INTRODUCTION TO
Nervous Systems

A Series of Books in Biology

EDITOR: *Donald Kennedy*

INTRODUCTION TO
Nervous Systems

Theodore Holmes Bullock
UNIVERSITY OF CALIFORNIA, SAN DIEGO

WITH THE COLLABORATION OF

Richard Orkand
UNIVERSITY OF PENNSYLVANIA

Alan Grinnell
UNIVERSITY OF CALIFORNIA, LOS ANGELES

W. H. Freeman and Company
SAN FRANCISCO

The illustration on the cover is an enlargement of the lower part of the dendritic arborization that appears on the right side of page 348. Reproduced from *The Postnatal Development of the Human Cerebral Cortex,* by J. L. Conel, Harvard University Press, Cambridge, 1959.

Library of Congress Cataloging in Publication Data

Bullock, Theodore Holmes.
  Introduction to nervous systems.

  Bibliography: p.
  Includes index.
  1. Neurobiology. I. Orkand, Richard, joint author.
II. Grinnell, Alan, 1936–   joint author. III. Title.
[DNLM: 1. Neurons—Physiology. 2. Neurophysiology.
WL102 B938i]
QP361.B92   612′.8   76-3735
ISBN 0-7167-0577-X

Printed in the United States of America

1 2 3 4 5 6 7 8 9

*To the students
who inspired this book
and the loyal associates
who made it possible*

CONTENTS

LIST OF BOXES

PREFACE

The goal of this book is to provide for advanced undergraduates, early graduate students, and medical students an introduction to the nervous system with emphasis on its systems aspects.

The nervous system, together with its close companion, the endocrine system, is unlike the other organ systems in that it is primarily concerned with signals, information processing, and control rather than the manipulation of substances and energy; it is a communication device. Its components, of course, do use substances and energy in the processes of signalling, recognizing, choosing, and commanding, as well as in developing and learning.

An insight that has slowly dawned, and is still not acknowledged in many expositions of neurobiology, is that the central questions about nervous systems, "How do they work? What's going on? What's the principle of operation?" have no single answers. Instead, the mechanisms, the constituents, and the principles are there to be uncovered, layer upon layer, from levels below the stereochemistry of membrane molecules to levels above the consolidation of en-

grams. The explosively expanding branch of knowledge called "neuroscience," still in its relative youth, already claims to embrace a wider spectrum of complexity, a greater range of levels of explanation, than any other science. None of the levels, submolecular to interhemispheric, is adequate by itself; none is the key level or the queen of the science. The span is too great even for an elementary text such as this, hence choices had to be made. The result is that this book deals with the first question, "How does it work?" It emphasizes the middle levels and says less about the molecular below and the psychological above.

The core problems we take up are relations and transactions between the cells and among the assemblages of cells in the nervous system. We deal with the encoding and decoding of neural signals, the evaluation and weighting of inputs, the formulating of outputs, and even the simpler elements of behavior. The book therefore embraces neuroanatomy and neurophysiology from the cellular level to that of subsystems of the brain. It cannot do justice to neurochemistry, neuropharmacology, energetic and nutritive metabolism, or to the mechanisms of learning.

The main reasons for offering this book are two. The first is to redress the relative neglect of the integrative aspects in current textbooks. The second reason is to present a major segment of the impossibly wide span of neuroscience in a logical series of levels, more or less comprehensively.

Following an introductory perspective (Chapter 1), the next four chapters deal with cellular componentry, first structurally (Chapters 2 and 3), and then functionally (Chapters 4 and 5), first within the cell (Chapters 2 and 4) and then between cells (Chapters 3 and 5). The next block of chapters (6, 7, and 8) considers integrative mechanisms at neuronal, intermediate, and behavioral levels. The last two chapters survey the development of the nervous system in the life of the individual (Chapter 9) and its evolution in the animal kingdom (Chapter 10).

A Glossary that is more than merely indicative is provided. We feel that familiar terms are often used in contemporary writing with insufficient care and that understanding is poor unless a definition can be given that is both inclusive and exclusive. The Glossary is intended to be used continually, not merely for reference in extremis.

We have tried to distill in order to keep the size of this volume compatible with the intention that it be used in parallel with others that deal more fully with the cellular mechanisms on the one hand and the details of the human brain on the other.

The treatment within the succession of chapters is not homogeneous. We hope the reader will benefit from, rather than be distracted by, the diversity of outlook among three authors who have quite different perspectives. As the title page and table of contents indicate, my collaborators have had an asymmetry of roles in the building of the book, hence they cannot be held responsible for the philosophy or treatment of chapters other than their own.

It is a major accompaniment of the growth and specialization of this field that neuroscientists today diverge widely in "slant" and emphasis. The brain as seen by one scientist may be hardly recognizable by another. The approach of one may be basically to explain observations in terms of understood mechanisms, whereas another may be more impressed by the gap between the few phenomena we can explain and the many we can presently only characterize. Some neuroscientists are primarily general physiologists, impressed by commonalities and attracted by the lure of broadly applicable principles; others are essentially comparative, impressed by the range within the most differentiated system nature has evolved.

It has been my credo that we must be eclectic if we are to do justice to nature's achievement and be ready for the changes that tomorrow may bring. This book, by example, advocates that each inquirer, whether student or investigator, sample widely even among the more subjective "slants" of authors, and rethink frequently even the settled dogmas.

This book is the combined effort of many more than the three of us. Each has been inspired by his students and aided incalculably by staff and colleagues, and it is to them that my collaborators join me in gratefully dedicating this book.

Theodore H. Bullock

December 1976

INTRODUCTION TO
Nervous Systems

1
THE SPECIAL NATURE OF
NERVOUS SYSTEMS AND NEUROSCIENCE

I. INTRODUCTION

To study the wellsprings of human nature should be challenge enough, but it isn't. At no time in his history has *Homo sapiens,* the wise one, been quite satisfied with an egocentric vision. To achieve peace on earth, human justice, health and well being—first order though they be—will never suffice as the only human goals. Man has a need to know, to understand, to experience. He probes the stars, the atoms, the heights of esthetic and creative feeling; he paints, he composes, and he walks on the moon. Despite a tendency to accept simplistic or supernatural answers, he nevertheless exhibits a drive to probe beyond the limited understanding that such answers offer.

In the tradition of inquiry and the need to know, the nervous system is a proper study of mankind—the nervous system in all its manifestations, the mysteries behind a true understanding of behavior, the origins of humanity. What is the nervous system's relation to behavior? How does it govern thought?

What is it like in its simplest forms? How has this behavior machine evolved, specialized, acquired new capacities and transformed the old in the course of evolution?

In addition to satisfying our desire to know, the discovery of answers to such questions (even though the answers may be far from definitive) is often of profound **humanistic significance.** The brain makes us err, makes us selfish, makes us altruistic and rational. The human brain, source of the world's most serious problems, is also the world's principal hope. Seeking a thorough understanding of ourselves—probing to uncover the intricacies of the brain—is certainly one of the most promising human activities. Such study has a long and dramatic history, but the accumulation of information about brain and behavior has accelerated enormously lately.

Building on the triumphs of the past few decades, life science is already well into the age of **neuroscience**—and still adjusting to an unprecedented kind of challenge. The familiar strategy of seek-

Figure 1.1
Levels of inquiry.

A specific cation may relieve severe depression.

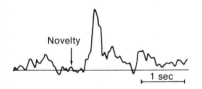

A specific wave of synchrony in a population of cells may signal the experience "What's that?" in response to a novel stimulus in a boring series.

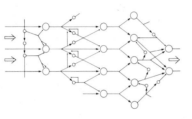

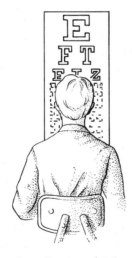

A specific array of active neurons may recognize E.

ing the common denominators of life is unequal to this challenge. It is behavior and its neural substrate that has most evolved, that makes higher animals higher and the human species highest of all. It seems most unlikely that there is a *single* code to be broken to explain love and hate, the pianist and the perjurer. Nevertheless, single-gene, single-enzyme deficiencies can cause devastating brain disease, and simple lithium salts can dramatically relieve some subtle psychiatric disorders. The challenge is to integrate widely disparate levels of inquiry (Fig. 1.1) and disciplinary approaches. What are these approaches?

It has been said that "everything comes down to molecules," and the most basic of the approaches can be called molecular neurobiology. This is itself a combination of biochemistry and biophysics, and grades imperceptibly into the next higher level of cellular microstructure and function—the one that deals with membranes and organelles. Still higher levels of anatomy and physiology deal with the layers upon layers of ever more intricate organization that constitute the nervous system. Study at such levels grades into physiological psychology and neuroethology.

The outstanding feature of this **array of disciplines** is its breadth. We don't expect to understand speech by limiting our investigation to cells, let alone to molecules or atoms. Nor can we wait to deal scientifically with motivation, drive, and emotion until we have systematically discovered all the fundamentals of enzyme dynamics, and then of cell organelles, cellular differentiation, and tissue organization.

Practicality and scientific strategy both demand that biology **advance at the same time on many levels** and many fronts, always struggling toward the integration of approaches and disciplines. It is obvious from the intricacy of the common object of study, the brain, that the subject matters, vocabularies, and problems of such disparate approaches as the chemistry of subcellular particulates and the systems analysis of constellations of cells in the cortex will be far apart for many years, given the relative primitiveness of our essential understanding of the whole. Great effort is required to integrate even closely related approaches.

In this book we limit our scope primarily to the signalling and **systems aspects**—that is, to the intermediate levels, which depend particularly for their significance on the organization of the cells. We will not deal in detail with intracellular componentry. Nor will we attempt, at the other end of the possible array, a full treatment of the organization of the human brain. That is thoroughly treated in many other books. But the principles of organization and integration, encoding and decoding, line-labeling, recognition, command, and the use of pattern are not. To discuss those subjects is our central objective. But in defining our scope, we must make other choices in another dimension.

As **evolution** from humble origins to lofty achievement is a universal feature of life, the nervous system—from diffuse nets in jellyfish to the human brain—is the outstanding consequence of evolution. The nervous system truly represents a biological phenomenon, and it is our contention that it cannot be understood simply as a part of general physiology or of human biology without the evolutionary perspective. Far more than for other systems, we are dependent, for any claims to adequate appreciation, upon the comparative view. To make this real now, before we are immersed in details, it suffices to point to the accomplishments of nervous systems—that is, to behavior in all its repertoire, nuances, and malleability. Contemplate the gulf between jellyfish and human (Fig. 1.2). It is almost as great as that between synapse and thought. Once more the sweep of our subject is so wide that we must make a choice. But for this choice, instead of cutting off one end of the pos-

sible range, as in the previous choice, we shall select animal groups at various points on the phylogenetic tree, aiming for at least a modest introduction to the indispensable biological perspectives.

II. NERVOUS SYSTEMS

The subject and theme of our book is nervous systems as systems. What, then, is a nervous system? It may be defined as an **organized constellation of nerve cells** and associated nonnervous cells; it includes receptors, but not most effector cells. Nerve cells—which we shall hereafter synonymously call **neurons**—may be defined as cells specialized for the generation, integration, and conduction of excited states, including most sensory but not effector cells. A corollary of this definition of nerve cells is that they derive their excitation intrinsically or from the environment, from special sense cells, or from other neurons and deliver it to other excitable cells or to effectors, such as muscle cells. A corollary of the definition of nervous systems is that they differ both quantitatively and qualitatively from other organ systems, because they deal only incidentally with materials and energy. Their function and specialization is to process information, and their organizational complexity greatly exceeds that of any other system.

Besides the defining features, certain **common attributes, of neurons,** though not universal, are usually helpful in distinguishing them. Such attributes include a brief impulse and refractory state, a local form of response at junctions, called a postsynaptic potential (further defined in the Glossary and in Chapter

Figure 1.2
Lower and higher animals differ primarily in elaboration of the nervous system.

50
10
4
6

5), and some electron-microscopic and biochemical specializations associated with these junctions. An important general statement about nerve cells, however, is that compared to other types of cells they are outstanding in the degree of differentiation among themselves (Fig. 1.3). A major theme of Chapter 2 is the variety of types of nerve cells. Heterogeneity and specificity are the hallmarks of neurons—not only in their forms, branches, and connections but in their distinctive chemical, physiological, pharmacological, virological, and other properties, and in their reactions to injury, disease, and external agents. Neurons are the least interchangeable of cells.

Nervous systems are easy to recognize in higher animals, but the defining criteria are difficult to apply in the lowest groups. As is true for the neuron, it is helpful to know certain **common attributes of nervous systems.** A central nervous system can be distinguished from a peripheral nervous system more or less clearly in all but the simplest forms (i.e., from flatworms upward). The **central nervous system** (CNS) contains most of the motor and internuncial cell bodies —that is, the nucleated parts of the neurons that innervate muscles and other effectors and of the neurons that are between sensory and motor cells. The **peripheral nervous system** (PNS) contains all the sensory nerve cell bodies, with rare exceptions, plus local **plexuses** (diffuse tangles of nerve cells and fibers) concerned with the body wall or viscera, local **ganglia** (knots of nerve cells) of either sensory or motor-and-internuncial composition, plus the peripheral **axons** (long processes of neurons), which make up the **nerves** (bundles of axons). We believe there are no isolated peripheral plexuses or ganglia without connection to the rest of the nervous system. All receptor cells are nerve cells except those of a few special sense organs of vertebrates, including taste buds, one form of touch corpuscle, acousticolateralis systems, and, according to the usage of some, rods and cones of the eye. These receptors are connected to the axons of the first-order afferent (incoming) neurons. Most sensory axons go all the way into the central nervous system, but a small number of them relay in peripheral plexuses, the remainder of the connection with the central nervous system being made by second-order afferent (entering) fibers. Similarly, most effectors are innervated by motor axons originating in the central nervous system, though some central motor neurons relay with peripheral motor neurons.

As we observed for neurons, these simple rules give only a hint of the elaboration achieved in the systems of higher animals. We know all too little of the essential achievements that have occurred during the **evolution of organized systems** of neurons. But a great deal is known of the trends in gross anatomy and in microscopic differentiation of nervous tissue. Higher groups (see the Glossary for a definition of "higher") in general have more differentiated receptors, more kinds of neurons and textured masses of neurons. In the lower phyla the distinction between peripheral and central nervous systems is less distinct; the increased distinction in higher phyla is called **centralization.** With increase in the complexity of these higher groups, there is repeatedly and independently a tendency toward **cephalization**—the greater concentration of neural masses and functional responsibil-

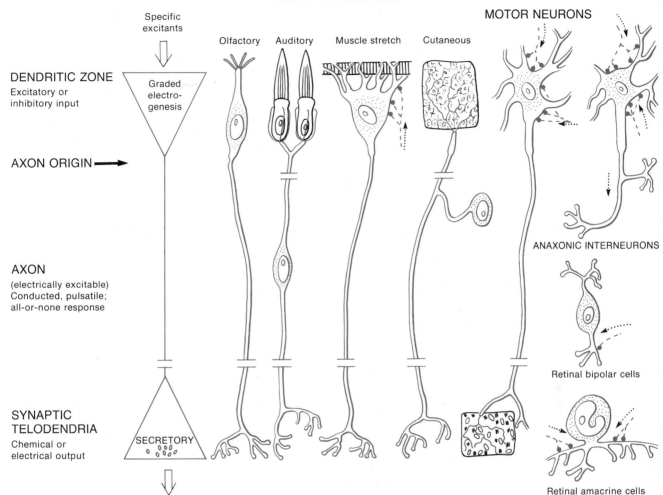

RECEPTOR NEURONS   INTERNEURONS
                              MOTOR NEURONS

Specific
excitants

Olfactory Auditory Muscle stretch Cutaneous

DENDRITIC ZONE
Excitatory or
inhibitory input

Graded
electro-
genesis

AXON ORIGIN

AXON
(electrically excitable)
Conducted, pulsatile;
all-or-none response

ANAXONIC INTERNEURONS

Retinal bipolar cells

**SYNAPTIC
TELODENDRIA**
Chemical or
electrical output

SECRETORY

Retinal amacrine cells

Figure 1.3
Diagram of a variety of afferent neurons, efferent neurons, and interneurons, arranged to bring out
the basic agreements in functional and structural features. The position of the soma or nucleated
mass of cytoplasm does not have a constant relation to the functional geometry in terms of impulse
origin. In axon-bearing neurons, the four major zones of interest in terms of neural processing (dendritic zone, zone of axon origin, axon zone, transmitting or synaptic zone) conform approximately to
the functional diagram of the generalized neuron proposed by Grundfest. But some axons do not
conduct impulses, some dendrites transmit as well as receive, and some telodendria receive as well as
transmit. Anaxonic neurons, in which the impulse-conducting region is absent, may be regarded as
having processes similar in nature to dendrites and telodendria. [Bodian, 1967.]

6

ity toward the head of the body (Fig. 1.4).

Figure 1.4
Types of central nervous systems and peripheral nervous systems.

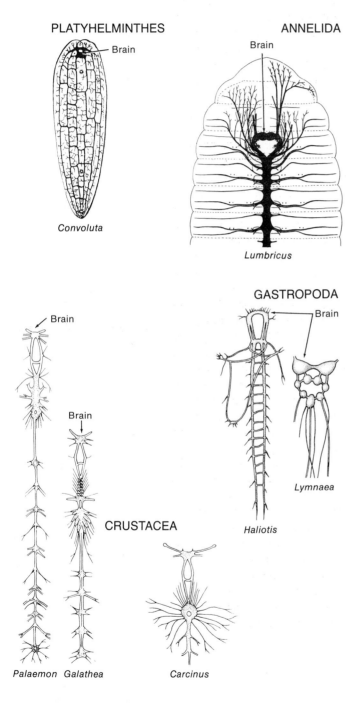

PLATYHELMINTHES

— Brain

Convoluta

ANNELIDA

Brain

Lumbricus

GASTROPODA

Brain

Lymnaea

Haliotis

Brain

Brain

CRUSTACEA

Palaemon Galathea   *Carcinus*

These trends are but superficial signs of the essential structural and functional aspects of evolution, which are much more difficult to specify. The **two major roles of the nervous system** are both discernible from the lowest to the highest groups. These roles are, in short, to counteract and to act (Fig. 1.5). In the first, the role of **regulation,** the nervous system acts homeostatically—that is, serves to preserve the status quo by making compensatory responses to stimuli that displace or perturb some condition of the organism. In the second, **initiation,** it acts to alter the status quo by replacing one mood or phase of behavior with another. Both roles show astonishing evolutionary development from the simpler invertebrates to the mammals; the initiating role has probably shown the most.

Learned behavior can be superimposed on either the regulating or the initiating category, but it pertains mainly to the second. For most animals, learning primarily promotes a more adaptive aiming, combining, and timing of species-characteristic acts that tend to occur anyway. Certainly there has been remarkable evolution in the degree and perfection of this form of plasticity.

Viewing the significance of the nervous system from a different standpoint, we may note that it performs in such a way as to extend the range of **speeds,** and therewith the **intricacy** of behavior: witness, for example, the capabilities of a pianist! Speed allows for an increase in the number of intervening steps between sensory input and motor output, the integrative transactions, and hence for numerous units of information handling. Thus the rich repertoire of behavior. Think not only of the fast reactions in conversational repartee, but of the controlled, slow motions in ballet and the myriad alternatives available.

If we compare lower invertebrates and higher vertebrates, we find that the higher forms have much greater **numbers of cells** in their nervous systems, even if we select animals of the same body size, such as a jellyfish and a rat. The difference for these two animals is about five orders of magnitude; a conservative estimate is that the higher form has at least 10,000 times as many cells in its nervous system as the lower one. The number of junctional contacts per neuron, afferent and efferent, has also increased vastly; an estimate that could be regarded as typical is even more difficult to make than the preceding one, but the numbers for common neurons in the species mentioned differ by a factor of four to five orders of magnitude. Thus the total number of junctions may have increased 10^9 or 10^{10} times. **The number of impulses per second** in each neuron and junction probably differs by an additional one to three orders of magnitude, but many neurons and junctions may function without impulses.

Along with these quantitative changes, a great flowering of **qualitative developments** has accompanied evolution. These have occurred at all levels—from the biochemical to the organizational. We will detail in subsequent chapters some examples of the differentiation of cell types, of integrative neuronal properties, of specified systems of cells and higher-level strategies that, taken together, must represent the principal achievement of evolution. Contrary to the view sometimes taken that large brains work to a great extent with masses of poorly specified cells having quasi-randomly distributed properties, we will argue that the remarkable advances in our knowledge of specific elaborations—advances reported in each issue of the research journals—are likely to continue for a long time. Particularly primitive must be our appreciation of the emergents, those phenomena not predictable from even a rather extensive knowledge at the next lower levels. Common to all the natural sciences, the emergents at each level have unpredictable consequences for function at the higher levels, and therefore may be expected to concatenate very large numbers of elements. Unanticipated properties have contributed much excitement to the study of neuroscience; examples chosen from various levels include the nerve impulse, synaptic potentials, trophic influence, specific regeneration, brain waves, contingent negative variations, ultradian 90-minute rhythms, and the localization of such functions as arousal, docility, speech and shape recognition. Such discovery—phenomenology, it is sometimes called—is one of the main hopes of gaining new insight, along with the reductionistic analysis of the mechanisms of such phenomena.

The higher nervous systems are far and away the most complex systems in nature (barring social systems devised by them). While this may discourage and certainly should awe and forewarn us, it promises a long continuation of the excitement that so marks research in neuroscience today. In such complexity there must surely lie a reservoir of discoveries, even new principles. It seems highly probable in terms of the advance of

Figure 1.5
Two roles of nervous systems.

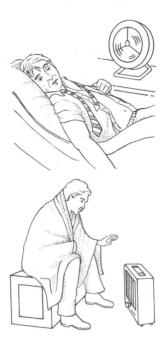

Regulation: short-term compensation to maintain status quo.

Initiation: longer-term change in the sphere of action.

science that major revolutions are in store, but they will probably not be like those that have taken place in the sciences of simpler systems—that is, not in the form of discrete, convulsive, datable overturns. Rather, it may be ex- pected that many quiet and gradual but drastic revolutions will take place, piece- meal and simultaneously, in disparate levels of the many-tiered science of nervous systems.

SUGGESTED READINGS

Eccles, J. C. 1977. *The Understanding of the Brain* (2nd ed.). McGraw-Hill, New York. [This presents a currently orthodox, yet quite personal, view of the whole sweep from cells and impulses to speech and consciousness.]

Fentress, J. C. 1976. *Simpler Networks and Behavior.* Sinauer Associates, Sunderland, Massa- chusetts. [A collection of essays that provides a sampling of current thinking on a wide array of issues.]

Kandel, E. R. 1976. *Cellular Basis of Behavior: An Introduction to Behavioral Neurobiology.* W. H. Freeman and Company, San Francisco. [A closely reasoned and highly integrated account of studies done chiefly on a single species, the mollusc *Aplysia;* simpler forms of learning are explained in cellular terms.]

Kuffler, S. W. and Nicholls, J. G. 1976. *From Neuron to Brain.* Sinauer Associates, Sunderland, Massachusetts. [Another excellent introduction from a different perspective. The authors "illustrate the main points by selected examples, preferably from work in which we have first-hand experience."]

Neuroscience Abstracts, Sixth Annual Meeting of the Society for Neuroscience. Volume two. Society for Neuroscience, Bethesda, Maryland. [These annual collections of more than 1600 abstracts provide a convenient cross section of current activity, as well as an appreciation of the range, diversity, and explosive progress in the field.]

2
MICROANATOMY OF NERVOUS ELEMENTS

I. INTRODUCTION

No other system in the animal body approaches the nervous system in the degree of dependence of its functions upon precise anatomical arrangements among its elements. This dependence can be seen in the distribution of such specialized structures as spines and vesicles, in the form of cell contacts, in the architecture of groups of cells, and in their connections over long distances. This chapter presents the main cytological and histological features of the cellular elements and their organization into nervous tissue. In the following chapter we consider the connectivity that links these elements into the more-or-less specific circuits of functional value to the organism.

In contrast to the cells of other tissues, most of which are much less varied in form, neurons show great diversity in shape. They have one or several processes (extensions) that may arborize (branch) profusely in characteristic ways and form synapses (establish contacts) with other neurons—receptor cells or effector cells. The cell body, from which the processes arise, is called the soma. If it has one process, it is unipolar; two processes, bipolar; three or more, multipolar (Fig. 2.1). The processes may be differentiated into axons and dendrites. Between the nerve cells are nonnervous cells of several kinds, called neuroglia, or glia. Some form sheaths around axons. Neuroglia and sheaths are the subject of Section III.

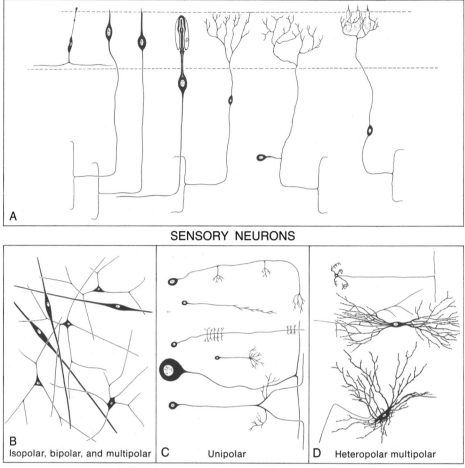

SENSORY NEURONS

B	C	D
Isopolar, bipolar, and multipolar	Unipolar	Heteropolar multipolar

INTERNEURONS AND MOTOR NEURONS

Figure 2.1

Types of neurons, distinguished on the basis of number and differentiation of processes. **A.** Sensory neurons. The most primitive (far left) send axons into a superficial plexus. In animals with a central nervous system, the commonest form is a similar bipolar cell in the epithelium with a short, simple or slightly elaborated (arthropod scolopale) distal process and an axon entering the central nervous system and generally bifurcating into ascending and descending branches. A presumably more derived form (third from right) is that with a deep-lying cell body and a long, branching distal process with free nerve endings. In vertebrates, such cells secondarily become unipolar and form groups in the dorsal root ganglia. Shown on the far right is a vertebrate vestibular or acoustic sensory neuron that has retained the primitive bipolar form but has adopted (presumably secondarily) a specialized nonnervous epithelial cell as the actual receptor element. **B.** Isopolar, bipolar, and multipolar neurons in the nerve net of medusa. These may be either interneurons or motor neurons or both; no differentiated dendrites can be recognized. **C.** Unipolar neurons representative of the dominant type in all higher invertebrates. Both interneurons and motor neurons have this form. The upper four are examples of interneurons and the lower two of motor neurons. Afferent branches may be elaborate but are not readily distinguished from branching axonal terminals. The number and exact disposition of these two forms of endings and of major branches and collaterals are highly variable. **D.** Heteropolar, multipolar neurons—the dominant type in the central nervous system of vertebrates. The upper two are interneurons; the lower one, a motor neuron. [Bullock and Horridge, 1965.]

Figure 2.2
Amacrine cell or neurons without a definite axon.

From the optic lobe of Sepia, *a cuttlefish.* [*Cajal, 1917.*]

From the optic medulla of Calliphora, *a fly.* [*Cajal and Sanchez, 1915.*]

II. THE NERVE CELL AND ITS PROCESSES

A. The Soma

That part of a neuron which contains the nucleus and surrounding cytoplasm is the **soma.** It may be a gentle swelling along the course of an axon, as in bipolar neurons; or a quasi-detached appendage of the axon, as in unipolar cells with long stem-like processes; or terminal or subterminal, as in superficial sensory neurons; or at the junction of dendrites and axons, as in vertebrate multipolars. The position of the soma is not a fundamental feature, especially in distinguishing between axon and dendrite. We can think of the nucleus and the surrounding cytoplasm as a trophic apparatus forming a swelling that may lie within axonal membrane, within dendritic membrane, or where the two kinds of membrane meet. The soma may be large ($>500\,\mu m$ in diameter) with a watery nucleus, small ($<2\,\mu m$) with a chromatin-rich nucleus and very little cytoplasm, or intermediate in size and nuclear composition. In some neurons the volume of the soma may be as little as 1/10,000th that of the axon; in others the volume may nearly equal or even exceed that of the axon.

The contents of the soma and the diverse types of neurons are treated in Section II-H. First it will be helpful to discuss the two kinds of processes that project from nerve cells.

B. Axons

The **axon** is usually, but imprecisely, considered functionally to be a process specialized for conduction of excitation over considerable distances. The term is better defined on histological grounds. Microscopically an axon can be recognized as a long fiber that is relatively uniform in diameter, smooth surfaced, has branches at long intervals, is usually covered with a sheath of closely applied cells, has synaptic endings upon it at only a few spots, and generally lacks some cytoplasmic constituents of the soma (Fig. 2.9,B). Most neurons have one axon. Even with these distinguishing features, it is sometimes impossible to say whether a given process is an axon or whether a given cell has an axon. Nerve cells without demonstrable axons are called amacrine cells (Fig. 2.2). A general term for processes, noncommital as to their type, is **neurite.**

The main functional characteristic of the axon is its ability to conduct excitation nondecrementally over some distance—that is, to support the all-or-none nerve impulse (Chapter 4). But this ability is not strictly a defining criterion. Impulses sometimes arise a millimeter or more downstream from the beginning of the axon, or they may arise upstream in the dendrites. It is not certain whether

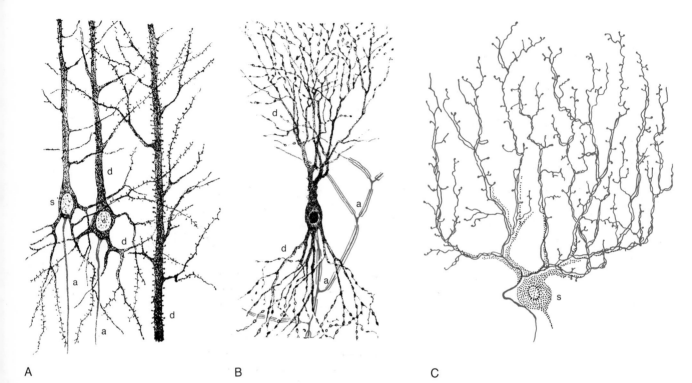

A   B   C

Figure 2.3

Axons and dendrites. Several types of neurons from the mammalian brain, impregnated by the Golgi method. The axons (*a*) are shown in color; dendrites (*d*) and somas (*s*) are in black. **A.** Pyramidal cells of the cerebral cortex, each with an axon extending from the lower pole of the soma, three or four basal dendrites extending sideways and downwards, and a thick apical dendrite extending upwards and out of the picture; at the right an apical dendrite whose soma is below and out of the picture. **B.** Pyramidal cell of the hippocampus with two sets of dendrites. **C.** Purkinje cell of the cerebellar cortex, faintly stained, with the axon of a distant soma, probably from the inferior olive of the medulla, entwining the dendrites; this axon is called a climbing fiber. **D.** Purkinje cell with its dendrites well stained; they form a flat fan. **E.** Basket cell of the cerebellar cortex with dendrites extending upwards and an axon forming terminal baskets around a row of faintly shown Purkinje cell somas. **F.** Cell from the cerebral cortex, with short axon, often called Golgi type II or intrinsic neuron because it does not project beyond the local region. Golgi impregnation. **G.** Motor neuron of the electric lobe of the medulla of the electric ray, *Torpedo*, teased out after dissociating the tissue, therefore with the processes broken off. Also shown are the bases of the dendrites and the axon (axon hillock) leading into the initial segment of myelinated fiber. [Cajal, 1909.]

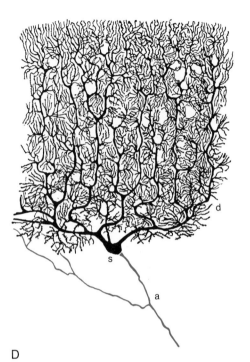

D

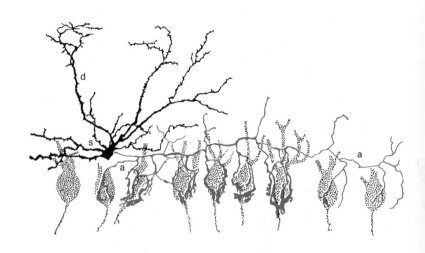

E

F

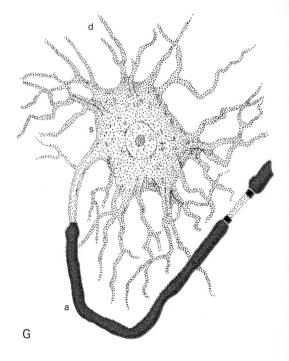

G

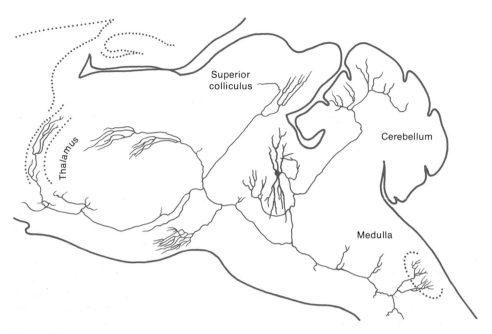

Figure 2.4
Neuron with extensive system of branching axons. A cell in the locus coeruleus, a nucleus in the
floor of the fourth ventricle, in the pons; Golgi impregnation; mouse. Collaterals distribute very
widely in many parts of the brain. [Courtesy of M. A. and A. B. Scheibel.]

impulses arise at all in neurons with very
short neurites (see Glossary). Some
axons—for example, many of those in
the vertebrate retina—function normally
without impulses. Terminals of axons in
many parts of the body probably fail to
conduct all-or-none impulses to the very
end but revert to graded local potentials.

The term "axon" therefore continues
to be based primarily on histological criteria (Fig. 2.3). Even these, however, are
not always unambiguous; the diversity of
neuron form is, after all, due partly to the
variety among axons. Some are relatively
short; others give off relatively numerous

collaterals (Fig. 2.4). In certain places in
the central nervous system, axons are
found without any sheath cell investment. Some axons receive synaptic endings not only at the origin and near the
terminals but at the node of Ranvier.
Some neurons have more than one axon;
for example, bipolar neurons, T-shaped
unipolars, and some multipolars.

The distinction between axon and
dendrite is usually clear in vertebrates,
but in invertebrates the distinction can
be difficult, suggesting that the two kinds
of processes are less differentiated in
lower forms.

C. Dendrites

Dendrites are often defined as processes specialized to act as the receptive regions for the neuron. They *are* the principal receptive apparatus, but these processes have recently been found to be transmissive (presynaptic) in some places as well; conversely, some axons are known to have receptive (postsynaptic) junctions. We are faced with a historic term and an evolving concept. The term "dendrite" was established for and should still be defined as a cytological category of processes that are relatively short, frequently branched, irregular in diameter, tapering, unsheathed, often beset with spines (see Fig. 2.3), containing cytoplasm resembling that in the soma, and receiving large numbers of synaptic endings over much of their surface. The distinction between axon and dendrite is more basic than that between either process and the soma (Fig. 2.5). The finer and more distal branches of dendrites are probably incapable of conducting all-or-none impulses, but instead are specialized to produce synaptic potentials in response to adequate presynaptic events. Reception is not the *only* role of the dendrites. They integrate converging input, generate spontaneous slow changes of state, both send and receive impulses at reciprocal synapses, and at some points initiate impulses (see Chapters 5 and 6).

The boundary between dendrite and soma is often indefinite. Dendrites branch in a wide array of patterns; these may be systematized in a simplified scheme for the vertebrates (Fig. 2.6).

In invertebrates the term "dendrite" can rarely be applied properly on the basis of known optical and electron-microscopical features. Perhaps there is a real difference in the degree of differentiation of processes in invertebrates compared to vertebrates (see also p. 436). Whenever it is necessary to speak of the "receptive processes" or the "proximal branches" of invertebrate neurons, those terms should be used instead of the anatomical term "dendrite."

D. Formed Elements in Nerve Cells

Neurons contain the structural components common to all cells, such as nuclei, nucleoli, mitochondria, and endoplasmic reticulum (Fig. 2.7). Some components are especially modified. We will deal here only with a few that are particularly important or unique to nerve cells (Fig. 2.8).

Synaptic vesicles and the junctional specializations associated with them are probably as diagnostic of neurons as any other component, although confined to limited parts of the cell surface and hence not always readily found. They are treated in Section II-G, below.

Fibrillar Structures. These are nearly ubiquitous in nerve cell cytoplasm—

16

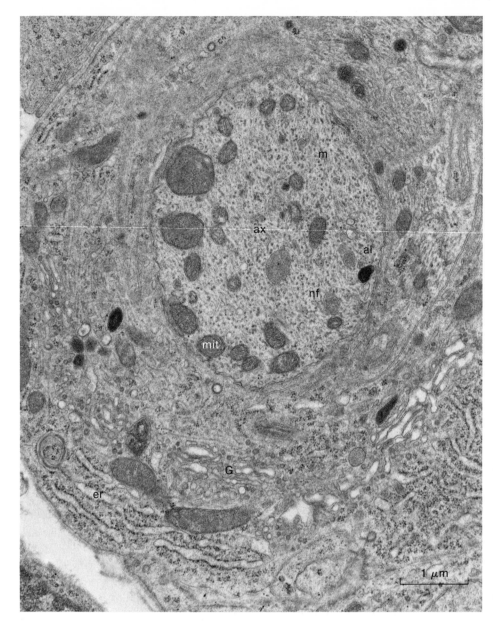

A

Figure 2.5
Axon and dendrite as seen in electron micrographs. **A.** The initial segment of an axon of a dorsal root ganglion cell from a rat. **B.** A dendrite (probably of a motor neuron) in the ventral horn of the spinal cord of a rat. *al,* axonal surface membrane; *At,* axon terminal; *ax,* axon; *den,* dendrite; *er,* gran-

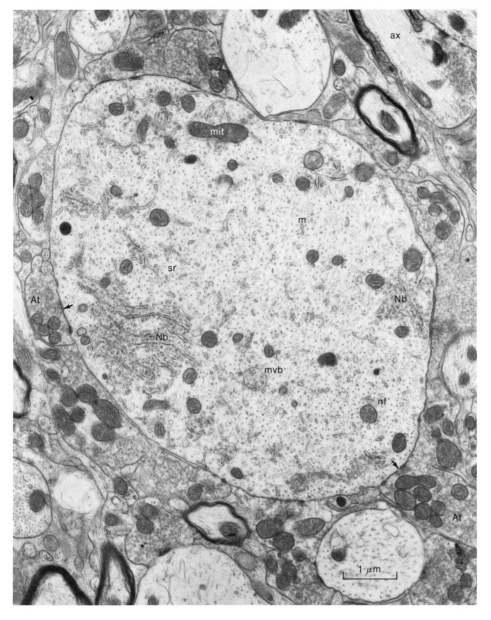

B

ular endoplasmic reticulum; *G*, Golgi apparatus; *m*, microtubules; *mit*, mitochondrion; *mvb*, multivesic-
ular body; *nb*, Nissl body; *nf*, neurofilaments; *sr*, smooth endoplasmic reticulum. The arrows in
part B show synapses. [Peters et al., 1970.]

18

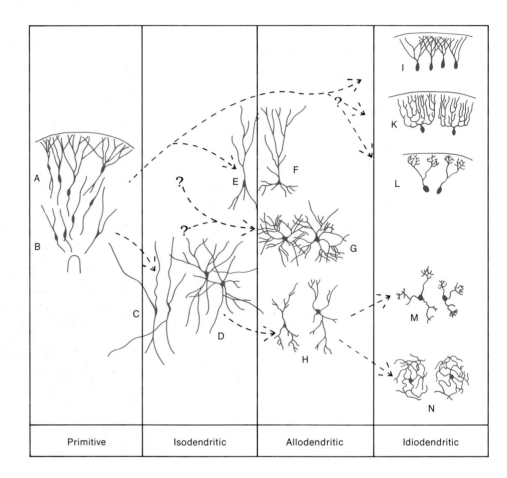

Figure 2.6
Diagram illustrating the probable phylogeny of vertebrate dendritic patterns. Two basic primitive
types are found in lower vertebrates (left). Neurons with subpial tufts (*A*), are seen intermingled with
periventricular leptodendritic neurons (*B*). The undifferentiated isodendritic pool is represented by
the neurons labeled *C* and *D*, and probably by *E*. As a result of morphological differentiation, various specific dendritic patterns (allodendritic) begin to appear in *F* (pyramidal neurons with basilar
dendrites), *G* (allodendritic neurons of the diencephalon), and *H* (allodendritic neurons of rhombencephalon). Eventually, certain highly differentiated forms (idiodendritic) appear; *I*, tufted granule cells
of gyrus dentatus; *K*, Purkinje cells; *L*, mitral olfactory neurons; *M*, tufted neurons of various secondary sensory centers; *N*, wavy precerebellar neurons. [Ramón-Moliner, 1968.]

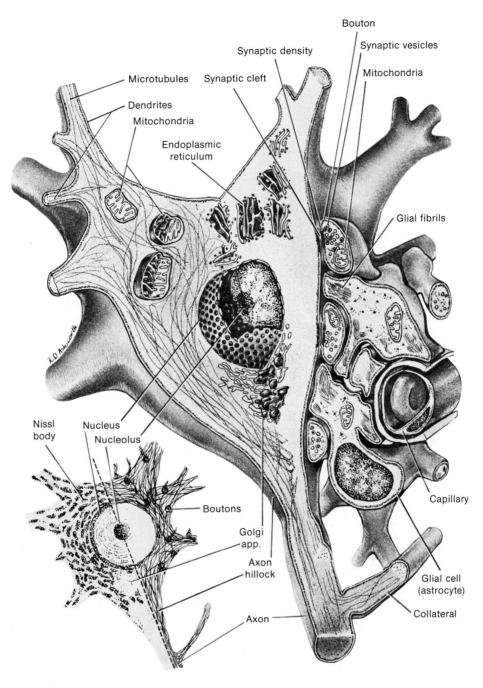

Figure 2.7
Nerve cell soma, showing organelles. The small view shows details seen at the light-microscope level. A stain selective for Nissl bodies was used on the left half; one selective for neurofibrils, on the right. The large diagram shows details seen at the electron-microscope level. [Willis and Grossman, 1973.]

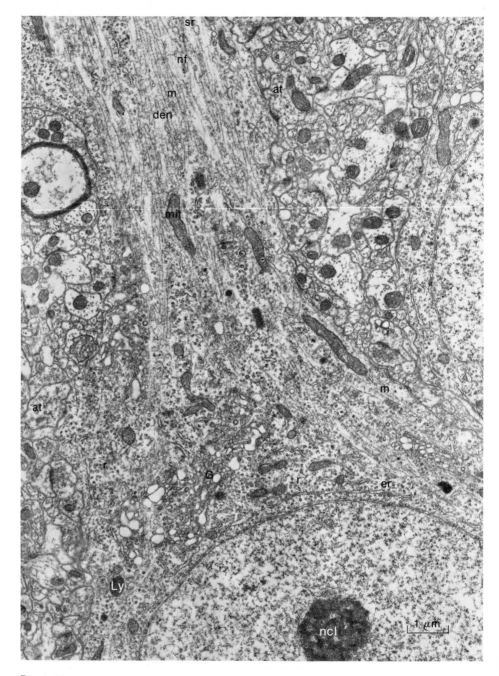

Figure 2.8
Formed elements in the neuron as seen in low-magnification electron microscopy; pyramidal cell soma in the cerebral cortex of a rat. *at,* axon terminals; *den,* dendrite; *er,* endoplasmic reticulum; *G,* Golgi apparatus; *ly,* lysosomes; *m,* microtubules; *mit,* mitochondria; *nf,* neurofilaments; *ncl,* nucleolus; *r,* clusters of free ribosomes; *sr,* smooth-surfaced endoplasmic reticulum. [Peters et al., 1970.]

axonal, dendritic, and somatic. There appear to be at least two basic types, neurofilaments and microtubules. **Neurofilaments** are fine threads 60–100 Å thick that are oriented roughly lengthwise in the processes and are commonly distributed uniformly throughout the cross section of a process. They are unbranched and of indefinite length. Some details of their ultrastructure (Fig. 2.8, 2.9,A) and composition are known, but their function is entirely unclear.

Microtubules, similarly, are longitudinal, unbranched organelles of indefinite length but with an outside diameter of 200–300 Å and a thick wall composed of spiral strings of bead-like globular proteins (Fig. 2.9,A). Their function is uncertain, but they may be involved in some form of transport of material along the axon. The distribution of both kinds of elements is not entirely uniform either among phyla or within the neurons of a given animal, and generalizations about their occurrence are not yet secure, let alone interpretable in terms of function.

The classical neurofibril (Fig. 2.10) of light microscopy is at least an order of magnitude thicker than the neurofilaments and microtubules, which can be seen only in the electron microscope. Although this classical element is generally regarded as an artifactual clumping of filaments or tubules, it may represent a natural concentration, as in the center of the 75-μm earthworm giant axon. Neurofibrils are usually described on the basis of impregnation with silver, which certainly distorts the normal structures. Nevertheless, fibrils *are* visible in the living state in the axons and somas of earthworms, as well as in those of jellyfish, lobster, fish, chick, and other forms.

Nissl Bodies. Components of many nerve cells, these chromophilic bodies are rendered conspicuous under the light microscope by means of basic dyes, but they can also be seen in some unstained, living cells with the aid of phase-contrast microscopy or microspectrophotometry. Their ultraviolet absorption spectrum and their reaction to ribonuclease digestion show that one of the principal constituents of the Nissl bodies is ribonucleoprotein. The electron microscope shows that an individual Nissl body may consist of large or small stacks of rough-surfaced or granular endoplasmic reticulum (ER) (Fig. 2.9,B). Ribosomes are arrayed along the outer surface of the membranes and in clusters or rosettes between them. These bodies are believed to represent the sites of protein synthesis, a process that is particularly active in nerve cells. The shape, size, number, and distribution of Nissl bodies vary greatly, but are characteristic for each type of nerve cell. In many of the larger vertebrate neurons, Nissl bodies are large and have distinct shapes. They provide a sensitive index to the condition of the cell, since a host of pathologic causes (including damage to the axon, asphyxia, toxic substances and possibly extreme demands on the metabolic and synthetic capacities) can change their form and position. The most common reaction to damage is *chromatolysis,* in which the Nissl granules disappear (in the light microscope) or form poorly outlined masses near the surface of one side of the cell while the nucleus undergoes

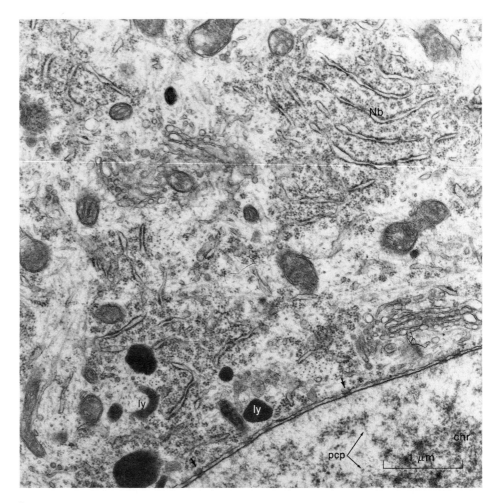

A

Figure 2.9
Formed elements in the neuron at high magnification; spinal cord of a rat. **A.** Nissl bodies (granular endoplasmic reticulum), nuclear envelope, Golgi apparatus (smooth cisternae near center and right edge.) **B.** Microtubules, neurofilaments. *ly*, lysosomes; *m*, microtubule; *Nb*, Nissl body; *nf*, neuro-filament; *sr*, smooth endoplasmic reticulum; the arrows in part A indicate pores in the nuclear

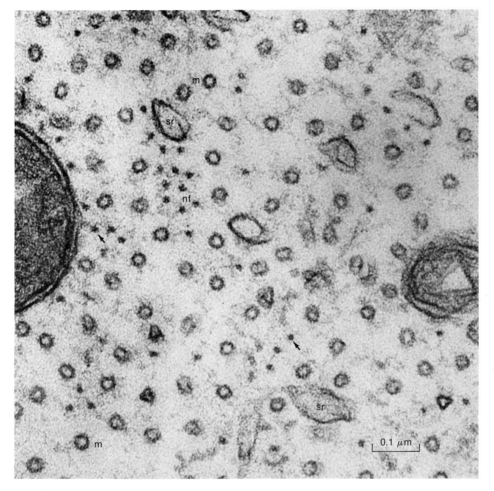

B

membrane; the arrows in part B indicate spoke-like appearance due to the grouping of neurofilaments into fascicles in this type of dendrite. [Peters et al., 1970; micrograph B by Dr. Raymond B. Wuerker.]

Figure 2.10
Neurofibrils: organelles seen in the light microscope, usually after special metallic impregnation. [Cajal, 1909.]

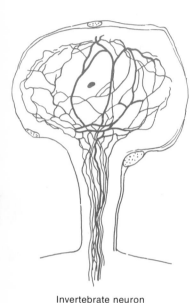

Invertebrate neuron

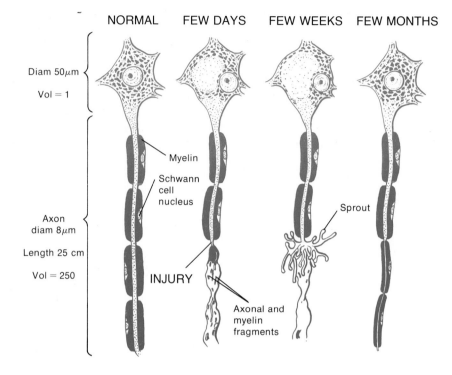

Diam 50μm

Vol = 1

Myelin

Schwann cell nucleus

Axon diam 8μm

Length 25 cm

Vol = 250

INJURY

Sprout

Axonal and myelin fragments

Figure 2.11
Regenerative stages of motor neuron after axon is severed. [Bodian, 1947.]

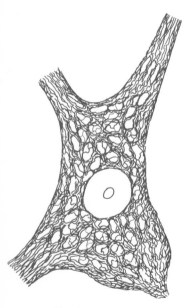

Vertebrate neuron

definite changes as well (Fig. 2.11). In some arthropod neurons, injury to the axon causes the ribonucleoprotein in the soma to concentrate in a ring around the nucleus. Changes of either kind may precede death of the cell or presage a gradual recovery. In many neurons, especialy among invertebrates, the corresponding rough-surfaced endoplasmic reticulum ("rough ER") is not collected into large bodies visible in the light microscope, but occurs in very small masses.

It is of great functional importance that rough ER is characteristically scarce or absent in axons but present in dendrites as well as in the soma. Granular endoplasmic reticulum is the chief protein-making machinery of the cell. How, then, does the axon, which may be a meter or more long and comprise the bulk of the neuron, get the proteins needed for normal function and replacement? The answer appears to be by movement of materials from the soma, a transport so formidable and important that we devote the next section to it. Even more characteristic of nerve cells than their Nissl bodies is their large endowment of cytoplasmic RNA, which is associated with a high rate of protein synthesis, particularly in the soma. This high rate of protein synthesis is, in turn, associated with the voluminous transport of materials along the axon.

Other general cytological elements of nerve cells are standard and do not require special comment in this work; these include the Golgi apparatus (which consists of clumps of smooth-surfaced endoplasmic reticulum), the mitochondria, a centrosome, and lysosomes. In addition, various inclusions may be present in widely varying numbers; these include lipoidal globules of several kinds, pigment granules of several kinds, glycogen granules, iron-containing granules, spiral whirls of dense lamellae, and modified ciliary structures, especially in the dendrites of sensory neurons.

E. Axoplasmic Transport

Axoplasmic transport is a term for a complex of widely divergent phenomena that contribute toward the movement of materials from the nerve cell soma to and along the axon or in the reverse direction (Fig. 2.12). A wide array of substances and methods have been used to study these phenomena in various nerves of many animals. At least two rates of transport coexist, one at about 1–10 mm/day (Fig. 2.13) and another at about 100–1000 mm/day. Intermediate rates of about 50–70 mm/day have also been reported. Most observations have been on proximo-distal transport (away from the soma), but slow distal-proximal transport has been measured as well. The composite picture from the various studies is consistent with the time-lapse cinephotomicrographs of axons in tissue cultures and in vivo, showing concurrent slow and fast streams, often sporadic but nevertheless simultaneous, in the same and in opposite directions. The materials moved include insoluble proteins, soluble proteins, glutamate, catecholamine-containing granules, monoamine oxidase, and phospholipids. Several amino acids and inert sugars are said not to be transported. Apparently some substances and cytoplasmic inclusions are quite selectively transported, some at the slow and others at the fast rate. Certain of these are transported more rapidly if the axon carries many impulses during the test period. The selectivity of materials and inclusions transported is consistent with the long-known striking differences in texture, staining properties, and inclusions shown by the cytoplasm of the soma (somatoplasm) and that of the axon (axoplasm), commonly with a sharp boundary at the axon hillock (Fig. 2.14).

The mechanisms of these slow and fast translocations are not understood. There is evidently at one and the same time a slow proximo-distal bulk transport of the whole cylinder of axoplasm (Fig. 2.15) and a complex fluid transport of some axoplasmic components relative to others in both directions and at various rates within the axoplasm. The former is equivalent to continuous growth; the latter is a mechanodynamic puzzle. A hydrodynamic model that was once postulated, in which materials were propelled by the pressure of their own rapid synthesis in the soma, has been tested and discarded. Two other ideas are still favored by some authors. One is that periaxonal forces generate a peristaltic pressure wave that squeezes the axoplasm ahead of it. The other is that struc-

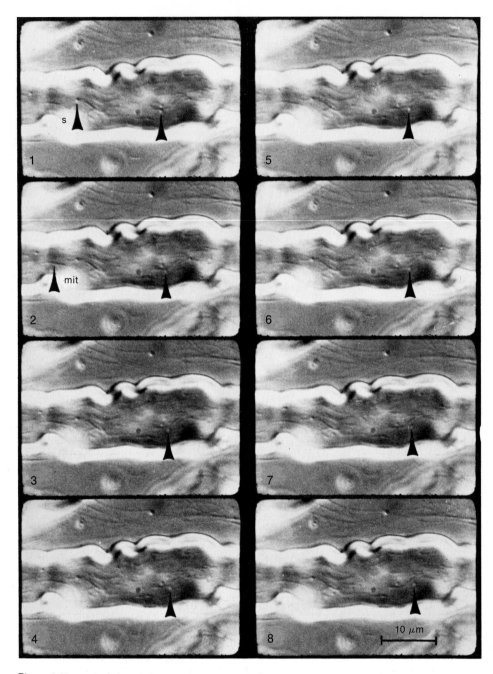

Figure 2.12

Movement of particles in the axoplasm. Frames from a cine film sequence of axoplasmic transport in peripheral axons isolated from adult chicken sciatic nerve; 120 frames/min; Nomarski differential interference microscopy. This technique makes it possible to visualize fast, slow, and jerky movements, mainly distally but also centrally. Here the large myelinated axon contains mitochondria (*mit*) and spherical particles (*s*). The particle marked with an arrowhead moves about 5 μm in the 4-sec sequence shown here. [Kirkpatrick et al., 1972.]

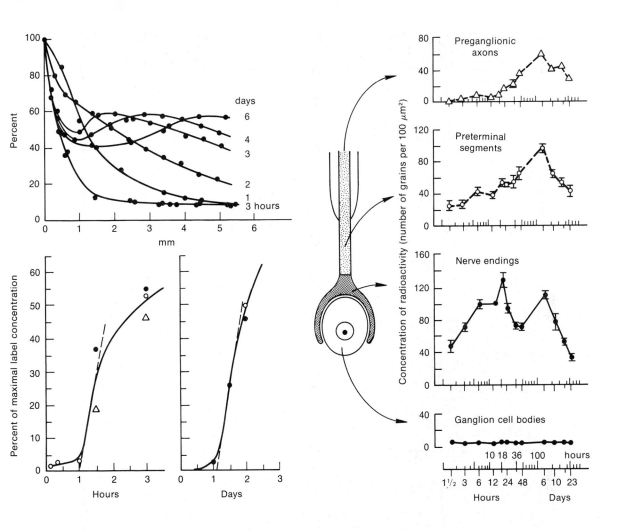

Figure 2.13

Axonal transport of materials. **Upper left.** Tritiated leucine was microinjected into the eyes of mice, where it became incorporated into retinal cells. On successive days the counts of silver grains in autoradiographs—normalized to 100% for each highest count (at "0 mm" = exit of optic nerve from bulb)—show an advancing crest traveling at about 1 mm/day. [Taylor and Weiss, 1965.] **Lower left.** Tritiated amino acid was injected into the chick brain stem; from there it moved out the parasympathetic preganglionic axons to the junctions (presynaptic calyces) with postganglionic cells in the ciliary ganglion. The graphs show the time of arrival of the first labeled molecules transported, respectively, by the fast and the slow flow. The ordinate is the percent of maximal label concentration recorded in presynaptic calyces by autoradiography (see Fig. 2.22) at various intervals. Open circles, [3H] leucine; filled circles, [3H] lysine; triangles, [3H] fucose. The first labeled macromolecules arrive at the nerve ending by fast flow in about one hour; by the slow flow, after one day. **Right.** Data from the same experiment, plotted to show the increase in concentration of radioactivity at several places as a function of time. Note the sudden rise that begins in preganglionic axons after 24 hours; two peaks at 18 hours and 6 days are seen in nerve endings, but only the 6-day peak in preganglionic axons and preterminal segments. [Droz et al., 1973.]

Figure 2.14

The contrast between axoplasm and somatoplasm.

Figure 2.15

"Damming" of axoplasm in constricted nerve fibers. One of the first discoveries indicating axoplasmic flow. [Weiss and Hiscoe, 1948.]

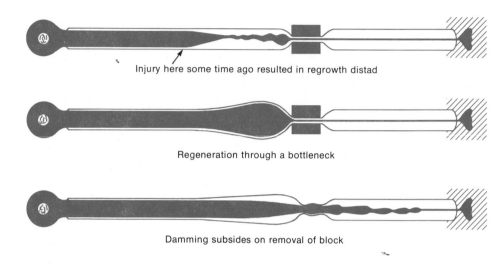

Injury here some time ago resulted in regrowth distad

Regeneration through a bottleneck

Damming subsides on removal of block

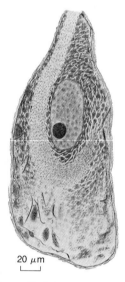

20 μm

Ganglion cell of Astacus, *a crayfish, showing intrasomatic origin of the axoplasm. Note also that strands of the sheath and its glial cell penetrate into the soma in some (= trophospongium) nerve cells. [Ross, 1922.]*

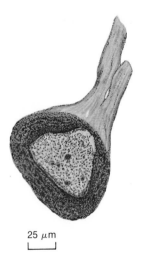

25 μm

Ganglion cell of Helix, *a snail, with similar features, though the magnification is too low to look for trophospongium. [La Croix, 1935.]*

tural changes in macromolecules in the axoplasm propel material selectively. Most interesting are new ideas that neurofilaments and/or microtubules contribute to axoplasmic transport by means of some kind of molecular cooperativity.

The significance of axoplasmic transport is doubtless multiple. It is critical to the maintenance of the axon. The substances synthesized in the soma are needed not only by the axon but by certain of the peripheral cells that it innervates and perhaps by the Schwann cells that surround it. Axoplasm probably serves as a route for information about the end organ to which the axonal terminal is attached—information that must find its way to the soma in order to influence the plastic properties of the soma-dendrite complex. In case of injury to the axon, axoplasm may provide the route by which the soma feels the effect of distal damage and inaugurates the chromatolytic reaction. Synthesis is not confined to the soma; it also takes place to some degree in the axon and its terminals. The soma is not the only source of materials for axonal metabolism; it has been shown experimentally that the axon is capable of direct uptake of some molecules through extracellular space, Schwann cells, and the myelin sheath. The existence of rapid transport from soma to nerve endings and of local synthesis in nerve endings makes it possible to think of the presynaptic membrane as a site for some plastic changes, as in rapid learning. Before the new findings on rapid transport became available, the presynaptic site hád seemed an unlikely one for such changes because of the remoteness of the soma (see also p. 363).

Figure 2.16

Some types of synapses distinguished on the basis of topographic relations. [*Bodian, 1972.*]

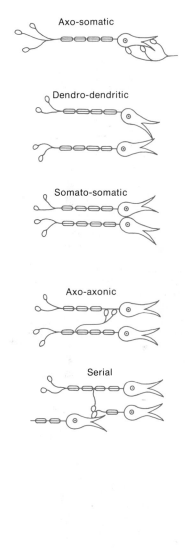

Axo-somatic

Dendro-dendritic

Somato-somatic

Axo-axonic

Serial

F. Physical Properties of Axons

Our knowledge of the physical properties of axonal materials, especially the changes in properties that accompany activity, is best treated here, although it is not entirely based on visible features of structure.

The viscosity of axoplasm varies widely among animals. In the giant axon of squids it is fluid enough to be expressed from a cut end by gentle squeezing, and this opportunity to obtain pure cytoplasm in quantity has been exploited in many studies. Viscosity is known to change measureably with electrical stimulation. Turgor has been measured in squid and cuttlefish axons, and tensile properties in crab nerve. A longitudinal orientation of more fluid channels is indicated by movements and shapes of droplets and vesicles. Stimulation of nerves has been found to alter staining character, volume, tension, and light-scattering in different preparations. Recent work has put some of these on a firm footing, especially changes with impulse conduction in light-scattering, in birefringence, and in phosphorescence after application of suitable dyes. It is still too early to know what these changes mean, but the hope is that they will give clues that may be interpreted in terms of alterations of molecular configuration during excitation or recovery.

On page 213 some evidence is cited of physical movements of nerve endings consequent to excitation.

G. Synapses

A **synapse** (Fig. 2.16) is the anatomical site of contact at which one nerve cell (the presynaptic one) can transmit a signal to another nerve cell (the post-synaptic one). This definition (a) implies that the neurons are separated by intact cell membranes, (b) distinguishes synaptic transmission from diffuse influence at some distance—for example, "field effects," or influence via neurosecretion, and (c) implies that each specialized site of contact is treated as *a* synapse. However, a single axon may synapse at several loci upon a single neuron and/or upon several different neurons. Likewise, a postsynaptic neuron often receives terminals from two or more presynaptic neurons within a small fraction of a micrometer—and in a characteristically organized manner, forming a triad or a compound synapse. Thus the definition also implies that there must be functional communication relevant to the information-processing role of neurons. There are many regions of so-called casual contact where, without exhibiting any visible specialization, neurons lacking the usual interposed glia are apparently incidentally juxtaposed; these regions are generally thought to have no functional influence, and are considered nonsynaptic. It may be only our ignorance, however, that permits the present ready dichotomy between synaptic regions and those nonsynaptic regions in which neuronal membranes may also lie

Figure 2.17
Diverse synapses upon giant neurons.

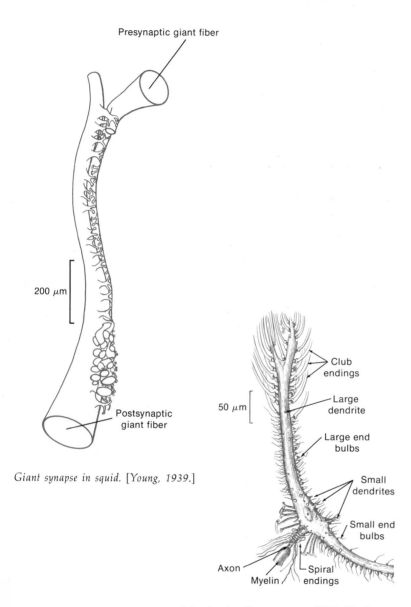

Presynaptic giant fiber

200 μm

Postsynaptic
giant fiber

Giant synapse in squid. [*Young, 1939.*]

Club
endings

Large
dendrite

50 μm

Large end
bulbs

Small
dendrites

Small end
bulbs

Axon

Myelin

Spiral
endings

Mauthner's cell synapse in goldfish. [*Bodian, 1952.*]

separated by only a few hundred Ångstrom units of intercellular space. We cannot exclude the possibility that weak influences are mediated through such "nonsynaptic" regions.

The concept of the synapse and its function is discussed further in Chapter 5. In the same chapter it is explained that a common, though not necessary, property of synapses is polarized, or one-way, transmission. The idea that the neuron is usually polarized, normally receiving excitation at its dendritic side and conveying it to its axonal terminal, was originally inferred by Cajal from his histological studies. He enunciated it as the "doctrine of dynamic polarization" (see also p. 103). Like his neuron doctrine, it has received abundant confirmation from modern electron microscopy as well as neurophysiology, though with significant refinement and qualification. Polarization is not, however, a universal characteristic of synapses as a class. The class is basically heterogeneous, both functionally and morphologically (Fig. 2.17). As is discussed in Chapter 5, there are electrically and chemically transmitting synapses, excitatory and inhibitory synapses, and each of the four possible permutations as well as a great range of morphological varieties.

Rather than describe an arbitrary "typical" or "normal" synapse, we shall acknowledge the significance of variety by selecting for illustration a number of the better known configurations. Because of differences in the scale of observation and the techniques used for the light and electron microscopes, there is a natural division between description at low magnification, in which emphasis is on topographic relations, and description at high magnification, in which emphasis is on ultrastructure and cell contacts.

Varieties of Contact at Low Magnification. The fabulous variety of characteristic forms of nerve fiber endings and synaptic contacts is one of the most impressive findings of anatomical science, as well as a distinguishing feature of the nervous system. It compels us to recognize a whole new field of study, the microarchitecture, or **architectonics,** of nervous tissue—a field that emphasizes spatial relations, geometry, and complex patterns. The first problem in attempting to deal with a welter of detail presented by nature is to discern **criteria for categorizing** that are not arbitrary or naive. Because our understanding of the functional significance of microarchitecture is still rudimentary, it is possible that the criteria used for classifying external form characteristics at low magnification, may not in fact include some of the most important features.

Synapses may be classified according to which parts of the presynaptic and postsynaptic neurons are involved: the contacts may be between presynaptic axonal terminal and postsynaptic soma, hence **axo-somatic,** or they may be **axo-dendritic** or, more rarely, **axo-axonal, dendro-dendritic,** or **soma-somatic.**

Another criterion is the general form of the whole array of axonal terminals that make contact with a given postsynaptic neuron (Fig. 2.18). We present here an incomplete list of examples modified from Cajal; more detail is given in Bullock and Horridge (1965).

1. Simple **contact-in-passing** between presynaptic axons and postsynaptic receptive processes (Figs. 2.18,A,C,D; 3.2,B; 3.3). This is possibly the dominant type of synapse in invertebrate neuropile (defined on p. 68 and in Glossary) (Figs. 2.54; 10.20,A; 10.25; 10.41; 10.51). Sometimes the simple contacts are single, whereas in other cases the two fibers touch, separate, and touch again several times before continuing on their courses.

2. Axo-dendritic connections by **climbing fibers** (Figs. 2.18,B; 2.19,B; 3.5,B; 3.16). This system resembles a vine entwining a tree. It offers extensive contact, serially ordered junctions, and private connection between two cells, contrasting in all these features with most of the other types. The best known example is the climbing fiber on the cerebellar Purkinje cell dendrite, which is also well established functionally as a powerful form of excitatory junction.

3. Axo-dendritic connection by **interdigitation.** In many places finger-like terminals of axons mesh, like gears, with corresponding dendritic projections. The most beautiful and diversified examples are found in the optic ganglia of insects (Figs. 10.41; 10.42; 10.43). Cajal, who himself analyzed this system in great detail, said "it seems that nature has attempted to show us in the insect nervous system . . . how in minimum space it is possible to organize a maximum of fine and subtle structures. . ." (1954, p. 81). Actually, for the sake of simplification, we are forcing into one rubric a multifarious array of endings in arthropods, cephalopods, vertebrates, and other groups (Figs. 2.29; 3.2; 3.3; 3.8; 10.31; 10.56). These exhibit the forms of brushes, tassels, tufts, taproots, shrubs, clubs, panicles, and other excrescences, each consistent and characteristic of its cell type. Moreover, specific localized terminal arborizations are known to arise at definite sites along the single stem process of unipolar cells, with a characteristic form of branching at each site.

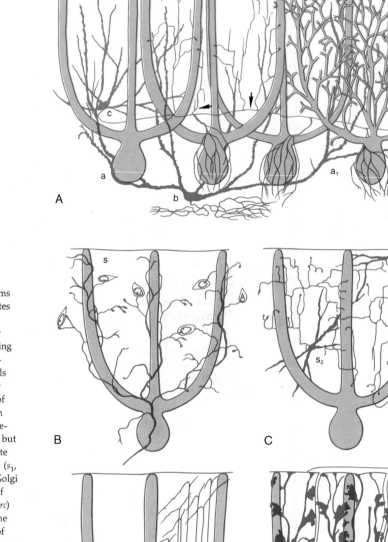

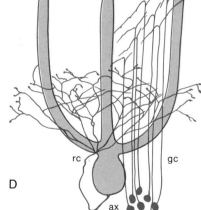

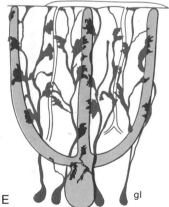

Figure 2.18
Variety of synapses on one neuron. Diagrams
of the surroundings of Purkinje cell dendrites
in the cerebellar cortex; the dendrites are
shown greatly simplified, even in the upper
right. **A.** basket cell axon (*c*) sends descending
collaterals to make baskets (a_1) around Purkinje cell somas (*a*) and ascending collaterals
(arrows) to make synapses in the molecular
layer. Large Golgi type II ganglion cell (*b*) of
the granular layer sends dendrites widely in
the molecular layer. **B.** Climbing fiber is presynaptic not only to the Purkinje dendrites but
to many adjacent elements, especially stellate
cells (*s*). **C.** Axonal plexuses of stellate cells (s_1,
s_2), which are short-axoned and therefore Golgi
type II cells, lie within the dendrite arbor of
one Purkinje cell. **D.** Recurrent collaterals (*rc*)
from Purkinje axon (*ax*) form a plexus in the
lower third of the molecular layer. Axons of
granule cells (*gc*) ascend and bifurcate to run as
parallel fibers at right angles to the plane of
the flattened Purkinje dendrite arbor. **E.** Stalks
of neuroglia cells (*gl*) make extensive contacts
with the dendrites and with cortical blood vessels (*bv*). All elements are simultaneously present about each Purkinje dendrite system, and
these systems extensively overlap. [Scheibel
and Scheibel, 1958b.]

4. Axo-dendritic connections by **right-angled arrays,** with axons of great length (Fig. 2.19,A). This curious and provocative synapse is best known in the cerebellum, where smooth, free endings of granule cell axons run for about two millimeters as unbranched, unmyelinated, terminal filaments in the molecular layer parallel to the surface and to each other, while Purkinje cell dendrites richly ramify in a more-or-less perfectly flattened espalier (plane) at right-angles to them. Thus each granule cell axon makes minimal passing contact with about 45 Purkinje cells in an ordered sequence, and each of the latter is touched by about 90,000 parallel granule cell axons.

5. Axo-dendritic connections by **laminar plexuses.** Especially in the retina of vertebrates, cephalopods, and insects, terminal axon arbors are found in a narrow stratum, interlacing with similarly confined postsynaptic receptive processes (Figs. 2.37; 2.53; 2.56; 3.4; 3.20; 10.41; 10.42; 10.50; 10.51; 10.56). Laminar ramifications may be repeated at as many as eight levels and extend to defined and characteristic distances in each stratum.

6. **Multineuronal functional complexes** of different types. One axon may embrace a circumscribed cell cluster. In certain places the axon ends by exploding into a thicket of fine twigs that intertwine with a discrete population of postsynaptic cells. Even more elaborate knots of axons and dendrites of several neurons may form a defined unit called a glomerulus (Figs. 2.33; 2.34).

7. Axo-somatic connections by **thick nests.** This class is exemplified by the cerebellar basket, which is a form of axonal ending that fits neatly over a single Purkinje cell soma (Figs. 2.3,E; 2.18,A; 3.17). A single presynaptic neuron may send axonal baskets to many Purkinje cells (Fig. 3.19). These are one of the best established forms of inhibitory junctions.

8. Axo-somatic connections by **sparse nests.** In contrast to the preceding class, these resemble early entwinement by growing vines (Fig. 2.19,D,F).

9. Axo-somatic connections by **calyces** (a kind of cup) (Figs. 2.19,C,H; 2.29,A). In certain places the postsynaptic cells receive broad, flat, petal-like axon endings that virtually engulf the soma. Such synapses made a special contribution to the establishment of the neuron doctrine because of the direct evidence they offer of the discreteness and independence of neurons.

10. Axo-somatic connections by **thickened terminal tubercles** (Figs. 2.19,G; 3.2,A; 7.13). This term covers a heterogeneous assortment of specialized endings with large or small expansions that make contact with the soma and large dendrites. The axon may end in a warty tubercle or a swollen suction-cup-like body or a terminal button.

11. Axo-axonal connection by simple **end-to-end contact.** The giant fibers of earthworms (Fig. 10.19) and the lateral giants of crayfish (Fig. 10.27) are actually chains of segmental units, each bounded by a complete cell membrane. Contact is by a simple, symmetrical apposition of membranes. These synapses are nexuses and electrical.

12. Axo-axonal connections by **postsynaptic protuberances.** These are found in the giant synapse of the squid (Fig. 2.17), in the giant-to-motor synapse in crayfish (Fig. 10.27), and in many invertebrate neurons where small tufted projections from the postsynaptic axon act as specialized receptive appendages.

13. **Somato-dendritic connections.** This is a relatively uncommon synapse, but is known at least in sympathetic ganglia.

14. **Neuroeffector junctions by free branched endings.** Most neuromuscular junctions in invertebrates (Fig. 2.20), the junctions on smooth muscle in vertebrates, most neuroglandular junctions, and the junctions on ciliated cells, luminescent organs, and chromatophores appear to be simple terminals, or contacts-in-passing, of moderately to extensively branched axons.

15. Neuromuscular junctions with **postsynaptic folds.** The axon terminals on vertebrate skeletal muscle fibers of the common-

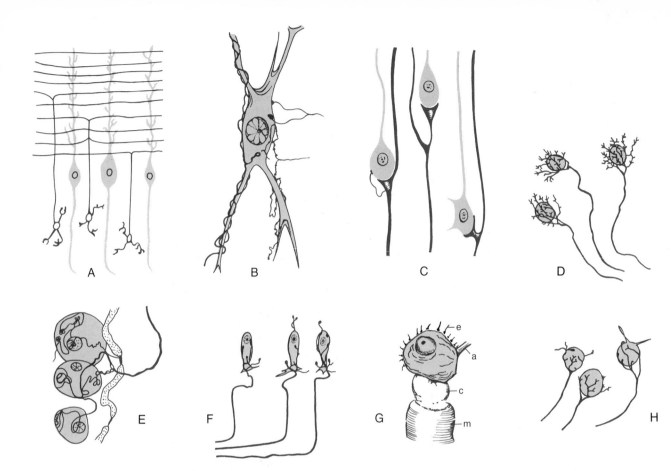

Figure 2.19
Types of synaptic endings observable with light microscopy and special stains; vertebrate central and peripheral nervous systems. **A.** Parallel fibers of cerebellar cortex making right-angle synapses with Purkinje cell dendrites, which lie in the plane vertical to the page. Golgi method. [Cajal, 1954.] **B.** Diagrammatic cell in Clarke's column of the spinal cord with its three types of synapses: (*i*) the "giant synapses" are seen on the large dendrites at the left, above and below, as entwining fibers from muscle spindle afferents; (*ii*) end feet on the right come from excitatory interneurons under the influence of skin afferents; (*iii*) the meshwork of extremely fine fibers on the soma comes from inhibitory interneurons in the pathway from antagonistic muscle afferents. [Szentágothai, 1961.] **C.** Cells of the tangential nucleus of a bird (young kite) receiving synapses from fibers of the vestibular nerve. Cajal method. [Cajal, 1954.] **D.** Terminal ramifications of afferent fibers in the lateral nucleus of the thalamus of a mouse. Golgi method. [Cajal, 1954.] **E.** Endings on intraparietal neurons of the auricle of the heart of a fish. Note the thick unmyelinated fiber terminating in special formations and the myelinated fiber giving off an unmyelinated collateral from the node of Ranvier. Method of Gros. [Laurent, 1957.] **F.** Nests formed by centrifugal fibers that reach the retina and surround cells, called associational amacrines. Methylene blue. [Cajal, 1954.] **G.** A cell in the reticular formation of the goldfish, which, besides small end bulbs (*e*), receives a large club ending (*c*) from a myelinated fiber (*m*); *a*, axon of postsynaptic neuron. [Bodian, 1942.] **H.** Calyces of Held in the nucleus of the trapezoid body of a kitten. Golgi method. [Cajal, 1954.]

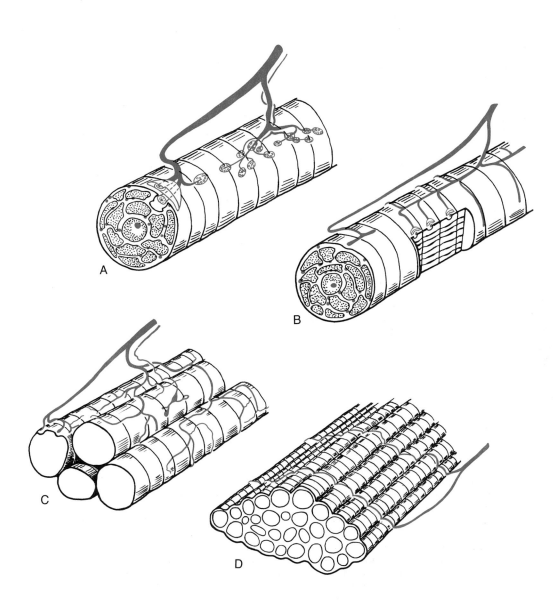

Figure 2.20
Insect neuromuscular endings. **A.** The end-plate type; general in leg muscles of coleopterans and hymenopterans (also in the proboscis of dipterans; note fibril columns, mitochondria, and glial cell covering nerve ending with vesicles. **B.** The metameric type with circular end branches approximately between every two Z-bands; found in *Dytiscus* (Coleoptera) leg muscle. **C.** The diffuse type; general in Orthoptera; note fast and slow nerve fibers. **D.** The simple transverse type, with rarely ramifying circularly disposed end branches; general in flight muscles of coleopterans and hymenopterans. Based on metallic impregnations and electron micrographs. [Courtesy J. Szentágothai and J. Hámori.]

est or twitch type are unique in the form of the expanded terminal apparatus (Fig. 2.21). The terminals spread more or less like a hand and involve a considerable area of the muscle surface; there is usually only one such ending, or end plate, on a given muscle fiber from a given axon, in contrast to the many in the next category. More remarkable is the specialization of the postsynaptic membrane; the muscle membrane is thrown into deep folds in which the presynaptic terminals do not follow, so that some space and intercellular material is left.

16. **Neuromuscular junctions with simple end plates.** In contrast to the previous class, end plates in muscle with multiterminal and polyneuronal innervation (see Glossary) are without subsynaptic folds (Figs. 7.2; 7.4; 7.5).

17. **Dendo-dendritic** contacts. Regarded as questionable until recently, these turn out to be common and are looming in importance (Fig. 2.58). They are often reciprocal (see p. 48 and Fig. 2.28).

18. **Soma-to-soma** connections are known and may have a similar interest.

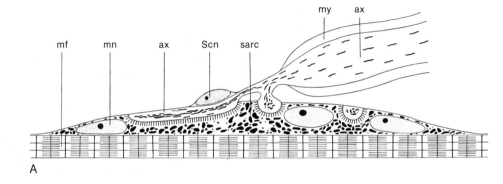

A

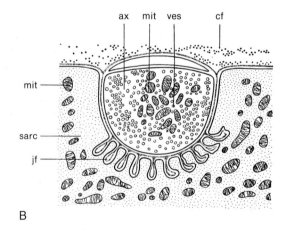

B

Figure 2.21
Vertebrate neuromuscular junction as revealed by the electron microscope. **A.** Schematic drawing of a motor end-plate at low magnification. **B.** Diagram of a cross section of one synaptic gutter and axon branch. *ax*, axoplasm; *cf*, collagen fibrils; *jf*, junctional fold; *mit*, mitochondria; *mf*, myofibrils; *mn*, muscle nuclei; *my*, myelin sheath; *sarc*, sarcoplasm; *Scn*, Schwann cell nucleus; *ves*, vesicles. [Couteaux, 1958, after Robertson.]

19. Contacts between **sensory terminals** and nonnervous receptor cells. These are found in the ear (Figs. 2.61; 2.65; 2.69), lateral line (Figs. 2.68; 2.72), taste buds (Fig. 2.71) and Merkel's and Iggo's skin receptors. They are properly called synapses, and show considerable variety. Doubtless their arrangement and form are functionally significant.

20. **Sensory endings that are themselves receptors.** Included strictly for comparison, this heterogeneous group includes sensory corpuscles, muscle spindles, olfactory neurons, free nerve endings in skin, and apparently all invertebrate receptors. Since these lack presynaptic elements, they are not synapses, but they are of interest here because of their form. The geometry of the endings is characteristic of each type and, we presume, related to function (see lower right margin). Section VI adds some details.

The list of synaptic forms by external configuration given above is not complete and includes several heterogeneous categories. In other words, there are a large number of presently distinguishable forms. Moreover, they do not occur at random but are consistent for given pre- and postsynaptic cells. Year by year, advancing anatomical information establishes in more detail and at additional sites the existence of consistent precision in the characteristics of the synaptic relations between specified neurons (see p. 99).

Varieties of Synapses at High Magnification. To the small handful of light-microscope methods available, the electron microscope (EM) adds a new and powerful method for revealing details of the processes of neurons and neuroglia. Previously they have been revealed only by very special metallic impregnations, by methylene blue staining, and recently by injected Procion yellow. In many animals no method yet attempted

has made nerve fibers visible in the light microscope. Electron microscopy makes it possible to trace processes to the ends of their finest branches; with the light microscope, it was never certain whether we could see them to the very end. By means of the powerful but laborious method of reconstruction from serial sections, the forms, branches, destinations, and connections can be mapped completely in volumes of tissue a few score of micrometers on a side and as much as twenty or so micrometers in depth (Fig. 10.43). EM techniques have been combined with light-microscope observation and staining—for example, by Golgi impregnation or Procion injection, sometimes using alternately thick and thin sections to advantage. Techniques of filling the neuron and its processes have been extended to the electron opaque precipitate of a cobalt salt (Fig. 10.61).

Autoradiography combined with the EM can reveal how axoplasmic transport distributes proteins made from radioactive amino acids taken up by a neuron (Fig. 2.22). The problem of following branches and visualizing form in three dimensions has been reduced somewhat by applying the technique of scanning electron microscopy to surfaces created by tearing, cutting, or fracturing nervous tissue along cleavage planes at different stages of preparation. (See further, Section VII, p. 93.)

Of the utmost importance has been one of the simplest findings of electron microscopy—namely, that neurons and all their processes are bounded by continuous cell membranes, thus remaining distinct and separate from each other. This confirms elegantly the basic tenet of the neuron doctrine.

Criteria for recognizing synaptic junctions in the EM and for identifying the

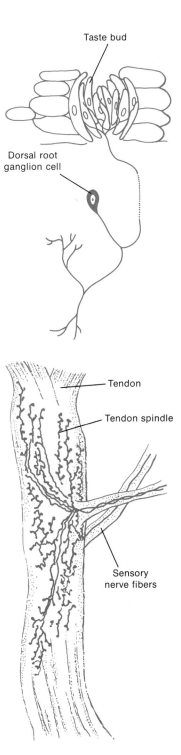

Taste bud

Dorsal root ganglion cell

Tendon

Tendon spindle

Sensory nerve fibers

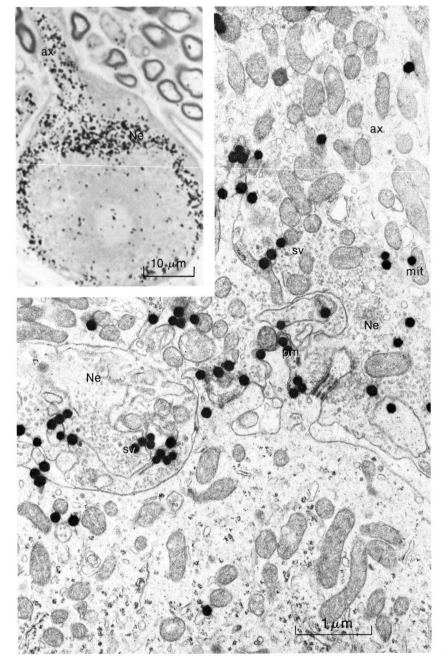

Figure 2.22
Autoradiographs visualizing axoplasmic transport of protein to synapses. ³H-labeled amino acid was injected into the vicinity of the nerve cells in the brain and was taken up and synthesized by them into protein. After 18 hours the ciliary ganglion in the orbit was prepared and sectioned. Both pictures show the radioactivity of proteins transported in the axons of preganglionic parasympathetic neurons in the IIIrd nerve. The inset, taken by light microscope, shows the dense accumulation of silver grains over the terminal part of the axon (*ax*) and its expanded ending (*Ne*), which embraces the postsynaptic soma (called a calyx and referred to in Figure 2.8, B. The large view, taken by electron microscope, is from an area similar to part of the inset; most silver grains are seen to be concentrated over the areas also occupied by synaptic vesicles (*sv*) and presynaptic plasma membrane (*pm*). Only a few are seen over mitochondria (*mit*) or postsynaptic cytoplasm; these proteins are presumably concerned with renewal of membrane components of the synapse. [Courtesy of B. Droz.]

pre- and postsynaptic sides have been gradually established by a patient process starting with cases in which the identification of pre- and postsynaptic sides was already clear on other grounds. We cannot be certain that our criteria are infallible, especially whether they include all the regions of functional interaction between neurons. Unspecialized junctional regions may go unrecognized.

The principal feature that characterizes synapses in the EM is an apposition of the pre- and postsynaptic cell membranes. This means that the barrier of neuroglial processes that generally separate neurons has windows. Appositions without gaps ("tight junctions") function as high-resistance seals but probably not as synapses. Those with 20–50 Å spaces ("gap junctions") forming an array between points of direct contact are called nexuses; they function as low-resistance electrical synapses. More commonly there is a cleft of 200 Å or more, which is even wider than some common, supposedly normal intercellular spaces. The synaptic cleft is specialized by its uniform width and by an ordered arrangement of a macromolecular material, mucopolysaccharidal in character, that occupies the intercellular spaces of nervous tissue generally.

The regions regarded as synaptic on electron-microscopic grounds generally have several other characteristics as well (Fig. 2.23). The cell surfaces appear thickened due to a greater electron density and condensation of the cytoplasm subjacent on the postsynaptic side or on both sides. The most typical specializations are on the presynaptic side. There may be an array of local thickenings of the presynaptic membrane, forming a hexagonal pattern of dense inward projec-

Figure 2.23
Variations of synaptic structure and topography. Arrows indicate direction of transmission. [Bodian, 1972.]

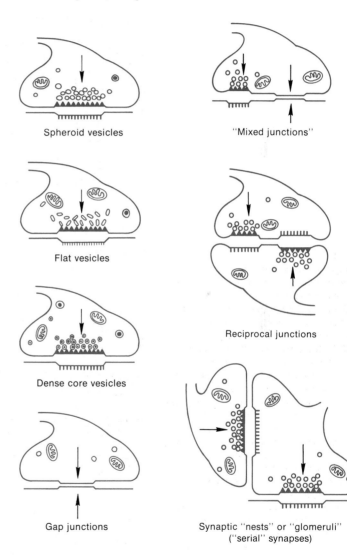

Spheroid vesicles

"Mixed junctions"

Flat vesicles

Reciprocal junctions

Dense core vesicles

Gap junctions

Synaptic "nests" or "glomeruli" ("serial" synapses)

Figure 2.24
Left. Low-power diagram showing a presynaptic terminal within the rectangle. **Center.** An enlarged view of the same terminal, showing the vesicular grid, "synaptophores" and vesicles incorporated into the cell membrane, a single mitochondrion, neurofilaments, and the cleft, but no postsynaptic detail. **Right.** Diagram of terminal at lower magnification than the preceding, as seen with a different fixative: *dv*, dense core vesicle; *sv*, spherical vesicle; *pr*, presynaptic projection; *po*, postsynaptic cytoplasm. [Akert et al., 1972.]

tions, with a period of 800 Å, enclosing thin spots large enough for one synaptic vesicle (see below) and occupied in some proportion by such vesicles (Fig. 2.24). Neurofilaments are typically absent but in certain terminal buttons may form a whorl (Fig. 2.31). One or a few mitochondria are often present near the junction.

Most characteristic of the synaptic features seen in the EM is a clump of synaptic vesicles, near the membrane. These are generally confined to the presynaptic side, but sometimes vesicles are seen on both sides. There are at least four kinds of synaptic vesicles.

1. Light-core or clear spheroidal vesicles 200–500 Å in diameter. In certain junctions examined, these occur in excitatory synapses and contain acetylcholine, but they also occur in others that do not involve acetylcholine, including some probable inhibitory endings.
2. Light-core ovoidal or pleiomorphic vesicles (easily flattened by certain preparative procedures) of about the same diameter. These are typical of certain synapses known physiologically to be inhibitory. Generalization is not justified, however.
3. Granulated vesicles of about 500 Å. These may contain catecholamines.
4. Dense core vesicles of 800–1000 Å. These may also contain catecholamines.

These classes and their physiological correlates are being modified with increasing knowledge, but it is clear that synaptic vesicles are not all alike, and that morphological and chemical-functional differentiation may go together. Therefore, it may be possible to use morphological criteria to infer the functional type or state of a synapse (Fig. 2.25). We must be prepared, however,

to learn that the criteria are not universal.

One technical development of particular promise in this area is the discovery of means of breaking off the axonal terminals near their ends. The postsynaptic membrane also breaks away from its neuron and adheres to the presynaptic membrane, forming a composite little body, the synaptosome. These bodies can be concentrated by centrifugation, permitting chemical as well as EM characterization of synaptic structures collected from different parts of the brain and collected during different functional states. Further treatment can isolate synaptic vesicles and permit chemical analysis of accumulated samples.

New developments are making it possible to count all the synapses, defined by EM, in a given area of a section. Although it is premature to make general statements, there seems to be evidence of changes in the count with physiological state and with age. Counts can also be made of the dendritic spines, and these have been found in certain cases to increase with age and with stimulation by so-called enriched environments —those full of objects and variety (Fig. 2.26; see also Chapter 9).

Other specialized synaptic structures occur commonly but not generally. We should not expect that the several types of junctions recognized above on the basis of light-microscopy would have a one-to-one correspondence with any classification of synapses based on EM evidence, since the scale of the former is much larger. Quite diverse configurations on that scale might have a common ultrastructural basis. Several types have, however, been distinguished on EM cri-

Figure 2.25
Change in synaptic morphology with change in state.

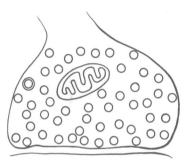

Nembutal

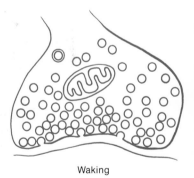

Waking

Synapses from anesthetized animals tend to be less arched, vesicles less aggregated and rarely opening to the cleft. [Akert and Livingston, 1973.]

Figure 2.26
Changes in synaptic substratum—that is, in dendrite morphology—with changes in age and experience.

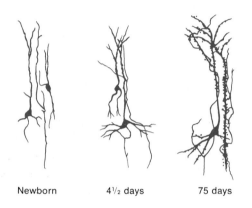

Newborn   4½ days   75 days

Development of apical dendrites in neurons of kitten cortex. [Scheibel and Scheibel, 1963.]

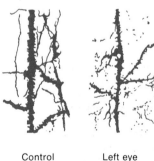

Control   Left eye blinded 30 days ago

Rabbit visual cortex apical dendrites in normal and unilaterally blinded adults. [Scheibel and Scheibel, 1968b.]

teria (Figs. 2.23; 2.27). Some examples are given in the following list.

1. **Symmetrical nexuses, or gap junctions.** Two apposed membranes with an overall thickness of 150 Å come within 20–40 Å of each other. This gap, as visualized by filling it with lanthanum, is actually a hexagonal array of 20 Å channels with a period of 100 Å; the channels communicate with the general intercellular space. Structures bridge the gap and put the two membranes into contact. Penetrating these bridges are channels perhaps 10 Å in diameter that allow ions and small molecules to pass between the cytoplasms of the two cells. The formation and dissolution of nexuses and low-resistance electrotonic coupling between cells are normal physiological and developmental processes. Nexuses correlate well with electrical transmission. Such junctions are distinguished from occluding, or tight, junctions, which have no gap or intercytoplasmic channels, and act as seals between cells to block the low-resistance pathway for current flowing along intercellular clefts. (Figs. 2.27 A,B,C; 2.28A).

2. **Axo-dendritic synapses** with narrow cleft. This category cannot be lumped with the preceding because, though transmission is electrical, it is polarized. The first example that was well studied both physiologically and ultrastructurally is the synapse between central giant and segmental motor axon, in the abdominal cord of the crayfish (category 12, p. 33; Fig. 10.27). The large postsynaptic axon, which supplies flexor muscles, sends short receptive processes, which may be regarded as microdendrites, into indentations of the presynaptic fiber (Fig. 2.27,F). The cleft is narrow, 15–20 Å. Vesicles may occur on either or both sides but are not numerous or in large clumps. Several examples are known in the vertebrates, including the excitatory electrical endings on the giant Mauthner cell of fishes. Not uncommonly, cells exhibit mixed electrical and chemical synapses, even in the same presynaptic fibers (Figs. 2.23; 2.29).

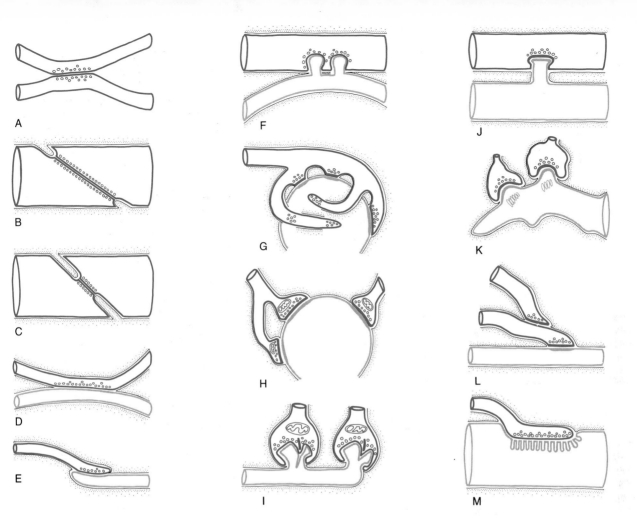

Figure 2.27

Types of synapses revealed by electron microscopy. Brown, presynaptic; gray, postsynaptic; black, the innermost extent of the sheath. **A.** Example of coelenterate axon-axon synapse in jellyfish ganglion, with vesicles on both sides and without adhering sheath cells. **B.** Earthworm septal synapse with close apposition of the membranes of the two cells and sparse vesicles on both sides. **C.** Crustacean septal synapse, as in B except that the synaptic area is restricted. Examples B and C have electrical transmission in either direction. **D.** Axon-axon synapse-in-passing typical of neuropile in many invertebrate ganglia, often with and often without sheaths. **E.** Axon terminal arborization ending on a fine dendrite in invertebrate neuropile. **F.** Crustacean giant-fiber-to-motor-neuron synapse, with postsynaptic motor fiber invaginated into the giant fiber. This example also has electrical transmission, but only in one direction. **G.** Axon arborization-to-soma synapse typical of vertebrate brain cells, but in invertebrates so far only clearly known as inhibitory endings on crustacean peripheral sensory cell of muscle receptor organ. **H.** Terminal buttons of axon arborizations typical of certain central neurons in vertebrates. **I.** Ribbon synapses between rod cell endings and dendrites of ganglion cells of vertebrate retina, with presynaptic specialization. **J.** Synapse between giant fibers of squid stellate ganglion, postsynaptic invaginated into presynaptic. **K.** Spine synapse (axon-dendrite) from cerebral cortical dendrite of vertebrates with postsynaptic specialization. **L.** Serial synapse, found so far in spinal cord, cerebral cortex, and plexiform layer of retina in vertebrates, but offering many potentialities for presynaptic inhibition and other complex interaction in neuropile. **M.** Specialized neuromuscular endings found in vertebrate skeletal muscle, with postsynaptic grooves. [Bullock and Horridge, 1965.]

44

Figure 2.28
Electrical and dendro-dendritic synapses. **A.** Medulla oblongata of a goldfish. In the center of the field is a synapse between an axonal club ending (*at*) and the lateral dendrite of a Mauthner cell. At the synaptic junction, the pre- and postsynaptic membranes come into close apposition and are separated by a distance of only about 20 Å. Thus a seven-layered structure is formed. This is an electrical synapse. At the site of apposition, striations (arrows) occur in the gap, giving a ladder-like appearance to the junction. On each side of this "gap" junction, the pre- and postsynaptic membranes diverge from each other to form punctate adhesions (triangles). Within the axon terminal are synaptic vesicles (*sv*) bounded by triple-layered membranes. **B.** Olfactory bulb from a rat. In the lower half of the field is the secondary dendrite (*den*) of a mitral cell. This dendrite synapses with a gemmule (*gem*) protruding from a granule cell dendrite. At the synaptic junction between these two dendritic components are two synaptic complexes with opposite polarities, as indicated by arrows. Where the direction of transmission, as judged by the grouping of the synaptic vesicles, is from the mitral dendrite to the gemmule, a dense filamentous material (*f*) underlies the postsynaptic membrane. Where the direction is from the gemmule to the mitral cell dendrites, the polarity is not marked by a postsynaptic density. [Peters et al., 1970; micrographs by T. Reese, courtesy of M. W. Brightman.]

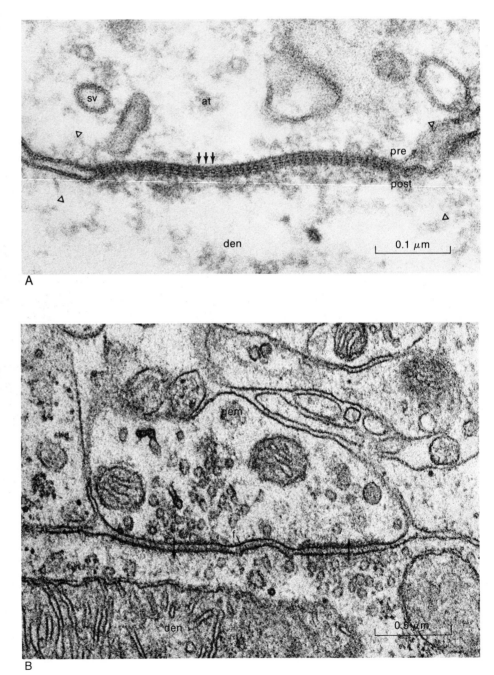

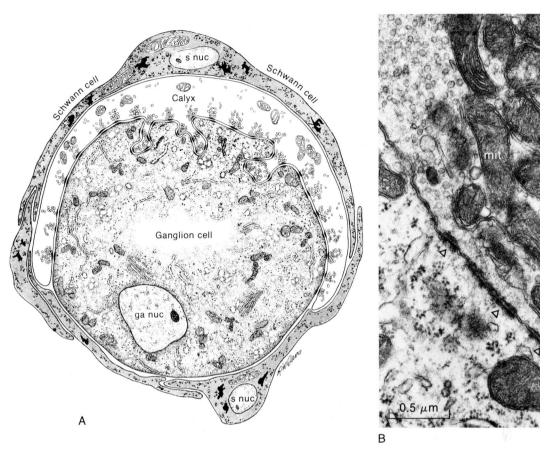

Figure 2.29

Mixed synapses. **A.** Calyciform synapse in the ciliary ganglion of the chick, as revealed by the electron microscope. Note the locally dense regions of the opposed synaptic membranes, the clusters of synaptic vesicles at these sites on the presynaptic side, and the uniform cleft width (300–400 Å).
ga nuc, ganglion cell nucleus; *s nuc,* Schwann cell nucleus. [De Lorenzo, 1960b.] **B.** Lateral vestibular nucleus of a rat. Occupying the center of the field is a large axon terminal containing many synaptic vesicles (*sv*) and mitochondria (*mit*). This terminal forms a mixed synapse with the perikaryon of a Deiters neuron. At the synaptic junction, complexes of three types occur. (1) Complexes in which there is a close apposition between the pre- and postsynaptic membranes (asterisk). These are considered as probable electrical synapses. (2) Complexes in which there is a cleft between the pre- and postsynaptic membranes and a prominent postsynaptic accumulation of dense material (arrow). Such junctions usually have synaptic vesicles associated with them, so that they have the form of typical chemical synapses. (3) Complexes in which there is a space between the pre- and postsynaptic membranes and symmetrically disposed dense material (triangles). These are typical puncta adhaerentia—points at which cells adhere, but which are not considered to be synapses. [From Peters et al., 1970; micrograph B by C. Sotelo.]

Figure 2.30
Two subtypes of chemical synapses that differ in dimensions. [Akert et al., 1972.]

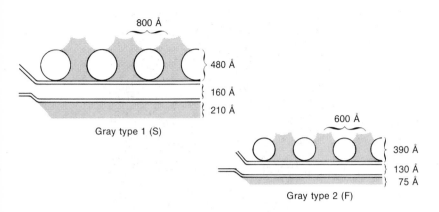

Gray type 1 (S)

800 Å

480 Å

160 Å

210 Å

600 Å

390 Å

130 Å

75 Å

Gray type 2 (F)

Figure 2.31
Varieties of axosomatic and axodendritic synapses.

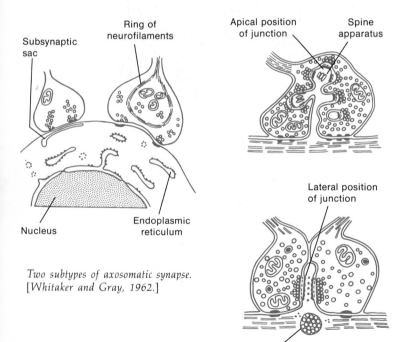

Subsynaptic sac

Ring of neurofilaments

Apical position of junction

Spine apparatus

Nucleus

Endoplasmic reticulum

Lateral position of junction

Subjunctional body

Two subtypes of axosomatic synapse.
[Whitaker and Gray, 1962.]

Two subtypes of axodendritic synapse.
[Hamlyn, 1962.]

3. **Axo-axonal synapses with short receptive processes, clustered vesicles, and wide cleft.** Represented by the squid giant synapse (Figs. 2.17; 2.27; see also p. 434) this type is well known physiologically. Electrical transmission has been ruled out, and some evidence for glutamate as the transmitter is reported.

4. **Unspecialized axo-dendritic connections** between fine fibers in invertebrates. In earthworm, insect, and other fine-textured neuropiles, numerous localized areas are presumed to be synapses solely on the grounds that they make contact without intervening glia and that they have clusters of vesicles on one side, which is thought therefore to be presynaptic. These quite unspecialized synapses may occur as contacts of fibers crossing at right angles (and therefore of small area), as longitudinal contacts of large area, or as end-knob contacts (Fig. 2.27,D,E).

5. **Axo-somatic and axo-dendritic synapses.** These wide-cleft (200–300 Å) junctions have been subdivided into two categories (Figs. 2.30; 2.31). One is simple and small in diameter (ca. 0.25 μm); the other is large (0.5–1 μm) and has more postsynaptic densification. The first category includes terminal button synapses on spinal motor neurons, endings on postganglionic autonomic neurons, and inhibitory junctions on the crayfish stretch receptor cell—obviously a heterogeneous set. The second includes many junctions of the mammalian cortex. These synapses may also be subdivided by vesicle shape (see above) and on other grounds (Fig. 2.31).

6. **Axo-dendritic synapses on spines.** Dendritic spines are a late-developing specialization of certain cells in higher animals, and they exhibit some special features in fine structure (Figs. 2.27,K; 2.31; 2.32).

7. **Invaginated ribbon synapses.** In the vertebrate retina and elsewhere, there are junctions similar to those of category 5 that exhibit a special presynaptic ribbon of unknown function (Fig. 2.27,I).

8. **Serial synapses.** These are the supposed bases for presynaptic inhibition; axon *A* is presynaptic to axon *B*, which in turn is presynaptic to neuron *C*, the junctions being within a few micrometers (Figs. 2.23; 2.27,L).

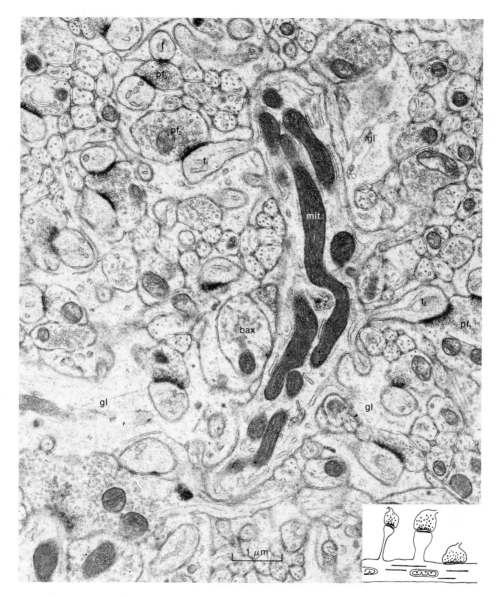

Figure 2.32

Synapses on spines of dendrites. In the center of the electron micrograph is the spiny branchlet of an adult rat cerebellar Purkinje cell; its covering is a layer of neuroglia (*gl*), and in its shaft are mitochondria (*mit*). A basket cell axon (*bax*) synapses directly onto the spiny branchlet. Two thorns (t_1, t_2) emerge from the shaft to synapse with parallel fiber axon varicosities (pf_1, pf_2); another parallel fiber varicosity (pf_3) is presynaptic to two Purkinje cell thorns (*t*). [Palay and Chan-Palay, 1974.] Inset at lower right is a low-magnification diagram of two synapses on spines contrasting with one on the dendritic shaft. [Scheibel and Scheibel, 1968b; courtesy of L. Westrum.]

Figure 2.33
Two types of synaptic glomeruli.

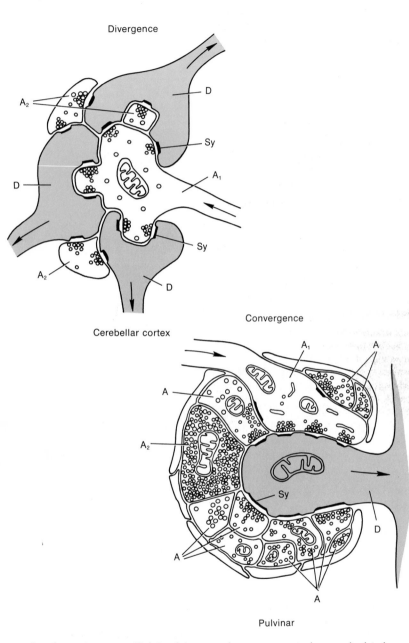

Divergence

Cerebellar cortex

Convergence

Pulvinar

Complexes of synapses (Sy) involving several neurons seem to have evolved independently many times. Some, as in the cerebellum, center on a single arriving axon terminal (A); others, as in the pulvinar of the thalamus, center on a single departing dendrite (D). [Steiger, 1967.]

9. **Neuromuscular junctions with post-synaptic grooves** (Fig. 2.27,M). This class corresponds to the light-microscope category 17. Both postsynaptic membrane and cleft material give positive cytochemical tests for acetylcholinesterase.

10. Neuromuscular junctions without grooves. This corresponds to the light-microscope category 15.

11. **Reciprocal synapses.** Wide cleft junctions that resemble those in category 5, above, are sometimes not polarized exclusively in one direction, so that a certain dendrite is solely postsynaptic, but instead lie side by side, polarized in opposite directions, so that each neuron is both pre- and postsynaptic. They are usually dendro-dendritic synapses (Fig. 2.28).

For the neurophysiologist it is important to note that one dendrite may bear morphologically diverse synapses; one presynaptic cell may produce morphologically different kinds of endings upon other fibers; one dendrite can be both pre- and postsynaptic.

Furthermore, one pre- or postsynaptic terminal specialization may make contact with several other neurons. That is, synapses are not always sites of intimate contact between one pre- and one postsynaptic unit, as we have already noted (p. 29). **Synaptic complexes** involving one or two enlarged central endings—in some examples, dendritic, in other examples, axonal—may be surrounded by an array of small terminals (Fig. 2.33). These end on each other as well as on the central element, and include both axon and dendrite terminals traceable to a number of cells, near and distant. The whole complex may be delimited by glia to form a characteristic knot, or **synaptic glomerulus** (Fig. 2.34). No doubt this is the site of rather complex, functional integrative transactions.

The extent and the specialization of dendritic surface available for synaptic contacts are vastly increased by the development of **spines and crests,** especially in some neurons. These vary widely in the profusion in which they are developed in different parts of the brain and in different animal groups. Of special significance is the recent finding that they can be more or less developed according to environmental factors or experience (Chapter 9).

The number of synaptic endings on a neuron varies widely among neurons, and not just because neurons differ greatly in extent of dendritic surface. It is estimated that a Purkinje cell of the cerebellar cortex receives some 90,000 synaptic endings. Counts of synaptic junctions in electron micrographs of the outer layer of the rat cerebral cortex indicate more than $10^9/mm^3$. Such figures should certainly be added to the commonly cited estimate of 10^{10}–10^{11} neurons in the nervous system of man (which later workers regard as low by perhaps a factor of ten). It must also be appreciated that the dogma that functional contacts are all visibly recognizable rests on limited evidence.

H. Diversity of Nerve Cells

Nerve cells can be **unipolar, bipolar,** or **multipolar** according to the number of processes emerging from the soma. In some primitive neurons the processes are indistinguishable; those neurons are called **isopolar** (Figs. 2.3; 10.2). Neurons in which axon and dendrite can be distinguished are called **heteropolar** (Figs. 1.3; 2.1; 2.3). The dominant type of neuron in vertebrates is a multipolar, heteropolar neuron (Fig. 2.53); the dominant type in invertebrate central nervous systems is unipolar with a branching stem process (Fig. 2.37; 2.53).

Another classification is functional, though usually demonstrated anatomically: nerve cells are **sensory, internuncial** (interneurons), or **motor.** Interneurons are often classified according to their connections: **projection** neurons send an axon a considerable distance rostrally or caudally in the central nervous system; **commissural** neurons send an axon to corresponding structures on

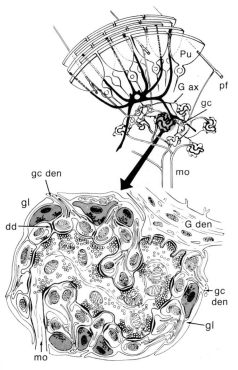

Figure 2.34
Synaptic glomeruli in the cerebellum. **Above.** A diagram at light-microscope magnification showing seven glomeruli in the layer of granule cells (*gc*). **Below.** One glomerulus shown at EM magnification. *mo,* mossy fiber afferents to the cortex; *G ax* (and shaded fibers below), axon from Golgi cell (dark shading signifies an inhibitory influence, in this case upon the dendrites in the glomeruli); *G den,* Golgi cell dendrite; *gc den,* granule cell dendrite; *gl,* glial capsule; *pf,* parallel fiber granule cell axon; *Pu,* Purkinje cell; *dd,* desmosomoid dendro-dendritic contacts. [Szentágothai, 1970.]

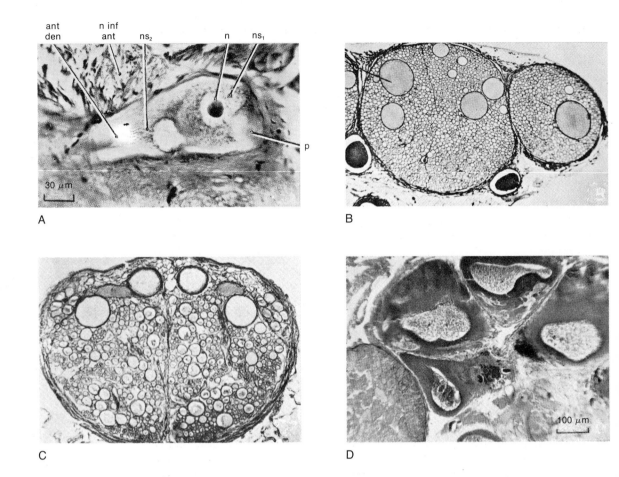

A

B

C

D

Figure 2.35

Giant cells and fibers. **A.** First-order giant cell in the brain of *Loligo*. Picroformol, hematoxylin, and eosin. [Young, 1939.] *ant den,* anterior dendrite; *n, nucleus; n inf ant,* anterior infundibular nerve; ns_1, central irregular granules of Nissl substance; ns_2, longitudinally arranged peripheral granules of Nissl substance; *p,* pathway from axon to dendrites. **B.** Transverse section of nerves of the cuttlefish *Sepia,* showing giant and small fibers. [Courtesy J. Z. Young.] **C.** Transverse section of the ventral cord of a crayfish fixed by the method of vom Rath. The four giant fibers are seen dorsally, and a motor axon is en route out of the cord, on each side, just about to pass dorsal to the lateral giants and to make synaptic contacts with them. [Robertson, 1961.] **D.** The somas of four large cells of the visceral ganglion of the gastropod *Aplysia.* The neuropile is below, right: a connective, which looks like a nerve, is forming on the left. Note the pigment in the large cell cytoplasm and especially the islands and trabeculae of glial tissue penetrating the cytoplasm (trophospongium). [Bullock, 1961.]

the opposite side; **intrinsic** neurons confine their axons to one side and level. Projection fibers, as well as entering sensory axons and exiting motor axons, may **decussate** (cross the midline) to other levels of the contralateral side or may remain ipsilateral with respect to their cell bodies.

Nerve cells are diverse in size and nuclear-cytoplasmic ratio. **Globuli** (Figs. 10.14; 10.20,B; 10.30; 10.31) and **granule cells** (Figs. 3.15; 10.75) in invertebrates and vertebrates, respectively, are the smallest nerve cells (e.g., soma less than 3 μm in insects; 10 μm in mammals) and typically have sparse cytoplasm and relatively large nuclei (ratio of diameters, nucleus : soma $\simeq$ 1) with dense chromatin. Large cells may range from 10 μm in soma diameter in small worms and insects to about 50 μm in whales. In some animals, **giant cell** somas and axons are found that reach maximum diameters of about 800 μm and 2000 μm, respectively (Fig. 2.35). Giant axons are generally larger than their cell bodies and are usually discontinuously larger than the next largest fibers, rather than being merely the end of a continuous spectrum. But *Aplysia* cell somas 800 μm in diameter have axons only about 50 μm in diameter. Conversely, giant axons may not have giant cells: squid axons 800 μm in diameter come from many cells about 40 μm in diameter by fusion of their processes (Fig. 10.52). The ratio of axoplasm to somatoplasm volume is therefore widely divergent, from less than one in small, short-axon globuli cells to more than 10,000 in serpulid polychaete giants; the figure 250 has been given for a spinal ventral horn motor neuron in a monkey. There is no modern theory for this great range or for the meaning of giant cell bodies. Giant axons are thought to be of value in two ways. One is their high

conduction velocity, though we should note that this is expensive in terms of volume, synthesis, and metabolism. The other is the large extracellular current and hence spike voltage, as felt by neighboring units; this may be of significance in influencing their excitatory state.

The **evolution of nerve cell types** can be inferred, to a degree, from their distribution among the phyla (Fig. 2.1). Cells with only one type of process (isopolar) are common in coelenterate nerve nets, where these processes may be quite long and axonal. Differentiated integrative processes and dendrites appear to be a later achievement than axons.

Sensory cells are generally bipolar, and the soma is situated primitively in the epithelium (Figs. 2.36; 2.61; 2.63), with a long axon extending into the nerve net, or, in most places in animals from flatworms to higher groups, into the central ganglia. This bipolar type persists as the common sensory neuron of invertebrates and is represented in vertebrates by the olfactory receptors and perhaps the rods and cones of the retina. But a trend begins in lower invertebrates toward deeper-lying bipolar cells, longer distal processes, more branching of those processes, and, eventually, clustering of the cells into ganglia. The extreme is the vertebrate dorsal root ganglion whose cells, originally bipolar, have become unipolar, with a T-shaped stem process that sends long axons both centrally and peripherally. Cell bodies of sensory neurons are nearly all outside the central nervous system.

Interneurons and motor neurons are not clearly distinguished in the most primitive systems, but distinct interneurons can already be seen in the earliest forerunners of ganglia—the nerve rings and marginal ganglia of medusae

Figure 2.36
Primary sensory neurons in lower animals.

Sensory bud in earthworm

Possible water-current receptor in flatworm

Specialized, presumed mechanoreceptor in polychaete

Photoreceptor in leech

Photoreceptors in earthworm

Eye in flatworm

52

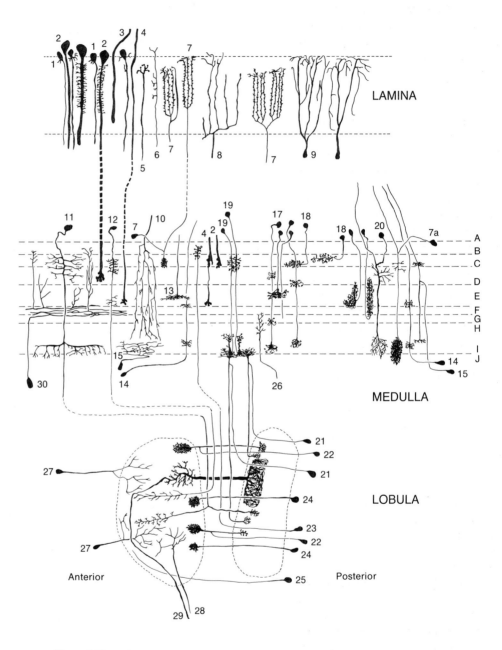

Figure 2.37
A variety of unipolar neurons, from the optic ganglia of a fly. The lamina, medulla, and lobula are successively more centrally situated neuropile masses in the optic ganglia. The numbers distinguish recognized types of neurons and represent only some of the known types (see also Figs. 10.41 to 10.43). [Cajal and Sanchez, 1915.]

(Figs. 10.2; 10.6). In the Platyhelminthes —the lowest group with a central nervous system—there is an abrupt change to an abundance of relatively advanced, central unipolar cells, but there are still many of the primitive cell types (Fig. 10.13).

In all the higher invertebrates an overwhelming majority of the interneurons and motor neurons are unipolars (Fig. 2.37); this cell type is far and away the most widespread form of central neuron in the animal kingdom. Literally scores of types of unipolar interneurons can be distinguished on the basis of significant differences in the character, extent, layering, and destination of their processes, reaching a peak in the optic ganglion of higher arthropods and cephalopods. The afferent and integrative branches are not always easy to recognize microscopically; they generally come off the axon, and may do so at many widely separate points along an axon, as may the efferent terminals. Therefore, in such neurons impulses may arise at different sites nearer or farther from the soma and conduct both ways, even meeting and cancelling. Unipolars are rarely found outside the central nervous system in invertebrates or inside the central nervous system in vertebrates. Somas of interneurons and motor neurons are nearly all inside the central nervous system in higher animals, except for the nerve supply to viscera.

Multipolar heteropolar cells are nearly universal among vertebrate interneurons and motor neurons, but are found in invertebrates only in a few special places (Fig. 2.38). In vertebrates there are multipolar isopolar cells in several special places, including some that appear to lack an axon; these are called **amacrine cells** (Figs. 2.2; 3.20). Differentiation of

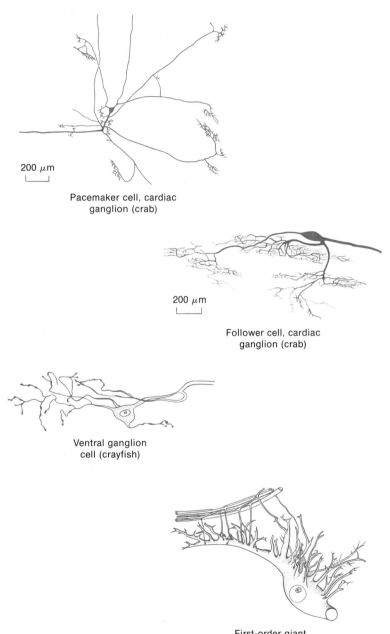

Figure 2.38
Multipolar neurons in invertebrates.
[Bullock and Horridge, 1965; after various authors.]

200 μm

Pacemaker cell, cardiac
ganglion (crab)

200 μm

Follower cell, cardiac
ganglion (crab)

Ventral ganglion
cell (crayfish)

First-order giant
cell (squid)

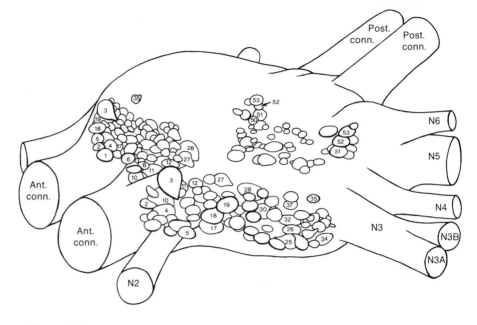

Figure 2.39
Identified cells. Motor neurons in the metathoracic ganglion of a cockroach. The numbered somas are some of the known and consistent cells identifiable in every individual. *Ant. conn.,* connectives to next anterior ganglion; *Post. conn.,* posterior connectives; *N2 to N6,* peripheral nerve trunks. [Cohen, 1970.]

vertebrate multipolars is much advanced; a large number of special forms can be distinguished. Indeed, even if we knew nothing about physiological, chemical, pharmacological, immunological, developmental, and pathological specificities, we would still be forced to recognize here the most elaborated specificity of cell types from the arrangement, distribution, orientation, and connections of the processes. The development of dendrites in extensive and characteristic arbors (branching patterns), resembling distinctive species of trees and shrubs, reaches a peak in vertebrates.

Identifiable neurons are a prominent feature of hirudineans, crustaceans, insects, and gastropods (Fig. 2.39). These are neurons so consistent as to be recognizable in each individual animal, characterized by input and output connections, coupling functions, principal branches, cytologic texture, pharmacology, and often the position of the soma. Such unique cells are most common among motor neurons; many interneurons, but fewer sensory neurons, are also identifiable. Examples are referred to in Chapters 3, 5, 6, 7, 9, and 10. This phenomenon is an indicator of the high degree of specification in many nerve cells and animal groups. It does not yet tell us how far specificity goes into the finest branches and endings, nor does it quantify consistency of transfer functions. The recent history of the subject has shown rather dramatically that this once-rare kind of consistency among neurons actually extends to many types and groups. We do not yet know how far it

Box 2.1 Origin of Neurons

In what structures did nervous functions arise? Many nonnervous cells, undifferentiated with respect to nervous functions, nevertheless carry out these functions, both in lower animals and in parts of higher animals. That is, many unspecialized cells have a limited capacity to be excited, to select adequate stimuli, to propagate changes, to correlate with the existing state or with other stimuli, and to respond adaptively. For example, Protozoa and Porifera seem to have no visible structures (apart from certain effectors) that are differentiated for such functions; nevertheless, those activities are carried out (see Bullock and Horridge, 1965, Chapter 7). Even the spread of excitation from cell to cell can occur to an important degree in some epithelia (e.g., *Hydra* and amphibian larvae) and in some smooth muscles. Some special cell-to-cell contacts and channels, possibly mediating such conduction, have been shown in the EM.

Preceding the development of specialized receptor cells and neurons were intracellular effector organelles and **independent effector cells.** Cilia, gland cells, and smooth muscle cells, as well as effectors of more limited distribution, can function without a nerve supply. Secondarily, they often receive innervation superimposing control on their intrinsic spontaneity.

It is generally believed that the first nerve cells were sensorimotor, derived from an epithelium and connected to effector cells. This hypothesis is attractive, mainly for lack of alternatives, but it has not been substantiated by a clear demonstration of the existence of such cells. The coelenterates possess the most primitive, unequivocal nervous system, but they are apparently already too far along in evolution. Even these simple forms have a diversity of neurons, including what may represent the second stage. Presumably, from primitive sensorimotor nerve cells there first differentiated a combined internuncial-motor nerve cell, from which pure interneurons and pure motor neurons later evolved.

Effectors believed to be independent of the nervous system.

Pedicellariae of sea urchin

Cilia of gut in *Anodonta*

Nematocysts (left, ready; right, discharged) from *Hydra*

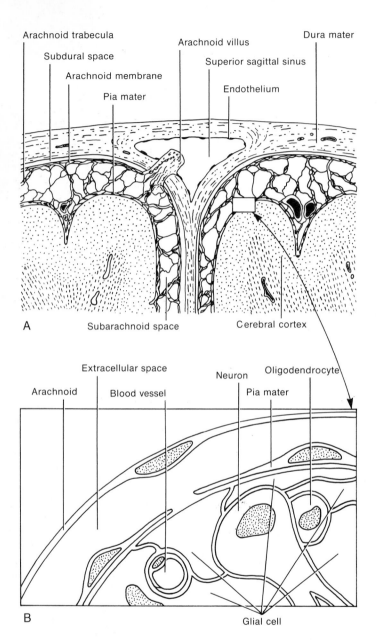

Figure 2.40

The surface of the brain and its coverings. **A.** Low-magnification diagram of the meninges in man. The small streaks in the brain tissue are sections of blood vessels. [Weed, 1923.] **B.** High-magnification diagram of a portion of the same, showing relatively open space between neurons and glial cells. This intercellular space is continuous with both ventricular and extraventricular cerebrospinal fluid space. The barrier to the passage of protein out of or into the blood is known to be at the endothelium lining the blood vessels. [Bunge, 1970.]

will extend or where to set the upper limit upon the fineness of the specificity, upon the animal groups or central levels that possess unique or nearly unique neurons or upon the hierarchial position of such units in recognition and command. But it seems inevitable that advance in anatomical and related physiological and chemical knowledge will increase the known degrees of specificity in the vertebrates as well as in the invertebrates.

III. NEUROGLIA AND SHEATHS

Nonnervous cells having some special relationship to nerve cells or fibers are called **neuroglia,** or simply glia, except that the term is not applied to the cells of blood vessels, trachea, muscle fibers, glands, epithelia, and connective tissue, though these are all associated in certain places with nervous tissue. Where the coverings of ganglia and nerves are attributable to ordinary connective-tissue elements, the term should not be used, but it is often difficult to decide, especially in invertebrates. The cells that form intimate sheaths around individual axons are glial. Most types of neuroglia share a common ectodermal origin with nerve cells (see Chapter 9).

The **coverings of ganglia** and central nervous tissue evolve from simple to complex in the animal series. Lower forms have nothing more than a basement membrane of secreted extracellular lamella. Higher invertebrates and lower vertebrates have simple cellular coverings. Mammals have complex meninges of several layers (Fig. 2.40). A tough outer membrane, the **dura mater,** contains inelastic connective tissue fibers in a dense, strong sheet. Under this is a delicate, complex investment made up of the leptomeninges—the **arachnoid,** next to the dura, and the **pia mater,** attached firmly to the brain tissue. Trabeculae (strands) of wispy "arachnoid" tissue cross a space occupied by a special cerebrospinal fluid and major blood vessels. The pia mater receives adherent end feet of deeper-lying neuroglial cells.

The coverings of major nerves are thick and subdivided in higher forms (Fig. 2.41). Indeed, only in higher invertebrates, in which well-formed wrappings evolve, do true nerves (discrete bundles of nerve fibers) become distinct. Note that inside these wrappings of bundles, most axons are individually ensheathed by glial cells.

The neuropile, or fiber core, in invertebrate ganglia is usually bounded and penetrated by neuroglial cells and their ramifying branches (Fig. 2.42). The rind of neuron somas is also permeated to a varying degree by similar cells and processes. In the vertebrates, both gray and white matter are infiltrated with ubiquitous glial cells. Many nerve cells and fibers, especially the larger ones, have their own private sheaths. **Trophospongium** is a branching network of glial processes that penetrate deeply into in-

Figure 2.41

The coverings of a nerve in a higher vertebrate. [*Shantha and Bourne, 1968.*]

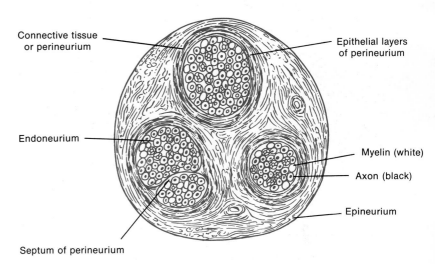

Figure 2.42

Neuroglia fibers in the octopus. **A.** The meshwork of glia and blood vessels in the subesophageal ganglion, as seen in a Bielschowsky silver impregnation. Note the greater density in the upper part, the rind, where the meshes contain large ganglion cells. **B.** Supporting tissue in nerves. At the left, a general view; at the right, a higher magnification, showing the glial processes wrapped around axons. Rio-Hortega method. **C.** Meshwork of satellite glia in layer of ganglion cells, showing the capsule for each neuron. Bielschowsky method. **D.** A blood vessel, on the left, showing the glial envelope outside the perivascular space (without muscle fibers); the glia project laterally as pseudovessels. On the right, a vessel is seen, indicated solely by its rich glial envelope. Rio-Hortega method. [Bogoraze and Cazal, 1944.]

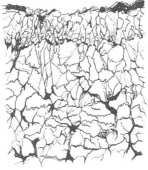

A   100 μm

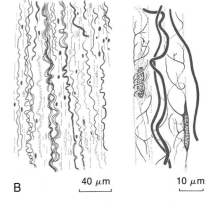

B   40 μm   10 μm

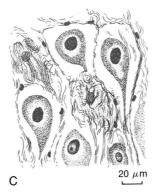

C   20 μm

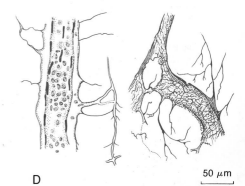

D   50 μm

Figure 2.43
Trophospongium. A giant nerve cell and neuroglial processes. [*Dehorne, 1935.*]

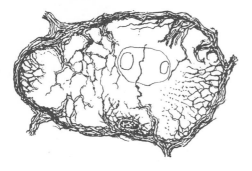

Figure 2.44
Various types of sheaths around axons.

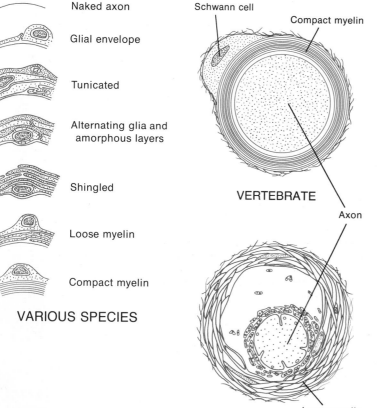

Naked axon

Glial envelope

Tunicated

Alternating glia and amorphous layers

Shingled

Loose myelin

Compact myelin

VARIOUS SPECIES

Schwann cell

Compact myelin

VERTEBRATE

Axon

Loose myelin

KURUMA SHRIMP

vaginated channels of the surface of some large nerve cell bodies (Fig. 2.43). It is thought to have nutritive significance and to facilitate ion exchange and the tolerance of stretch.

Axons with sheaths are the rule, but many smaller fibers, even in higher vertebrates, are called **naked axons** because they are not immediately surrounded by a glial cell. In sheathed axons glia may surround an axon as a single cell layer, or in loose folds, or as overlapping shingles of multiple cells (Fig. 2.44). Collectively, the foregoing varieties are called unmyelinated fibers because they lack the following feature. Glia may instead form tightly wrapped spirals around the axon, and such fibers are called myelinated (see also Glossary).

Unmyelinated fibers are actually of quite different kinds. Naked axons are the simplest; all nerve fibers in coelenterates, plus limited bundles in certain places in higher animals, including mammals, are naked axons. Axons covered by a single cell layer are more common and actually lie in an invaginated groove in each of a chain of elongated glial cells called Schwann cells. Such a glial cell usually has several grooves, and some have hundreds of parallel axons in bundles, each with its own invagination (see below and Fig. 2.45). Several types of loose sheaths several cells thick are characteristic of and confined to higher invertebrates. Earthworm giant fibers have the most elaborate sheaths of the unmyelinated category, with scores of lamellae of uniform thickness, only some pockets of glial cytoplasm here and there, and a total thickness of about 10% of the fiber diameter. Even thicker in proportion and probably more specialized, to judge from the velocity of nerve impulses, which is

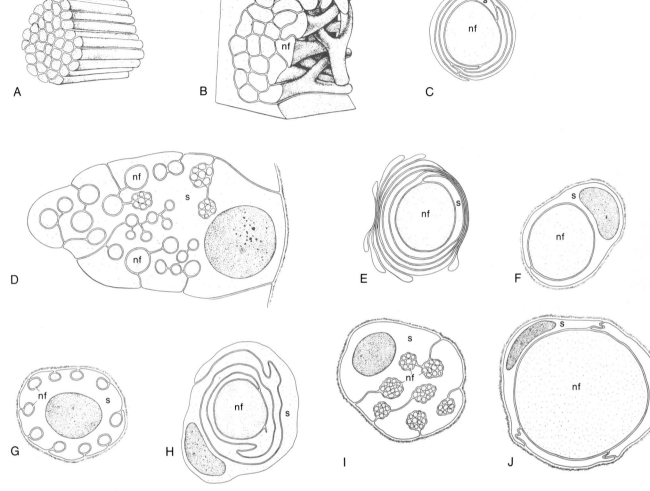

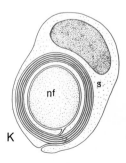

Figure 2.45

Schematic diagrams of various types of glial ensheathment. *nf,* nerve fiber; *s,* sheath cell. The glial component of the sheath is emphasized rather than connective tissue elements. **A.** In some forms nerve fibers have no sheaths at all (as in coelenterates). **B.** In many forms nerve fibers (or neurons) are ensheathed as a group, with a glial sheath situated between them and their source of nutrient (as in vertebrate CNS). **C.** Large nerve fibers (or neurons) may be surrounded by multiple layers of glial processes (as in insects). **D.** In some forms, large glial cells ensheath axons singly or in groups (as in the leech). **E.** Multiple layers of glial processes may have some compaction between them (as in the earthworm). **F.** Large nerve fibers may be surrounded by a single Schwann cell, as in insect peripheral nervous system (PNS). **G.** Small undifferentiated nerve fibers may be enclosed in individual troughs of glial membrane (as in mammalian PNS). **H.** In some cases, loosely applied glial processes are irregularly spiraled around the axon (as in insect PNS). **I.** Very fine fibers may, in some places, be enclosed as a group in glial membrane troughs (as in vertebrate olfactory nerve). **J.** Some large axons may have a mosaic of Schwann cells applied to their surface (as in squid). **K.** In vertebrate PNS (and CNS) systematic spiralization and compaction lead to the formation of myelin. [Bunge, 1968.]

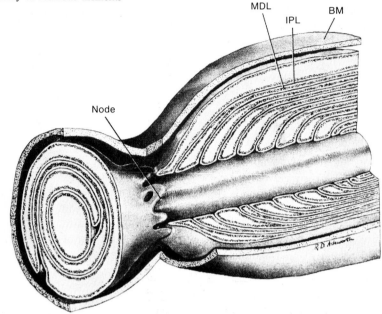

Figure 2.46
The node of Ranvier. The spiral double membranes formed by the invagination of the glial cell surface (shown in Fig. 2.45,**K**) make loops where they end, resting on the axon, the innermost first, the outermost last. Adjacent glial cell units form the internodes and meet at the nodes. *MDL*, major dense line, the result of two glial surface membranes coming together, inner cytoplasmic sides in contact; *IPL*, interperiod line, the result of the two outer faces of glial surface membrane coming together; *BM*, basement membrane. [Willis and Grossman, 1973.]

the highest known (190–210 m/sec at 20°C), are the sheaths of giant fibers in large shrimps. These may reach 50% of the total fiber diameter. In marked contrast, squid giant fibers have a glial cell sheath that can be less than 1 μm thick (much less than 1%) only a few cells deep. The numerous folds in these membranes form contorted intercellular channels that provide ion pathways from the axon, via the space (70 Å) between axon and glial surfaces, to the basement membrane that lies outside the glia and thence to the loose connective tissue outside of that.

Myelinated fibers are defined by the myelin sheath, a wrapping of many layers of doubled neuroglial cell membrane, tightly spiralled such that all the cytoplasm is squeezed out. The myelin sheath is thus formed of part of the membrane of the glial cell. Myelin sheaths vary from about 20 to 70% of total fiber (axon plus sheath) diameter. Figure 2.46 shows the important structural details of the myelin sheath and its relationships at the end of the elongated glial cell. A single glial cell accounts for a sheath length of about 0.5 to 2 mm, at least in peripheral axons, in which such cells are called <u>Schwann cells.</u> In the central nervous system the myelin is formed by cells called oligodendroglia, and they wrap the axon in a different way to form the spiral, as shown by comparing Figures 2.45,K and 2.48 (see also p. 351).

Where one sheath cell ends, the next one begins, leaving a short gap called the node of Ranvier (Fig. 2.46). We shall see in Chapter 4 that this gap is of crucial importance in the mechanism of impulse propagation in myelinated axons, which thereby gain several marked advantages over unmyelinated axons. True myelin is peculiar to vertebrates. Even in the lowest vertebrates, the cyclostomes, none has been found. But in elasmobranchs it is abundant and seemingly as well developed as in mammals. If this is true it would appear that in the perspective of nervous evolution this significant advance was a one-step invention. True nodes with a similar functional meaning have not been demonstrated in invertebrates, but functional spots acting as nodes are strongly suggested in earthworms and shrimps.

The **types of neuroglial cells** are well known in vertebrates, poorly known in invertebrates. Five cell types are distinct in vertebrates: Schwann cells, astrocytes, oligodendroglia, microglia, and ependymal cells, which line the ventricles. Schwann cells have already been discussed.

Astrocytes (Fig. 2.47) have many processes that radiate more or less evenly, branch profusely, and apply themselves to a blood capillary or to the pia mater. These processes expand into cell feet that, together, form a virtually continuous sheet-like membrane, so numerous are the astrocytes. This membrane of interdigitating foot plates separates the nervous tissue proper from the blood vessels and pia. Functional contacts among astrocytes, and between astrocytes (or other glial cells) and neurons, are probably abundant. This is difficult to assess, however, because they may include rather unspecialized areas of con-

tiguity. In the optic nerve of amphibians, and perhaps generally, there is a low electrical resistance between neuroglia that are perhaps astrocytic, but a high, normal membrane resistance between glial cytoplasm and intercellular space or neuronal cytoplasm (see also Chapter 4).

Oligodendroglia (Figs. 2.47; 2.48) are even more numerous, probably the most abundant cell type in the central nervous system, but they are smaller and, except

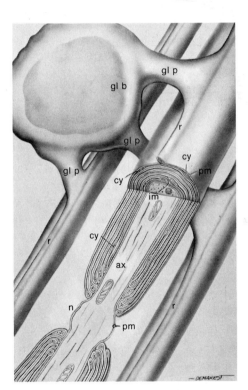

Figure 2.48
The myelin sheath in the CNS. Each oligodendroglial process forms the internode for one axon. *ax*, axon; *gl p*, glial cell processes; *cy*, cytoplasm of glial cell; *im*, inner mesaxon; *gl b*, glial cell body; *n*, node; *pm*, plasma membrane; *r*, ridge of cytoplasm. [Bunge et al., 1961.]

Figure 2.47
The principal types of neuroglia in the central nervous system of vertebrates.

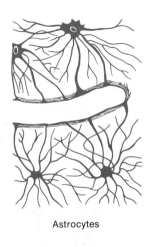

Astrocytes

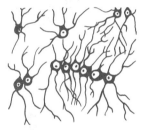

Oligodendroglia

Microglia

Chapter 2
Microanatomy of Nervous Elements

Figure 2.49
The oligodendroglial myelin sheath of a single glial process, unrolled to show its structure. [Hirano and Dembitzer, 1967.]

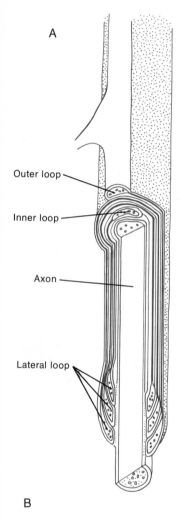

A

Outer loop

Inner loop

Axon

Lateral loop

B

for the myelin sheaths that they form (Fig. 2.49), less conspicuous. Sparse, dense cytoplasm hugs a rounder, darker nucleus; processes are few and thin. In tissue culture, which has proved to be a most useful tool for studying neuroglia, these cells are highly active; their processes move, bend, wave, and squirm ceaselessly in the unrestricted space. It remains an important question just how much movement these and nerve endings undergo in vivo.

Microglia (Fig. 2.47) are small cells whose nuclei and long branched processes are both bent in an angular manner. They are abundant and specialized in nervous tissue, but their roles, embryological origin, and relations to other cells are still obscure.

Ependymal cells (Fig. 2.55,A) form the epithelial lining of the spinal canal and brain ventricles. They contribute foot plates (like those of astrocytes) to capillaries and, if it is nearby, to the pia mater. In certain places, together with tufts of capillaries and the pia mater, ependymal cells make up a local specialization—the choroid plexus, which participates in producing the cerebrospinal fluid.

Invertebrate glial cell types cannot be directly identified with any of these five cell types, and categories have not been agreed upon that embrace several major groups of animals. In lower forms there may be only one general glial cell type, but in insects, cephalopods, and probably other higher groups, glia are differentiated into numerous kinds.

The functional significance of neuroglia is discussed in Chapter 4, page 174.

IV. NERVOUS TISSUE

Nerve cells and the nonnervous components associated with them occur in certain places as isolated elements, but most are organized into tissues—epithelial layers, plexuses, peripheral ganglia, nerves, central ganglia, cords, brains, connectives, tracts, and commissures. The general features of each of these can now be considered.

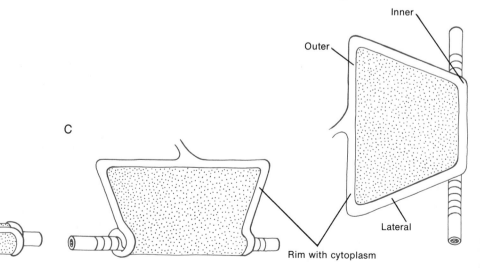

C

D

Inner

Outer

Lateral

Rim with cytoplasm

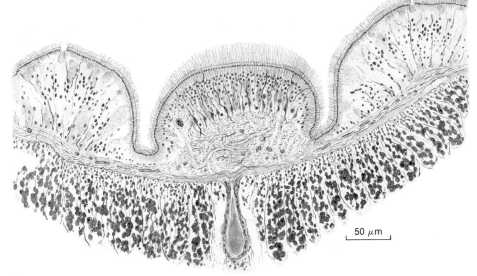

Figure 2.50
Example of a basiepithelial nervous system. The general plexus and dorsal cord of the trunk in a protochordate (Enteropneusta). In the epithelium over the cord can be seen bipolar sense cells and one giant cell; in the general plexus, very fine fibers contrast with giant (5 μm) axons. Coelomic muscles and dorsal blood vessel, below. Bodian protargol impregnation. [Bullock and Horridge, 1965.]

The primitive form of nervous tissue was probably intraepithelial, and this is its main form in lower phyla (Fig. 2.50). Cell bodies lie among the ordinary epithelial cells of the body surface. Neurites form a more-or-less thick layer of nerve fibers that lie deep to the cells and just above the basement membrane. Early in evolution, before the development of true nerves as well-defined, ensheathed bundles, plexuses, defined as tangles of deeper-lying strands of cells as well as fibers, were prevalent (Fig. 2.51). They are common even in higher animals in certain regions, such as the wall of the gut.

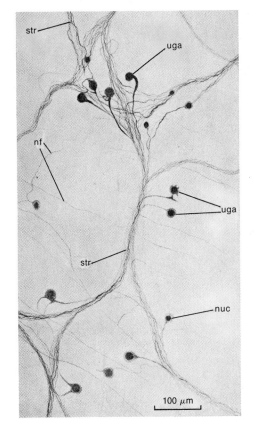

Figure 2.51
A peripheral plexus in the wall of the posterior stomach of the snail *Helix*. Bielschowsky-Gros silver stain. [Ábrahám, 1940.] *str*, strand of many nerve fibers forming the meshes of the plexus; *uga*, unipolar ganglion cells, which might be interneurons or motor neurons; *nuc*, nucleus; *nf*, nerve fibers.

Figure 2.52
*Varieties of central nervous tissue composed
mainly of axons en route. [Upper, Zawarzin, 1925,
after Retzius, 1892; lower, Cajal, 1909.]*

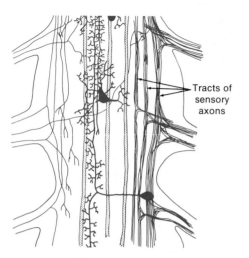

Tracts of
sensory
axons

*Tracts, ventral cord
of an earthworm*

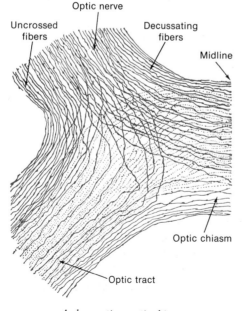

Optic nerve

Uncrossed
fibers

Decussating
fibers

Midline

Optic chiasm

Optic tract

*A decussation, optic chiasm
of a cat*

Ganglia evolved very early as circumscribed collections of somas and exist in all levels of complexity. The term is used for some large subdivisions of the central nervous system (Fig. 10.49; p. 480) and for small peripheral knots of sensory, motor, or mixed function (Fig. 10.44).

Nerves, which are usually, but not always, free of neuron somas, connect the central nervous system with some peripheral receptors or effectors or ganglia. Each of the larger named nerves is characterized by its own composition in terms of numbers of axons of different diameters and functions.

The **connectives** and **commissures** of invertebrate central nervous systems are likewise ensheathed bundles of axons, but they connect central ganglia, either longitudinally (connectives) or transversely (commissures) (Figs. 10.13; 10.16; 10.44; 10.45; 10.87). They may be quite free of somas or more or less invaded by cells or neuropile. **Tracts,** or fascicles, are smaller bundles and may be more-orless well-defined, compact or loose, homogeneous or mixed in axon composition, pure or invaded by cells and synapses (Fig. 2.52, top).

A feature common in all bilateral animals is **decussation** of axons, or the crossing of a median plane by projection fibers (Fig. 2.52, bottom). Usually the plane is in the long axis of the body, and decussation from side to side. But in optic systems of arthropods and cephalopods, there are decussations behind each eye across a plane in the axis of the eye. Some decussations are complete, others partial. Developmental and physiological explanations of this tendency have been suggested but are at best speculative, and it seems probable that no one explanation accounts for decussations in general.

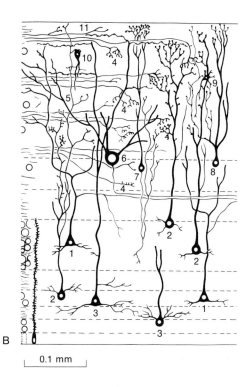

0.1 mm

Figure 2.53

Invertebrate versus vertebrate synaptic and nerve cell regions. **A.** Optic lobe of a crab. Golgi method. Only a selection of the cells and afferent fibers is shown. Note cell somas at periphery, without synapses; neuropile masses inside dotted lines include cell synapses but no cell somas. [Hanström, 1924.] **B.** Optic lobe of a frog. Golgi method. Only a selection of the cells and afferent fibers is shown. The numbered cell types are described in Székely et al., 1973; briefly they may be listed as (1) large pyramidal neuron with type 1 dendrite, (2 and 3) large pear-shaped neurons with types 2 and 3 dendritic arborization patterns, (4) optic nerve terminals, (5) ascending axon, (6) large ganglion cell, (7) small pear-shaped neuron with descending axon, (8) small pear-shaped neuron with beaded axon-like process, (9) stellate neuron, (10) amacrine cell, (11) assumed endings of diencephalic afferent fiber.

In all animals with a central nervous system, there is a division of nervous tissue into two zones—one zone with nerve cell bodies and one without. But this primary differentiation takes place in two quite different ways, one characteristic of invertebrate groups and the other of vertebrates. In invertebrates (Fig. 2.53) the nerve cell bodies are gath-ered into a rind, or superficial layer, of the ganglion, or cord, leaving a central core of fibers. The main feature, however, is that there are few or no fibers ending in the cell rind; the synaptic field and the long pathways are both in the fiber core (Fig. 2.54). In vertebrates (Fig. 2.55) the two zones are called gray matter and white matter. The gray matter

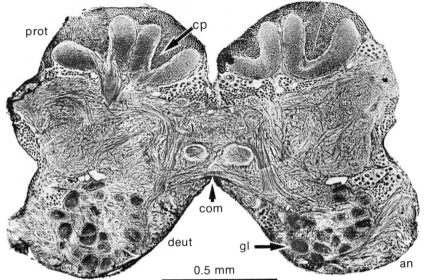

A

Figure 2.54
Invertebrate brain tissue, low and high magnification (see also Fig. 10.56). **A.** Frontal section of the brain of a cockroach. Bodian reduced silver method. [Boeckh et al., 1970.] *com,* commissure between left and right sides of the deutocerebrum (*deut*); *cp,* corpora pedunculata; *gl,* glomerulus; *prot,* protocerebrum; *an,* antennal nerve. **B.** Electron micrograph of synaptic region of an ant brain. Ek_1 (dark) and Ek_2 (light) synaptic bulbs; *mit,* mitochondrion; *egr,* electron-dense granules; *sy,* synapse. [Steiger, 1967.]

B

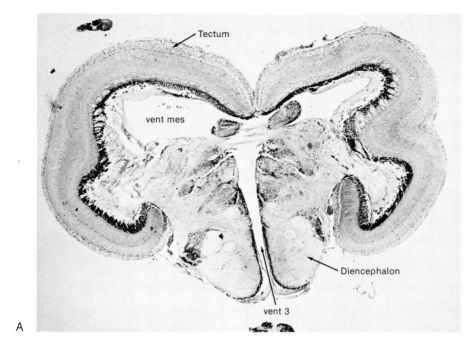

A

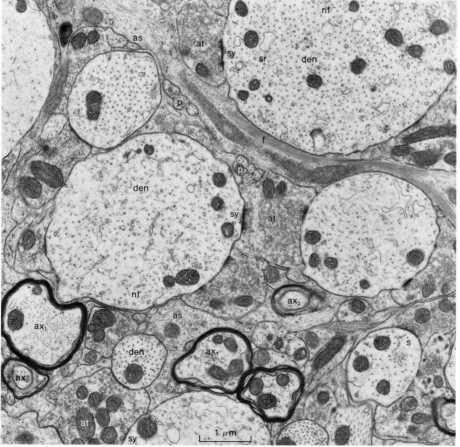

B

Figure 2.55

Vertebrate brain tissue, low and high magnification (see also Fig. 10.56). **A.** Midbrain, including optic lobes, of a flatfish, *Hypsopsetta,* stained by the Klüver method, which colors myelin dark. *Dienceph alon,* portion of posterior hypothalamus; *Tectum,* roof of the optic lobe; *vent 3,* third ventricle, the cavity of the diencephalon; *vent mes,* the ventricle of the mesencephalon. **B.** Electron micrograph of neuropile in the ventral horn of the spinal cord of a rat. *at,* axonal terminals; $ax_1,$ $ax_2,$ myelinated axons; *as,* astrocytes; *den,* dendrites; *f,* fibrils in cytoplasm of astrocytes; *p,* preterminal, unmyelinated axons; *nf,* neurofilaments; *sy,* synapses; *sr,* smooth endoplasmic reticulum. [Part B, Peters et al., 1970.]

consists of the nerve cell bodies plus the dendritic and axonal endings, hence the synapses; the white matter consists only of axons en route, and is white because of its large proportion of myelin sheath material. Neuroglia pervade all zones.

This difference between vertebrates and invertebrates, besides being a general morphological feature, may perhaps have physiological significance as well. It accords with the prevalence of dendrites that spring from, and of synapses on, the soma in vertebrates—features that are extremely exceptional in invertebrates. (Some of the exceptions are well known: the receptor cells in the muscle receptor organ of crayfish, the cardiac ganglion cells in lobster, the first-order giant cells in squid.) It may also accord with a basic difference in the nature and properties of the afferent, integrative processes of vertebrates and invertebrates. The dendrites of vertebrates are more fleshy, bushy, and thorny, and perhaps more defined or regimented in their branching patterns for each neuron type, than the corresponding processes of invertebrates. Possibly these differences play a role in the activity and synchrony behind the slow smooth brain waves that are more general and conspicuous in vertebrates (pp. 228, 434).

The fiber core of the invertebrate ganglion is differentiated, except in the lowest forms, into tracts and neuropile. The tracts, when they are pure tracts, are equivalent to vertebrate white matter (without true myelin)—that is, fibers en route, without endings or synapses. **Neuropile** (Fig. 2.56) is a mass of fibers with afferent endings, axonal terminations, and synapses, but without nerve cell bodies. Vertebrates have local regions of neuropile in the gray matter, both between nerve cells and in regions relatively free of nerve cells. Neuropile is therefore a concentrated region of integrative nervous events. It is still largely a terra incognita in terms of structural pattern and functional detail but deserves attention by new methods.

Degrees of differentiation of the neuropile can be found and are sensitive indicators of complexity and hence the level of advancement. Neuropiles in the lowest forms are loose and coarse in texture and lack signs of local textural differentiation. The higher the nervous systems, the more signs of textural differentiation (Fig. 2.56). Certain specialized forms of neuropile may be mentioned. **Glomeruli** are knots of tighter weave than the surrounding neuropile; they may involve several kinds of neurons and be internally ordered. **Stratified** neuropile is well seen in the optic ganglia of cephalopods and insects; layers of different density and grain are clearly segregated, manifesting the planes in which many fine nerve fibers ramify and end (Fig. 2.56; 10.41; 10.42; 10.51). The calyces and stalks of the **corpora pedunculata** are additional specializations, distinguished from coarser neuropile by their very dense, smooth texture, which is due to their fine, short, packed processes and many synapses (Fig. 2.54).

Differentiation in the cell rind of invertebrates is relatively modest, but nevertheless pronounced. The simpler rinds are by no means without structural specification, for in them we find the best studied examples of individually recognizable cells. The highest centers in invertebrate brains are believed to be the globuli masses, which are composed of tiny tightly packed globuli cells. These are found already in polyclad flatworms,

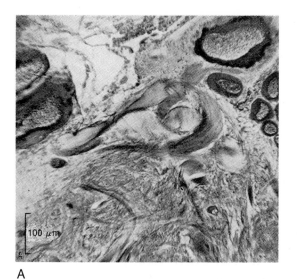

A

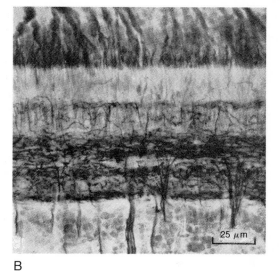

B

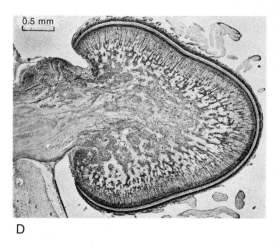

C

D

Figure 2.56

Variety of differentiated forms of neuropile in invertebrates. **A.** *Aplysia,* visceral ganglion with unspecialized fiber core, here traversed by the stem processes of giant cells, which can be seen at left.
Reduced silver. [Bullock, 1961b.] **B.** *Octopus,* plexiform layer of the cortex of optic lobe; above and
below are the granular layers. The pure neuropile is highly differentiated into strata. [Young, 1971.]
C. *Harmothoë* (Polychaeta, Polynoidae), transverse section of the brain, showing common neuropile in
center, three specialized stalks of the corpora pedunculata surmounted by "globuli" cell masses,
clumped neuropile or glomeruli, in lateral lobes and tracts entering the circumesophageal connectives
below. [Courtesy B. Hanström.] **D.** *Octopus,* optic lobe, low-power view, showing the great differences
in the neuropiles of the cortex (same as B), the outer and inner portions of the lobe and the main
tracts connecting with the brain (left). [Young, 1971.]

as well as in the more advanced poly-chaetes, arthropods, and molluscs. In cephalopods there is one more sign of high development—islands of nerve cell somas scattered throughout a large neuropile mass (Figs. 2.56; 10.50).

Differentiation of gray matter in vertebrates is even more pronounced in texture and architectonics (cell architecture). This is due in part to the presence of nerve cell somas, which besides their wide range of forms, branching, and orientation (Fig. 2.57) are diversely disposed in the tissue. It is noteworthy that, as in invertebrates, many higher levels of the brain are characterized in part by relatively smooth, dense neuropile—for example, the **molecular layer** of the cerebral and cerebellar cortices—and by small, compact cells. These so-called **granule cells** are much like globuli cells, although multipolar. Within the brain, a relative concentration of nerve cells or a discrete mass of gray matter may be called a nucleus or a ganglion. In contrast, gray matter of a certain relatively unspecialized texture and crossed by many anastomosing strands of white matter is called **reticular formation;** it is developed in places at the margin between gray and white matter in the spinal cord and brain stem. Other types of gray matter are seen in the retina (Figs. 3.20; 3.21), the substantia gelatinosa, the inferior olive, the red nucleus, the tectum (Fig. 2.58), and the hippocampal cortex, to mention a few of the innumerable sites with distinctive features of architectonics. The details are beyond the scope of this book but may be found in Cajal, Bumke and Foerster, and works on special parts of the brain.

The **functional meaning** of such specialized architecture must be great, but is little understood. Columns and strata, glomeruli and reticula, conspicuous order and apparent disorder must have a biological significance. Chapter 3 deals with principles and examples of connectivity among organized systems of neurons. Diversity in neural histology must depend on diverse patterns of graded, electrotonically summing transactions, with significant functional aspects distributed along spatial continua. If we are to make physiological sense of tissue differentiation, we must strive to understand connectivity in terms of coupling functions and interactions. The existence of strata and other forms of cell architecture suggests that one of the prime questions is: "What parameters in the neural process are being spatially represented, first by being segregated from other parameters and then by being distributed systematically?" This is one way of asking a fundamental question about how the brain works. It emphasizes emergent properties above the level of single unit properties, and focuses, with the perspective of this chapter, on the nonrandom, specified order as a domain of approachable functional anatomy.

The **extent of potential interaction** of neurons has one measure in sheer numbers of countable nerve cell bodies within the area of ramification of the processes of a single neuron. Omitting as uncountable the number of other neurons represented only by their axonal terminations in this area, Scheibel and Scheibel counted 4125 cell bodies within the subarea of the *dendritic* arborization of one cell of the magnocellular nucleus of the kitten medulla and 39,375 cells within the subarea of terminal branching of one afferent *axon*, probably a so-called nonspecific fiber, to the cortex.

The **intercellular space** in nervous

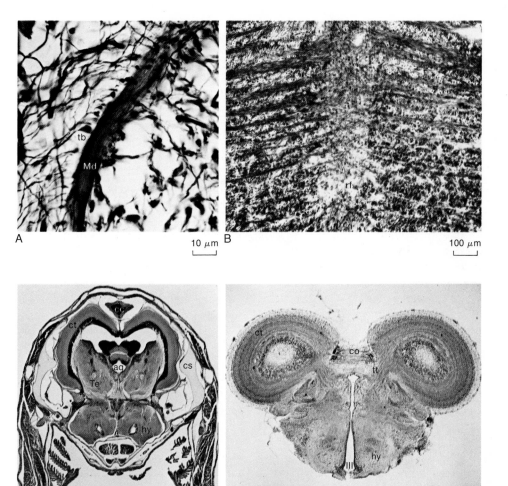

Figure 2.57
Variety of differentiated forms of neuropile in vertebrates. **A.** Neuropile around the dendrite of the giant cell of Mauthner (*Md*), including many axonal endings on it in the form of terminal buttons (*tb*); goldfish, Bodian reduced silver stain. [Courtesy D. Bodian.] **B.** Reticular formation (*rf*), scattered cells and axon bundles in the neighborhood of the midline of the medulla; cat, Bielschowsky reduced silver stain. **C.** Whole head of young flatfish showing the cartilage of the skull (*cs*), the mesencephalon above with the cortex of the tectum (*ct*) and of the cerebellum (*cc*), the diencephalon below with its hypothalamic gray matter (*hy*) and various textures of neural tissue for example beside and below the aqueduct (*aq*) of the midbrain, in the tegmentum (*Te*); Masson trichrome stain. **D.** Similar level in larger specimen of a goldfish, showing the stratified cortex of the tectum (ct), commissural fibers (co) between the two sides, tracts from the tectum to the thalamus (*tt*), gray matter of different texture in the diencephalon surrounding the third ventricle (III) and including the hypothalamus (*hy*); Bodian reduced silver stain.

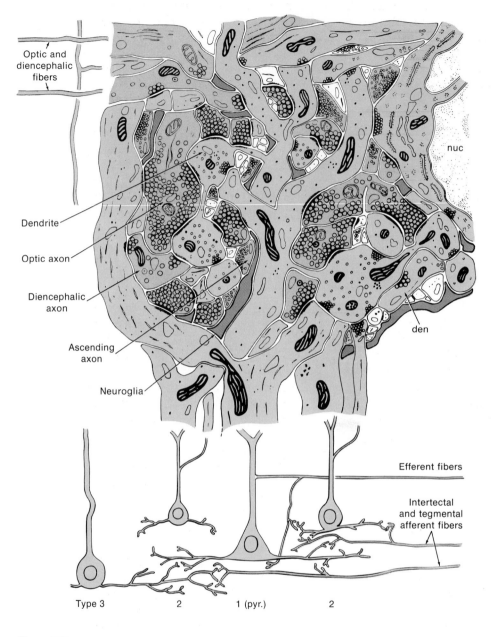

Figure 2.58

Above. Diagrammatic electron-microscope analysis of some relations among cells in the optic tectum of a frog (see also Fig. 2.53,**B**). **Below.** Low-magnification summary of the three types of dendritic arborizations. The shaft dendrite of the neuron with type 3 pattern ascends, and its collaterals enter the upper view from the left, together with the optic and diencephalic afferent fibers. The shaft dendrites of the three other neurons and the two ascending axons continue from below into the upper view as indicated, and they arborize in a cellular sheet. The groups of endings compose two glomeruli, each with the four types of terminals. Dark brown represents glia; light brown represents dendrites (*den*); gray represents axonal terminals; *nuc,* nucleus of a stellate neuron. [Székely, et al., 1973.]

tissue is small, at least in higher animals. Since the neurons and their processes have, in general, ordered and presumably obligatory geometric relations with each other, it is likely that glial elements serve to conform to the residual spaces and fill them. The cell membranes of adjacent cells are commonly reported to be about 200 Å apart, in what are usually believed to be well prepared EM sections. This would aggregate to a total intercellular compartment of less than 5% of the tissue volume. A body of evidence is gaining acceptance in some quarters for the interpretation that the cells in these classical micrographs are artifically swollen during fixation and that intercellular volume is actually about 15–20%. Moreover, this view holds that the spaces grow larger and smaller depending on functional state, a very pregnant degree of freedom since it might be expected to alter the physiological properties of the cell mass. The intercellular space is thought to contain a complex mucopolysaccharide material that can take on or give up water to account for the volume changes and probably influence interactions between cells in other ways as well.

V. DEGENERATION AND REGENERATION

Degeneration of various kinds can be distinguished. We are not concerned with pathological conditions but only with certain features of general biological interest or useful as tools in studying the organization of nervous systems.

When an axon is severed and cannot reunite, death of the distal moiety (the portion cut off from the nerve cell body) appears to be inevitable. If it is myelinated, the myelin sheath also degenerates and its fragments gradually disappear (Fig. 2.59). This is an example of intercellular trophic dependence: part of the Schwann cell suffers when the axon dies. Sheath degeneration has not been found in any of the loosely sheathed invertebrate nerve fibers, although in insects the sheath reacts by thickening and granulating.

The distal degeneration of axon and sheath is called **Wallerian degeneration** and is useful in tracing tracts in the central nervous system because special stains that react with the degenerating fibers make it possible to follow them for long distances.

Information is meager about axon degeneration in invertebrates. In some nerve fibers, especially motor axons, degeneration is not visible, either functionally or microscopically, for weeks or months after a transection. In others, especially in sensory axons and in tracts of the central nervous system, both kinds of signs are obvious within a few days.

Proximal to the interruption of an axon, a reaction of variable degree may be observed. It is most noticeable in the soma of vertebrates, which may exhibit **chromatolysis**—a fragmentation and dispersion of the Nissl granules into fine particles with an eccentric shift in position of the nucleus away from the axon hillock. This gradually disappears over a few weeks, but some cells die merely from transection of the axon (**"retrograde degeneration"**). In a few situations, following Wallerian degeneration, the next cell in the pathway also dies. Such **transsynaptic degeneration** is known from the loss of cells of the lateral geniculate nucleus after optic nerve section in early life, the loss perhaps being due merely to

74

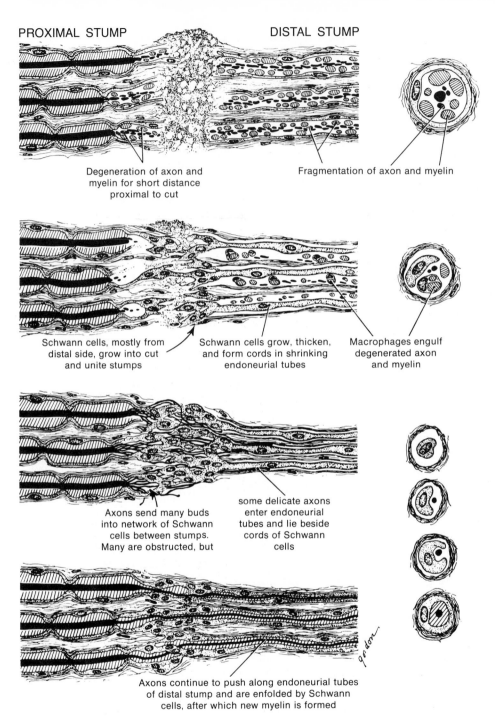

PROXIMAL STUMP   DISTAL STUMP

Degeneration of axon and
myelin for short distance
proximal to cut

Fragmentation of axon and myelin

Schwann cells, mostly from
distal side, grow into cut
and unite stumps

Schwann cells grow, thicken,
and form cords in shrinking
endoneurial tubes

Macrophages engulf
degenerated axon
and myelin

Axons send many buds
into network of Schwann
cells between stumps.
Many are obstructed, but

some delicate axons
enter endoneurial
tubes and lie beside
cords of Schwann
cells

Axons continue to push along endoneurial tubes
of distal stump and are enfolded by Schwann
cells, after which new myelin is formed

Figure 2.59
Diagram showing the changes that occur during regeneration of a severed peripheral nerve.
[Ham, 1969.]

the quiescence of input. Apart from disease, the chief cause of natural degenerative changes is ageing. The brain, however, is more resistant to changes from this cause than other organs. Loss of cells proceeds in advanced age but may be hardly appreciable. A common sign of change with old age is an accumulation in some neurons of lipochrome pigment of unknown significance.

Regeneration of cells or at least of fibers may be a universal tendency, successful in some neurons and not in others. It can be very rapid. Coelenterates can re-establish conduction across cuts or between host and graft in less than a day, possibly by fusion of processes. Brains and cords regenerate in many lower metazoa; those of certain closely related forms are unable to regenerate, and one suspects that a suppressing factor supervenes. In higher invertebrates regeneration is little known in the central system. Peripheral nerves can often regenerate quite well. The study of crustacean motor axons suggests that regeneration takes place here by fusion of undegenerated proximal and distal stumps instead of by new outgrowth.

Peripheral regeneration in vertebrates has been studied extensively (Fig. 2.60). Axons sprout numerous fine outgrowths from the proximal stump; these branch, probe their way through the tissues, and take off optimally if they encounter an empty Schwann cell tube of a degenerate axon to guide them. Most of the sprouts do not become established and are resorbed. In motor axons especially, many new branches survive after a nerve lesion, so that each axon supplies far more muscle fibers than normally. Under favorable conditions new fibertips advance at 1 to 2 mm a day in mammals.

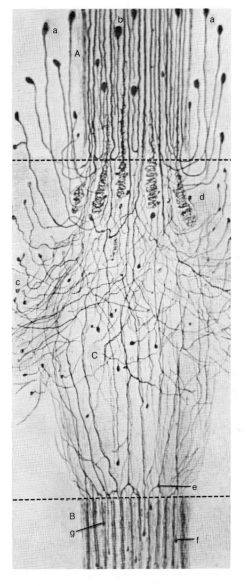

Figure 2.60
Regeneration in a peripheral nerve, showing the disposition of the outgrowing fibers from the central stump (A), through or deflected by the scar (C) into the distal stump (B). a, b, axon branches growing back toward CNS; c, branches growing laterally; d, coil formed by some recurrent axons; e, bifurcating fiber; f, g, axons growing into the peripheral stump. [Cajal, 1928.]

Figure 2.61

Some receptor neurons. [Bodian, 1967.]

Auditory
(human)

Olfactory
(human)

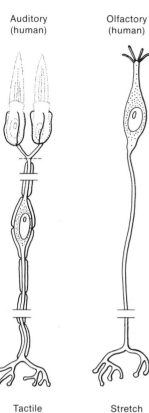

Tactile
(human)

Stretch
(crayfish)

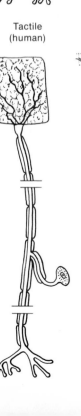

Regeneration is said to be virtually nil in the central nervous system of homeotherms. In certain places in lower vertebrate classes, it is spectacularly successful, as in the complete regeneration of the retina and central optic connections in urodeles. Recent work suggests that limited regeneration does occur in mammals but that certain factors frustrate it, including the formation of a glial scar that acts as a barrier. It may be significant that oligodendrocytes and Schwann cells differ in their accomodation of sprouting tips. Optic nerve fibers, and others, sprout after certain kinds of injury to themselves or after loss of neighboring terminals. This can result in increased invasion of areas normally reached by very few axons. The ability to sprout is much greater in the developing mammal.

Many important questions that pertain to the development of the nervous system, such as the way in which regenerating fibers find their ultimate destination, are dealt with in Chapter 9.

VI. RECEPTORS AND SENSE ORGANS

The microanatomy of sensory elements is much more varied (Figs. 2.61; 2.62) than might be supposed from a list of the classical five senses (sight, hearing, taste, smell, and touch) or even a somewhat more modern list (adding the senses of pain, temperature, position, and the unconscious reception of blood pressure, blood chemistry, lung stretch, visceral distension, and others). One reason is that many of these major modalities have several subclasses of receptors—for example, cold and warm receptors; phasic (adapting) and tonic (nonadapt-

ing) or mixed receptors; mechanoreceptors for light touch, vibration, light pressure, deep pressure, muscle stretch, tendon stretch, and joint position; and taste buds specialized for different sensations. Other less familiar subdivisions are also known. Another reason is that distantly related animal groups possess quite different structures for the performance of analogous functions. For example, the ears, eyes, and other organs of insects, are fundamentally unlike those of vertebrates. Finally, we may recognize a third reason for the unexpectedly wide variety of sense receptor among animals. That is the existence of many unfamiliar forms of reception and many sense organs of unknown function. Examples are infrared receptors in certain snakes, electroreceptors of several classes and subclasses in many fish, wind speed indicators in locusts, vibrating-gyroscope strain-gauges in flies, turbulence and water displacement detectors in the lateral-line system of aquatic vertebrates, and a host of organs for which the only functional inference is based on a rich afferent innervation.

In the known diversity, grounds for classification are difficult to establish. Elsewhere in this volume we have explained the currently used modality-classes, named according to the form of stimulation—for example, photo-, phono-, chemo-, mechano-, thermo-, and electroreceptors. An older distinction is between exteroceptors, interoceptors, and proprioceptors (see Glossary).

A comprehensive account would be out of place here, but the principal features of the neural aspects of a limited sample of receptors will be briefly described. The essential anatomy of the ultimate transducers is still beyond us; not even the electron microscope has

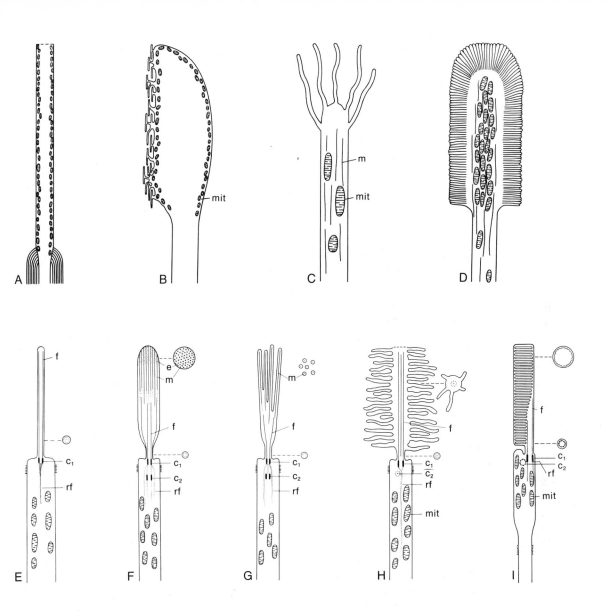

Figure 2.62

Receptive portions of receptor cells for various modalities. Upper row shows receptors with non-ciliary organization; lower row, receptors with ciliary organization. **A.** Pacinian mechanoreceptor of mammals. **B.** Crayfish muscle stretch receptor. **C.** Reptile chemoreceptor of Jacobson's organ. **D.** Photoreceptor of the planarian *Dendrocoelum lacteum.* **E.** Insect scolopidial mechanoreceptor. **F.** Insect hair, or campaniform, mechanoreceptor. **G.** Arthropod olfactory receptor. **H.** Photoreceptor of the hydromedusan *Polyorchis penicillatus.* **I.** Rod photoreceptor of vertebrates. Diagrams are only roughly to scale; magnification of A and B about four times less than for all other diagrams. c_1, c_2, centrioles; e_1, e_2, intensely stained material associated with microtubules; f, ciliary filaments; *mit*, mitochondria; *rf*, root filaments; *m*, microtubules. [Thurm, 1969.]

Figure 2.63
*Primitive receptors.
Solitary sense cells in the
epithelium of a coelenterate.*
[*Hanström, 1928b.*]

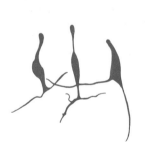

revealed the molecular architecture that is altered by the adequate stimulus. But it is possible to narrow greatly the search to a probably relevant region of small dimensions (Fig. 2.62).

A. The Spectrum from Solitary Cells to Complex Organs

In addition to solitary sense cells (Figs. 2.1,A; 2.63; 2.64), proper organs (defined as ordered arrangements of two or more types of tissues) are present in animals as simple as jellyfish, especially hydrozoan medusae. Even before organs of the digestive, reproductive, or any other system have evolved, these animals possess fairly complex eyes and statocysts. Throughout the vast spectrum of animal types, ranging from such simple types on up to the vertebrates, there is a conspicuous correlation between the degree of

development of such organs of special sense and the organism's habit of life. Sense organs are meager or absent in parasitic and sessile or sedentary species but well developed in free-ranging, especially predatory forms. The richness of the behavioral repertoire, the histologic level of differentiation of the central nervous tissue, and the number of discretely controllable movements are all correlated with the degree of development of sense organs.

Scattered, solitary receptor cells diffusely distributed in epithelia are present not only in the lowly species (simpler phyla and sessile or parasitic species in advanced phyla); they are widespread also in the most complex animals. Insects, crustaceans, and cephalopods—the highest invertebrates—have great numbers of solitary sense cells. A trend toward sinking of the cell body beneath the epithelium, clumping into ganglia,

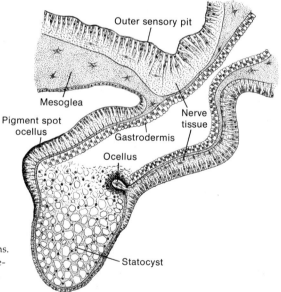

Figure 2.64
Primitive sense organs in addition to unspecialized sense cells. The marginal body and sensory pit of a medusa, *Aurelia.* The statocyst and ocelli are specialized sense organs. Numerous generalized, solitary sense cells are in the epithelium of the outer sensory pit. Nerve tissue shown in color. [Schewiakoff, 1889.]

and branching of the distal process is illustrated in its fullest development in the abundant dorsal root ganglion cells with "free nerve endings" in all vertebrates.

B. Primary Sensory Neurons and Secondary Nonnervous Sense Cells

Probably all receptors in invertebrates, and most receptors in vertebrates, are neurons. These are called primary sensory neurons (Fig. 2.65). Some receptors in the vertebrates are apparently nonnervous cells that synapse with sensory nerve endings from first-order afferent neurons. These so-called secondary sense cells are the taste cells, the hair cells of the inner ear, and the sense cells of Iggo's and Merkel's skin receptors, which actually transduce the chemical or mechanical stimulus into a physiological event controlling transmission across the sensory synapse to the nerve endings. They are considered not to be neurons because they lack well-developed processes, but in origin, ultrastructure, and function they resemble nonspiking neurons. Rods and cones are sometimes treated as neurons, sometimes as secondary sense cells.

C. Receptors in and near the Skin

A distinction is made between the common **somesthetic senses** and the special senses like vision and hearing. "Somesthetic" refers to the senses of the body wall, especially in the skin, and is nearly equal to the first three of the following categories. We shall use functional categories below, although each is anatomically heterogeneous. For some of the commonest somesthetic senses, we still have only tentative identifications of histological endings with local sensations.

Mechanoreceptors. Free nerve endings (Fig. 2.66)—that is, unencapsulated terminals—certainly serve as mechanoreceptors in certain places where there is no other type of ending (pinna of the ear, cornea). Meissner's corpuscles (Fig. 2.67) may be mechanoreceptors, as are Merkel's and Pacinian corpuscles, each with distinctive physiological properties. More than a dozen types of cutaneous mechanoreceptors are distinguished in mammals, only partially identified anatomically. There are hairy skin types I, II, G_1 (guard hair$_1$), G_2, D (down hair), F_1 (field$_1$), F_2, C (unmyelinated, C fiber), and PC (Pacinian corpuscle). Sinus hairs (like vibrissae) are of at least three types. Glabrous skin (like that on the palms) is of at least two types. A wide variety of generalized and specialized mechano-

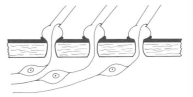

Figure 2.65
Primary versus secondary sense cells.

Primary sense cells are true neurons. Arthropod mechanoreceptors have a distal dendritic process distorted by a cuticular hair.

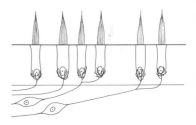

Secondary sense cells are not true neurons. They transduce the stimulus into a signal transmitted to the first-order afferent neurons. Hair cells in the vertebrate acousticolateralis system are such transducer cells. [Bullock and Horridge, 1965.]

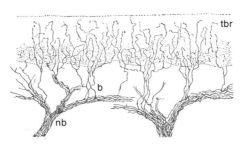

Figure 2.66
Free nerve endings. Intraepidermal sensory terminals in the sole of the foot of a dog. Golgi method. The stratum corneum of the skin is shown above. *b*, bifurcations of axons; *nb*, nerve bundle; *tbr*, terminal branches. [Cajal, 1933.]

receptors are known in invertebrates, especially arthropods (pp. 420–422), among them campaniform sensilla, hair sensilla, and chordotonal organs.

A large class of mechanoreceptors with diverse functions is included among the vertebrate receptors, collectively called the **acousticolateralis receptors** (Fig. 2.68). The name refers to the fact that they embrace the organ of hearing of higher vertebrates and the lateral line organs of lower vertebrates, as well as such derivatives as the vestibular organs for senses associated with equilibrium. The lateral line organs include special derivatives for electroreception (p. 84). Others are differentially influenced by certain chemicals, and some are quite sensitive to temperature changes, though neither of these stimuli is likely to be the normally adequate stimulus. The proper designation of any receptor should be

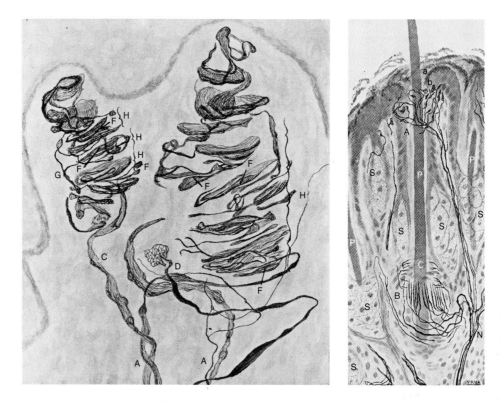

Figure 2.67
Endings in hair follicles and sensory corpuscles. Reduced silver stain. **Left.** Meissner's corpuscle in the skin of the finger. A, afferent nerve fiber; C, transition region between capsule of corpuscle and sheaths of nerves; D, branch of nerve going to a nearby corpuscle; F, lens-shaped expansions of nerve fibers; G, fine vertical fibers; H, planes, or strata, of fibers. **Right.** Hair of young rat. A, special forms of nerve fiber termination in loops; B, the same in palisades; C, nerve ring around hair base; N, small nerve emerging from the plexus in the dermis; P, hair; S, sebaceous gland; a, stratum corneum of epidermis; b, deep transitional epidermis. [Cajal, 1933.]

Figure 2.68
Sensory epithelia of the acousticolateralis system.

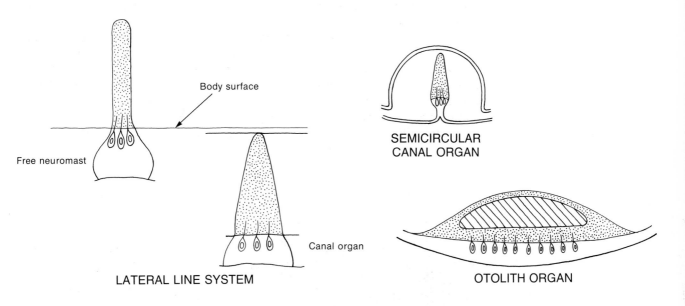

Body surface

Free neuromast

SEMICIRCULAR
CANAL ORGAN

Canal organ

LATERAL LINE SYSTEM

OTOLITH ORGAN

The afferent nerve fibers are not shown. [Dijkgraaf, 1952.]

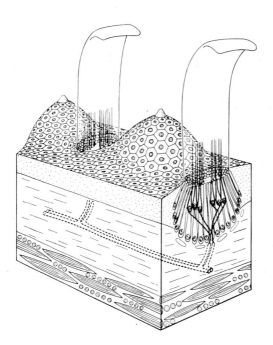

*Detail of neuromasts, with nerve fibers (broken lines)
and cupulae projecting from sense hairs. [Dijkgraaf,
1963; modified from Görner, 1961.]*

based on evidence of the biologically significant stimuli that normally excite the receptor. This criterion is not satisfied for many histological or physiological varieties.

Acousticolateralis organs have three features in common. (a) The receptors are so-called secondary sense cells typified by hair cells—specialized epithelial cells with cilia at the free end and nerve endings touching the base and sides (Fig. 2.69). (b) The supply of sensory nerve

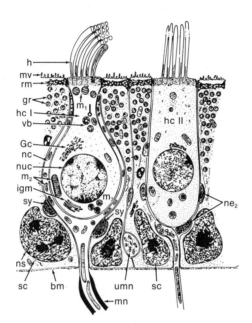

Figure 2.69
Two types of hair cells and nerve endings in the vestibular sensory epithelium; diagram from electron micrographs. *bm*, basilar membrane; *Gc*, Golgi complex; *gr*, granules in a supporting cell; *h*, hairs; *hc I*, hair cell of type I; *hc II*, hair cell of type II; *igm*, intracellular granulated membranes; *m$_1$, m$_2$, m$_3$*, mitochrondria; *mn*, myelinated nerve; *mv*, microvilli; *nuc*, nucleus; *nc*, nerve chalice; *ne$_2$*, granulated nerve ending at the base of a hair of type II (colored, indicating efferent inhibitory influence); *ns*, nucleus of a supporting cell; *rm*, reticular membrane; *sc*, supporting cell; *sy*, granulated synapse; *umn*, unmyelinated nerve; *vb*, vesiculated bodies. [Engström and Weršäll, 1958.]

fibers comes from the brain, at the level of the rhombencephalon, essentially the medulla, via cranial nerves, with the somas in ganglia of these nerves. Some, but not all of these receptors receive an efferent nerve supply from the brain, exerting, mainly, an inhibitory influence on the sensitivity at the hair cell. (c) There is usually, but not in the primitive lateral-line organs, an enclosed space into which the cilia project. Accessory structures, such as a cupula, basilar membrane, or otoliths, are associated with selecting and filtering certain stimuli and enhancing the receptor's sensitivity to them. These accessories define sense organs such as the semicircular canals and cochlea.

Five major subclasses of lateral-line organs are distinguished, including (a) primitive, free neuromasts on the surface, (b) canal organs within an extensive system of cutaneous canals over the body, especially along the lateral line (Fig. 2.68), (c) pit organs, (d) ampullae of Lorenzini, and (e) tuberous organs (Fig. 2.72). The last three are described below under Electroreceptors. These major windows on the world of lower vertebrates were lost as amphibians came out of the water. The inner ear remains as a homologue in terrestrial amphibians and in reptiles, birds, and mammals.

Thermoreceptors. These are probably mainly free nerve endings, but some special corpuscles have been thought to serve "warm spots" (warmth-detecting); another class, "cold spots" (cold-detecting). These identifications are too uncertain to perpetuate here, but there is no question that some afferent units increase their impulse discharge upon warming and that others do so upon cooling. The "warm fibers" concentrated in special organs of pit vipers and boas, adapted to receive infrared radiation from objects only very slightly warmer or cooler than their surroundings, are remarkably richly branched, mitochondria-packed, free nerve endings.

Nociceptors. There are probably two distinct classes of receptors that discharge impulses especially in response to noxious stimuli—that is, such as would cause damage or pain in man. The discreteness of such receptors from other modalities has often been doubted but can now be supported with direct evidence (*see* Altman and Dittmer, 1974, vol. 2, Table 138, pp. 1142–1149). The nociceptors are free nerve endings of thin unmyelinated axons belonging to the size category of "C" fibers (see p. 209). Some medium-sized, or "A-delta," fibers are also nociceptive. In addition, there is probably a contribution to pain from extraordinary forms of firing of tactile, pressure, or other afferents.

Chemoreceptors. Under this heading is a wide variety of structures, including free nerve endings in special visceral organs, such as the carotid body, which detects and signals blood O_2 level and CO_2 level; special cuticular pegs and hairs, in arthropods, supplied by the distal processes of primary sensory neurons, which respond to flower odors, sex attractants, and the like, and the receptors for taste and smell in humans (Fig. 2.70). These,

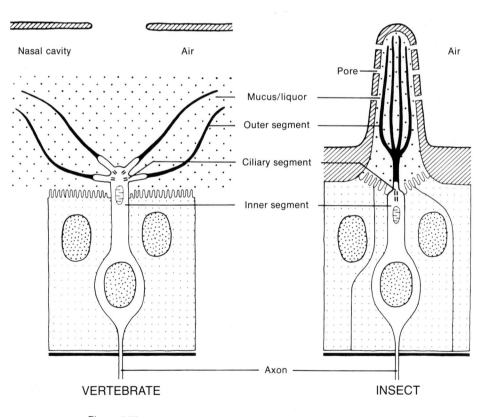

Figure 2.70
Olfactory receptors, vertebrate and invertebrate. [Steinbrecht, 1969.]

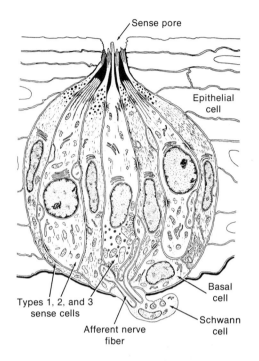

Figure 2.71
A taste bud. [Murray, 1973.]

Sense pore

Epithelial cell

Types 1, 2, and 3 sense cells

Afferent nerve fiber

Basal cell

Schwann cell

of course, differ profoundly. The olfactory organ (Figs. 2.61; 2.62) resembles primitive sensory epithelia in that the receptors are primary sensory neurons, are superficial, and have derivatives of cilia. The gustatory organ (Fig. 2.71) is the other extreme in being far from the primitive form. Like the acousticolateralis series, with which there may well be some evolutionary relationship, the taste organs are based on specialized, so-called secondary sense cells that lack neuronal processes. One of many remarkable features of the taste cells is their rapid turnover; the typical cell lives ten days in man.

It should be recalled in contrasting vertebrate organs of taste and smell that the difference is not basically between an airborne stimulus and a waterborne (saliva) stimulus. Both organs evolved in early aquatic vertebrates, and were possibly specialized for the reception of different classes of chemicals.

Electroreceptors. In our sampling of somesthetic receptors, the recently discovered modality of electroreceptors requires a few more functional comments than is normal for this chapter, since it is unfamiliar. Anatomically the two main classes of electroreceptors, each with subclasses, are part of the lateral-line system of lower vertebrates (Fig. 2.72). One of the two classes is called **ampullary,** and its most highly developed subclass is the ampulla of Lorenzini in sharks and rays. This organ is marked by a long canal running in the plane of the skin; the canal has a very high wall resistance, contains a jelly that has a high electrical conductance, and connects a surface pore with a terminal ampulla (bag). The lining of the ampulla includes the modified hair cells that act as receptors. Ampullary organs are highly sensitive to low-frequency (0.1–20 Hz) electric fields; a small shark can detect a potential difference equal to the voltage generated by one flashlight cell (1.5 V) across 1500 km of sea water. This ability is useful in passive detection of prey and in orientation to feeble low-frequency fields of inanimate origin. The other class is called **tuberous,** and includes several subclasses, all sensitive to high-frequency (50–5000 Hz) AC. This ability is useful in detecting the discharges of electric organs (organs in electric fish

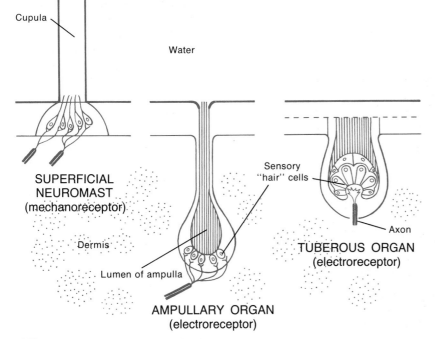

Figure 2.72
A mechanoreceptor and two electroreceptors of the lateral-line system of fishes: a superficial neuro-mast and two classes of specialized organs, in schematic form. [Dijkgraaf, 1967, and Szabo, 1974.]

that generate pulses or waves of current in the water). An electric fish receiving its own discharges is using an active system to detect perturbing objects ("electrolocation"). Receiving the discharges of other electric fish, the same receptors form a passive system for detecting social signals ("electrocommunication"). Ampullary organs and low-frequency electroreception occur in many nonelectric fish; tuberous organs, only in electric fish. The subclasses of tuberous organs differ in the way they

encode stimulus intensity (Fig. 6.11) and in their filter properties, sensitivities, adaptations, and structures.

D. Receptors in Muscle and Connective Tissue

A large variety of receptors, principally mechanoreceptive, is found within muscle tissue, in associated tendons, fascia, and joint capsules, and in mesenteries and other visceral membranes. Some are

Figure 2.73
A Pacinian, or lamellated, corpuscle. [*Cajal, 1933.*]

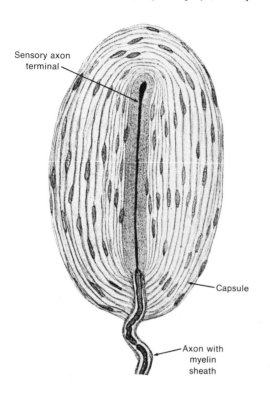

Sensory axon terminal

Capsule

Axon with myelin sheath

free nerve endings; others are elaborately encapsulated, like the Pacinian, or lamellated, corpuscles.

One significance of the **Pacinian corpuscle** (Figs. 2.62; 2.73) is that after careful study, the capsule cells that distinguish it have been shown not to be receptors. The ending of the afferent axon is the receptor, transducing small, fast deformations into receptor potentials. The capsule cells filter the stimulus before it reaches the sensory ending. The Pacinian corpuscle also represents an example of extremely rapid adaptation, normally firing a single impulse for each sufficiently rapid pressure wave.

Arthropods have a variety of stretch-sensitive organs made of connective tissue strands that bridge a joint and contain primary sensory neurons—cell bodies as well as afferent endings.

In higher vertebrates, stretch receptors in muscle tissue have evolved into elaborate little organs called muscle spindles (Figs. 2.74; 4.29). A small group of muscle fibers encapsulated by a sheath have the spindle-like form and are hence called intrafusal fibers. They receive both motor and sensory endings. The motor endings ("fusimotor endings") come from a special size class of nerve fibers called the gamma efferents. These fibers are separately activated by spinal cord mechanisms, from the motor fibers to the main mass of muscle, the alpha efferents, as will be described on page 267. The afferent endings are of two types, serving slightly different functions and terminating in different ways on the intrafusal muscle fibers. Their important central effects will be treated on pages 267–269.

Crustaceans have analogous **muscle receptor organs** (MRO's). They add another feature—inhibitory efferent fibers that reduce the response of the receptor to stretch (Fig. 2.75). Some MRO's are relatively more phasic, and others are tonic in response to stretch. The most conspicuous anatomical feature of MRO's is the sensory nerve cell body, situated right in the organ and offering a highly accessible multipolar heteropolar neuron for study.

E. Receptors in the Viscera

A large class of receptors is often neglected, partly because they rarely contribute to our sensations, functioning in important ways entirely below our

Figure 2.74
The vertebrate muscle spindle and the Golgi tendon organ. Left. A whole muscle and its tendon.
Right. A spindle and its innervation.

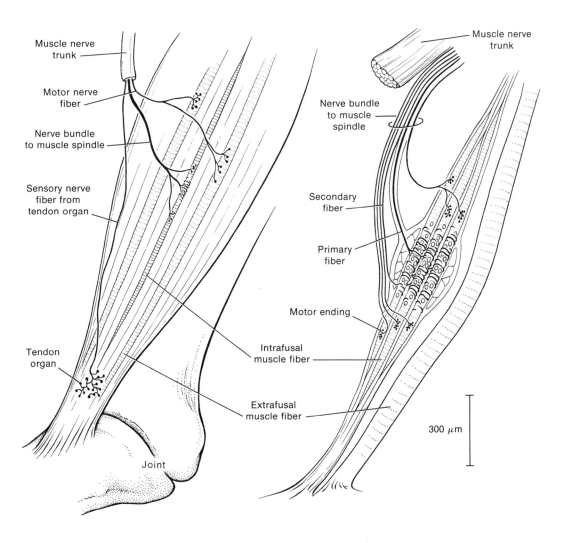

Muscle nerve trunk

Motor nerve fiber

Nerve bundle to muscle spindle

Sensory nerve fiber from tendon organ

Tendon organ

Joint

Muscle nerve trunk

Nerve bundle to muscle spindle

Secondary fiber

Primary fiber

Motor ending

Intrafusal muscle fiber

Extrafusal muscle fiber

300 μm

awareness. These are the ubiquitous receptors in blood vessels, deep tissues, and the visceral organs. Chiefly free nerve endings, these are diverse in geometric disposition and probably in function. They are mainly fine, unmyelinated fibers in mammals, quite numerous and powerful in influence. Besides controlling the activity of the viscera and the circulation, they modify the reflexes of skeletal muscle and, in humans at least, determine in various and little-known ways our general affect or feeling of well being.

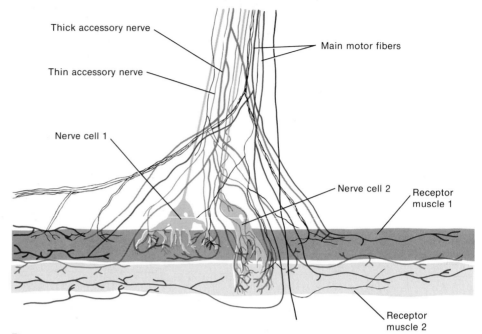

Figure 2.75
Crustacean muscle receptor organ. Diagram of the abdominal organs of one side in the lobster *Homarus*. Dark brown fibers, motor neurons; light brown, sensory; medium brown, the thick and thin inhibitory (accessory) neurons; black, small fibers of unknown function accompanying the inhibitory and main motor nerves. The thick and the thin inhibitory axons each supply the two nerve cells and both muscle fibers. The thick black fiber on the right runs to the median portion of the superficial dorsal muscle, sending a small branch to receptor muscle 2; the dark brown fibers on the extreme left are branches from axons that also serve normal muscles. Nerve cell 1 is the "slow" cell; nerve cell 2 is "fast." [Alexandrowicz, 1951.]

F. Receptors for Light

Unicellular primary sensory neurons specialized for photoreception are found scattered in the body wall of sea anemones, earthworms, clams, and many other forms. At the other extreme of complexity are the eyes of higher (more complex) groups. Two basic types of photoreceptor cells have been distin-guished (Fig. 2.76) each found in simple and complex eyes and suggesting two independent inventions. One is called **the ciliary type** because it contains a more-or-less modified ciliary apparatus. This type includes the rods and cones in vertebrate eyes, which are especially dis-tinguished by the development of a stack

Figure 2.76 (*facing page*)
Photoreceptors evolved along two main lines, the ciliary and the rhabdo-meric. A schematic representation of selected examples. [Eakin, 1968.]

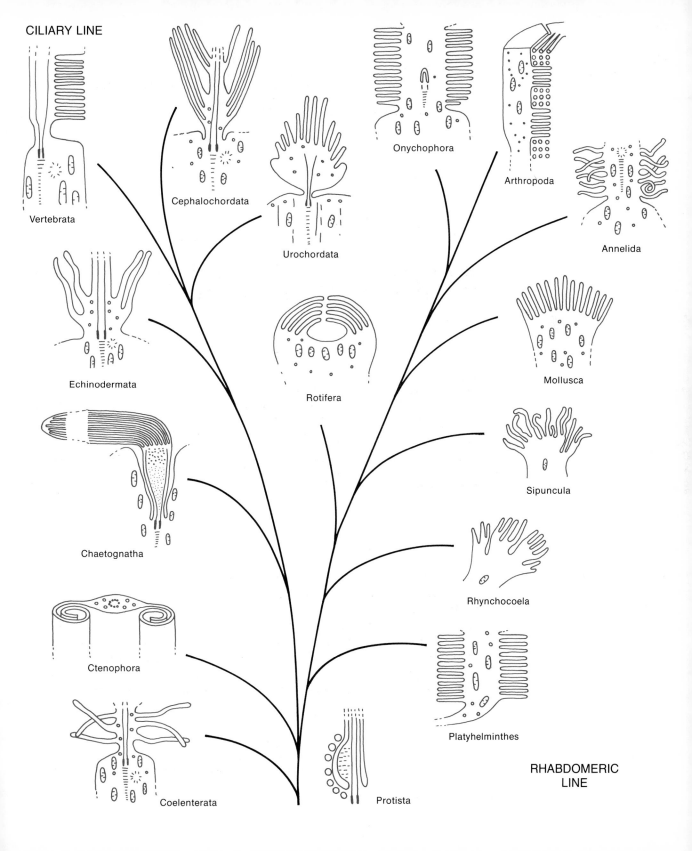

CILIARY LINE

Vertebrata

Cephalochordata

Urochordata

Onychophora

Arthropoda

Annelida

Echinodermata

Rotifera

Mollusca

Chaetognatha

Sipuncula

Ctenophora

Rhynchocoela

Coelenterata

Protista

Platyhelminthes

RHABDOMERIC
LINE

Figure 2.78

The two basic types of photoreceptors.
[Eakin, 1963.]

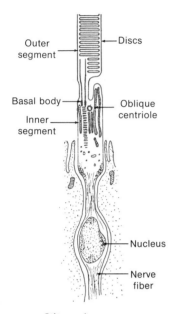

Ciliary photoreceptor

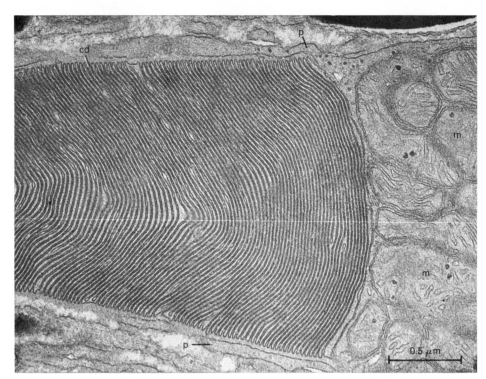

Figure 2.77

The sensitive part of a photoreceptor in high magnification. Base of the outer segment of a cone from a frog, *Hyla,* reared in darkness. The cone discs (*cd*) do not show vesicular breakdown; *m*, mitochondria in inner segment of cone; *p*, two processes of inner segment extending along outer segment. [Eakin, 1965.]

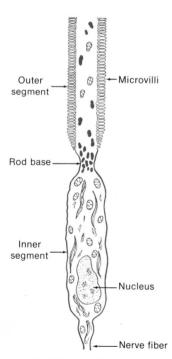

Rhabdomeric photoreceptor

of membrane folds that contain the photoreceptor pigment (Fig. 2.77). Here, it is presumed, the membranes transduce absorbed photons, down to as few as one quantum, into an event that becomes highly amplified and propagated to the distal end of the cell to control transmission at the synapse. The folds retain a connection with the cell membrane in cones but become free inside the cell in rods.

The other type of photoreceptor, called **the microvillous type,** is chiefly developed in arthropod and mollusc eyes, both simple and compound. In this type the pigment is also contained in processes of the cell membrane, but these take the form of finger-like villi, ultramicroscopic in size, densely packed and systematically oriented, forming the rhabdoms of light microscopy (Fig. 2.78).

In the context of this book it is the neural or organizational aspect of the more advanced eyes that is of interest. Fortunately, excellent introductions to the structure and function of the mammalian eye and retina are readily available in standard texts. The connections of the several cell types in the retina are treated in the next chapter.

G. Receptors for Sound

The ear is distinguished from other mechanoreceptors because of its complexity in higher vertebrates. Quite independently, several arthropod groups —crustaceans, insects, and spiders—have each invented their own auditory specialists from more general displacement transducers (see Fig. 2.79). Molluscs, including cephalopods, curiously have not, and the lowest vertebrates—cyclostomes—have hardly achieved a differentiation of sound receptors from their antecedents. This evolution occurs, as it were, before our eyes, in the vertebrate spectrum ranging from sharks to mammals: in fish, patches of sense cells evolved acoustical sensitivity in the parts of the internal ear or membranous labyrinth called the sacculus and lagena; in mammals, the spiral cochlea evolved from an elongated lagena. The remainder of this section deals only with mammals.

The anatomy of the external, middle, and inner ear, including the cochlea and the sound path, is readily available in standard texts. The organ of Corti within the cochlea includes the sense cells and nerve endings (Fig. 2.80). We have already seen the hair cells—the receptors for sound—in other parts of the acousticolateralis system (p. 82), where they are specialized for the stimuli of water movement, substrate vibration, steady position, tilt, acceleration, and electric current. The main features of interest in the present context are the two poles or faces of the hair cell, the outer as presumptive transducer and the inner as first encoder into transmitter release. As in other hair cells, the specialization for reception of mechanical events appears to be associated with the cilia; these are deflected by

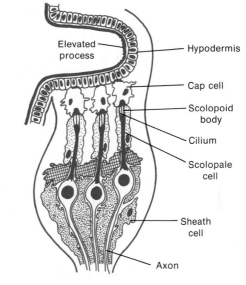

Figure 2.79
Tympanal organ of a locust. [*Gray, 1960.*]

Elevated process — Hypodermis — Cap cell — Scolopoid body — Cilium — Scolopale cell — Sheath cell — Axon

a shearing motion between the hair cells and the overhanging tectorial membrane, which moves relative to the hair cells, dragging the cilia with it. Just how these motions, amounting to no more than a few Ångstroms, cause a change in electric current through the hair cell is not known. The selectivity of each receptor, as measured by the response area (or receptive field, in the frequency domain) of each auditory nerve fiber, is determined primarily by the physical properties of the basilar membrane, under the hair cells. The consequence of this is that, since the basilar membrane cannot change properties in a very short distance, nearby hair cells and the nerve fibers excited by them have extensively overlapping response areas or effective bands of frequencies.

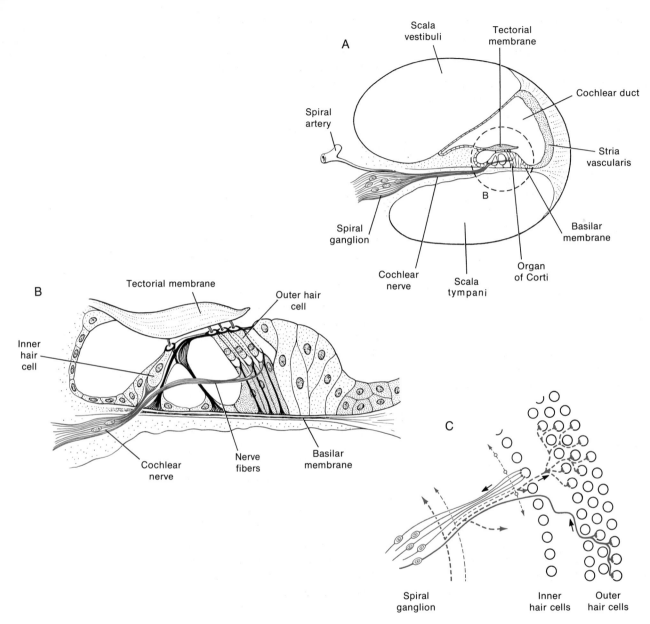

Figure 2.80
Phonoreceptors. The organ of Corti of mammals in cross sections and in plan view. **A.** Low-magnification vertical section through the cochlea. **B.** Section at medium magnification. **C.** The pattern of innervation of the organ of Corti, with afferent (solid lines) and efferent (broken lines) nerve fibers (high magnification). The fibers destined for the outer hair cells are represented by thick lines and those for the inner hair cells by thin lines. Relative numbers of the fiber types are approximately correct. [Spoendlin, 1968.]

The sense cells in the organ of Corti are of two kinds, inner and outer hair cells (Fig. 2.80), the latter far more numerous but receiving only about 5000 of the afferent fibers (in the cat). The inner hair cells seem built to be more discriminating, with 45,000 afferent fibers, each ending unbranched on one hair cell, which it shares with 10 or 20 other fibers. In contrast, each afferent to the outer hair cells ramifies and contacts many of them.

The cochlea receives a bundle of about 500 efferent or centrifugal nerve fibers arising from cells in the brain and exerting an effect at least largely inhibitory upon reception. These axons branch profusely and make some 40,000 endings —mainly upon the sense cells in the case of the outer hair cells, mainly upon the afferent terminals in the case of the inner hair cells. The efferent endings in the latter case are relatively few, in the former case extremely abundant, just the opposite of the numerical comparison of afferent endings.

VII. METHODS OF VISUALIZING NEURONS AND TRACING CONNECTIONS BETWEEN THEM

Tremendous effort has gone into the development of methods for making nerve cells and their processes visible, for tracing nerve pathways, for discovering the origins of all inputs to a given cell, and for tracing all outputs to other cells. Many of these techniques were perfected and used around the turn of the century, especially by the premier neuroanatomist, Santiago Ramón y Cajal (see p. 103). Others have become available more recently, especially through the introduc-

tion of the electron microscope and the development of electrophysiology and microneurochemical techniques. Some of the most important of these techniques are the following.

Classical Morphological Techniques

Methylene Blue. One of the first—and still one of the best—for selectively staining nerve cells, this technique is based upon a "supravital stain," one that is taken up by living tissue even though it is itself toxic. It works on invertebrates and vertebrates and on both myelinated and unmyelinated axons, often even staining to the terminal. It is most useful in the peripheral nervous system or in smaller ganglia. (See Figs. 2.3,B; 2.4; 2.75; 10.18; 10.25; 10.32; 10.56.)

Golgi Stain. A powerful silver stain, invented by Golgi and perfected by Cajal, that—for unknown reasons—selectively stains only an occasional nerve cell, leaving all others (more than 98% of the total) untouched. The cell that is stained, however, is stained reddish brown or black throughout the full extent of its processes. Because so few cells are stained, and these darkly, thick sections can be used and the whole cell visualized. This stain has been used more than any other in determining form of nerve cells. (See Figs. 2.3,D,E; 2.26; 2.37; 3.7; 3.18; 10.13; 10.31; 10.41; 10.42; 10.56.)

Reduced Silver Stains. These important techniques, developed by Cajal, Bielschowsky, Bodian, and others, depend on chemical reduction of silver ions (as in photography)—a reaction that can be made selective for nerve cells. Normal axons, myelinated or unmyelinated, are stained all the way to the terminal. With such stains, fiber pathways can be traced and synaptic endings visualized; but because all fibers are stained, it is difficult to follow individual fibers. (See Figs. 2.10; 2.50; 2.51; 2.54,A; 2.56,A,B,D; 2.57, A,B,D; 2.60; 2.67; 10.50; 10.56.)

Myelin Stains. Often associated with the name Weigert, certain ways of applying the dye hematoxylin selectively color myelin sheaths a dark blue. Unlike the foregoing, these are reliable routine procedures and very useful for low-magnification anatomy of major tracts, but are not good for establishing actual connections. (See Figs. 10.59,left; 10.74; 10.79; 10.91,B.)

Marchi Stain. This early technique is used for selectively staining degenerating myelin, which permits the tracing of fiber pathways following section of myelinated fiber bundles. It does not show terminals, however, and hence is not adequate for tracing of final connections.

Modern Morphological Techniques

Nauta-Gygax and Fink-Heimer Degeneration Stains. Products of long development, these silver methods selectively stain degenerating axons and terminals. They are much used in "experimental" anatomy—that is, after lesions are carefully placed in animals.

Chromatolysis Stains. Based on changes in concentration and distribution of RNA in the soma, these techniques identify the cell bodies of neurons whose axons have been cut (Fig. 2.11).

Electron Microscopy. The introduction of the electron microscope revolutionized the study of nerve connections, making it possible today to visualize the precise nature of the contacts between cells (Figs. 2.28; 2.29; 2.32; 3.17; 3.21) and to study cell organelles (Figs. 2.5; 2.8; 2.9) and sheaths (Figs. 2.54; 2.55; 2.58). Identification of processes can be difficult, however, so techniques have evolved for identifying degenerating processes, or for doing electron microscopy on cells previously identified in light microscopy with reduced silver or Golgi impregnation, or after labeling the cells in life by any of the following techniques.

Neurochemical Labeling. Living neurons will take up certain substances, including amino acids, and distribute them throughout their processes, right to the terminals. If the substance is electron dense or radioactively labeled, the molecules that have become incorporated into the rapidly transported axoplasm can be visualized. At the right time, the tissue is prepared and examined by electron microscopy and autoradiography. This technique has proved particularly useful for studying the incorporation of precursors of potential transmitters into specific neuron types. (See Fig. 2.22.)

Dye Injections. Two important new families of stains have been developed for the visualization of individual cells, especially ones that have been studied electrophysiologically. Procion yellow and related compounds, which fluoresce when exposed to UV light, are nontoxic, can be electrophoresed or pressure-injected into a cell, and will spread rapidly throughout the cell without crossing the membrane and leaking out. Another injectable substance, cobalt chloride, although toxic, will also spread rapidly throughout a cell, and can be precipitated with ammonium sulfide to form a black, electron-dense image of the cell. These stains have been used to good advantage to study the shape and connections of cells in a variety of preparations. (See Figs. 9.7,B; 10.26; 10.61.)

Endocytosis at Chemical Synapses. Taking advantage of the fact that during and after transmitter release, vesicle membrane appears to be pinched off from the terminal membrane, enclosing small amounts of extracellular fluid during the process (endocytosis), it has been possible to put markers in the extracellular medium and show them going into terminals that have been repetitively active. Especially useful is the enzyme horseradish peroxidase, which is picked up in vesicles and can then be exposed to hydrogen peroxide and other reactants to produce a dense, black precipitate in the terminal that was active. Alternatively, the preparation can be left longer and the movement of peroxidase traced to the cell, permitting one to determine where the relevant cell bodies are located. This is probably the best technique currently available for tracing pathways.

Physiological Methods Useful in Mapping Connections

Lesion and Ablation. In conjunction with the preceding and following methods, damage to specific brain areas can be used to determine connections. The results are difficult to interpret where the lesion can damage not only the intended neurons but also neighboring structures or fibers passing through the area. Though indispensable, and in many situations the only method available, it must be used with great care lest the findings be ambiguous.

Electrophysiological Recording. Carefully placed stimulating and recording electrodes—often positioned with the aid of special instruments that permit selection of sites according to X, Y, and Z coordinates from atlases (stereotaxy)—are the basis of widely used techniques for recording from single cells or groups of cells. By analyzing latency, waveform, and distribution of responses, connections can be traced with considerable accuracy. Postmortem verification of exact electrode position is sometimes the weakest link.

Stimulation. Coupled with recording or lesions, stimulation of receptors with "physiological" (= the normal modality of) stimuli or of different parts of the nervous system with electric shocks can provide important information about the nature of connections and the role of stimulated structures.

Injection or implantation of drugs, etc. Drugs, hormones, or labeled amino acids or precursor molecules can be inserted into specific locations of the nervous system and their distribution followed or the sites of influence recorded.

SUGGESTED READINGS

Bourne, G. H. 1968–1972. *The Structure and Function of Nervous Tissue* (6 vols.). Academic Press, New York. [These volumes contain articles on topics of a wide range of specialization.]

Bullock, T. H., and G. A. Horridge. 1965. *Structure and Function in the Nervous Systems of Invertebrates.* W. H. Freeman and Company, San Francisco. [Chapter 2 is a detailed treatment of microanatomy with emphasis on the comparative aspects.]

Kater, S., and C. Nicholson. 1973. *Intracellular Staining in Neurobiology.* Springer-Verlag, New York. [This is a fascinating little volume, with twenty highly technical chapters and startling illustrations on recent techniques based upon the distribution by the living cell of foreign substances throughout its processes.]

Nauta, W. J. H., and S. O. E. Ebbesson. 1970. *Contemporary Research Methods in Neuroanatomy.* Springer-Verlag, New York. [A practical and evaluative treatment of a wide array of techniques, many of them new, from electron microscopic to fluorescence and experimental degeneration methods, this book should be examined even if it is only to see the remarkable armamentarium of new tools available and the examples of them in the figures.]

Peters, A., S. L. Palay, and H. deF. Webster. 1970. *The Fine Structure of the Nervous System: The Cells and Their Processes.* Springer-Verlag, New York. [This is a small book, chiefly a magnificent set of full-page electron micrographs with an excellent, condensed text.]

Ramón y Cajal, S. 1952. *Histologie du système nerveux de l'homme et des vertébrés* (2 vols.). Instituto Ramón y Cajal, Madrid. [This is a translation in French from the Spanish, of a classic in neuroanatomy first completed in 1904. It is well worth browsing through even for those who read no French, for it contains a wealth of drawings that are still considered to be authoritative.]

3
STRUCTURAL BASIS OF CONNECTIVITY

I. INTRODUCTION

To a degree that is remarkable, considering the diversity of function involved, each nerve cell in the brain responds in a similar fashion. The cell in all cases sums all inputs and provides a resulting output ranging from cessation of activity (if the net input is strongly inhibitory) to a prolonged burst of spikes (if the input causes a prolonged depolarization above threshold). This output, in turn, is but one input among many to a number of higher-order cells. The action potentials of one cell are usually indistinguishable from those of other cells. The synaptic potentials, too, despite the wealth of subtle differences between synapses documented in the previous chapter, are generated by similar or identical mechanisms in vast populations of cells having different functions. Indeed, it cannot be emphasized too strongly that the complicated operations performed by the nervous system depend primarily on the connections between cells—that is, on which cells affect which others and in what ways—whereas the small number of different types of action potentials and synaptic potentials are merely general

mechanisms for accomplishing the interaction.

Two major questions are raised in this chapter: (1) How precise are the connections between neurons, and how important is the precision? (2) What specific anatomical circuits occur in well-studied parts of the CNS, and how do these connections account for the function of the cells involved?

In this chapter we first cite some examples, at the cytoarchitectural level, of anatomical regularities among arrays of units that help determine the functional value of connections. Next we jump to a much higher level of structure and consider some examples that show principles of topographic organization of large functional systems. Then we look more closely at the detailed circuitry of some relatively well worked out subsystems of connected neurons and examine the principles of neural connectivity that they illustrate. Many of the physiological features of connectivity that determine its effects are considered in Chapters 6 and 7 (e.g., convergence, divergence, redundancy, lateral inhibition, feedback, and filter properties).

II. CYTOARCHITECTURAL FEATURES GOVERNING THE EFFICACY OF CONNECTIONS

Most neurons have many synaptic inputs. In a spinal motor neuron (Fig. 3.1) the number of contacts is estimated to be about 5000; in a cortical pyramidal cell, perhaps 30,000; in a cerebellar Purkinje cell, as many as 80,000. Some inputs are excitatory, some inhibitory; some strong, some very weak. Obviously, in cells with thousands of inputs, any one synapse is unlikely to be capable of exciting the cell above its threshold for firing. Indeed, it appears that many of these synapses release no more than one or two quanta of transmitter, and often fail to release any transmitter upon receiving a presynaptic spike. Some individual synapses, however, may invariably cause a cell to fire. The activation of different populations of these synapses

under different conditions not only permits a wide range of different outputs from a given cell, but also enables that cell to function in many different neural pathways. Thus it is important to know what governs the efficacy of a given input.

There are a discouragingly large number of **ways in which two synapses may differ in effectiveness:** in quanta of transmitter released per given depolarization (itself a function of recent activity), in proximity to the subsynaptic membrane, in ease of diffusion of transmitter from the subsynaptic surface, in rate of hydrolysis or uptake of transmitters, in amount of receptor present, in the nature of the postsynaptic conductance change, and in the distance of the synapse from the site of postsynaptic impulse generation. Any of these variables could be subdivided into separate component processes, and probably all of them will be found, eventually, to be important in determining synaptic efficacy. Moreover, many of the component processes may vary in subtle ways to account for changes in neural function with experience. Several are treated in detail in Chapters 2, 5, and 6.

Dominant synaptic inputs tend to form many or larger synapses, with many synaptic vesicles and postsynaptic potentials (PSP's) of high quantal content; inputs with fewer or smaller synaptic endings release fewer quanta per impulse. Because of these differences, even different branches of the same fiber may have a strong effect on one postsynaptic cell and a weak effect on another.

There are also well-documented cases in which the tightness of glial wrapping around a synapse strongly influences its effectiveness. Since synapses on dendrites are often enclosed in a glial wrapping, it is possible that such insulation is

Figure 3.1
Synaptic terminal boutons on the surface of a motor neuron, reconstructed from serial electron micrographs. [Poritsky, 1969.]

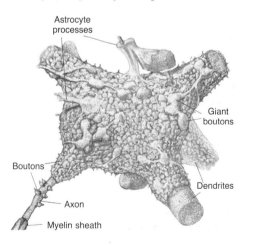

Astrocyte processes

Giant boutons

Boutons

Dendrites

Axon

Myelin sheath

of widespread importance. When the axon of a neuron is cut, the synaptic inputs onto the neuron frequently decline sharply in effect. There is some evidence that this loss of effectiveness is associated with a slight displacement of the presynaptic endings from the cell surface and with the interposition of glial cell processes between them and the cell surface.

Describing the morphology and physiology of a given synapse does not sufficiently describe its role in affecting the interaction of two cells; it is necessary to analyze a given synapse in the context of all others (in the same subsystem). Tracing the effective connections between neurons thus becomes extremely complicated, for function depends not only on the existence of synapses, excitatory or inhibitory, but also on such relationships as the proximity of other synapses and their shapes and terminal diameters. These spatial relationships are central to the importance of cytoarchitecture. (See also Box 6.4, p. 234.)

A. Consistency of Neuron Form and Connections

Since the days of Cajal, we have known that there are many distinguishable neuron shapes, always reliably found in characteristic areas in the central nervous system. Many of these consistent variations in neuron structure are described in Chapter 2. A large proportion of the motor neurons, many interneurons, and some sensory **neurons are individually indentifiable** in annelids, arthropods, and molluscs. Even in vertebrates, a few single cells are individually identifiable by their shape, position, and connections (e.g., giant electromotor neurons in certain electric fish, Mauthner cells in fish

and tailed amphibians, and giant spinal interneurons in lampreys). In the invertebrates, in which many fewer cells govern behavior, individually recognizable neurons have long been known. Only recently, however, with the advent of methods (e.g., procion yellow) for visualizing the whole structure of a given neuron, has it been appreciated that such cells are commonplace and that the consistency that they exhibit in shape, structure, inputs, and outputs extends even into the complex ganglionic neuropile and to the dynamics of connectivity (pp. 54, 257, 264, 274, 409, 413, 417, 432, 441). It seems likely that in the vertebrates, in which such consistency is much harder to find, all degrees of equivalence of cells will be found, including cells that are either unique or that overlap with only a few others.

For only a few cell types has the precision of their connectivity been pursued beyond determining that the site of synapse with a target cell is, for example, predominantly on the apical dendrite or on the secondary or tertiary branches. Accurate analyses of the number, location, size, and shape of one particular cell's terminals on another are not easily accomplished. At least one such analysis has been done, however, in an elegant electron-microscopic study of the connectivity between the receptor cells of the *Daphnia* eye and the second-order cells in the brains of these tiny crustaceans. *Daphnia* can be parthenogenetically cloned, so that each individual is genetically identical to every other. The eye consists of 22 ommatidia, each with 8 receptor cells. These 176 cells synapse with 110 second-order neurons in the brain, and the axons of each group of 8 receptor cells synapse with 5 neurons in a tightly organized "optic cartridge." The whole system is small enough to be reconstructed by computer from electron micrographs. This has now been done by Levinthal and his colleagues, who have shown that each of the 176 receptor cells and 110 second-order neurons is recognizable in the eyes of all individuals of the clone on the basis of their location, morphology, and connections. The position of the cell body and the branching patterns of homologous cells are highly uniform, and so are the connections. Each receptor cell synapses with only certain of the 5 second-order neurons in the optic cartridge, and then always more strongly (more synaptic terminals) with some than with others. Typically, there is no contact with 2 or 3 of the second-order neurons. Thus the connections between cells, and their relative strength, are reliably determined genetically. Nevertheless, the exact number, geometry, and distribution of endings all vary within quite broad limits for connections between corresponding cells in the eyes of different individuals (or the two sides of a given individual). These differences, which may not affect function significantly, apparently reflect the genetic noise level. (Macagno et al., 1973). Consistent connectivity among the cells in the complex optic cartridges in insects is mentioned on page 424.

B. Segregation of Inputs

We now know, from combinations of morphological and physiological studies, that in many cell types of characteristic form, **the synapses made by different afferents are consistent in position and structure.** For example, hippocampal pyramidal cells at a given site have a

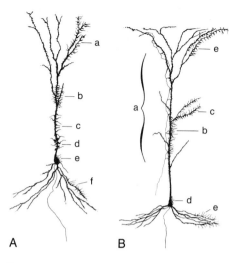

Figure 3.2
Segregation of inputs on a hippocampal pyramidal cell (A) and a large fifth-layer pyramidal cell of the cerebral cortex (B). Each letter represents a separate area of synaptic termination receiving input from separate, distinct afferent systems. [Scheibel and Scheibel, 1970b.]

quite consistent dendritic structure, such as that shown in Figure 3.2,A. From studies of degeneration, it has been found that there are six major categories of inputs and that they are segregated, those of each category making contact on different parts of the dendritic tree. In pyramidal neurons of the visual cortex, sectioning of specific afferent fibers from the lateral geniculate causes a loss of synaptic contacts (and dendritic spines) in one area of the central shaft of the apical dendrite (input b, Fig. 3.2,B). Other inputs innervate basal dendrites and secondary or tertiary dendrites. Many other examples could be cited. For example, Figure 3.3 shows segregated inputs from different sources contacting different dendrites of a brainstem reticular neuron and a spinal cord motor neuron. In the olfactory bulb, mitral cell

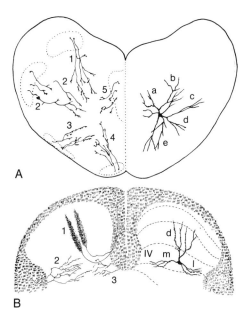

A

B

Figure 3.3
Schematized drawings showing segregation of input onto **(A)** a brainstem reticular neuron (right) by collaterals from afferents of five different tracts or nuclei within the brainstem (left) and **(B)** second-order sensory neuron in lamina IV of the spinal cord (right) and the three major sets of afferents that innervate its three dendritic regions. [Scheibel and Scheibel, 1970b.]

structure of frog retinal ganglion cells is another good example (Fig. 3.4). On the basis of dendrite structure, five distinct cell types can be distinguished, and they probably correspond to the five types of ganglion cells that are distinguished on the basis of physiology (see pp. 250–255, 374). In cell types *A* and *B* (Fig. 3.4), the two separated parts of the dendritic tree receive input from different sources in ways that are almost certainly important to the function of the cells.

Perhaps the best example of input segregation, however, is shown by the cerebellar Purkinje cell. These huge cells, with their broad, flat arborizations (Figs. 2.18; 2.29; 2.32), receive input from four different sources. (1) General weak excitation comes from as many as 80,000 granular cell axons (known as parallel fibers), which end on as many dendritic spines. (2) Weak inhibition comes from the processes of stellate cells, which make contact with primary and secondary dendrites. (3) Strong inhibition comes from the combined terminals of 10 to 20 basket cell axons, which are clustered in a basket shape around the axon hillock of the Purkinje cell (Fig. 3.20). (4) Finally, overriding all other inputs is the strong excitation exerted by a single climbing fiber that runs along the soma and entire dendritic tree of the Purkinje cell, apparently synapsing over much of its area of

dendrites are so completely separated that the primary afferent input can be shown to be localized entirely on apical branches and input from sources within the bulb entirely on side branches. The

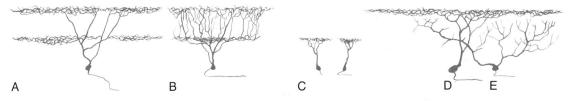

A   B   C   D   E

Figure 3.4
Different forms of retinal ganglion cells in the frog, showing their different dendritic distributions and geometries. Based on drawings of Cajal from Golgi-stained material. [Scheibel and Scheibel, 1970b.]

Box 3.1 Chronology and Background on the Interpretation of the Brain in Terms of Cells

A selection of highlights in the history of ideas. Earlier background is given in Box 7.4 on page 286, together with references for further reading.

1660 Malpighi founded microscopic anatomy, using improved and even compound lens systems—products of the first wave of development of microscopes from the simple magnifiers that had been known for centuries. Useful magnification of about 10X was achieved in the 16th century, and probably only several times that by the mid-17th. One of the earliest to publish observations in short papers—most were sent to the Royal Society of London—Malpighi described all kinds of plant and animal structures, fresh, dried, and cooked. He discovered capillaries and saw cells ("utricules," "globules") in plants. In the cerebral cortex he saw pyramidal cells and considered them glandular—secreting the fluid whereby muscles are moved to contract via the hollow nerves.

1665 Hooke at Oxford saw cavities in cork and called them "cells."

1672 Grew concluded that plants consist of microscopic chambers, or cells. His and Malpighi's detailed analyses of bark, wood, pith, buds, leaves, and flowers were not improved upon for a century. That these materials hold basic clues to the cellular basis of all living tissue was not recognized till the 1830's.

1668–1685 Leeuwenhoek made simple lenses that were perhaps unsurpassed till the 19th century. It is said that he attained (exceptionally) usable magnifications as high as 270X. Examining everything to hand, animate or inanimate, and without theorizing, he described bacteria, blood corpuscles, spermatozoa, and striated muscle, but nothing notable in nervous tissue.

1740 Swedenborg, in Uppsala, described pyramidal cells as being connected by threads to all parts of the body and to each other! He placed the soul in the cortex, which is an apparatus for sensation and its translation into motion.

1795 Bichat, in Paris, classified 21 tissues, including arterial, venous, glandular, dermal, serous, mucous, somatic and visceral muscle, and somatic and visceral nervous tissue.

1810 Gall, in Vienna, considered nervous tissue to be built up of nerve fibers, ganglia to be junctions of the fibers, the spinal cord to be a series of ganglia with tracts of nerve fibers running through it to the cortex, where intellect, soul, and hereditary traits reside.

1833 Ehrenberg saw myelinated tubular and nontubular strands and nerve cells in dorsal root and sympathetic ganglia.

1837 Remak recognized the true relation of nerve cells to nerve fibers for the first time, in observations on sympathetic ganglia.

1838 Purkinje described nerve cells from the cerebellum. He developed a microtome, an ophthalmoscope, and other devices.

1839 Schwann, building on Schleiden, proposed the Cell Theory—a sweeping new generalization and simplification, stating that tissues consist of cells. This was rapidly accepted and elaborated.

1841 Mohl proposed that the cell has living contents, which he called protoplasm, not merely slime or moisture.

1842 J. Müller and many pupils, including Henle, Reichert, Remak, Kölliker, and Helmholtz, developed animal cytology. The second wave of improvement of the microscope began, reaching a high level in the last quarter of the century.

1842–1849 Hannover saw the true relation of nerve cells to fibers in the cortex, Helmholtz in invertebrates, von Kölliker in dorsal root and V ganglia. Hannover concluded in 1849 that all fibers are connected with cells; Remak agreed in 1855; Deiters agreed in 1862.

1871 Schultze saw neurofibrils in living nerve cells in *Torpedo;* he had introduced good fixation (OsO_4) in the 1860's, as well as the idea that a cell is not so much a space with walls as a living substance without any necessary skin.

1872 Gerlach introduced metallic impregnation ($AuCl_3$) and described gray matter as a diffuse nerve net with undreamed-of complexity due to the fusion of fine dendrites. Meynert, Hensen, and others agreed.

1873 Golgi introduced his bichromate-silver method, which gave the first good pictures of the general form of whole nerve cells. It was ignored for 14 years, though later recognized as "the find of the century."

Box 3.1 (*continued*)

1884–1886 Weigert invented a special stain for myelin; Nissl invented a special stain for cell bodies; Marchi, one for degenerating myelin; and Ehrlich, the intra-vital methylene blue stain for whole neurons. These are four of the handful of most important stains.

1886–1888 Cajal, Nansen, and Kölliker began to use the Golgi method. Forel struck the first effective blow to the nerve net theory by finding that retrograde degeneration is confined to the damaged cell. The same conclusion was supported in·embryos by His, who said nerve cells behave as centers giving origin to fiber outgrowths, just as was claimed in 1857 by Bidder and Kupffer.

1888 Cajal found that the axis cylinder always terminates freely in proximity, not continuity, with other cells or dendrites. Examples of this finding are contained in his first descriptions of basket and climbing fiber relations to Purkinje cells in the cerebellar cortex.

1891 Waldeyer reviewed the literature and coined the term "neuron." He supported Cajal. Many workers were now using Golgi and other methods, including gold chloride. Some objected to the neuron theory, including Weigert, Held, Ápathy, Bethe, Bielschowsky, and Nissl.

1894 Cajal's third book, *New Ideas on the Fine Anatomy of Nervous Centers,* summarized more than 40 of his own papers on neurons.

1897 Sherrington introduced the term "synapse" for a concept put forth at least 12 years before by Romanes (1885). Cajal had developed the law of dynamic polarity by 1891.

1898 Nissl announced that the neuron theory was dead. Cajal took neuronal unity as a fact, and did not actively debate. He had not yet studied fibrils, whose "continuity" was Nissl's chief objection.

1901–1903 Cajal improved his reduced silver techniques, studied fibrils, and bombarded the journals with findings (22 papers in 1903) on fibrillar discontinuity. His work was immediately confirmed by more than 25 other authors. Nevertheless, some reports of continuity persisted even into the 1950's.

1906 Cajal convincingly showed that embryonic axons grow out from nerve cells, terminate in a growth cone, and later become sheathed, instead of arising by fusion of a row of cells, as proposed by Schwann (1839). Many workers confirmed his findings, including Harrison (1907), using tissue culture. The 1906 Nobel Prize in Physiology and Medicine was shared by Cajal and Golgi. This merely served to heat up the controversy; vigorous debate continued for more than 25 years, a few authors objecting even in 1958 (Tiegs; Boeke; Stöhr, Jr.; Held).

1912 Cajal's 2-volume monograph on degeneration and regeneration settled a long dispute over autoregeneration of axons distal to a cut (by Schwann cells, without the help of the center) versus outgrowth from the center, which he favored.

1900–1933 Scores of workers, including many pupils of Cajal, produced hundreds of papers. Cajal's summary of the "irrefutable evidence" appeared in 1933 in his *Neuronismo o reticularismo?*

1955 Electron-micrographic evidence of the generality of a bounding membrane gradually accumulated, and only special exceptions have been found. A quiet revolution took place in our understanding of the nature of the neuron, its ·processes, its membrane mosaic, its diverse relations with neighboring cells.

T.H.B.

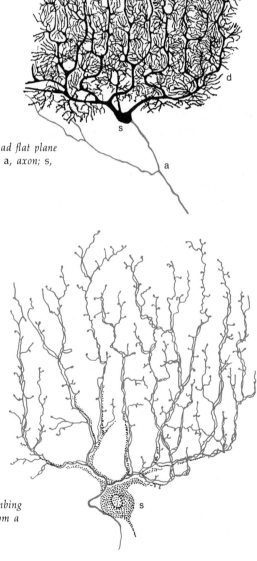

The dendritic arbor is a broad flat plane of richly branched dendrites: a, axon; s, soma; d, dendrite.

One of the inputs is the climbing fiber, a terminating axon from a soma (s) in the medulla.

contact (Fig. 3.5). The significance of these different inputs is further discussed in Section IV below (see also pp. 31, 49, 345, 347, 453).

This segregation of inputs may not mean that different types of afferents specifically recognize and grow to a certain patch of dendritic membrane. Instead, it may mean that synapses form sequentially during development (see Chapter 9). Whatever the mechanism, it seems probable that most neurons receive input that is segregated to some degree.

What, then, is the **importance of this segregation?** Too little is understood about dendritic properties to answer this question completely. Nevertheless, there is convincing evidence that the position of a synaptic input along a dendrite determines its interaction with other synapses and its effect on the firing pattern of the postsynaptic cell.

Although there are a growing number of exceptions, it is generally felt that the **dendrites of most cells do not generate spikes;** that is, they serve as passive conductors of synaptic potentials to the cell soma, where all potential changes are integrated and the net effect read out at the axon hillock. This conclusion is based upon experimental evidence, but makes good theoretical sense. If dendritic membranes were capable of generating spikes, the summation of inputs would be highly nonlinear. If the dendritic membranes are, in effect, passive conductors, then it follows that synaptic inputs near the soma will be more effective for any given conductance change than more distal inputs. Potentials generated in an apical dendrite will be less attenuated than potentials arising on a branch of the apical dendrite. Segregation of inputs thus has the effect of weighting each type of input differently in its influence on the cell's readout.

It is also evident that inhibitory endings, which usually act by increasing membrane conductance to K⁺ or Cl⁻ (see Chapter 5), are more effective when they make contact near the site of impulse generation rather than far out on dendrites, where their short-circuiting action would only affect nearby excitatory synapses on the same dendrite. Consistent with this expectation is the finding that many inhibitory inputs end predominantly on the cell soma or axon hillock or proximal parts of the dendritic tree. Nevertheless, it is perfectly reasonable to expect that inhibitory synapses will be found to occur at strategic locations far out on dendrites as well, where they could function to suppress specific excitatory inputs.

Dendritic spines may be morphological substrates for highly specific interactions between excitatory and inhibitory inputs on a dendritic branch (Fig. 3.6). Many large cortical neurons receive as much as 95% of their input via synapses onto spines. The stalk of a dendritic spine probably represents a significant resistance to flow of current generated by a depolarization of the terminal bulb. This resistance would attenuate the depolarization seen in the dendrite branch. Thus it is of interest to note that in some large neurons, the dendritic spines are short and broad near the cell soma but become progressively longer and thinner farther out on the dendritic tree. This could simply be a way of exaggerating the difference in influence between synapses near the cell body and those farther out. A more important function of the spines, however, may be to permit plastic changes in a synapse's efficacy. If there were a mechanism for elongating or

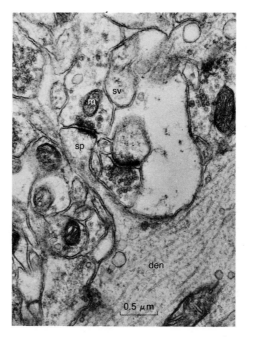

Figure 3.6
Electron micrograph of a dendritic spine (*sp*) from a rat cerebral neuron dendrite (*den*), showing synapses with mitochondria (*m*) and synaptic vesicles (*sv*) formed both on the bulbous spine terminal and on the stalk. The synapses on the stalk may represent a mechanism for specific inhibition of inputs onto particular spines. [Diamond et al., 1970.]

shortening a spine, this could greatly alter the degree of its attenuation of a signal. W. Rall has pointed out that, if one makes reasonable assumptions about the increasing input resistance of dendritic branches with distance from the cell soma, it is clear that progressively longer spine stalks would be required to keep the spine resistance in a range where minor changes in spine length would still affect synaptic efficacy. The problem is essentially one of impedance-matching. Because the input resistance of a distal dendritic branch is very high,

Figure 3.7
Golgi-stained dendritic branches of a human Purkinje cell, showing the numerous dendritic spines.
[*Eccles et al., 1967.*]

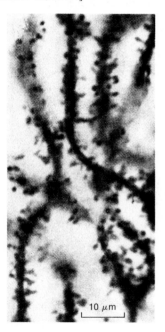

10 μm

the stalk resistance must also be high, so that part of the potential drop would have to be across the stalk—and subject to increase or decrease by plastic changes in stalk length or diameter. That dendritic spines have this function is pure speculation, (see also pp. 41, 388); nevertheless, the possibility that they do is of great importance.

The number and shape of specific populations of spines (Fig. 3.7) are dependent on the presence of specific afferent inputs (see Chapter 9, p. 347), which argues that they may be uniquely involved in certain types of synapses. Often synapses are seen both on the tip of the spine and on its stalk. One suggestion made on the basis of recent EM and electrophysiological studies is that the synapse commonly found on the stalk is inhibitory (Fig. 3.6) and that shunting of the membrane conductance at that point can suppress excitation coming into the synapse at the tip of the spine without affecting excitatory inputs elsewhere on the dendrite. These authors studied the synapse between the fish Mauthner cell and the motor neurons it innervates. They found that the excitatory input to a population of motor neuron dendritic spines from one Mauthner cell could be rendered ineffective if the contralateral Mauthner neuron was activated within 0.16 msec after the excitatory input. The inhibitory input is thought to end on the stalks of the spines. Thus there may be a large resistance built into the stalk that slows the spread of activation from spine to dendrite, permitting specific inhibition within a narrowly defined time. This still-unsubstantiated mechanism remains an exciting possible consequence of highly organized, segregated inputs onto

different portions of the dendritic tree.

The dendrites thus provide important and complicated sites of interaction as well as channels of variable effectiveness for excitatory inputs, depending on their position relative to the cell body. It seems inevitable that further research will reveal a great number of further refinements of integration accomplished by special dendritic properties.

III. TOPOGRAPHIC ORGANIZATION OF THE NERVOUS SYSTEM

There is also precise organization at the level of assemblies of neurons. We have seen, in Chapter 2, how assemblies can be arranged to form laminae, glomeruli, and nuclei. Not only are cells grouped together functionally into "nuclei" that possess inputs of the same modality or control aspects of the same sensory analysis or efferent outflow; within these cell groups, and in the fiber pathways between them, there is usually a further organization called topographic localization. Cells are located next to other cells that have overlapping or adjacent receptive or motor fields. For example, in the mammalian auditory system, the primary receptor neurons are located along the length of the basilar membrane of the cochlea. Because the basilar membrane continuously changes properties from basal to apical ends of the cochlea, neurons along it are excited by different sound frequencies at different positions —the highest audible frequencies at the base of the cochlea, the lowest frequencies at the apex. These first-order neurons innervate approximately three times as many second-order cells in the

Box 3.2 Maintenance of Temporal Information

The importance of accurate timing of inputs—and the maintenance of this timing across several synapses—is obvious in sensory pathways of the nervous system. Among the best examples are the cells of the medial superior olive in the auditory pathway. These cells have two major dendritic trees, extending from opposite poles of the cell soma. One receives synaptic input from the ipsilateral cochlear nucleus, the other from the contralateral cochlear nucleus, as shown in the figure. Normally, the contralateral input is excitatory, the ipsilateral input inhibitory. These cells appear to be involved in the determination of the azimuth of a sound source. When a sound is presented from a contralateral side, most cells are excited; ipsilateral sounds cause inhibition. In elegant experiments, it has been shown that the cells could go from strong excitation to full inhibition with a change of only a few hundred microseconds in arrival time of the two inputs. Human observers can behaviorly distinguish arrival time differences as small as 10 μsec. In certain cells, inhibition was observed when the ipsilateral signal preceded the contralateral by any amount between 0.2 and 1.4 msec, not with greater or smaller differences. Different cells in this nucleus show sensitivities to different delays and help constitute a mechanism for accurate determination of acoustic signal angle in space.

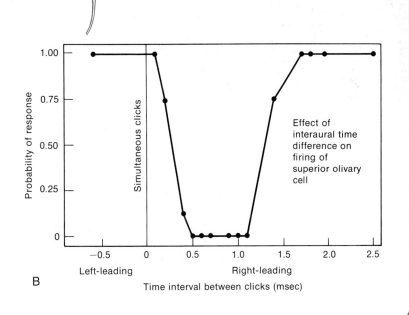

Synaptic geometry associated with precise temporal interaction in the auditory system. The medial superior olivary cell shown schematically in **(A)** receives input from both ipsilateral and contralateral auditory pathways. In the example shown in **(B),** the ipsilateral (right) ear input is strongly inhibitory and suppresses excitatory input from the contralateral (left) ear when its input precedes the contralateral by 0.4–1.1 msec. It is effective only within this narrow time slot. [Part A, Stotler, 1953; B, Galambos et al., 1959.]

Effect of interaural time difference on firing of superior olivary cell

Figure 3.8

Diversity and order among terminals of a single type of axon. Cochlear nerve fibers (A) ending in two parts of the ventral cochlear nucleus of a dog. [Cajal, 1909.]

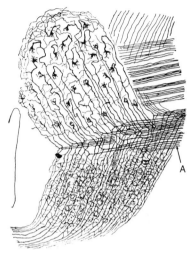

cochlear nuclei, but their divergency is even greater than that (Fig. 3.8). Each primary fiber is thought to terminate on 75 to 100 second-order cells. Despite this large amount of divergence (of first-order branches), and a like degree of convergence (onto second-order cells), the cochlear nuclei are "tonotopically" organized; that is, recordings from successive cells in a given microelectrode penetration show that adjacent cells are driven by adjacent best frequencies (Fig.

3.9). There is not a single "map" of the cochlea, but rather three or more (one estimate places the number as high as 13); nevertheless, within each subgroup of second-order cells, the cochlear input is distributed in an orderly way. At still higher neural levels, all the way to the auditory cortex, tonotopic localization is maintained.

Similar topographic organization is well documented in all of the other major sensory pathways. It has been

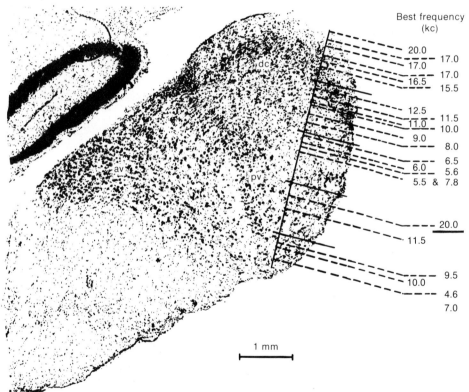

Best frequency (kc)

20.0
17.0 — 17.0
16.5 — 17.0
— 15.5

12.5
11.0 — 11.5
— 10.0
9.0
— 8.0
6.5
6.0 — 5.6
5.5 & 7.8

— 20.0
11.5

10.0 — 9.5
— 4.6
7.0

1 mm

Figure 3.9

Tonotopic organization of the cat cochlear nucleus. A saggital section is shown through the left dorsal cochlear nucleus (*dc*) and anterior (*av*) and posterior (*pv*) ventral cochlear nuclei. A microelectrode was advanced along the line shown, and cells recorded at the various depths. The frequency to which each was most sensitive is shown at the right. Note that along this penetration there were two separate maps of the cochlea. [Rose et al., 1959.]

most thoroughly studied in the **somato-sensory pathways.** Each dorsal root brings into the spinal cord a large number of sensory fibers from many different types of receptors. Within a segment or two in the spinal cord, these are roughly segregated according to modality. Many decussate immediately to the opposite side of the cord, and synapse before sending second-order axons to the brain in the spinothalamic tracts. The rest send ascending collaterals to the brain on the same side, without synapsing. Thus two major populations sort themselves out —one ascending to the cord in the contralateral lateral column, the other in the ipsilateral dorsal column. In the lateral columns run afferents providing information primarily about the body's condition: warmth and cold; sexual sensations; tickle, itch, and pain from skin, tendons, joints, viscera, and muscles; including the sense of muscle fatigue (Fig. 3.10,A). These afferents tend to be mainly small-

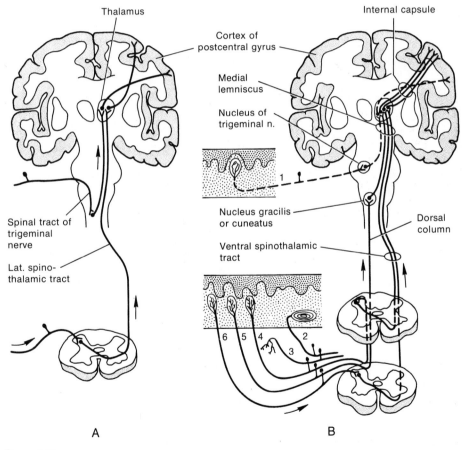

Figure 3.10
A. Schematic drawing of major afferent pathways for pain and temperature receptors. **B.** Pathways for major tactile inputs to the cerebral cortex. Shown are Meissner's corpuscles (1, 4, 5, 6), Pacinian corpuscles (2), and joint receptor endings (3). [Gardner, 1968.]

diameter myelinated and unmyelinated fibers. In the dorsal columns run afferents carrying information of a more specific, perceptual nature: muscle, tendon, and joint sensibility used in appreciating body position and degree of movement; and touch and pressure sensitivity used in locating sites of contact with an object, or the weight, shape, and texture of an object (Fig. 3.10,B). Some touch and pressure fibers from both cutaneous and deep receptors are present in the lateral columns as well.

Both dorsal and lateral columns are arranged in **laminae** containing fibers arising in successive spinal ganglia (innervating different areas of the body) and entering the spinal cord via separate dorsal roots. Contributions from the most caudal inputs lie most superficially in the spinal tracts. Thus an electrode that penetrates a dorsal column on a transverse axis from lateral to medial will record activity first from the most posterior parts of the body, then from successively more anterior segments, making a continuous map of the animal's body. When the electrode passes from one lamina to the next in the lumbar part of the cord and the lower thoracic segments, there is an abrupt backward jump in the receptive field locations, reflecting the fact that each dorsal root innervates a field that overlaps slightly with the one in front and in back. Interestingly, however, this abrupt transition in receptive field area disappears at higher spinal levels. Apparently, the fibers coming from different parts of the body sort themselves out so that the spatial distribution is truly continuous. (Such sorting out of fibers is seen in extreme form in the vertebrate optic nerve. In this brain tract, the axons of retinal ganglion cells appear to be mixed randomly, but as

they approach their target—for example, the optic tectum—they segregate themselves into two major divisions (one from the dorsal part of the eye, the other from the ventral) and then grow over the surface of the target nucleus in such a way that they restore a precise topographic representation of the retina (see Chapter 9 for a more complete discussion of this phenomenon).)

As we have seen, the fiber pathways in the vertebrate spinal cord are **segregated by modality** and are topographically **organized with respect to body position.** The topographic organization is maintained, and even refined, at the level of the primary **somatosensory cortex.** Recordings of evoked or single-unit activity at different points on the cortex confirm anatomical evidence that the afferents project to and drive cortical cells in a broad band of cortical tissue immediately posterior to the large central sulcus that runs mediolaterally across the top of the cortex. The somatosensory cortex of each cortical hemisphere maps the opposite side of the body, reflecting the fact that virtually all of the afferent fibers have decussated somewhere along their pathway. Running from the lateral to the medial extremes of this somatosensory projection area is an orderly progression of receptive fields, from the face to the tail. The map is distorted, however. The amount of cortical tissue devoted to a given part of the body is not proportional to the area of that part but to the density of sensory innervation. Thus areas devoted to tongue, face, soles of hands and feet, and genitalia are disproportionately large, and areas devoted to trunk or upper limb innervation relatively small. Within each anterior-posterior band associated with a particular part of the body, there is another constant progression of receptive fields; for example, within the foot and hand areas, the progression is from lateral to medial. (Just anterior to the central sulcus, the primary motor cortex is similarly mapped, but it is distorted in a different way, most cells in those areas being devoted to controlling the finest movements, especially of the hands.) A careful mapping of the somatosensory cortex (Fig. 3.11) shows a complex picture, but one which makes sense if it is assumed that the primary organization is established in the embryonic anterior-posterior axis, before the limbs develop. In the final map there is a continuous trajectory from the face around the forelimb (proximal to distal in the anterior side, then back on the posterior side), down the trunk, around the hind limb, and out the tail. (Details of multiple representation are omitted here.)

The organization, however, is even more sophisticated than this map would indicate; Mountcastle and his colleagues were the first to show that the **sensory cortex is arranged in a series of columns,** perpendicular to the surface, with a diameter of about 0.5 mm. Within each column, all cells recorded are concerned with the same sensory submodality (e.g., light touch, joint movement, or bending of hairs) and have approximately the same receptive field. Adjacent columns may deal with the same sensory submodality but be slightly displaced in receptive field; or they may deal with a totally different submodality. A similar and perhaps even more ordered organization has been shown by Hubel and Wiesel in the **visual cortex** (Fig. 3.12). Here, however, the columns are not devoted to different sensory modalities. All cells in a 50 μm column are most sensitive to images of straight lines or bars of the same orientation and approximately the same receptive field in space, though they have different sensitivities to brightness, bar thickness, contrast, and the like. Adjacent columns, as we move along the cortex in one axis, show alternating ocular dominance, but approximately the same orientation specificity. Moving along the cortex in the other axis, the orientation specificity gradually changes, in an orderly way (Fig. 3.13). Ocular dominance is strong only where the lateral geniculate axons enter the cortex, in layer IV. In deeper and more superficial layers, there is increasing lateral interaction between ocular dominance columns, so that most cells are driven by both eyes, with only slight dominance by one eye.

The result of this fine-grained organization is that cells involved in the processing of information about a given kind of stimulus, received at a given location, are located close together. Since most information processing depends on the subtle interaction of cells receiving much the same information, this proximity greatly facilitates the interaction. The morphology of the sensory cortex clearly supports a primarily vertical organization. Each column contains a few thousand cells, and there are many vertical connections between cells in the same column, as well as lateral connections between cells in different columns. Differentiation of function of cells within a column is gradually becoming known, so that the apparent redundancy of cells grows less as anatomy and physiology progress.

An extreme and convincing example of complex cortical columnar organization is seen in the conspicuous cell assemblies of the mouse somatosensory cortex, which processes input from facial

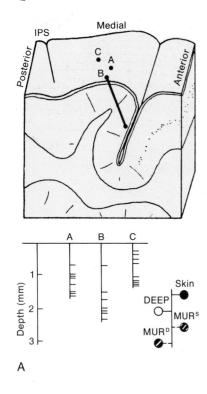

A

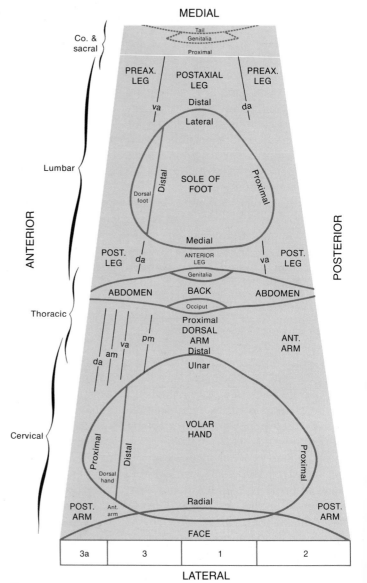

B

Figure 3.11 *(facing page)*
Topographic organization of the somatosensory cortex of a primate, *Macaca mulatta. (IPS,* intraparietal sulcus.) **A.** Reconstruction of three microelectrode penetrations, at *A, B,* and *C,* through the cortex of the right hemisphere in a region called *S1* (first somatic area). The modality and receptive field *(RF)* of each neuron encountered is marked on the representations of body parts. The positions of the neurons along the penetration are indicated at the lower left: broken lines mean a multiunit record *(MUR),* solid lines a single unit (symbols indicate stimuli to the skin (ˢ) or to deep cutaneous sense organs (ᴰ). All neurons depicted were responsive to hair movement or mechanical stimuli on the skin. Such detailed maps permit the following overall map. **B.** The postcentral gyrus is shown schematically unfolded, with the level of spinal origin of input shown to the left. Note the regular progression from face and head on the ventral-lateral surface to the tail at the top of the cortex, and the distortion of the map that results from emphasizing the analysis of input from the face, hands, and feet. *da,* dorsoaxial line; *va,* ventroaxial line; *am,* anterior midline; *pm,* posterior midline. The numerals (1, 2, 3, 3a) along the cut lateral edge of the map indicate the cytoarchitectural fields that make up S1. [Whitsel et al., 1971.]

Figure 3.12
Above. Postulated columnar organization of the primary visual cortex in the macaque. **Below.** Different layers of the lateral geniculate (*i* = ipsilateral, *c* = contralateral) send their output to adjacent parts of the cortex at the level of layer IV. Here they innervate interneurons that distribute their influence both vertically up and down "ocular dominance columns"—really slabs—and laterally to adjacent columns, so that most cortical neurons can be driven binocularly, albeit with slightly stronger effect by the eye that provides primary input to that column. Perpendicular to these wider (0.25 mm) columns are narrow (*ca.* 25–50 μm) "orientation columns" (actually slabs), all cells of which are selectively excited by signals having a certain orientation. Adjacent orientation columns differ by only a few degrees in their selectivity. A family of these orientation columns would contain cells responsive to all possible orientations, all with essentially the same receptive field. Nearby would be cells organized into similar orientation and ocular dominance columns having a slightly displaced receptive field. The cellular circuitry shown represents the input to an upper-level complex cell from two neighboring ocular-dominance columns, but from the same orientation column. [Hubel and Wiesel, 1972.]

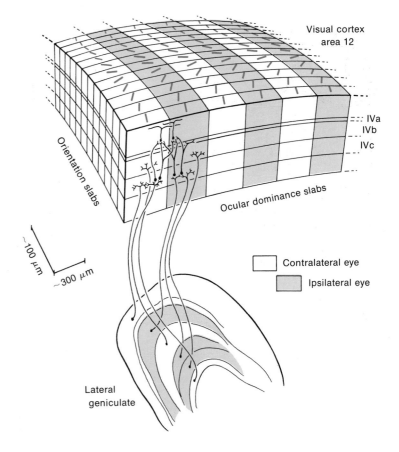

114

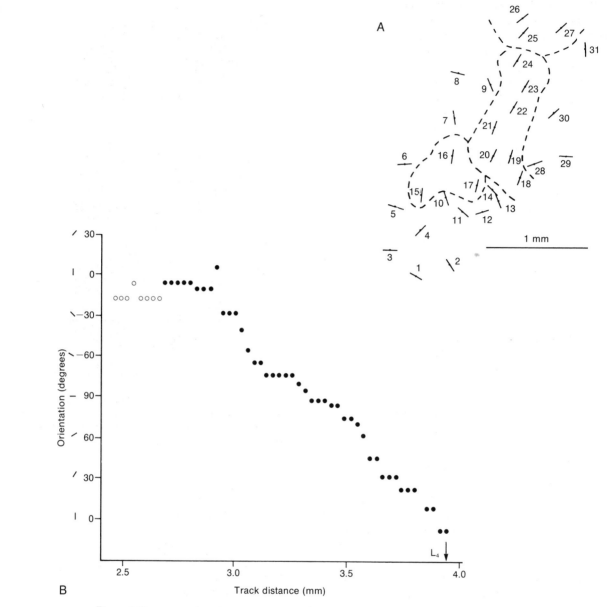

Figure 3.13

Distribution of best orientation sensitivity in the visual cortex. **A.** Plan view of part of the cat cortex, showing the most effective orientation of a bar of light for the cells in a column under that symbol. Broken lines partially outline three areas of common orientation preference. [Hubel and Wiesel, 1963.] **B.** Graph of the best orientation of cells encountered in a single track of a microelectrode pushed at 3° to a tangent of the monkey cortex (left hemisphere), showing the steady progression of orientations at the rate of 165°/mm. Open circles from contralateral eye, filled circles from ipsilateral eye; arrow "L4" is an anatomical landmark. [Hubel and Wiesel, 1974.]

Thus the **role of cerebellar integration** is presently conceived by many scientists as being one of sampling input from proprioceptive and other receptors that signal relations of the body in space, first by enhancing differences in receptor activity via lateral inhibitory and feedback circuits that sharpen the topographic localization of activity in the cerebellar cell populations, and then by translating the patterns of Purkinje cell excitation into similarly patterned inhibition of the deep cerebellar nuclei. The role of the cerebellar output in coordinating motor activity is thought to be somewhat as follows. The deep cerebellar nuclei are themselves excited by collaterals of the mossy and climbing fibers and by collaterals of the cerebral pyramidal cell axons that form the main channel of outgoing commands to the spinal motor neurons. The cells of the deep cerebellar nuclei in turn project excitatory input (a) back onto the pyramidal cells in the cerebral cortex via one or two synapses, forming a positive feedback loop, and (b) onto other brain stem neurons in descending motor pathways other than the pyramidal. The Purkinje axon input to the deep cerebellar nuclei interrupts this loop in a specific topographically patterned way, modulating the efferent outflow to avoid maladaptive motor neuron activity.

For the purpose of this chapter, our main objective has been to use the cerebellum as a real example of the complexity, types of integration, and importance of specific synaptic geometries encountered in a brain structure of only moderate complexity.

V. THE VERTEBRATE RETINA

Vertebrate retinas vary greatly in complexity. One of the simplest, with relatively few and large cells, is that of *Necturus*. This is the first retina to be thoroughly analyzed by both electrophysiological and ultrastructural methods. As a result, we have a fairly complete picture of the interactions that take place. Figure 3.20 is a diagrammatic survey of the structural relationships.

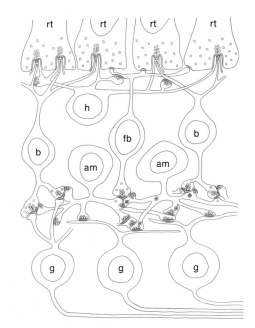

Figure 3.20

Summary diagram of synaptic contacts in a vertebrate retina, based upon investigations in *Necturus*. In the outer plexiform layer, receptor cell terminals (*rt*) interact with both bipolar (*b*) and horizontal (*h*) cells in "triad" synapses. Flat bipolars (*fb*) are excited via less complex synapses. Horizontal cells interact with each other (not shown) and with bipolars, as well as feeding back inhibition onto receptor cells. In some vertebrates the receptor cells are known to have excitatory contact with each other. In the inner plexiform layer, bipolars innervate amacrine cells (*am*) and ganglion cells (*g*); amacrines interact with each other, with bipolars, and with ganglion cells. (See text for more details concerning the nature of the interactions.) [Dowling, 1970.]

In the retina, as in the cerebellum, there are currently recognized **five types of neurons,** one of which provides the output of the system. These five types are organized in three cell layers, separated by two "plexiform," or neuropile, layers in which synaptic interactions take place. In the outer cell layer are the photoreceptor cells. In the middle cell layer are horizontal cells, bipolar cells, and amacrine cells. The innermost cell layer contains the retinal ganglion cells whose axons constitute the optic nerve.

The **receptor cells** provide input to both **bipolars** and **horizontal cells** (H-cells) via highly complex synapses in which there are typically three postsynaptic processes—two H-cell processes and one bipolar process—protruding into invaginations in the base of the receptor cell and apposed to a presynaptic ribbon (Fig. 3.21). (Such presynaptic "ribbons," characterized by a row of vesicles on either side of a dense bar, are a specialization found so far only in a few other synapses.) Occasionally,

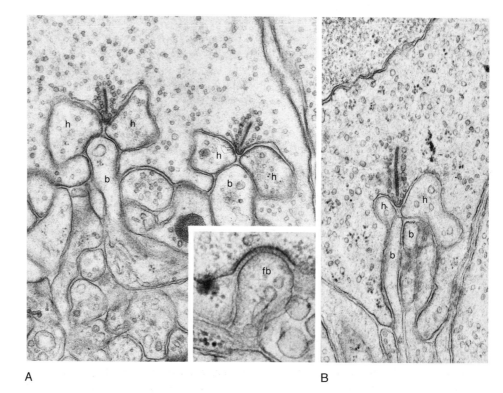

A   B

Figure 3.21
Electron micrographs of monkey cone (**A**) and cat rod (**B**) terminals synapsing with horizontal cell (*h*) and bipolar (*b*) processes. In both cone and rod terminals, transmitter is apparently released from the presynaptic "ribbon." *Inset:* frog flat bipolar (*fb*) synapse on a cone. [Dowling, 1970.]

one of the H-cell processes in this synaptic **"triad"** can be seen to be synapsing on the adjacent bipolar process. There is now convincing evidence, both morphological and physiological, of interaction between receptor cells, at least in some species. The bipolars tend to have narrow receptive fields, receiving input from only a small number of receptor cells in one region of the retina. The H-cells, on the other hand, extend broadly across the retina, summing input over a much larger area, extending perhaps ten times as far from side to side. They interact with bipolar cells over this whole area, both in the triads just described and directly onto separate bipolar cell processes. The H-cells mediate lateral inhibition of the bipolars.

The bipolar cells constitute the vertical pathway for carrying input from the outer to the inner plexiform layer. There the bipolars interact with **amacrine cells** (A-cells) and **ganglion cells,** in a maze of different types of synapses. In some synapses, reminiscent of the foregoing triad, the bipolars innervate both the A-cells and the ganglion cells, again with a ribbon synapse in the bipolar terminal (see Fig. 3.20). Often there are two A-cell processes postsynaptically, rarely two ganglion cell processes. The A-cells extend over considerable distances laterally and synapse on bipolar terminals, other A-cells, and ganglion cells with conventional (as opposed to ribbon) synapses. In many instances, the A-cell process that receives input from a bipolar makes an adjacent feedback synapse onto the bipolar process (a **"reciprocal" synapse**). This feedback may be responsible for the transient nature of some A-cell response (see below). Also commonly observed are chains of synapses between A-cell processes, from one process onto another for three or four consecutive synapses (serial synapses). There is, in fact, a good correlation between the preponderance of amacrine cell synapses seen and the complexity of integration in the retina. In the cat and the monkey, for example, all but about 20% of the bipolar terminals synapse on only one amacrine cell and one ganglion cell. The ganglion cells respond with simple concentric receptive fields, and lateral inhibition appears to be the principal integrative function of the retina. The retina of the frog or of *Necturus,* on the other hand, has a much larger fraction (50–75%) of bipolar terminals that synapse with more than one amacrine cell and not with ganglion cells. In these animals, the ganglion cells receive most of their input from amacrines; and although some show concentric receptive fields, others are much more complex in stimulus requirements (e.g., for borders, rate and direction of movement, and small objects; see pp. 250, 374, 464).

It appears that at least two different synaptic pathways of significantly different function exist in the inner plexiform layer. In the most direct pathway, bipolars directly excite ganglion cells, with parallel lateral input from amacrine cells. In the second pathway, bipolars excite amacrines, which, in turn, after interaction among themselves, excite ganglion cells. These morphological differences can be correlated with two types of ganglion cell responses: one in which sustained activity reflects the slow time course of bipolar cell responses, and another in which phasic on-off responses are typical of some amacrine cells.

The **electrophysiology of the retina** is extraordinarily interesting, especially in

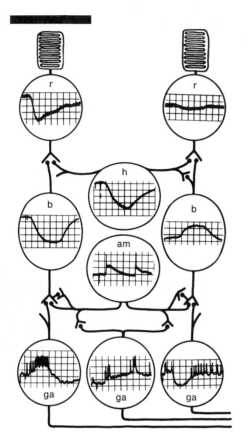

Figure 3.22
Summary of the electrophysiological responses
recorded from sample cells in a vertebrate
(*Necturus*) retina when the rod on the left is
stimulated. Postsynaptic structures are depicted
by rounded terminals. Note that receptors (*r*),
bipolars (*b*), and horizontal cells (*h*) do not have
spikes, but that amacrines (*am*) and ganglion
cells (*ga*) do. (See text for further explanation.)
[Dowling, 1970.]

relation to the anatomical connectivity.
The first three cells in the network, the
receptor cells, bipolar cells, and horizontal cells, all respond to light with slow,
graded responses, and do not produce
spikes (Fig. 3.22). In the receptors the
response is always hyperpolarization,
associated with a light-induced decrease
in membrane conductance to Na^+, reducing a depolarizing "dark current." The
horizontal cells are also hyperpolarized,
but the bipolars may be either depolarized or hyperpolarized (depolarized in
"on-response" pathways to ganglion
cells). Probably the ribbon synapses of
the receptor tonically release transmitter that depolarizes the horizontal cells.
With receptor hyperpolarization, transmitter release is reduced, allowing the
postsynaptic horizontal cells to hyperpolarize also. The first evidence of lateral inhibition is seen at the level of the
bipolars. On-center stimulation causes a
large slow depolarization in the center,
and hyperpolarization in the surround.
Stimulation by a bright annulus surrounding a dark center produces a response of the opposite polarity, usually
after a significant delay. The surround
effect is apparently mediated by horizontal cells.

Amacrine cell physiology is less well
understood. In *Necturus*, most are reported to respond with transient depolarizing postsynaptic potentials and
spikes to both the "on" and the "off"
of a stimulus anywhere in their receptive fields. In other animals, it has been
reported that amacrines respond only
with graded depolarizations. Ganglion
cells, on the other hand, have been well
studied and respond either with transient depolarizations and spikes at the

"on" and "off" of a stimulus, or in a way clearly resembling the responses of bipolar cells, with a sustained slow depolarization and steady spike discharge. Stimulation of the surround in the latter case causes inhibition of activity.

The ganglion cells that respond to both the "on" and the "off" of a stimulus often show maximal responses to stimuli moving in a certain direction. Two plausible synaptic arrangements that are consistent with the known morphology and could explain directional specificity in a ganglion cell are shown in Figure 3.23. In the first example, there are directionally polarized inhibitory synapses by horizontal cells on bipolars. In the second, it is assumed that bipolars excite amacrines and that amacrines excite ganglion cells but inhibit other amacrine ganglion cell synapses. The observed serial synapses between amacrines could then account for large responses to a wave of stimulation passing in one direction across the retina and no response (null) to movement in the other direction (see Fig. 7.8).

Box 3.3   Color Vision

In animals with color vision, the role of retinal neurons is much more complicated. For example, in fish retinas, which also have large cells, the horizontal cells appear to receive input from three populations of cones having different wavelength specificity. Some horizontal cells receive similar input from all, responding with hyperpolarization at all wavelengths; the response amplitude is mainly a function of luminosity. Other types of horizontals respond with hyperpolarization to some wavelengths and depolarization to others (see figure), thus yielding "opposite color" responses. Bipolar cells driven by cones typically show concentric field organization in which the center is either hyperpolarized or depolarized by one wavelength, whereas stimulation of the surround by a different wavelength has the opposite effect. Amacrine and ganglion cells show correspondingly complex behavior.

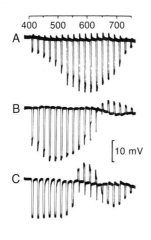

Responses of three different types of horizontal cells in the carp retina to flashes of light of different wavelength (top). **A.** Luminosity type, responding with hyperpolarization at all wavelengths. **B.** Biphasic chromaticity type, reversing polarity once as the wavelength is altered. **C.** Triphasic chromaticity type, reversing polarity twice as the wavelength is altered. [Tomita, 1965.]

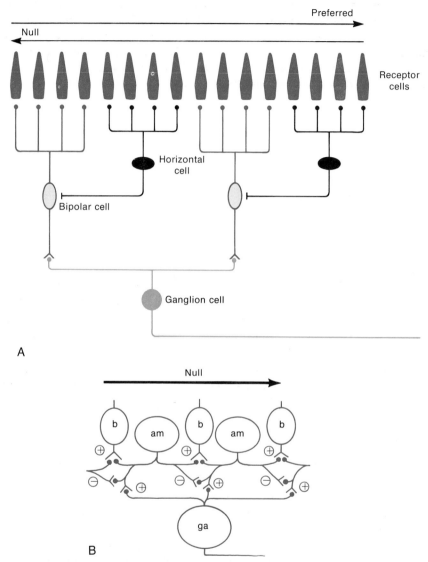

Figure 3.23
Possible synaptic circuitry, consistent with known morphology, that could explain directionally sensitive ganglion cell responses. **A.** Each bipolar receives excitatory input from one group of receptors, inhibitory input from another group of receptors, to one side, via a horizontal cell. A stimulus moving to the right would tend to excite the ganglion cell; stimuli moving to the left would first excite the horizontal cells, which could then inhibit the bipolars, inhibiting direct response to the receptors. [Michael, 1969.] **B.** Amacrine cells (*am*) are excited by bipolars (*b*), and they in turn excite adjacent amacrines and ganglion cells (*ga*) in one direction, but inhibit ganglion cells (or amacrine cell inputs to ganglion cells) in the other; there could be excitation to a stimulus moving in one direction and no response (null) to movement in the other. [Dowling, 1970.]

Thus it is possible to recognize in the particular connections between retinal neurons the morphological substrate for lateral inhibition (the major function of the horizontal cells in the outer plexiform layer) and for directional specificity and discrimination of other dynamic aspects of stimulation (both outer and inner plexiform layer). The role of cells having no spike potentials is especially interesting, since it may turn out that a large number of similar interneurons of the brain function in a comparable way.

SUGGESTED READINGS

Dowling, J. E. 1968. Synaptic organization of the frog retina: An electron microscopic analysis comparing the retina of frogs and primates. Proc. Roy. Soc. London, Ser. B, **170**:205–228. [Excellent, readable description of retinal circuitry and its functional implication.]

Eccles, J. C. 1973. *The Understanding of the Brain.* McGraw-Hill, New York. [Introduction to basic neurophysiology and to the principles of neuronal integration necessary to understand synaptic circuitry. Goes into detail in the context of best known systems.]

Eccles, J. C., M. Ito, and J. Szentágothai. 1967. *The Cerebellum as a Neuronal Machine.* Springer-Verlag, Berlin. [Good statement of morphological and electrophysiological properties of the cerebellum, with educated hypotheses about how the organ works, as of publication date.]

Kandel, E. R., ed. 1976. *Cellular Biology of Neurons* (Handbook of Physiology. The Nervous System, vol. 1). Williams and Wilkins. [Up-to-date review papers on a wide variety of problems in neurobiology, with especially relevant chapters on neuronal organization.]

Palay, S. L., and V. Chan-Palay. 1974. *Cerebellar Cortex: Cytology and Organization.* Springer-Verlag, Berlin. [Authoritative description of the ultrastructure and organization of the cerebellum.]

Schmitt, F. O. 1970. *The Neurosciences: Second Study Program.* Rockefeller Univ. Press, New York. [This volume and the next contain excellent collections of papers by leaders in the field on many topics, including synaptic function, organization, development, and integration.]

Schmitt, F. O., and F. G. Worden. 1974. *The Neurosciences: Third Study Program.* M.I.T. Press, Cambridge, Mass.

Shepherd, G. M. 1974. *The Synaptic Organization of the Brain.* Oxford Univ. Press, New York. [Careful analysis of synaptic circuitry in the best studied neuronal systems: spinal cord, olfactory bulb, retina, cerebellum, thalamus and cortex. Good introductory chapters on the principles involved.]

4

EXCITATION AND CONDUCTION

I. INTRODUCTION

From anatomical studies it is apparent that the nervous system consists of an enormous number of cells woven into a functioning network. The role of this network is to analyze the external world and the internal environment and to coordinate the behavior appropriate to the animal's needs. At first view the task seems so complicated that a great variety of nerve cells functioning in basically different ways would seem to be needed to accomplish it. Present evidence suggests that this is not the case. Despite the complexity of shapes and types of neurons, as we have seen in the preceding chapters, their basic physiological properties are similar. Superficially, the situation is analogous to that found in an electronic computer. Simple components with the same properties are intricately and precisely connected to represent complex parameters in the form of stereotyped electric signals.

In this chapter and the next the physiological properties of the components of the nervous system will be considered.

Studies of the electrical signs of neuronal function have advanced rapidly due to the availability of techniques for measuring this kind of activity. As a result we can explain, on a relatively sophisticated level, how the electrical signals are generated in individual cells.

It is hoped that by understanding how the individual elements work and the way in which they are connected, one can then appreciate the operation of the network. This approach has been quite successful in the analysis of the behavior of simple neuronal populations. It must be recognized, however, that new approaches will probably be required for an understanding of more complicated functions. Many networks are likely to have properties that cannot be ascertained solely from an appreciation of the function of the individual cells and the ways in which they are connected. Moreover, many of the most interesting problems of nervous system function, such as the specificity of neuronal connections during embryological development, the mechanisms of storing information in the nervous system, and the long-term

When a nerve is stimulated electrically the potential change in each fibre is of fixed magnitude and duration; it depends only on the state of the fibre and not on the strength of the stimulus which set it in motion. . . . The [sensory] message consists merely of a series of brief impulses or waves of activity following one another more or less closely. In any one fibre the waves are all of the same form and the message can only be varied by changes in the frequency and duration of the discharge. In fact the sensory messages are scarcely more complex than a succession of dots in the Morse Code.

Adrian, 1932

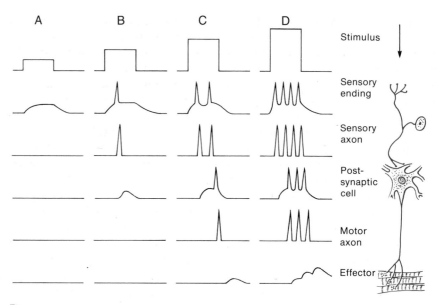

Figure 4.1
Diagram of a simple reflex arc. A stimulus is applied to a sensory nerve terminal with increasing intensity from **A** to **D**. Initially, only a sensory generator potential is produced locally in the nerve terminal. With increasing stimulus, the generator potential increases in amplitude, giving rise to nerve impulses in the sensory axon. Such action potentials propagate to the nerve terminals, where transmitter is released, producing an excitatory postsynaptic potential **(B)**. With repetitive impulses **(C and D)**, the synaptic potentials sum, giving rise to impulses in the motor axon. Such impulses travel the length of the motor axon and release excitatory transmitter, which activates the effector, in this case a striated muscle. The effector response is graded in accord with the number of impulses in the motor axon and the length of the interval between them.

chemical interactions between neurons and between neurons and their satellite cells, may not have any obvious electrical correlates. An understanding of the common properties of the basic elements is essential before attempting to analyze the behavior of the whole system.

A. Electrical Signals in the Nervous System

To illustrate the most common electrical signals generated by neurons, it is useful to consider the operation of a simple **reflex arc** consisting of a sensory neuron, a motor neuron, and an effector. Such a system is illustrated in Figure 4.1. External energy is absorbed by the sensory neuron. This leads to a voltage change across the membrane of the nerve terminals, called a **sensory generator potential.** The amplitude and duration of the potential are continuously variable with the magnitude and duration of the external source of energy. It is confined to the endings of the sensory neuron. Sensory generator potentials are the electrical analog of the external energy as transformed by the receptor.

If the sensory generator potential is sufficiently large it leads to a brief transient potential change, of constant duration and constant amplitude, across the membrane of the axon—the **nerve impulse,** or spike, often called the nerve action potential. The spike potential is much larger and briefer than the generator potential and is rapidly and faithfully conducted the entire length of the axon. The spike potential amplitude and duration are fixed by the properties of the axon; the magnitude and duration of the generator potential are represented in the intervals between nerve impulses and their number.

When the action potential reaches the nerve terminal it causes the release of a specific chemical, the synaptic transmitter. This substance diffuses across the narrow gap between the two neurons and produces a change in the postsynaptic cell; the change can either lead to spikes in that cell or decrease the likelihood of spikes. If the transmitter action leads to nerve impulses as shown in the figure, it is said to be **excitatory;** if it prevents nerve impulses it is **inhibitory.** Excitation is produced by decreasing the voltage difference across the postsynaptic cell membrane, and inhibition is usually produced by increasing that potential. These **synaptic potentials** are variable in amplitude, can add with one another, and are confined to the postsynaptic region. The manner in which the postsynaptic cell sums all the excitatory and inhibitory influences is termed **synaptic integration.** If the synaptic connection is excitatory it will lead to one or more impulses in the motor neuron, just like those in the sensory neuron. These impulses are conducted to the terminals, where a chemical transmitter is released that activates the effector. If the synaptic potential is inhibitory it decreases or stops activity in the postsynaptic cell, rendering the effector organ less active.

Neurophysiologists now consider, by convention, that the membrane potential of the cell is that of the inside with respect to the outside. The outside potential is considered "zero." With the introduction of intracellular recording techniques it was found that as the electrode enters the cell a negative potential is recorded. This stable **negative potential,** recorded across the cell membrane in the absence of activity, is termed the **resting potential.** A positive change in membrane potential is termed a **depolarization;** a more negative change, a **hyperpolarization.** These conventions were not well established until the mid 1950's, and so one must be aware of the varying conventions used by different authors when referring to the older literature.

These three types of potential changes —the sensory generator potential, the nerve impulse, and the synaptic potentials—are the major stereotyped electrical signals in the neurons. Their properties and occurrence are remarkably constant in the various parts of the nervous system. Knowledge of how they are produced is relevant to understanding all aspects of nervous system function.

II. IONIC PERMEABILITY AND MEMBRANE POTENTIAL

Nerve cells, like other cells, are surrounded by a surface membrane. This membrane presents a formidable barrier to the movement of ions between the aqueous phases inside and outside the cell. The permeability of the membrane

—that is, the ease with which a given ion species crosses it—differs even for rather similar inorganic electrolytes. The membrane is able to distinguish between ions according to their physicochemical properties; it allows some to pass more easily than others.

Most of the electrical signals in the nervous system are produced by changes in the relative permeability of the membrane to various ions, primarily sodium, potassium, chloride, and calcium. The changes in permeability may result from changes in the electrical potential difference across the membrane, from the reaction of specific chemical substances with reactive sites on the membrane, or from the absorption of external energy. Resting neurons are more permeable to K^+ than to other ions. However, the rate at which K^+ diffuses across the membrane is 10^8 times slower than its diffusion rate in free solution. The permeability to Na^+ is about 50 times less than that to K^+ at rest. The action potential in a nerve is produced by a reversal of this permeability pattern, initiated by a change in the potential difference across the membrane. That is, the action potential results from a change in Na^+ permeability from 1/50 to about 10 times that of K^+. Similarly, changes in permeability pattern are responsible for excitatory and inhibitory synaptic potentials and sensory generator po-

tentials. It is therefore of primary importance in the understanding of the electrical signals in the nervous system that the distribution of ions across the membrane, the permeability of the ions, and the factors that determine ionic permeability and distribution be clearly established.

Many of the experimental results to be considered have been obtained from studies of the giant axon of the squid *Loligo*. This remarkable axon was called to the attention of physiologists by J. Z. Young in 1936. He appreciated that its large size (up to 1 mm in diameter) would be of great advantage in physiological studies. Over the past 30 years two groups of neurophysiologists, one headed by K. S. Cole at Woods Hole, Massachusetts, and the other by A. L. Hodgkin in Plymouth, England, utilized this axon to analyze in remarkable detail the membrane processes that underlie the nerve impulse. The main results and conclusions have subsequently proved to be applicable to describing the behavior of nerve cells in other animal species.

A. Distribution of Ions Across the Nerve Membrane

Table 4.1 indicates the approximate concentration and transmembrane ratio of

Table 4.1
Ionic Concentrations Across a Squid Axon Membrane

	Intracellular (mM/kg H_2O)	Extracellular (mM/kg H_2O)	Ratio extracellular/intracellular
K^+	400	20	1/20
Na^+	50	440	9/1
Cl^-	50	550	11/1
Ca^{++}	0.4	10	25/1
Organic anions	350	0	—

ions inside and outside of a giant squid axon. The table contains the necessary information for our purposes but is incomplete; it should be remembered that the total of positive and negative ions on either side of the membrane is equal and that the membrane is permeable to water, so that the two solutions are isosmotic. In addition, the important parameter is ionic activity rather than concentration. Measurements of diffusion of radioactive K^+ in squid axons show that the activity coefficient of K^+ is similar inside and outside. Other ions, particularly Ca^{++}, have a greater tendency to bind to cellular constituents, and their concentrations do not accurately reflect their ionic activities. The basic principles of electrical activity can be explained most simply by using ionic concentrations only.

B. Origin of the Membrane Potentials

If the potential difference across a resting nerve membrane is measured directly by electrodes that respond to voltages inside and outside of a neuron, it is found that the interior of the cell is 60 to 70 millivolts (mV) negative with respect to the outside. This steady potential difference is the "resting potential" (Fig. 4.2). To see approximately how this potential arises, we will make some simplifying assumptions. From the table we find that the concentrations of Na^+ and K^+ are vastly different on either side of the membrane. Suppose a cell has a high concentration of K^+ and organic anion and a little Na^+ on the inside and a high concentration of Na^+ and Cl^- with a little K^+ on the outside, as shown in Figure 4.3. Suppose also that the membrane is permeable to K^+ but not to any of the other ions.

Under these conditions a chemical gradient would result from the difference in K^+ concentration on either side of the membrane, tending to make K^+ diffuse out of the cell (Fig. 4.3). Will the cell continue losing K^+ until the chemical gradient disappears and the K^+ concentration is equal on either side of the membrane? Certainly not, for as soon as a few K^+ ions diffuse outward, there will be organic anions inside whose negative charges will not be neutralized by nearby positive charges. The organic anions will migrate toward the membrane with K^+ but will be unable to pass through it. This excess negative charge on the inside will then tend to pull positive ions back in. As the membrane is impermeable to Na, the only cation to be pulled back is K^+. Quite rapidly the system will come into an equilibrium in which the chemical gradient forcing K^+ to diffuse outward is balanced by an electrical gradient tending to pull K^+ in. Thus it appears that the potential gradient is produced by the concentration gradient and that the concentration gradient is maintained by the potential gradient. The situation can be thought of as analogous to a weight hanging from a spring. The weight produces tension in the spring, and the tension in the spring keeps the weight from falling further. The weight represents the concentration gradient for K^+ and the tension in the spring the electrical gradient.

When the forces associated with the concentration gradient and the electrical gradient are equal and opposite there will be no net force causing K^+ to move; the system will be in **electrochemical equilibrium.** As the membrane is permeable to K^+, there will still of course be exchange of K^+ as a result of thermal agitation of ions, but there will be no net movement. The potential difference

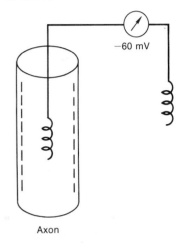

Figure 4.2
Resting potential.
At rest the interior of a nerve cell is found to be about 60 mV negative with respect to the exterior. This potential difference exists across the cell membrane.

Figure 4.3
In a cell permeable only to K^+, the diffusion of K^+ down its concentration gradient leaves the interior of the cell negatively charged.

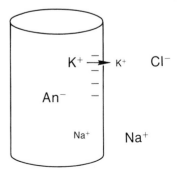

necessary to balance a given concentration gradient is called the **equilibrium potential.** It should be clear that the larger the concentration gradient, the larger will be the required electrical gradient to keep the ion in equilibrium (a larger weight produces more tension in a spring). The actual quantitative relation between potential and concentration gradient was worked out by physical chemists in the nineteenth century and is called the Nernst equation. For potassium it is

$$E = \frac{RT}{zF} \ln \frac{[K^+]_o}{[K^+]_i}$$

where E is the equilibrium potential, R the gas constant, T the absolute temperature, F the Faraday constant, z the valence of the ion species, and K_o and K_i the outside and inside potassium concentrations. The equilibrium potential is simply the concentration gradient stated in units of electricity. In 1902 Julius Bernstein knew that the potassium concentration differed greatly inside and outside of muscle cells. He suggested that the cell was permeable to potassium and that the difference in potassium concentration produced the resting potential. This hypothesis has been subjected to rigorous experimental testing and has proved to be substantially correct. The tests derive from a consideration of the Nernst equation. A good test is to measure the potassium concentration on either side of the membrane and compare the potential predicted from the Nernst equation with the actual membrane potential and then to vary these concentrations and observe the change in membrane potential. In the squid axon the internal K^+ is 400 mM and the external 20 mM. Substituting these values and those for the gas constant, temperature, Faraday constant, and

valence into the Nernst equation and converting to log base 10, we have

$$E = 58 \log (K_o^+/K_i^+)$$
$$= 58 \log 0.05$$
$$= -75 \text{ mV,}$$

which is in fair agreement with the value of -70 mV observed in the intact squid axon. Figure 4.4 is a graph of the relation between resting potential and external potassium concentration in a frog axon. The measured relation is close to that predicted by the Nernst equation for high external potassium concentrations, but deviates from predicted values at low concentrations. Later it will be shown that such deviation is entirely

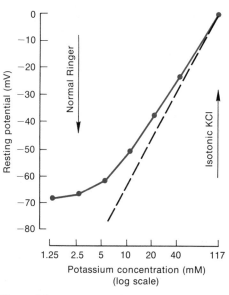

Figure 4.4

Effect of external potassium concentration on the resting potential of a frog axon. The broken line has a slope of 58 mV per tenfold change in potassium concentration, as expected from the Nernst equation. [Huxley and Stämpfli, 1951.]

expected if the nerve is not perfectly selective for potassium and also slightly permeable to sodium. In recent years a striking demonstration of the dependence of the resting potential on the difference in potassium concentration on either side of the membrane has been made in perfused squid axons. In these experiments the axoplasm is squeezed out of an axon and replaced with an artificial solution. When the artificial solution inside the axon is high in K^+ and the external solution low in K^+, as in the normal situation, the resting potential is normal. But if the concentration gradient is reversed by making the outside solution high and the inside solution low in K^+, the potential is reversed.

Now consider a cell in which the ionic gradients are the same as in Table 4.1 but with a different permeability pattern. Suppose that the cell is only permeable to Na^+. Since Na^+ will be more concentrated on the outside than the inside, it will tend to diffuse inward (Fig. 4.5). But because the membrane is impermeable to Cl^- and other external anions, the movement of a small number of Na^+ ions will create a potential difference across the membrane, the slight excess of negative charges on the outside tending to pull Na out. Looking at it another way, the inside will now be slightly positive with respect to the outside. There are no extra anions inside to neutralize the entering Na^+ ions. The factors limiting the inward movement of Na^+ are the same as those that limit the outward flow of K^+. Sodium will continue to enter until a potential is established that will balance the concentration gradient. The Na^+ equilibrium potential is given by the Nernst equation,

$$E = RT/zF \ln (Na_o^+/Na_i^+).$$

For the squid axon,

$$E_{Na^+} = 58 \log (440/50) = +55 \text{ mV}$$

(i.e., the inside is positive relative to the outside). In this way one can calculate the equilibrium potential for any ion (calcium, chloride, etc.), provided that one knows the concentrations, or, more correctly, the ionic activities, on either side of the membrane. We have seen that the transmembrane potential of a cell with the ionic concentrations given for a squid axon can vary from -75 mV to $+55$ mV, depending on whether the cell is permeable to K^+ or Na^+.

It is important in these considerations to appreciate that the actual movement of ions across the membrane necessary to produce the potential changes is so small as to leave the ionic concentrations essentially constant. To illustrate this, we shall calculate the number of ions necessary to cross the membrane to produce the observed potential. It can be recalled that the cell membrane consists of lipid and protein, is about 50 Å thick, and has a high electrical resistance. Thus we have a situation in which two conductors, the internal and external solutions, are separated by a high-resistance layer. This situation is analogous to an electrical capacitor. Direct electrical measurements have shown that the capacitance of the nerve membrane is about 1 microfarad/cm^2. The amount of charge that must be transferred across a capacitor to produce a given voltage change can be calculated from the relation between capacitance, voltage, and charge—a relation implicit in the definition of a capacitor, $C = q/V$, where C is the capacitance in farads, q is the charge in coulombs, and V is the potential in volts. To produce a 100 mV potential change, the charge that needs

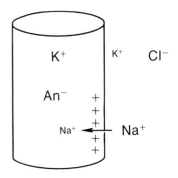

Figure 4.5
In a cell permeable only to Na^+, the diffusion of Na^+ down its concentration gradient leaves the interior of the cell positively charged.

to be transferred is given by $q = CV = 10^{-6} F \times 10^{-1} V = 10^{-7}$ coulomb/cm². One mole of a univalent ion is transferred by 96,500 coulombs (the Faraday constant). The number of moles of a univalent ion transferred by a coulomb of electricity can be determined from the Faraday constant: 96,500 coulombs/mole $\cong 10^5$ coulombs/mole. The moles transferred by

$$\frac{10^{-7} \text{ coulomb/cm}^2}{10^5 \text{ coulomb/mole}} = 10^{-12} \text{ mole/cm}.$$

To see how large a change in concentration would be produced by the efflux of 10^{-12} mole/cm², we must know how much K^+ is contained in the axoplasm surrounded by a square centimeter of squid axon membrane. Assuming an internal K^+ of 400 mM and a diameter of 1 mm, there is about 10^{-5} M of K^+ for each square centimeter of membrane. Therefore, only $10^{-12}/10^{-5}$, or one out of ten million of the K^+ ions in the cell—a fraction much too small to be measured by chemical techniques—must diffuse out of the axon to change the membrane potential by 100 mV. Thus the statement made earlier that the numbers of positive and negative ions on either side of the membrane are equal is not absolutely true. The ions that account for the disparity, however, are confined to the immediate vicinity of the membrane, and their number cannot be measured chemically.

The general principles from the above discussion can be summarized.

1. A concentration gradient of an ion species across the membrane can be balanced by a potential gradient. This potential gradient is called the equilibrium potential and is determined from the Nernst equation, $E = (RT/zF)(\ln \text{ion}_o/\text{ion}_i)$.

2. When the potential across the membrane does not equal the equilibrium potential for the ion, there is a force on the ion tending to cause it to diffuse across the membrane.

3. The ion will diffuse across the membrane only if the permeability of the membrane permits movement. Electric current across membranes results from ion movements.

4. The potential change produced across the membrane by the movement of the ion brings the membrane potential toward the equilibrium potential of the ion.

5. The number of ions that must move across the cell membrane necessary to produce short-term potential changes is so small as to leave the ionic gradients essentially unaffected.

6. The potential changes arise from the passive movements of ions down their concentration gradients; the energy source for these movements is the ionic concentration gradient. Later it will be shown that the metabolism of the cell provides the energy needed to maintain the concentration gradients.

So far the membrane has been assumed to be permeable to only one ion at a time. If more than one ion can permeate the membrane, each one will produce a transmembrane current; the relative current contributed by each ion species will determine the membrane potential. The factors leading to ionic movement are the net electrochemical force pushing the ion across the membrane and the resistance of the membrane to the ionic movement. The relation between ionic current I, membrane resistance R, and membrane potential V is given by Ohm's law, $V = IR$. For convenience we use the conductance of the membrane, g, which is the reciprocal of resistance ($g = 1/R$). The force tending to push the ion is simply the difference between the membrane potential and the equilibrium potential of the ion

Box 4.1 The Cell Membrane [*Kindly contributed by Francis O. Schmitt.*]

The typical cell membrane is now pictured as a highly dynamic structure composed of two regions of different properties, which together make up this highly integrated cell organelle. There is an inner component facing the cytoplasm, including a bimolecular layer of mixed lipids. Within this lipid matrix are embedded various kinds of specialized molecular machinery, chiefly proteinaceous (e.g., specific receptors) plus their associated enzymes (e.g., adenylate cyclase), pumps, microfilament attachments, and probably many more structures still unidentified. The outer region of the membrane is visualized as less homogeneous than the inner region; elongated fingers of molecules are thought to extend like antennae into intercellular space. Included among such molecules are glycoproteins whose elongate carbohydrate chains bristle with negative charges (e.g., sialic acid), which may interact with Ca^{++}. These molecules are thought to be important in providing specific recognition sites, important for establishing appropriate intercellular interaction, especially in development and in certain brain functions. Glycolipids also extend from the outer layer.

Dynamism is the leitmotiv of the modern membranology. Lipid molecules (remaining oriented) apparently can diffuse remarkable rapidly throughout the inner bilayer (according to electron-spin labeling experiments). Proteins also move about both as individual globular molecules and as complexes (receptors, enzyme trains, etc.); the slower variety of movement of large aggregates has been dubbed "continental drift."

Measurements have been made on nerve fibers (vagus nerve by McConnell), but little is actually known about the degree of fluidity of interaction of membrane-borne lipids, proteins, and complexes in the membrane limiting the axon, the dendrite, or synaptic surfaces.

Membranes are constantly being renewed by freshly synthesized substances (possibly containing biophysical or biochemical information). With an enclosing membrane provided by the Golgi apparatus, this material diffuses to the cell membrane and fuses with it. Passing in the reverse direction, from membrane into cell interiors, are other materials, some of which are degraded by liposomal enzymes for ease of reabsorption.

A truly exciting, everchanging, two-dimensional show!

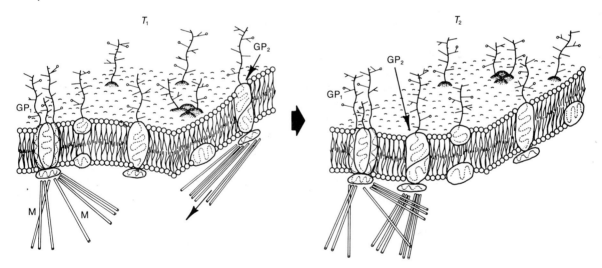

Modified version of the fluid mosaic model of cell membrane structure. T_1 and T_2 represent different points in time. Certain hypothetical integral membrane glycoprotein components are free to diffuse laterally in the membrane formed by a lipid bilayer, whereas others, like the integral glycoprotein–peripheral-protein macromolecular complex (GP_1), are impeded by membrane-associated components (M). Under certain conditions, some membrane macromolecular complexes (GP_2) can be laterally displaced by membrane-associated contractile components in an energy-dependent process. [Nicolson, 1974.]

$(V_m - E_{ion})$. It will be recalled that at the equilibrium potential, the ionic gradient and electrical potential are in balance. The greater the deviation of membrane potential Vm from E_{ion}, the greater is the force tending to make the ion move. The ionic current can therefore be calculated as $I = g(V_m - E_{ion})$. The equations for Na^+ and K^+ currents are as follows:

$$I_{Na^+} = g_{Na^+}(V_m - E_{Na^+});$$
$$I_{K^+} = g_{K^+}(V_m - E_{K^+})$$

When the membrane is permeable only to Na^+ and K^+ at a steady level of membrane potential, the Na^+ and K^+ currents must be equal and opposite. Therefore,

$$I_{K^+} = -I_{Na^+}$$

and

$$g_{K^+}(V_m - E_{K^+}) = -g_{Na^+}(V_m - E_{Na^+});$$

solving for V_m gives

$$V_m = \frac{g_{Na^+}}{g_{Na^+} + g_{K^+}} E_{Na^+} + \frac{g_{K^+}}{g_{K^+} + g_{Na^+}} E_{K^+}$$

where

$$\frac{g_{Na^+}}{g_{Na^+} + g_{K^+}} + \frac{g_{K^+}}{g_{K^+} + g_{Na^+}} = 1$$

Thus the membrane potential depends on the relative conductance of the membrane to the two ion species, g_{Na^+}/g_{K^+}. By substituting into the equation, we can calculate that if the membrane conductance to K^+ is, say, 1000 times greater than that to Na^+, the membrane potential will essentially be equal to E_{K^+}, and if much more permeable to Na^+ than K^+ it will be essentially at E_{Na^+}. Thus it can be seen that by varying the relative conductance of Na^+ and K^+, the membrane potential can be at any value between -75 and $+55$ mV. The relation between the conductance ratio of Na^+ and K^+ and

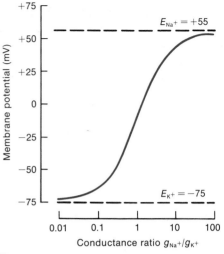

Figure 4.6
The relation between ratio of sodium conductance to potassium conductance (g_{Na^+}/g_{K^+}) and membrane potential for a cell permeable only to potassium and sodium. Broken lines indicate sodium equilibrium potential, E_{Na^+}, of $+55$ mV and potassium equilibrium potential, E_{K^+}, of -75 mV.

V_m is shown in Figure 4.6. It is well to keep in mind that these potential changes are produced solely by changes in the conductance ratio. The ionic gradients, and therefore the equilibrium potentials for the individual ions, remain unchanged.

A more complete description of how membrane potential is related to ionic gradients and membrane conductance should include all ions capable of permeating the membrane, the ionic gradients of each of the ions, and their relative permeabilities. The relevant ions are Na^+, K^+, and Cl^-, and by combining the Nernst equation for each ion and a factor for their relative permeabilities the following expression is obtained.

$$V_m = \frac{RT}{zF} \ln \frac{[K^+]_o + b[Na^+]_o + c[Cl^-]_i}{[K]_i + b[Na^+]_i + c[Cl^-]_o}$$

where b and c are the permeabilities of Na$^+$ and Cl$^-$ relative to K$^+$. Note that the chloride ratio is reversed because Cl$^-$ is negatively charged.

Although it appears that the terms **permeability** and **conductance** are being used as synonyms, they have different physical meanings. "Permeability" is a property of the membrane (and the term is also used for "capacity to permeate"), whereas "conductance" is a measure of the actual ion flux and depends on both permeability and concentration. For example, if one increases the potassium concentration of the aqueous medium that surrounds a cell, the membrane property (permeability) may remain constant, but the exchange of ions (conductance) would increase with the number of collisions between potassium ions and the membrane. Under a given set of conditions the two parameters usually change together.

An equation similar to that given above was formulated by Goldman and utilized by Hodgkin and Katz in their development of the ionic hypothesis. It is often referred to as the Constant Field equation because in its derivation it was assumed that the potential field through the thickness of the membrane is constant. Another way of describing the electrical behavior of the membrane is to draw an electrical model of the membrane in which each of the concentration gradients is represented by a battery whose voltage is equal to the equilibrium potential for a particular ion species and the internal resistance of the battery is a measure of the conductance of the membrane to that ion species. The model, shown in Figure 4.7, also includes the membrane capacitance, as previously mentioned. To utilize this model one needs to understand Ohm's law and to

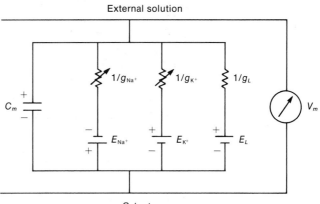

Figure 4.7
An equivalent circuit for an element of an excitable neuron membrane. C_m is the membrane capacitance; $1/g$ (conductance) is a measure of the resistance of the membrane to the movement in the various channels, Na$^+$, K$^+$; and the remaining ions (e.g., Cl$^-$) are shown as a constant-leakage (L) channel. The batteries (E) indicate the electromotive force produced as a result of the asymmetrical distribution of the ions. V_m is the membrane potential.

be able to calculate potential in circuits with batteries arranged in parallel.

Our consideration of the ionic hypothesis can be summarized by indicating that the concentration gradients of ions across the nerve membrane lead to membrane potentials that depend on the permeability pattern of the membrane to the various ion species. For each ion species there is a unique membrane potential —the **equilibrium potential** for that ion, at which there is no net force tending to push the ion across the membrane. If the membrane potential is not equal to the equilibrium potential, the ion will tend to diffuse across the membrane, provided that the membrane is permeable to that ion. The movement of the ion will bring the membrane potential toward the equilibrium potential for that ion. At rest the nerve membrane is predominantly permeable to K+, and the resting membrane potential is close to the potassium equilibrium potential.

The success of the ionic hypothesis is that one can account for potential changes within the boundaries of the equilibrium potentials solely by varying the permeability pattern of the membrane. The best evidence that such ionic movements account for potential changes in the nervous system is obtained by measuring the actual movement of the ions with the aid of radioactive tracers. This evidence will be presented below. Some potential changes in the nervous system cannot be

accounted for by changes in permeability pattern. They are produced by changes in the concentration gradients for some ions or by selective transfer of ions across the membrane by metabolic machinery in the membrane. These will be discussed later.

C. Ionic Basis of the Nerve Impulse

The necessary membrane property that provides a basis for the action potential is its ability to distinguish between Na+ and K+ and to control the relative per-

---

Box 4.2. Ionic Batteries in Parallel

Ohm's law, $E = IR$, expresses quantitatively the observed relations between the electromotive force E in volts, the current I in amperes, and the resistance R in ohms, in an electric circuit. In nerve and muscle cells the electromotive forces are derived from the concentration gradients of the ions, using the Nernst relation. The current is that which flows across the membrane as a result of ion movements, and the major resistance is that exerted by the cell membrane to the movement of the ions. For a cell permeable only to sodium and potassium we can consider that the sodium and potassium batteries E_{Na^+} and E_{K^+} are in parallel with the membrane resistance R_L. The ionic current I delivered by each battery depends on the resistance of the membrane to the flow of the particular ion species (see Fig. 4.7). The membrane potential V_m can be determined by using Millman's theorem, according to the following equation

$$V_m = \frac{E_K/R_{K^+} + E_{Na^+}/R_{Na^+}}{1/R_{Na^+} + 1/R_{K^+} + 1/R_L} \ldots$$

This expression of the theorem holds for any number of parallel branches, provided that there is no series resistance between the branches. For excitable cells this means that the resistances of the cytoplasm and of the external medium between the different membrane sites must be insignificant. For a cell with a resistance to potassium 100 times less than that to sodium we would find the following for a unit area of membrane:

$$E_{K^+} = -90 \text{ mV} \qquad R_{K^+} = 10^4 \text{ ohms}$$
$$E_{Na^+} = +50 \text{ mV} \qquad R_{Na^+} = 10^6 \text{ ohms}$$
$$R_L = 10^5 \text{ ohms (the resistance to all other ions,}$$
termed leakage resistance)

The above theorem gives $V_m = -81$ mV. If the resistance to sodium is decreased 1000 times to 10^3 ohms, $V_m = +37$ mV. In such a circuit the internal resistances of the batteries play a major role in determining the current supplied by each battery and therefore the membrane potential. In practice, membrane potential changes in cells are not instantaneous, because of the capacitance of the membrane. This factor will be taken into account when the electrical properties of the membranes are discussed.

meability of each. We have seen that at rest the membrane is predominantly permeable to K$^+$, and the membrane potential is therefore close to the potassium equilibrium potential. The brief electrical change that we call the action potential is brought about by a sudden influx of Na$^+$ ions moving down their electrochemical gradient and bringing the membrane potential toward the equilibrium potential of Na$^+$. The Na$^+$ influx results from a large increase in conductance to Na$^+$ produced by a decrease in the membrane potential. This constitutes a positive feedback loop, as shown in Figure 4.8. The influx of Na$^+$ is governed by the sodium conductance, which is controlled by the potential difference across the membrane and in turns alters the potential difference in the direction that further increases the influx of Na$^+$. Once initiated, the process goes to completion and the resulting potential changes are independent of the energy source producing the initial depolarization. Thus the action potential is all-or-none: under a given set of conditions it either occurs in full or does not occur at all. As the membrane potential approaches E_{Na^+}, the g_{Na^+} decreases toward its resting value (a process termed sodium inactivation), and there is an increase in g_{K^+}. As a result, K$^+$ ions flow out of the cell, and the membrane potential returns toward E_{K^+} (Fig. 4.9). The relations between g_{Na^+}, g_{K^+}, and V_m during the action potential are illustrated in Figure 4.10.

It has already been indicated that the membrane permeability pattern determines the membrane potential. We will now consider how the membrane potential controls the permeability pattern and how the interaction between these two —membrane permeability and membrane potential—leads to the action potential.

Figure 4.8
The regenerative nature of the depolarization phase of the nerve impulse arises from the positive feedback of membrane potential, sodium conductance, and sodium influx.

Figure 4.9
Repolarization of the membrane results when potassium efflux restores the internal negativity (negative feedback).

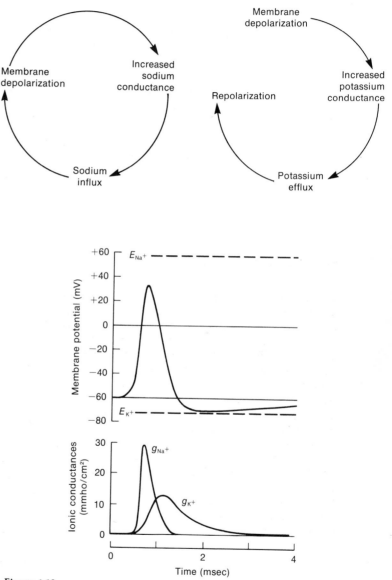

Figure 4.10
Theoretical solution of Hodgkin-Huxley equations for changes in membrane potential (*top*) and sodium and potassium conductances (*bottom*) as a function of time. Such changes would produce an influx of sodium of 4.33 pmole/cm^2 and an efflux of potassium of 4.26 pmole/cm^2. [Hodgkin and Huxley, 1952b.]

D. Experimental Evidence for Ionic Basis of Action Potential

Subsequent to suggesting the resting potential arose from the unequal distribution of K$^+$ across the membrane and the selective permeability to K$^+$, Bernstein proposed that stimulation made the membrane permeable to all ions and the resting potential transiently disappeared. This hypothesis predicts that during the action potential the conductance of the membrane increases and that the action potential amplitude is only as large as the resting potential. Although some experiments suggested that in fact the amplitude of the action potential exceeds that of the resting potential, the problem could not convincingly be solved until it was possible to measure the membrane potential and the conductance of excitable cells directly.

In 1939, the experiments of Cole and Curtis on the squid giant axon provided direct evidence that the permeability of the membrane increases during the passage of the nerve impulse. Figure 4.11 shows the decrease in resistance between two electrodes on either side of a squid axon during the passage of the impulse. The fall in membrane resistance shows that the permeability of the membrane to ionic flow is increased during the action potential. The resting membrane resistance was about 1000 ohm cm^2; at the peak of the action potential, 25 ohm cm^2. A second set of experiments performed simultaneously by Hodgkin and Huxley in England and by Cole and Curtis in the United States necessitated an important modification of the Bernstein hypothesis. In these experiments a capillary electrode about 100 μm in diameter was inserted two or three centimeters longi-

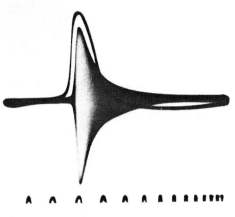

Figure 4.11
Superimposed records of the time course of an action potential (solid curve) and increase in conductance (indicated by broadening of band in impedance-measuring circuit) of a squid axon. Time marks are 1 msec apart. [Cole and Curtis, 1938.]

tudinally down a giant squid axon, and the potential between the internal electrode and a reference electrode in the seawater outside the axon was recorded. They observed that when the axon is stimulated and gives rise to nerve impulses, the resting potential not only disappears but the potential across the membrane is actually inside positive by about 40 mV. Because the action potential clearly did not result simply from the disappearance of the resting potential, as suggested by Bernstein, a new explanation had to be found. During World War II, biologists turned their attention to various pressing national needs. Following the war, work on the nature of the action potential resumed. In 1949 Hodgkin and Katz provided evidence for the "sodium hypothesis" of the nerve impulse. They found that the amplitude and rate of rise of the action potential depended on the external Na$^+$ concentration in a manner that would be expected if the action potential were the

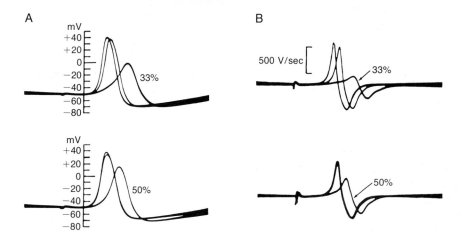

Figure 4.12
Effect of reducing external sodium on the overshoot, and maximum rate of rise of the action potential in squid axon. **A.** Effect of 33% sea water, 67% isotonic dextrose (*top*) and 50% sea water and 50% isotonic dextrose (*bottom*) on the overshoot of the action potential. **B.** Effect of the same solutions on the rate of change of the membrane voltage. The overshoot and maximum rate of rise of the action potential are both reduced when the concentration external sodium is reduced. [Hodgkin and Katz, 1949.]

result of a transient increase in the permeability of the membrane to Na⁺. We have already seen that the equilibrium potential for Na⁺ is about +55 mV. If the membrane were to become transiently permeable to Na⁺, the membrane potential would move toward the Na⁺ equilibrium potential. When the external Na⁺ is reduced, the Na equilibrium potential is reduced, hence it would be expected that the amplitude of the action potential's positive overshoot would also be reduced. In addition, as sodium ions are expected to carry the inward current to depolarize the membrane, one would also expect a reduction in the maximum rate of rise of the action potential, which is an indication of the intensity of the inward current. These results are shown in Figure 4.12.

For the simplest case the membrane potential could be considered to be deter-mined solely by the sodium equilibrium potential, and the relation between the peak of the overshoot and the sodium equilibrium level would be predicted by the Nernst equation,

$$E = (RT/zF) \ln (Na_o^+/Na_i^+),$$

which reduces to

$$58 \log (Na_o^+/Na_i^+),$$

meaning that the overshoot would change by 58 mV for a tenfold change in external Na⁺ concentration. Good agreement with prediction has been found in amphibian myelinated nerve fibers, as shown in Figure 4.13.

Hodgkin and Katz, in the introduction of the paper in which they state their Na⁺ hypothesis, refer to a series of experiments by Overton (published in 1902 in the same volume of Pflügers Archiv

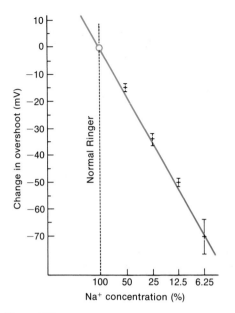

Figure 4.13
Effect of the reduction of external sodium on the overshoot of the action potential in frog myelinated axons. The line has a slope of 58 mV per tenfold change in external sodium. Sodium is replaced by the impermeant ion choline. [Huxley and Stämpfli, 1951.]

as Bernstein's hypothesis). Overton had found that a frog skeletal muscle loses its excitability if the sodium is removed from the bathing medium. He proposed that during activity Na^+ might exchange for K^+. He realized, however, that the internal Na^+ in muscle remains low even after years of activity. Failing to conceive of a mechanism whereby the ionic gradients were maintained, he suggested that this fact might rule out the possibility of an ionic exchange. In the past 25 years, however, experiments have revealed that nerve cells, like many others, utilize metabolic energy to remove Na^+ from the inside of the cell in exchange for K^+: **the sodium pump.**

The initial observations of Hodgkin and Katz that sodium was essential for the action potential in squid axon led to a sophisticated series of experiments by Hodgkin and Huxley, which culminated in a quantitative description of the time course and magnitude of the ionic conductance changes and currents during the action potential. Their analysis enabled them not only to reconstruct the action potential from a knowledge of the variables but also to explain a whole variety of phenomena that could previously be described only in rather vague terms, such as threshold, refractory period, afterpotentials, and propagation. Their analysis firmly established the ionic hypothesis and prepared the way for an understanding of ionic mechanism in synaptic transmission as well.

The problem in analyzing the magnitude and time course of the Na^+ and K^+ conductances was that these parameters depended on the membrane potential and varied with time. The solution of the problem was found by systematically isolating the variables by controlling the electrical activity of the cell and the ionic environment. In order to separate these interdependent variables it was necessary to control one of them—the membrane potential—experimentally.

The technique for electrically controlling the potential of the cell is termed a **voltage clamp** because it allows one to set the membrane potential at any level and to maintain that level despite membrane conductance changes that would otherwise tend to bring the membrane potential toward a new equilibrium level. This entails driving externally imposed electrical current through the resistance of the membrane in order to produce a voltage drop that counteracts the natural one. A diagram of the experimental

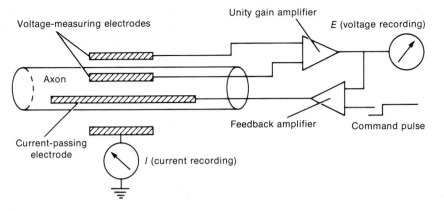

Figure 4.14
Experimental arrangement for "voltage clamping" a length of squid axon. A command pulse is applied to a high-gain feedback amplifier, which puts out a current proportional to the difference in voltage between the command pulse and the recorded membrane potential. The voltage is recorded across the axon membrane, and the current necessary to keep the membrane potential equal to the command voltage is recorded by an ammeter situated between the output of the feedback amplifier and ground. In this way one can set the membrane potential at a desired level and determine the membrane current that results from the change in membrane potential.

setup is shown in Figure 4.14. The technique has two compelling advantages. First, it allows one to measure the total current across the membrane in response to a change in membrane potential and, second, at constant voltage the membrane current only flows through the resistive element. Since current across the capacitance of the membrane is a function of the rate of change of voltage with time, when the membrane potential is held constant the current through the capacitance is zero. From the relation between membrane current, conductance, and potential, $I = gV$, it is apparent that if the voltage is known and constant and the total current across the membrane is determined by measuring the current necessary to hold the membrane potential at a given value, one can obtain measurements of the total conductance of the

membrane as a function of voltage and time. To determine which portions of the current and conductance change is due to Na^+ and which to K^+, it was then necessary to remove most of the Na^+, replacing it with a cation that cannot cross the membrane, and repeat the measurements. The potassium current and conductance were determined from measurements in low Na^+ solution and those for Na^+ by subtracting from the total currents and conductance from those for K^+.

Records from voltage-clamped squid axons are shown in Figure 4.15. These records illustrate how the sodium and potassium currents were determined for a given level of membrane potential. At the top it can be seen that the membrane was depolarized from rest by 56 mV. This depolarization produced a transient

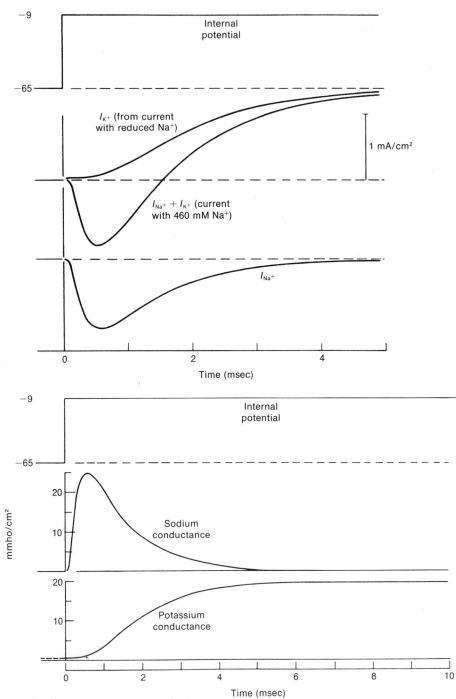

Figure 4.15 (*facing page*)
Analysis of data from voltage clamp experiments designed to determine the components of membrane current carried by sodium and potassium and the corresponding changes in sodium and potassium conductance. **Above.** A depolarizing voltage step was applied at $t = 0$. The sodium current is obtained by subtracting the total current ($I_{Na^+} + I_{K^+}$) from the potassium current, the latter recorded in a reduced concentration of sodium. [Hodgkin and Huxley, 1952a.] **Below.** Calculated changes in sodium and potassium conductances for step depolarizations from the relation $g_{ion} = I_{ion}/(V_m - E_{ion})$. [Hodgkin, 1958; based on Hodgkin and Huxley, 1952a,b.]

inward current followed by a prolonged outward current. Two lines of evidence indicate that the early inward current was carried by Na$^+$. First, the inward current disappeared when the membrane was clamped at the Na$^+$ equilibrium potential (about $+50$ mV), and at values above $+50$ mV the early current was outward. Under the latter condition the interior of the fiber is actually more positive than the E_{Na^+}, so that when the Na$^+$ conductance is increased Na$^+$ flows outward. Second, when the Na$^+$ is removed (curve I_{K^+}) there is no early inward current, only an outward current presumably carried by K$^+$. That the outward current is carried by K$^+$ was found directly by loading the axon with radioactive K$^+$ and measuring the efflux of K$^+$ when the membrane potential was clamped. Hodgkin and Huxley found excellent agreement between the actual number of K$^+$ ions that flowed outward and the number calculated from their measurements of membrane current as necessary to carry the outward current. By subtracting the total current in normal seawater from that in Na$^+$-free seawater, they obtained the magnitude and time course of the Na$^+$ current (I_{Na^+}). The time course and magnitude of the Na$^+$ and K$^+$ conductances can then be calculated as a function of time from the relation $g = I/E$; these are shown in the lower half of the figure. Note that the increase in Na$^+$ conductance is transient even though the depolarization is maintained, whereas that for K$^+$ is maintained. By repeating these measurements at various membrane potentials, a family of curves of the time course and magnitude of the conductances to Na$^+$ and K$^+$ was determined. Figure 4.16 shows the calculated Na$^+$ and K$^+$ conductances for a variety of membrane potentials and the relation between peak sodium conductance and membrane potential. An important feature of these results is that the magnitude of the sodium conductance varies with membrane potential and time in a continuous fashion; g_{Na^+} increases as the cell is depolarized. Later we will consider how this continuous relation can give rise to an explosive all-or-none electrical change—the nerve impulse. Previously we considered how the current will vary depending on the difference between the equilibrium potential and membrane potential ($V_m - E_{ion}$) if the conductance is constant. Now we must appreciate a system in which both conductance and current can vary with membrane potential. A second feature of Figure 4.16,A is that g_{Na^+} decreases with time even though the membrane is held in a depolarized state. This process of decreasing conductance is termed **Na$^+$ inactivation.** As a result of Na$^+$ inactiva-

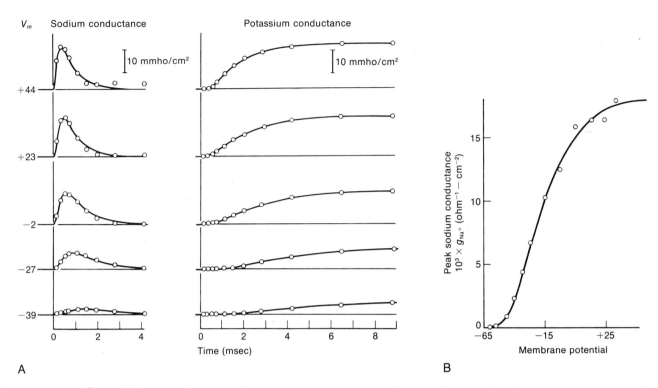

Figure 4.16
A. Changes in sodium and potassium conductance with time, following a step depolarization of the squid axon membrane from rest (-65 mV) to the indicated potential. The circles are experimental estimates derived as in Figure 4.15, and the smooth curves are the solutions to the equations used by Hodgkin and Huxley to describe the changes in conductance. [Hodgkin, 1958.] **B.** Relation between peak sodium conductance and membrane potential in the squid axon during a voltage clamp step. [Katz, 1962.]

tion, the increase in g_{K^+} leads to more rapid repolarization. It is clearly efficient if the g_{Na^+} decreases as g_{K^+} increases. In this way the membrane potential returns toward E_{K^+}, and the number ions exchanged is kept down.

When Hodgkin and Huxley sought to determine the time course of inactivation of Na^+, they began by assuming that depolarization of the membrane has a dual effect—that it increases g_{Na^+} with a rapid time constant and decreases it with a slow time constant. Stated another way, they assumed that the inactivation

process is the result of two parameters, the magnitude and duration of the depolarization. After determining the relation between membrane potential and Na^+ inactivation, they could then describe the way in which g_{Na^+} and g_{K^+} increase with time and the way in which g_{Na} decreases with time for each value of membrane potential. As shown in Figure 4.16, g_{K^+} does not appear to inactivate with time in the squid axon, although it is known to show inactivation with a slow time course in certain other cells, such as snail neurons.

Box 4.3   Ion Gates in the Membrane

The voltage dependence of ion conductances suggests that conductance changes involve a change in position or conformation of charged molecules fixed in the membrane. For sodium, such changes have been found to give rise to a measurable gating current. With a step depolarization (but not hyperpolarization) of a squid axon, a transient outward current is found to precede the inward sodium current and to coincide with the opening of the sodium channels. The gating current is capacitative in origin and very much smaller than the ionic current carried by sodium. It can be measured under conditions in which the ionic currents have been essentially eliminated by substituting impermeable ions for sodium and potassium. Procedures that inactivate the sodium current, such as prolonged membrane depolarization, also block the gating current.

E. Reconstruction of the Action Potential

In order to obtain mathematical expressions describing the action potential, Hodgkin and Huxley had to fit the observed curves with empirical expressions describing each of the parameters. When computations were completed it became clear how the apparently complex behavior of the axon in response to stimulation could accurately be predicted from knowledge of the relatively simple interactions of ionic conductance, membrane potential, and time. To describe the curve for potassium conductance it was necessary to use an equation with a fourth-power variable, suggesting that a path for K$^+$ ions is formed by four ions moving to a region of the membrane under the influence of the membrane potential. The activation of the Na$^+$ path was assumed to occur when three events, each with the same probability, happened simultaneously; and it was only

necessary to postulate that a single process was involved in the inactivation process. Appropriate constants were determined to account for changes in membrane potential, temperature, and calcium concentration. The equation for membrane current consists of four terms:

$$I = c(dV/dt) + (V_m - E_{K^+})g_{K^+}n^4$$
$$+ (V_m - E_{Na^+})g_{Na^+}m^3h + (V_m - E_L)g_L.$$

The first term is the current across the capacitance of the membrane, expressed in terms of the membrane capacitance C times the rate of change of potential with time. The second term is the K$^+$ current, expressed as the difference between the membrane potential and potassium equilibrium potential $(V_m - E_{K^+})$ multiplied by g_{K^+} and a constant n to the fourth power, which is necessary to fit the relation between V_m and g_{K^+}. The third term is the Na$^+$ current, in which m is a factor for activation raised to the third power and h is the inactivation. The final term is relatively unimportant and takes into account current carried by other ions, principally Cl^-, through conductance channels ("leakage") not previously considered. Using only this equation, expressions for the dependence of n, m, and h on time, and the assumption that the action potential was continuously propagating at constant velocity in a nerve fiber in a large fluid volume, Huxley calculated the action potential in Figure 4.10 using a hand-powered calculating machine. The laborious process was accomplished by setting the membrane potential below threshold and calculating the changes in the variables at intervals of 10 μsec. By determining what happened to g_{K^+} and g_{Na^+} and V_m for each successive 10 μsec interval, he obtained the expected time course of the voltage change and conductance changes shown in Figure 4.10. The success of these equations has been remarkable.

They describe the behavior of the axon under a variety of conditions, including maintained depolarization and variations in Ca^{++} concentration, and they account for refractory periods and threshold. This success justifies the conclusion that the electrical behavior of the axon can be predicted accurately from knowledge of the interrelations between membrane potential and ionic conductances.

F. Action Potential Threshold

In Figure 4.16 it can be seen that the Na^+ and K^+ conductances are both continuous functions of membrane potential. The question arises how such a system can give rise to an all-or-none electrical event that is initiated when the membrane is depolarized to a critical threshold potential. Consider a resting axon in which g_{K^+} is much greater than g_{Na^+} and

---

Box 4.4 Selectivity of Sodium and Potassium Channels

There are several lines of evidence indicating that the channels through the membrane for sodium and potassium movements are separate. First, the voltage-clamp experiments demonstrate that the conductance of each channel has a different dependence on membrane potential and time. Second, such experiments have shown the channels differ in the ease with which other cations can traverse the channel. The sodium channel is almost equally permeable to sodium and lithium, both of which are about 12 times more permeable than potassium. The potassium channel is 100 times more permeable to potassium and rubidium than to sodium. In addition, pharmacological agents are available that selec-

tively block one channel or the other. These agents have proved valuable in studying the channels independently and in studying the behavior of nerve cells in the absence of nerve impulses. Tetrodotoxin (TTX), obtained from puffer fish, selectively blocks the increase in sodium permeability with depolarization when applied to the outside of most axons in concentrations of 5×10^{-9} to 100×10^{-9} M. Tetraethylammonium (TEA) blocks the voltage-dependent increase in potassium conductance in many neurons. The compound usually has to be injected inside the neuron in concentrations of 1–10 mM to be effective.

the membrane potential is therefore close to the E_{K^+}. If the cell is slightly depolarized by passing current across the membrane (without a voltage clamp), there is an increase in g_{Na^+}, and Na^+ ions move down their electrochemical gradient, tending further to depolarize the membrane toward E_{Na^+}. This inward Na^+ current is opposed by an outward K^+ current tending to bring the membrane potential back toward the resting level. With depolarization, I_{K^+} increases because the difference between V_m and E_{K^+} increases even though g_{K^+} initially stays constant. I_{Na^+} increases because the instantaneous increase in g_{Na^+} with depolarization is greater than the decrease in $V_m - E_{Na^+}$. If I_{K^+} is greater than I_{Na^+} the membrane potential returns to rest level. At some critical amount of depolarization, g_{Na^+} will increase to a level at which I_{Na^+} just exceeds I_{K^+}, and the membrane potential moves toward E_{Na^+}. The membrane potential at which I_{Na^+} just begins to exceed I_{K^+} is the threshold, and the current required to bring the membrane potential to this level is the threshold stimulus (Fig. 4.17). Each slight increase in depolarization further increases g_{Na^+}, leading to an increased inward I_{Na^+} and increased depolarization. This regenerative, or positive feedback, cycle gives the action potential its explosive nature.

G. Refractory Period

As the interval between two suprathreshold stimuli decreases, the amplitude of the second action potential gradually decreases until eventually the second stimulus fails to produce an action potential. During the time when the second action potential is smaller, the threshold is also elevated. This period is termed the relative refractory period. At very short intervals the action potential can no longer be elicited by even very strong stimuli, and the axon is said to be absolutely refractory. This property is of interest in that it limits the maximum frequency possible for nerve impulses. The explanation for this phenomenon readily follows from the Hodgkin-Huxley explanation of the action potential. Sodium inactivation is greatest at the end of the action potential. At this time a depolarization of the membrane cannot sufficiently increase g_{Na^+} to the point at which the I_{Na^+} can exceed the increased I_{K^+}. At longer intervals a strong depolarization will increase g_{Na^+} sufficiently to trigger an impulse, but because g_{Na^+} is partially inactivated and g_{K^+} still elevated, the peak of the overshoot occurs further from E_{Na^+} than normal (i.e., the overshoot is smaller).

H. Accommodation

If the axon is depolarized at a sufficiently slow rate it will not give rise to a nerve impulse. This is expected by the Hodgkin-Huxley model, since Na^+ inactivation and the delayed increase in g_{K^+} can occur as rapidly as the slow increase in g_{Na^+} at this critical rate of depolarization. At rates just above the critical, an increase in the threshold is seen; this increase is called accommodation. Other parts of the neuron may show more accommodation than the axon. This use of the term must be distinguished from accommodation in sense organs that depend on changes in the incidence of stimulating energy to the receptor—for example, focusing the lens of the eye.

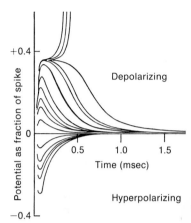

Figure 4.17
Extracellular records of the response of a crab axon to hyperpolarizing and depolarizing electric shocks.

The three strongest depolarizing shocks produce action potentials. (Amplitude of action potential is taken as 1.0 on the vertical scale.) [Hodgkin, 1938.]

Figure 4.18
Impulses and afterpotentials in twitch muscle of frog, recorded intracellularly (top); *single node of frog nerve, recorded extracellularly* (middle); *and intracellular spike from squid axon* (bottom). [*Tasaki, 1959.*]

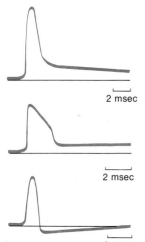

2 msec

2 msec

1 msec

I. Afterpotentials

After the peak of the action potential the membrane potential does not simply return directly to the resting level. Instead, it may first increase to a greater negativity than at rest, then decrease to less than rest, then increase a second time to greater than rest, and finally return to the resting membrane potential. The names of these various afterpotentials found in the literature can lead to confusion, because as neurophysiologists changed from external to transmembrane potential recording techniques, the conventions for positive and negative were changed. But we now have reasonable explanations for the observed potential changes, and the names have become less important; we therefore briefly consider the potentials and their bases.

As indicated above, the earliest afterpotential in squid axon and many other cells is an increase in the membrane potential to greater than the resting level (Fig. 4.10). This afterpotential ("undershoot") occurs because the potassium conductance is still somewhat greater than at rest, even though the sodium conductance has returned to rest level. The membrane potential is therefore even closer to E_{K^+} than at rest. As g_{K^+} decreases, the membrane potential returns toward rest level.

Following this early increase, the membrane potential may become less negative than rest level. It will be recalled that, after the impulse peaks, repolarization is produced by an increase in g_{K^+} and a small efflux of potassium from the cell. As neurons are closely surrounded by their satellite cells, separated by a narrow cleft of some 200–300 Å, the efflux of K$^+$ (amounting to a few p-moles/cm^2) increases the K$^+$ concentration in this narrow space. This decreases the ratio K_o^+/K_i^+. The potassium equilibrium level is reduced, and the membrane is slightly depolarized. With time, the potassium diffuses away or is taken up, and the K$^+$ in the cleft returns to normal values. This potential change is best seen after repetitive stimulation of the neuron, when the increase in external K$^+$ can be substantial, up to about 5 times normal concentration.

Finally, in many neurons, particularly small unmyelinated axons, repetitive stimulation will lead to a long-lasting increase in membrane potential. At this time the membrane potential exceeds the equilibrium potential for all known ions and hence cannot be explained solely on the basis of the pattern of membrane permeability to ions. This potential is greatly diminished by decreases in temperature or by metabolic poisons that inhibit the active extrusion of Na$^+$, which accumulates in the cell as a result of the action potentials. Later in the chapter we will see how the activity of the sodium extrusion process can lead to an increase in the membrane potential. In some excitable cells the sequence of afterpotentials differs from that in squid (Fig. 4.18). These have not yet been fully explained.

J. Pacemaker Potential and Spontaneous Impulses

Repeated spike potentials occur in some cells in the absence of external triggering. Recordings from such cells are shown in Figure 4.19. Following a spike the membrane potential increases, then slowly depolarizes and gives rise to another action potential. This slow depolarization results, in some cases, from a gradual decrease in g_{K^+} until the action potential threshold is reached. The phase of slow depolarization is termed the pacemaker potential, because its rate determines the rate of impulse activity in the neuron. The process underlying this potential change is somewhat similar to that discussed above for the initial increase in membrane potential following the action potential.

Some neurons normally exhibit spontaneous impulses; others do not. In some of these the cell may spontaneously depolarize to threshold, fire a burst of impulses, and then hyperpolarize before again depolarizing to threshold (B in Fig. 4.19). The interval between impulses and the duration of the bursts can be quite variable (from tenths of seconds to many seconds). The pattern of spontaneous impulse activity is highly dependent both on influences from other neurons and on the neuronal environment. Under some circumstances neurons may go through cycles of membrane potential changes without giving rise to impulses. Such changes in membrane potential appear to result either from cyclical changes in the permeability pattern of the membrane or from the activity of metabolic-dependent ion pumps (see below). Even many normally quiescent neurons can be made to give rise to spontaneous potentials. For example, the squid axon can be made to fire by lowering the external calcium concentration. Lowering external calcium decreases the depolarization necessary to increase the sodium conductance to threshold, and with a sufficient decrease in calcium the cell will give rise to impulses at the normal resting potential.

K. Calcium Action Potentials

Various neurons and muscle cells, in vertebrates and invertebrates, give rise to action potentials when the sodium is removed from the external solution and replaced by calcium. In these cells a decrease in membrane potential leads to an increase in calcium conductance, and calcium diffuses into the cell in a manner analogous to that for Na^+ in the squid axon. Under these conditions the peak of the action potential should vary with the change in Ca^{++} equilibrium potential given by the Nernst equation for Ca^{++},

$$E = (RT/zF) \ln (Ca_o^{++}/Ca_i^{++}).$$

As calcium has a valence of $+2$, the slope is half that for Na^+, or 29 mV for a tenfold change in Ca^{++}. Under normal conditions in these neurons, both Na^+ and Ca^{++} appear to contribute to the inward current during the action potential, and a decrease in the concentration of either ion will decrease the overshoot of the action potential. Even in the squid axon, where sodium carries almost all of the inward current, there is a small influx of calcium with each action potential. This can be detected using optical techniques following intracellular injection of an axon with aequorin, a protein isolated from jellyfish that emits light upon reacting with ionized calcium. A consideration of the role of Ca^{++} in action potentials is complicated by the influence of Ca^{++} on cell permeability, on resting potential, and on the change in Na^+ conductance during the action potential. In general, an increase in external Ca^{++} increases the resting potential, decreases cell permeability, and enhances the increase in g_{Na^+} during the action potential.

Figure 4.19
Pacemaker potentials and action potential in Aplysia neurons.

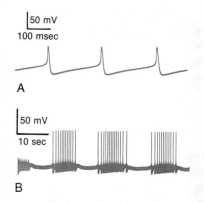

A

B

A. *Following each action potential, the cell spontaneously depolarizes to threshold, giving rise to another action potential. The slow depolarization is termed the pacemaker potential.* **B.** *The pacemaker potential gives rise to a burst of impulses. [Courtesy of L. Tauc.]*

Today there is growing appreciation of the role of Ca^{++}-spikes in excitation-secretion coupling in axon terminals and neurosecretory cells, as well as in the conducting processes of many cells.

L. Direct Measurements of Ionic Movements

Important confirmation of the ionic hypothesis comes from experiments in which the movements of Na^+ and K^+ across the membrane are directly measured with radioactive tracers or by chemical analysis of the axons. It was stated on p. 136 that in order to alter the charge across the $1\,\mu F/cm^2$ membrane capacitor by 100 mV, the movement of $10^{-12}\,M/cm^2$ of a univalent ion is required. Thus, during an action potential at least $10^{-12}\,M/cm^2$ of Na^+ should enter the fiber and a similar amount of K^+ should leave. Although it has not been feasible to measure the exchange during a single impulse, it has been possible to measure the resting inward flux of Na^+ and efflux of K^+, elicit a known number of impulses, and measure the changes in fluxes resulting from the impulses. The results show that 3×10^{-12} to $4 \times 10^{-12}\,M/cm^2$ of Na^+ and K^+ crosses the membrane per nerve impulse. These values are greater than the minimum necessary, presumably because the increases in conductance to Na^+ and K^+ partially overlaps in time. In Figure 4.10 it can be seen that at the peak of the action potential the conductance to both ions is increased.

M. Maintenance of Ionic Gradients

It has been noted above that the action potential and resting potential in nerves depend on the unequal distribution of ions, principally Na^+ and K^+, across the membrane, and that during activity there is a small increase in cell Na^+ and loss of cell K^+. Although the exchange in a single impulse is small, with prolonged repetitive activity the ionic gradients would disappear unless the cell were able to get rid of the accumulated Na^+ and restore the internal K^+. Sodium must be moved out against a large electrochemical gradient, and the process therefore requires expenditure of metabolic energy. A mechanism that moves an ion across a membrane against an electrochemical gradient is termed an ion pump. The mechanism in nerve responsible for moving Na^+ out of the cell is therefore termed the **Na^+ pump.** It should be pointed out that to give a process a name is not necessarily an indication that the process is understood. Nevertheless, although the chemical nature of the pump is not known, many of its properties have been well described. The pumping process is relatively slow; in the living animal there is a long-term steady state in which the Na^+ concentration is maintained within the limits necessary for electrical activity. The main evidence that the efflux of Na^+ from an axon depends on metabolism is that the process is highly temperature dependent and can be inhibited by a variety of metabolic inhibitors, including cyanide and dinitrophenol. It is of special interest that these inhibitors do not block the action potential, provided that the ionic gradients are already established. In the presence of inhibitors the gradients do gradually run down, and then impulse production is precluded. The energy for the electrical activity is derived from the concentration gradients, which can be considered as batteries. Dramatic evidence that metabolic sources of energy are not directly involved in the generation of the action potential was obtained from experiments in which the contents of a squid axon were squeezed out and the axon reinflated with a solution containing only inorganic ions and no sources of metabolic energy, such as glucose or adenosine triphosphate. Under these conditions it was only necessary to maintain normal intracellular concentrations of Na^+ and K^+ in order for the axon to continue having a resting potential and to generate action potentials when stimulated. Moreover, when the Na^+ pump is inhibited, the influx of Na^+ during the action potential is not decreased.

Normally, a metabolic source of energy is required to maintain the ionic concentration gradients. Evidence has been obtained indicating that high-energy phosphate-containing compounds can provide energy for Na^+ pumping. Figure 4.20 shows the rate at which radioactive Na^+ leaves a loaded squid axon. When the axon was bathed in cyanide-containing solution, the efflux markedly decreased. When ATP was injected into the axon, the efflux of Na^+ increased. If ATP was hydrolyzed to AMP and inorganic phosphate before injection, it was without effect. Other high energy phosphate compounds capable of leading to the synthesis of ATP, such as arginine phosphate, also increase Na^+ efflux.

N. Performance of the Na^+ Pump

The job of the Na^+ pump is to maintain the Na^+ concentration inside the cell at a low level by removing Na^+ from the interior of the cell as fast as it enters. This task is made easier by the low resting permeability of the membrane to Na^+. However, because the electro-

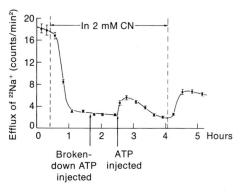

Figure 4.20
Evidence that high-energy phosphates can support the active extrusion of sodium from squid axon. A squid axon was loaded intracellularly with radioactive sodium, and the efflux measured as a function of time (note time scale in hours). The efflux is markedly reduced by the addition of 2 mM CN. An intracellular injection of hydrolyzed ATP has no effect on the efflux, but the injection of ATP produces a transient rise in sodium efflux. [Caldwell and Keynes, 1957.]

chemical driving force for Na^+ is large, there is a constant inward Na^+ current that must be balanced by the activity of the pump. There appear to be at least two ways in which the pump does its job. In the first, Na^+ is pumped out in exchange for K^+. Since a positive ion inside is exchanged for a positive ion outside, no net charge is transferred by this process, and it is referred to as a neutral or Na^+-K^+ coupled pump. The second mechanism involves the pumping out of ionic Na^+ without exchange for K^+. This pump moves positive ions out, leaving the interior of the cell more negative; it is called an electrogenic pump. There is now good evidence for both mechanisms.

The Neutral Pump. The entirely neutral pump exchanges Na^+ for K^+, ion for ion. If in the resting state the internal Na^+

remains constant, then for each Na^+ diffusing into the axon a Na^+ ion must be pumped out. For each K^+ ion that diffuses out, a K^+ must be pumped in. If the membrane potential is constant, the inward Na^+ current must equal the outward K^+ current. The membrane potential will be at a level where the passive outward K^+ flux equals the passive inward Na^+ flux; this potential will be somewhat less negative than the potassium equilibrium potential. As we saw above, the membrane potential is determined in large part by the relative Na^+ and K^+ conductances and their concentration gradients. Since the resting cell is only slightly permeable to Na^+, the resting potential is very close to E_{K^+}; but it differs sufficiently to provide a driving force for outward K^+ diffusion to balance the resting inward Na^+ flux. Recall that the current for an ion is a function of both the conductance and the driving force, $V_m - E_{ion}$. Evidence for the coupled pump is that metabolic inhibitors decrease not only the Na^+ efflux but also the K^+ influx, as measured with radioactive tracers, and that when the external K^+ concentration is reduced the Na^+ efflux decreases by about 70%.

Electrogenic Pump. With this mechanism the Na^+ that diffuses inward is pumped back out without exchange for K^+. At rest the active efflux of Na^+ balances the passive influx, there is no potential generated, and the membrane potential is essentially at E_{K^+}, as there is effectively no inward Na^+ current. Following a period of impulse activity, however, Na^+ may be pumped out faster than it diffuses passively inward, because of the temporary increase in intracellular Na^+,

so that the interior of the cell becomes more negative. Electrogenic pumps have been found in neurons from invertebrate ganglia as well as in mammalian axons. Evidence that the Na^+ pump can produce an increase in membrane potential has been obtained in a variety of ways. First, if Na^+ is injected into a neuron through a micropipette, the membrane potential increases to a greater extent than if an equivalent amount of another cation, K^+ or Li^+, is injected. Second, this increase in membrane potential can be inhibited by such pump inhibitors as ouabain. When the electrogenic pump is stimulated by an increase in internal Na^+, the membrane potential of the cell is greater than E_{K^+} and cannot be predicted from a knowledge of the distribution of the ions across it by either the Nernst or the Goldman equation. Following prolonged activity in small unmyelinated fibers, the axons undergo a large hyperpolarization—the "positive afterpotential." This potential is blocked by metabolic inhibitors of the Na^+ pump. In some neurons the resting potential is decreased by metabolic inhibitors. In these cases the membrane potential approaches, or even becomes more negative than, E_{K^+} because of outward Na^+ flux via an uncoupled pump. Inhibition of the pump blocks the outward flux, and the membrane potential assumes a value dependent on the relative inward Na^+ current and outward K^+ current.

Many neurons appear to utilize both mechanisms of Na^+ pumping, depending on the needs of the cell. In addition, there is evidence that other ions, particularly Ca^{++} and Cl^-, are also pumped in some neurons. These pumps are especially important in those neurons in which Ca^{++} plays some role in carrying the inward current during the action potential and in which conductance changes to Cl^- are involved in synaptic transmission.

III. CONDUCTION OF THE NERVE IMPULSE

A. Passive Electrical Properties

In the previous sections the nerve membrane was considered to behave in a uniform manner, as if the membrane potential and permeability were the same throughout the cell. Normally, this is not the case. A major function of neurons is to obtain information from one part of the body or nervous system and transmit it to a distant part. The accuracy with which information is transmitted is dependent on how well the electrical signals are relayed from one part of the neuron to another. In order to understand this process, we need first to know how good an electrical conductor the neuron is. Passive electrical properties of the neuron do not involve an active response (i.e., a change in ionic conductance). The axon consists of an electrical conductor, the ionic fluid axoplasm, bounded by a high resistance surface membrane, which is itself surrounded by at least a thin layer of conducting medium, the extracellular fluid. The whole system is analogous to an undersea cable —copper wire surrounded by an insulating cover and immersed in seawater. Thus the passive electrical properties of axons are sometimes referred to as cable properties. In comparison to a submarine cable, the axon is very poorly designed for long-distance transmission of electrical signals because the axoplasm is a poor conductor (restivity $\cong$ 30 ohm cm) and the cell membrane is leaky (resistivity $\cong 10^9$ ohm cm); the resistance across a square centimeter of nerve membrane about 50 Å thick is about 1000 ohms A voltage applied at one point decays almost to zero through resistive loss and leakage within a few millimeters. In addition to its resistive properties, the cell membrane is so thin that it has significant electrical capacitance. The high resistance membrane serves as the dielectric and the surface of the internal and external electrolyte solutions as the plates of the capacitor. The equivalent electrical circuit for an arbitrary patch of nerve membrane is therefore a resistor and capacitor in parallel. To complete an electrical analog of an axon, it is necessary to connect a resistor inside the membrane (representing the axoplasm resistance, r_i) to an outside resistor (representing the resistance of the external fluid, r_o). Such a circuit is shown in Figure 4.21.

In order to study the electrical properties of the axon, it is necessary to pass current across the membrane and to record the resulting voltage changes. The techniques for making these measurements were developed about 25 years ago and have proven so useful to neurophysiological analysis that they will be considered in some detail. We have already seen, in the study of the squid axon, the advantage of directly measuring the membrane potential and being able to control it. Other neurons, however, are not only much smaller than squid axons but also anatomically less accessible. For example, the larger motor

Figure 4.21
Electrical analog of a nonmyelinated axon membrane.

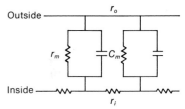

The specific resistance of the axoplasm, R_i, is about 200 ohm-cm; the membrane resistance, R_m, is about 200 ohm cm^2; and the membrane capacitance, C_m, is 1 $\mu F/cm^2$. The values for c_m, r_m, and r_i for a unit length of axon depend on axon geometry and are related to the specific values in an infinite cable by the relations $c_m = 2/2\pi a C_m$, $r_m = R_m/2\pi a$, and $r_i = R_i/\pi a^2$, where a is the radius of the axon. In saline solutions, the external resistance, r_o, is relatively low and can be ignored.

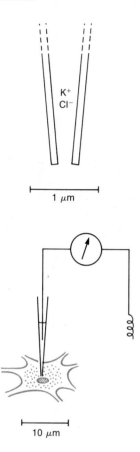

K+
Cl−

1 μm

10 μm

A glass capillary tube is heated in the middle and rapidly pulled apart so that the final tip diameter is about one micrometer. Such a pipette is filled with a solution of nearly saturated KCl, usually by boiling at reduced pressure. The electrode can then penetrate cells as small as about 10 μm in diameter without significant damage, and the membrane potentials can be recorded between the interior of the cell and a reference electrode connected to the extracellular fluid.

neurons in the mammalian spinal cord are only about 50 μm in diameter and are located deep inside the spinal column. These cells would readily be damaged by attempts to impale them with the sort of electrodes originally used on squid axons. The technical problems of directly measuring potentials from smaller cells were solved by the development of the **capillary microelectrode** (Fig. 4.22). This electrode consists of a glass tube pulled out after heating so that the final tip diameter is about 0.5 micrometers. The capillary is then filled with a nearly saturated solution of K^+Cl^-. Because the opening at the tip is small, the electrode has a high electrical resistance of about 10^7 ohms. Since the electrode punctures the cell to connect inside to outside, special amplifiers are needed that drain essentially no current from the cell and which assure that the recorded potentials faithfully represent the electrical activity of the cell. Such microelectrodes can be inserted into a cell with the aid of micromanipulators. The potential changes are recorded by measuring the potential difference between the microelectrode and an "indifferent," or reference, electrode in the extracellular fluid. While both electrodes are in the external fluid the amplifier is adjusted to balance out any voltage that exists between the electrodes themselves. When the microelectrode is advanced into the quiescent cell, the resting membrane potential is recorded. The microelectrode becomes negative with respect to the reference electrode. If a second microelectrode is inserted in the same cell close to the first, the same resting potential is recorded. This result suggests that the electrode is sufficiently small so that the cell can be penetrated without producing a significant change in the membrane potential.

In order to study the electrical properties of the cell, we can insert one microelectrode to record the potential and a second to pass current between the interior of the cell and external bathing solution. In this way the current passes across the cell membrane resistance, and the resulting potential change is recorded. With either the current-passing or the recording electrode outside the cell, one does not see an appreciable potential change when the same current is passed.

Passing current between the inside and outside of the cell is equivalent to passing current over r_m, between the internal resistance r_i and the external resistance r_o in the electrical model. It can be seen in Figure 4.23 that it takes some time for the recorded membrane potential change to reach a steady value. The reason for this is that at first most of the current goes to charge the capacitance. But, as it charges, more and more current must flow through the resistance. When the capacitance is fully charged, all the current goes through the resistance. The current through the resistance produces a voltage, $V = Ir_m$. The membrane voltage change hence increases with time to a steady level that depends on the charging rate of the membrane capacitance. The voltage across the capacitor always equals that across the resistor. This ability of the capacitance to accumulate charge is responsible for the delay in the system. After the capacitor is charged, the current flows through the resistor, and the membrane potential remains steady. When the imposed current is turned off, the charge stored on the capacitance leaks off through the resistor network, and there is a delay in the return of the potential to the baseline. The time course of both the increase and

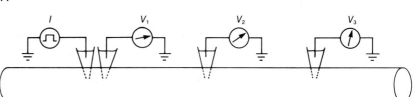

Figure 4.23
Passive spread of potential in an unmyelinated axon. **A.** The axon is considered as an infinite series of membrane patches, each consisting of a resistance, r_m, and a capacitance, c_m, in parallel and connected by an internal resistance, r_i, and a negligible external resistance, r_0. When a current is applied across the membrane at point 1, some of the current crosses the membrane at that patch, and the remainder leaves at more distant patches. Because of the internal resistance, less current flows at successively more distant patches. Thus the voltage change across successive patches decreases with the distance from the point at which the current is introduced. The decay of voltage along the axon is exponential. The space constant of the axon, λ, is given by $\lambda = \sqrt{r_m/r_i}$ is the distance for the potential to decay to $1/e$ of its value at patch 1. When r_i is low, as in a large-diameter axon, the potential spreads farther. **B.** The decay of potential with distance. The time course of the potential change will also vary with distance from the site of current injection. The time constant for the change in potential, $\lambda = r_m c_m$, is the time for the potential to rise to within $1/e$ of its final value at patch 1. At successively more distant patches, the time constant appears to increase, because the internal resistance is in series with the membrane resistance. **C.** The time course of the potential change at different patches. **D.** The internal resistance of the axon has two effects on the distribution of potential: first, to reduce the voltage change at more distant patches; second, to slow its time course.

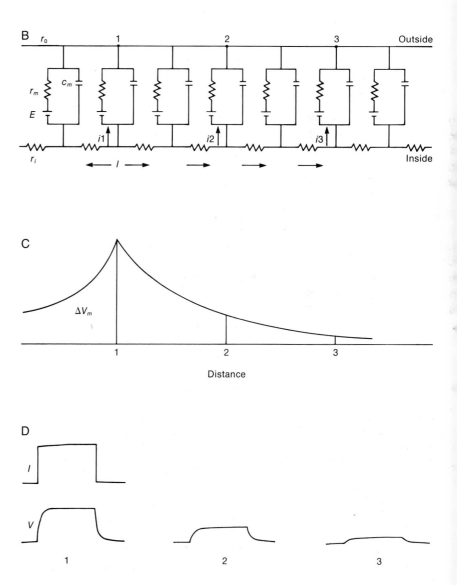

decrease of the potential is the same, determined by the magnitude of the resistance and capacitance. The **time constant** $\tau = r_m c_m$ is the time required to change the potential to within $1/e$ (37%) of its final value. This passive potential change is called an electrotonic potential (see Glossary).

So far it has been assumed that both the voltage-recording and the current-passing electrodes are close together in the middle of the axon. What happens when the recording electrode is moved some distance away? It becomes necessary to consider the current that flows longitudinally down the axon through the resistance r_i and then across the membrane. The resistance r_i exerts two effects on the recorded potential change further down the axon. First, as there is a voltage drop across this resistance, the voltage reaching the part of the membrane at the second recording position is decreased, and less current flows across this region, making the voltage change less. Another way to look at this is to consider that as some of the current has leaked out near the current electrode, there is less current leaking out at the second element. In addition to a decrease in voltage there is also a greater delay in the buildup of the potential change. This occurs because the effective resistance and capacitance of the circuit is increased. In Figure 4.23 it can be seen that this is a continuous process; the voltage change decreases and the time constant of the potential change increases as the distance from the current passing electrode increases. Under steady-state conditions the decay of voltage along the fiber is exponential and is determined by the relative value of r_m and r_i. The distance over which the potential decays to $1/e$ is called the **space constant,** $\lambda = \sqrt{r_m/r_i}$.

If r_i is low, as in large-diameter conductors, there is less voltage loss in the interior of the fiber, and a larger axon, therefore has a larger space constant than a smaller axon. Similarly, if the membrane resistance is large, little of the current leaks out across the membrane close to the current electrode, and the space constant is larger. In experimental situations, the external resistance r_o is usually so small in comparison to the other resistances that it can be neglected. Under some conditions it can be important; for example, if the axon is immersed in oil or lifted into air, there is only a thin layer of saline surrounding the axon and r_o becomes large. Since current flows in the circuit $r_i \rightarrow r_m \rightarrow r_o \rightarrow r_m$, an increase in r_o has the same effect as an increase in r_i. When r_o increases, the space constant— or length constant, as it is also called— decreases.

In the squid giant axon the length constant is only a few millimeters. In small vertebrate unmyelinated fibers the length constant is a few tenths of a millimeter. Thus electrical signals can be conducted passively only for distances of about that magnitude. Although the cable properties of axons are not adequate for passive long-distance signalling, they are nevertheless extremely important for conduction of the action potential and integration of information over distances of tens to hundreds of microns—for example, from the distal portion of a dendrite to the axon hillock of a neuron.

B. Local Current Spread and Propagation

In order to understand the role of local circuits and the passive properties of the

A

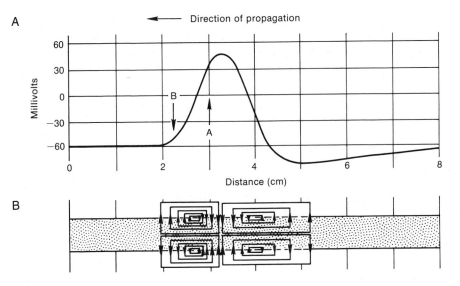

Figure 4.24
Nerve impulse as a function of distance along an axon. **A.** Membrane potential. **B.** Electric currents flow across the membrane and in the external medium as a result of the difference in potential. These currents depolarize the region ahead of the impulse, leading to changes in sodium permeability. [Keynes, 1958.]

membrane in the conduction of the action potential, it is useful to visualize the action potential as a function of distance along the axon. Since we know the conduction velocity and duration of the action potential in a squid axon we can draw the potential profile of the spike along the axon. Figure 4.24 is such a drawing, and it can be seen that the action potential occupies about 2 cm of axon. To simplify the factors involved in conduction, we will consider the events at two points along the axon, as shown in the diagram. At point A the membrane potential has decreased to a level at which g_{Na^+} is greatly increased and Na^+ flows inward faster than K^+ can leave. The inside of the membrane is becoming more positive; that is, there is a net increase in the number of positive ions

inside the cell. The increased positivity at this point tends to repel positive charges longitudinally down the axon. This ionic current will be carried by the intracellular ions in accordance with their relative mobilities and concentrations. As the mobilities of Na^+ and K^+ are not very different, and since K^+ is more highly concentrated internally, it will carry current in both directions away from the positivity. At point B this current provides more positive ions, and the membrane potential decreases. This situation produces an apparent paradox in that positive charge flowing inward at A depolarizes the membrane and at B positive charge is flowing outward and it is also depolarized. The essential reason why positive charge flows outward at B is that there has been an excess of positive charge built up in the immediate vicinity of the membrane, and this accumulation of charge provides the driving force for the outward current. When sufficient charge accumulates at B, so that the membrane potential is reduced to threshold, this region undergoes the regenerative increase in g_{Na^+} and gives rise to an action potential. In this way the action potential moves along the axon. An important part of this argument is that the action potential spreads as a result of the passive flow of current bringing each successive region of membrane to threshold. The mechanism for the action potential does not spread. It is contained in each patch and is activated by the depolarization produced by the spreading current. The rate of spread of an action potential thus depends both on passive electrical factors, the resistance and capacitance of an axon, and on active factors, the degree and rate of change of g_{Na^+} with depolarization.

Historically, two ideas evolved as to how the action potential is propagated. The first was that a chemical change spreads down the axon and that the electrical changes were a reflection of that chemical change. The second was that the action potential is fundamentally an electrical change that spreads along the axon. In the years 1937–1940, Hodgkin performed a series of experiments demonstrating that local electric currents determine the propagation of the impulse and that the action potential *is* an electrical event in the axon. In the above discussion of the spread of passive potential changes, it was noted that the electric potential should spread further along the axon when the external resistance r_o is low rather than high. This is true because at low r_o there is less potential drop in the external fluid and thus more

potential change across the cell membrane. If propagation depends on the flow of electric current, the conduction velocity should decrease when the axon is transferred from saline solution to either oil or air. A thin film of saline solution around the nerve has a higher electrical resistance than does a large volume of saline. Hodgkin found that this is the case: transfer of axons from saline solution to air produces a reversible decrease in conduction velocity of some 30%. In addition, decreasing the external resistance in saline solution by laying an axon on a grid of platinum wires shorted together produced the expected increase in conduction velocity. He concluded that when the external resistance is low the conduction velocity is high. Hodgkin's findings are entirely consistent with the view that ease of spread of electric current along the axon determines in part the conduction velocity of the nerve impulse.

In a second series of experiments the nerve conduction was blocked at one point by cooling. The action potential propagated to the point of cooling, but beyond that point only a graded potential was produced, which decreased in amplitude with distance from the block. The decrease in action potential amplitude was the same as the decrease in potential of an artificial depolarization that did not elicit any active response. By testing the threshold stimulus needed to excite the axon at various distances from the point at which conduction was blocked, he found that the change in excitability (change in required stimulus strength) was the same whether the axon was depolarized to a subthreshold level by the action potential or by the artificial depolarization. Thus there was no essential difference between the action potential and an applied artificial depolarization either in ability to spread along the unexcited axon or in effect on excitability.

The finding that the action potential propagates as a result of the ionic current flowing ahead of the impulse has important consequences in considering not only the speed but also the safety factor for propagation of the impulse. When the membrane area ahead of the impulse increases, as in the case of axon branching, the distribution of current may, under some conditions, be inadequate to depolarize all of the branches to threshold, leading to branch point failure. This may be important in determining depolarization of presynaptic endings, and therefore the amount of transmitter released (see Chapter 5).

Saltatory Conduction. Conduction velocity can be increased by decreasing internal resistance, increasing membrane resistance, or decreasing membrane capacitance. Under these conditions the electrotonic potential will more effectively spread down the axon. In the squid and other invertebrates, the internal longitudinal resistance is decreased by the increase in axon diameter. The penalty for using this mechanism is that the nerve trunks become very thick. A combination of rapid conduction and small size has been achieved in many of the axons in vertebrate nervous systems through a cooperative relation between axons and their satellite cells, the Schwann cells in the peripheral nervous system and the oligodendroglia in the central nervous system. These cells enwrap the axons, forming layers of compact myelin consisting of closely apposed cell membranes. The myelin insulates the axon and produces an effective increase in the resistance of the axon to outward current flow. In addition the layers of myelin add capacitance in series with the membrane capacitance, thus in effect decreasing the membrane capacitance. At intervals the myelinated axons are exposed (nodes of Ranvier), and current can leave, producing the depolarization to trigger an action potential. The gain in efficiency is remarkable. A frog myelinated axon 12 μm in diameter conducts at 25 m/sec, whereas a squid axon 350 μm in diameter conducts at the same velocity. This represents a 100-fold saving in cross-sectional area. Moreover, because the action potential of myelinated fibers is confined to the nodes, the ionic exchange is reduced and the metabolism necessary to maintain the ionic gradients is less.

Convincing evidence of saltatory conduction (i.e., that the action potential jumps from node to node in myelinated fibers) was obtained in the experiments of Tasaki and Takeuchi and of Huxley and Stämpfli. It can be seen in Figure 4.25 that there is an active inward current only at the nodes. At the internodal region they recorded a small outward current. The safety factor for conduction is so great in myelinated axons that if a node is artificially blocked, by application of a local anesthetic, the action potential can jump across the block to a more distant node. Saltatory conduction therefore provides additional evidence for the local circuit theory of conduction and emphasizes the importance of electrical factors in determining the spread of potentials in nerves.

The importance of the miniaturization possible with myelinated nerves becomes apparent if we consider the mammalian optic nerves. These nerves contain about 1 million axons. Without myelination, the diameter of the optic nerve would have to be increased from 2 to perhaps 100 mm to carry the same information at the same speed.

C. External Recording of the Nerve Impulse

During the action potential, current flows across the membrane, longitudinally along the axon, and completes the circuit through the external medium. According to Ohm's law the potential difference between two points depends on the current flow between the points and the resistance. When an axon is in saline solution, a propagating impulse causes current to flow in the external medium; the resistance between two electrodes on the surface of the nerve is low compared to the transmembrane resistance and the internal longitudinal resistance, hence only a small potential is recorded. If, however, the nerve is placed in moist air or in mineral oil the current will flow through the thin film of fluid adhering to the outside of the axon. This film has a high resistance, and under these conditions a larger potential change is recorded by the electrodes. The amplitude of the potential change recorded will depend on what fraction of the total resistance is accounted for in the circuit by the thin layer of saline. Although this technique does not give accurate data about potential changes across the membrane, the information can be quite valuable. For example, in many experiments one asks the question, did the nerve give rise to impulses, and if so, what was the frequency of the impulses? An important point to remember when using this technique is that the time course of the potential change re-

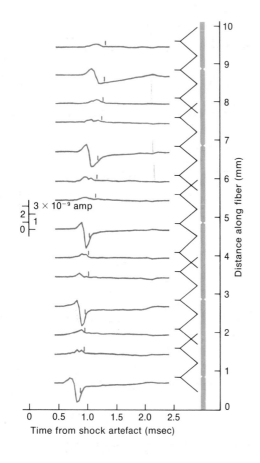

Figure 4.25
Membrane currents measured along the length of a single frog myelinated axon. Each trace shows the difference between the longitudinal currents recorded 0.75 mm apart, as indicated at the right. Inward current appears downward. Note that inward current is recorded only when an internode occurs between the recording electrodes. Tick marks show time of peak membrane potential. [Huxley and Stämpfli, 1949.]

corded externally indicates the time course of current flow in the external fluid rather than the potential change across the membrane.

Monophasic and Diphasic Recording. With external recording from two electrodes on the axon, the changes in potential are complicated in appearance. The potential first swings in one direction and then the other, thus the name diphasic action potential. This occurs because as the action potential reaches the first electrode, the current flows inward at that point and outward elsewhere along the fiber. When the action potential reaches the second electrode, the same series of events takes place. But as we record the potential difference between the electrodes, the potential first swings in one direction and then in the other. To remove this complication, one often prevents the action potential from reaching the second electrode. This can be accomplished by crushing the nerve between the electrodes, by blocking conduction with applications of highly concentrated K$^+$ or by using drugs, such as local anesthetics. The results of this procedure are shown in Figure 4.26. The potential change now occurs only near one electrode and is termed a monophasic action potential. In recent years a technique has been introduced that enables one to record almost the entire change in membrane potential with external electrodes. This is accomplished by increasing the external resistance between the electrodes so that it is much higher than the membrane or internal longitudinal resistances by perfusing the region between the electrodes with ion-free isotonic sucrose solution; the procedure is called the **sucrose-gap technique.**

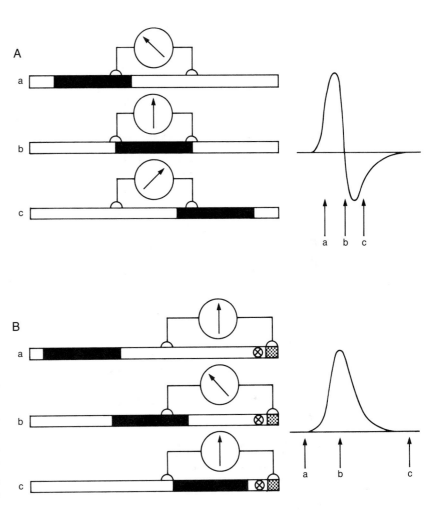

Figure 4.26

Recording the action potential with external electrodes. **A.** Diphasic recording. Two electrodes are placed on an axon, and the difference between them is recorded on the oscilloscope. When the impulse reaches the first electrode, a potential difference is recorded in one direction (*a*), and when the impulse reaches the second electrode the potential swings in the opposite direction (*c*). When the impulse is between the two electrodes, there is no potential difference recorded (*b*). **B.** Monophasic recording. If the impulse is prevented from propagating to the second electrode by crushing, cooling, or anesthetizing the region between the electrodes (indicated by ⊗), the impulse appears monophasic.

D. Recordings from Nerves with Many Axons

Records of impulses from whole nerves are complicated by at least two additional factors. First, the compound action potential amplitude is graded with stimulus intensity. The reason for this is that as the stimulating current is increased, more and more axons are depolarized to threshold and give rise to action potentials. The action potential in each axon is all-or-none, but as more axons are activated the current flow in the external medium increases and the recorded potential change increases. No matter how many axons are stimulated, the action potential recorded cannot exceed in amplitude that of a single transmembrane recording. The reason is that the parallel axons act as parallel batteries, and the voltage produced in such a circuit can only approach the voltage of the greatest cell. A second complication arises from differences in conduction velocity in the fibers composing the whole nerve. These arise from differences in diameter and myelination and from the associated effects discussed previously. It is worthy of note that the conduction velocities are not distributed at random; instead, each nerve has a characteristic fiber-diameter

Box 4.5 Optical Detection of Neuronal Signalling

The ability to record electrical activity in neurons by optical rather than electrical techniques has recently been realized. This has been accomplished by exposing neurons to a class of dyes known as merocyanines. The fluorescence or light absorption of some of these dyes is a linear function of membrane potential, so that optical records of changes in membrane potential may be obtained. The figure shows simultaneous records of changes in light absorption and membrane potential in a squid axon. The technique holds great promise for the recording of electrical activity from cells too small to be penetrated by microelectrodes and for simultaneous measurements of individual membrane potentials from relatively large populations of neurons.

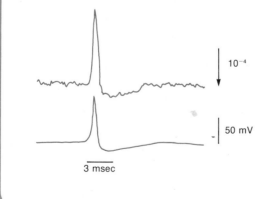

3 msec

The absorption change (top trace) of a dyed squid axon during the action potential (bottom trace). The absorption change, measured in a single sweep, is large enough to allow optical monitoring of action potentials in a 150 μm length of giant axon. The direction of the vertical arrow to the right of the trace indicates the direction of an increase in light intensity. [Ross et al., 1974.]

histogram and, hence, velocity distribution. For example, in the frog sciatic nerve there is a group of slower-conducting myelinated axons and a group of very slowly conducting unmyelinated axons in addition to the fast-conducting fiber group. The axons require different strengths of current to be stimulated and subserve different physiological functions. Each group of axons gives rise to a distinct hump or peak in the potential record, hence the name **compound action potential.** The properties of the different types of fibers were studied by Erlanger and Gasser who, in the 1930's, were the first to utilize the newly available cathode ray oscilloscope. Figure 4.27 shows some of these early records from whole nerves. It can be seen in these records that the greater the distance between the recording and stimulating electrodes, the greater the time difference between the peaks. By analogy, if two individuals traveling at different speeds leave a place together, the further down the road one records their arrival the greater will be the separation in time.

IV. SENSORY RECEPTORS

The output of information from a neuron is a function of various influences impinging on it from the environment. For most neurons it is the result of the various excitatory and inhibitory effects exerted on them at synapses. Some neural elements, however, are specialized to detect and translate particular forms of external energy of interest to the function of the organism; these are termed sensory receptors. As seen in the previous chapter, the anatomical specializations necessary to absorb external energy and to convert it into a meaningful signal

are varied and uniquely suited to their function. The sensory signal appears always to be a change in membrane potential. This can lead to a change in the amount of chemical transmitter released, which would affect an apposed nerve cell or if the sensory structure itself were a neuron with a long axon, the signal could lead to a change in the number or pattern of nerve impulses. In fact, most or all neurons will respond to a sufficiently strong nonneural stimulus such as pinching, cutting, or burning. Sensory receptors are distinguished by their structural and chemical adaptations, which render them capable of responding to external energy selectively and with remarkable sensitivity. The external influence may be chemical, mechanical, thermal, or electromagnetic (i.e., light, electric field, magnetic field). The kind of energy that produces an appropriate behavioral response by the organism is termed the **adequate stimulus** for the receptor. Some of the adaptations of sensory receptors are extraordinary and lead to extreme behavioral sensitivity of the organism. For example, visual receptors in many animals, including man, can respond to a single quantum of light; chemical receptors in insects, dogs, and probably many other animals, to a single molecule of an odorous substance; temperature receptors in snakes, to a rise in temperature of $0.003\,°C$; and electroreceptors in fish, to electric fields as small as 0.01 microvolt/cm. Additional modalities (see Chapter 6, Section III.B) that may be detected by sensory receptors include sound, pain, muscle tension, muscle length, joint position, touch, pressure, hydrogen ion concentration, and partial pressure of oxygen. In all cases the function of the receptor is to analyze the environment and convert the necessary

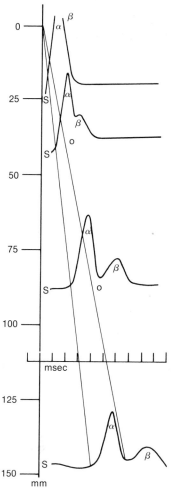

Figure 4.27
Monophasic compound action potential of frog sciatic nerve. These records illustrate the appearance of the first two (α,β) elevations of the compound action potential when recorded at increasing distances (from above downward) from the site of stimulation. Note that the relation between conduction time and distance is linear for both groups of axons, indicating a constant conduction velocity, and that, as expected, the delay between the two groups of fibers increases with distance. Each peak in the record is a combined result of the action potentials in a number of myelinated axons. [Erlanger and Gasser, 1937.]

information into a suitable physiological code. Although, as one might suspect, the physiological processes involved in the detection and transduction of the stimulus are varied, and in no case completely understood, there is some consistency in the ways in which impulses in sensory axons are initiated.

It was established in the 1920's by Adrian and his colleagues that sensory receptors somehow translate external energy into all-or-none nerve impulses for processing by the central nervous system. The impulses in all the sensory systems appear identical, and clearly the interpretation of these impulses by the nervous system depends on the central connections of the incoming nerve fibers (see Chapter 6, Labeled lines, p. 214). It follows from this that a punch in the eye gives one the sensation of light in the environment—a sensation that is independent of the true nature of the stimulus. The muscle spindle provided us with the initial insights into the performance of sensory receptors, and the general principles found remain applicable to a variety of sensory systems.

A. Generator Potentials in the Muscle Spindle

The function of the muscle spindle is important to the control of muscle contraction by the central nervous system. The muscle spindle consists of modified, or intrafusal, muscle fibers in parallel with the regular, or extrafusal, muscle fibers. The sensory nerve ending is wound around the middle of the fibers, forming an annulospiral ending. In addition, the muscle fiber has motor innervation. Stimulation of these axons makes the muscle fiber contract and, because

the region of the sensory endings is not itself contractile, this increases stretch on the annulospiral ending; stretching the muscle also stretches this sensory ending.

In 1931 Matthews studied the frog toe muscle; the preparation he used contained a single nerve ending that responds to stretch. His study showed that stretch of the muscle produces a discharge of all-or-none action potentials in the sensory nerve and that the frequency of the discharge (reciprocal of the interval between impulses) was roughly proportional to the logarithm of the weight on the end of the muscle. These early studies established the **first principle of sensory coding,** that intensity and duration of the sensory stimulus can be indicated by the spacing and duration of a discharge of identical nerve impulses. An additional indication of intensity is the number of receptors responding to a given stimulus.

In 1950 the work of Katz led to a major advance in our understanding of how the sensory stimulus is converted to nerve impulses. By recording close to the sensory ending of a frog muscle spindle, he found that a local potential change in the nerve terminal, whose amplitude and duration were related to the strength and duration of the stretch, led to nerve impulses in the sensory axon. This local potential change in the terminal was called the **generator potential** (see Glossary for usage). The summary of Katz's paper is reproduced in Box 4.6.

The muscle spindle gives information not only on the steady state stretch of the muscle but also on the velocity of the stretch (Fig. 4.28). Note also that at a given steady stretch, the frequency of the nerve impulses decreases with time. This decrease, termed **adaptation,** is a common property of many receptors,

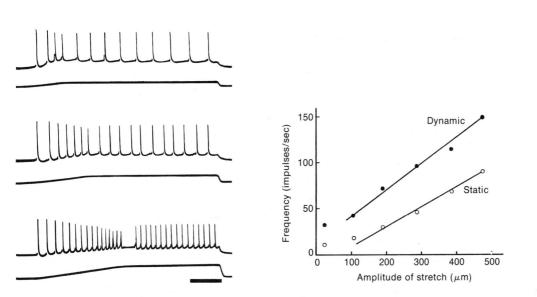

Figure 4.28
Left. Impulses in a sensory axon from a muscle spindle, recorded in response to progressively increasing stretch (shown on lower trace) to three different lengths. (Time bar 50 msec.) **Right.** Impulse frequency (ordinate) as a function of amplitude of stretch (abscissa) for sensory axon from a muscle spindle. The dynamic frequency is that measured during the linearly increasing stretch, whereas the static measurements were made 200 msec after the onset of the maintained stretch. [Ottoson and Sheperd, 1971.]

Box 4.6   Sensory Generator Potentials

1. When a frog muscle is stretched, its sensory nerve endings become depolarized, and a local potential change can be recorded from the sensory axon at a point close to the spindle.

2. This potential change varies with the rate and amplitude of stretching, and gives rise to repetitive impulses in the sensory nerve.

3. By applying a local anaesthetic it is possible to abolish the nerve impulses without affecting the local electric reactions to stretch. There is evidence for two distinct components of the potential change associated with the dynamic process of stretching and with static extension respectively.

4. When a stretched muscle is released, a transient potential change in the opposite direction (i.e. a positive variation at the nerve endings) is observed.

5. The local potential change appears to arise from a direct action of the mechanical stimulus on the sensory nerve endings and to be a link between the mechanical input and the rhythmic output of impulses from the sense organ.
[From Katz, 1950.]

and its time course—varying from extremely rapid (complete in a few msec) to extremely slow (minutes)—is one of the main features by which receptors are differentiated. Adapting receptors are called **phasic;** nonadapting receptors are called **tonic.** Many receptors are "**phasic-tonic,**" meaning that after an initial adapting phase the frequency settles to a plateau that is maintained as long as the stimulus condition is maintained. In the muscle spindle, adaptation results from a combination of factors. There is some mechanical adaptation due to the elasticity of the tissue and perhaps some changes in the conductance of the nerve terminal membrane with prolonged depolarization.

B. Ionic Basis of Generator Potential

How does deformation of the sensory terminal lead to the generator potential? One possibility is that stretch on the membrane opens selective membrane channels leading to the influx of sodium ions. But the experimental work does not support such a simple hypothesis. It has been found that the generator potential in muscle spindles is insensitive to tetrodotoxin, indicating that if the sodium channels are involved they have different characteristics from those in the axon. The amplitude of the generator potential is sensitive to reductions in both sodium and calcium. The situation is not clear. Perhaps the generator potential arises from an increase in permeability to all ions (including large organic ions like choline), as was originally suggested for the basis of the action potential by Bernstein in 1902.

C. Central Control of Sensory Sensitivity

The muscle spindle, especially in mammals, offers one of the clearest examples of how the central nervous system can set the sensitivity of the sensory organ and greatly extend its useful operating range. It is apparent from the diagram that when the extrafusal muscle fibers contract, the stretch on the muscle spindle decreases and the afferent discharge decreases (Fig. 4.29). If, however, the motor nerves to the spindle were active, it would cause contraction of the intrafusal fibers, as was indicated above, and increase the stretch on the sensory ending, restoring the discharge. The effect of stimulation of the motor axons to the spindle, termed **fusimotor fibers,** in determining the response of the spindle to stretch can be seen in Figure 4.29. The role of this system in the reflex behavior of muscle is discussed on p. 266ff.

In some other receptors the central control of sensitivity is even more direct. In the crustacean muscle receptor organ of the abdomen, for example, there is an efferent fiber from the central nervous system that acts on the receptive terminals of the sensory neuron, reducing the response to a given stretch (see p. 421).

The muscle spindle offers a good example of a receptor in which the stimulus is applied directly to the nerve ending of a primary sensory neuron (see Chapter 2, Section VI. B). As already indicated, in other cases, there is a specialized receptor called the sense cell, inserted as it were, ahead of the first-order afferent neuron. The sense cell detects the change in the environment and then secondarily excites the afferent nerve terminals by synaptic transmission.

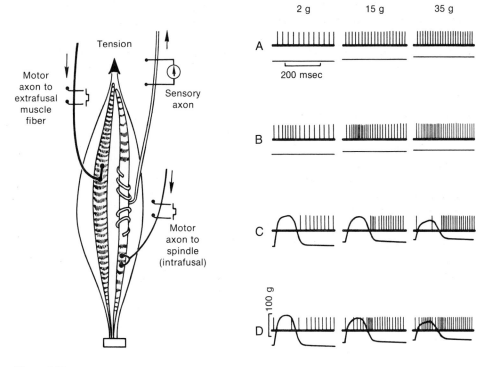

Figure 4.29
Left. Anatomical relations for intrafusal and extrafusal muscle fibers of a mammalian muscle.
Right. Efferent control of a stretch receptor in mammalian muscle. **A.** Impulses recorded from a sensory axon from a muscle spindle with three different loads on the muscle tendon 2g, 15g, and 35g. Lower beam indicates tension of muscle. **B.** Motor nerves to the muscle spindle are stimulated at 100/sec for 0.1 sec. This stimulation increases the discharge frequency in the sensory axon without measurably increasing the muscle tension. **C.** Effect of stimulation of the motor axons to the remainder of the muscle on the sensory discharge. Contraction of the muscle, as recorded on the lower trace, causes a decrease in the sensory discharge. **D.** Stimulation of both the motor axons to the motor fibers to the sensory spindle and the remaining muscle fibers. Under these conditions the sensory discharge is better maintained. [Parts A to D, Hunt and Kuffler, 1950.]

It may release a chemical transmitter substance in some proportion to the stimulus or it may employ electrical transmission (Chapter 5). Changes in membrane potential in the specialized sense cell are termed **receptor potentials** —a term that historically overlaps "generator potentials" and is often used as a special case of the latter (see Glossary). The anatomy and presumed functional behavior of one of these receptors—the hair cells in the lateral line of fish—is diagrammed in Figure 4.30 (see also pp. 78–85, 92).

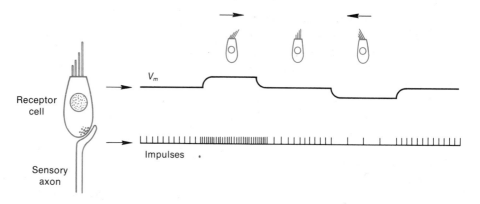

Figure 4.30
Diagram of the anatomical relation between a sensory hair cell and the sensory axon in a fish lateral-line organ, the presumed potential changes in the receptor cell, and the impulse discharge in the
axon. There is a chemical synapse between the sensory cell and the afferent axon. Bending of the
hair in one direction presumably depolarizes the receptor cell, increases the release of an excitatory
chemical, which depolarizes the nerve terminal, and increases the frequency of nerve impulses. Bending in the opposite direction produces the opposite effects. [Flock, 1965.]

V. PHYSIOLOGICAL PROPERTIES OF NEUROGLIAL CELLS

Neurons in both the peripheral and central nervous system are found to be
closely associated with nonneural cells.
In the periphery these cells in vertebrates
are called Schwann cells; in the central
nervous system they are known as neuroglial cells. In the vertebrates three main
classes of neuroglia have been described,
the oligodendrocytes, the astrocytes, and
the microglia. In vertebrate nervous systems neuroglial cells greatly outnumber
neurons and are generally smaller in size;
they constitute about 50% of the volume
of the nervous system.

Neuroanatomists first appreciated the
distinction between nerve cells and glial
cells about the middle of the last century.
This distinction was based on the form
of the cells; in particular, the glial cells
seem to lack specific connectivity with
neurons and have different reactions to
histological staining procedures from
those of neurons. The close apposition of
neurons and glial cells and the interposition of glial cells between blood
vessels and neurons suggested to these
anatomists that glial cells play some supporting nutritive function in nervous
activity. Although we know through
electron-microscopic observations that
many glial cells contain the necessary
organelles to support metabolic activity,
we still lack evidence that metabolism
within glial cells participates in neuronal
function. In fact, the only well-estab-
lished **role of glial cells** in the function
of neurons is that the oligodendroglia
form the myelin of central axons. It is
not known what role the large number
of astroglia and microglia perform in

the functioning of the nervous system or what else the oligodendroglia might be doing. One hopes that some of the sophisticated microchemical techniques currently becoming available might be useful in detecting changes in glial chemistry that accompany nerve activity.

A. Membrane Properties of Neuroglia

Resting Potentials. Glial cells in both vertebrates and invertebrates have large resting membrane potentials of up to 90 mV, significantly higher than those found in neurons (60–70 mV). The magnitude of the resting potential depends on the K^+ concentration of the surrounding fluid, as in neurons and muscle fibers. In Figure 4.31 the relation between membrane potential and K^+ is shown for both glial cells and neurons of amphibians. It can be seen from the graph that at normal concentrations of K^+ (2–3 meq/liter), the glial membrane potential is not only higher than that of the neurons but also more sensitive to variations in K^+. At K^+ concentrations above about 1.5 meq/liter, the glial cell membrane potential behaves as an accurate potassium electrode. That is, the slope of the membrane-potential–K^+-concentration curve can be predicted from the Nernst equation $E = RT/zF\ ln(K_o^+/K_i^+)$. Thus, at 25°C, a tenfold increase in the external potassium concentration decreases the glial membrane potential by 59 mV. This suggests that the resting glial cell is very much less permeable to sodium than is nerve or muscle. This graph can also be used to gain information about the internal composition of the glial cells. The Nernst equation predicts that the glial membrane potential would be abolished

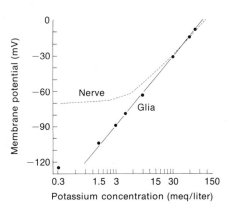

Figure 4.31
Relation between membrane potential and external potassium for amphibian myelinated axons and neuroglia. Solid line is drawn according to Nernst relation with slope of 59 mV. [Orkand, 1971.]

when the internal and external K^+ concentrations become equal. There would be no gradient for K^+ across the cell membrane, and hence no tendency for K^+ to diffuse in or out of the cell. From the graph we can see that the glial membrane potential would disappear at an external K^+ concentration of about 100 meq/liter. Glial cells therefore contain about 100 meq/liter of K^+ in their cytoplasm. Little else is known about the ionic composition of glial cell cytoplasm, but it appears that the concentration of sodium is much lower than that of K^+, as it is in neurons.

Electrical Properties. In contrast to neurons, when the membrane potential of a glial cell is either increased or decreased by the passage of current, the glial membrane behaves passively, as a fixed resistance and capacitance in parallel. This indicates that the permeability of the glial cell remains constant despite changes in membrane potential, at least for potential changes of up to 100 mV, depolarizing or hyperpolarizing. The shape of glial cells is so complicated that the membrane areas cannot readily be determined. Therefore, the specific membrane properties of glial cells, resistance and capacitance per unit area, have not yet been accurately determined.

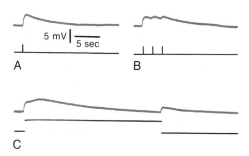

Figure 4.32
Effect of nerve impulses on the membrane potential of glial cells. Intracellular recording from a glial cell in the optic nerve of *Necturus*. Light flashes, as indicated on lower trace, were directed at the eye. **A.** Single light flash of 100 msec sets up a transient depolarization. **B.** Same flash repeated three times. **C.** Light stimulus maintained for 27 sec. The initial glial depolarization declines due to adaptation of nerve impulses. When the light is turned off, an "off" discharge of impulses produces an additional depolarization. [Orkand et al., 1966.]

Electrical Connections Between Neuroglia. Most neurons are electrically isolated from one another by the relatively low resistance of the fluid surrounding them and their own high-resistance membranes. They communicate at synapses usually by secreting chemical transmitters. Occasionally, they form electrical connections called electrical synapses (see Chapter 5). The glial cells have been found usually to be electrically connected to one another. Current injected into one glial cell can leave through adjacent cells and can spread from one cell to the next passively in a manner analogous to the electrotonic current spread along the nerve axon. The glial cells are therefore said to constitute an electrical continuum. Anatomically, glial cells are found to be connected by specialized contacts called gap junctions, and presumably it is at these points of attachment that the ionic current passes from one cell to the next. The physiological importance of these contacts is not known, nor is it known if other substances such as metabolites might move from one glial cell to another.

B. Neuron-Glia Interactions

If glial function is to be coordinated with neuronal function a mechanism must exist whereby the glial cell detects activ-

ity in neurons. Ionic currents do not pass from neurons to glial cells. That is, a potential change in a neuron is not detected by the adjacent glial cells. The action currents traverse the low resistance extracellular clefts rather than enter the glial cells. Nerve action potentials in unmyelinated axons have been found to depolarize the adjacent glial cells; the reason this depolarization is produced is actually quite simple. It will be recalled that during the nerve impulse a small quantity of Na^+ enters the axon, causing depolarization, and that a small quantity of K^+ leaves during repolarization. The efflux of K^+ from the axon transiently raises the K^+ concentration of the intercellular cleft just outside the active axon. The increase in K^+ has been found to depolarize the axon giving rise to the negative afterpotential. It has already been shown that the glial cell membrane potential depends on the ratio of K^+ concentration across the membrane. Thus, when the K^+ in the intercellular clefts increases as a result of nervous activity, the glial cell becomes depolarized. In Figure 4.32 it can be seen that impulse activity in the amphibian optic nerve produced by flashes of light shone in the eye produces a depolarization of slow time course in glial cells surrounding the unmyelinated axons of the nerve. This experiment shows that nervous activity is reflected by changes in membrane potential in glial cell. What is not known is how the glial cell responds to this depolarization. This kind of signal could serve to coordinate glial metabolism with the overall level of nervous activity. At best, this kind of signal is not very specific, because each glial cell may surround many thousands of axons. A hypothesis for specific transfer of information between neurons and glia would require a much more specific signal, but at present such a signal is not known to exist.

SUGGESTED READINGS

Adrian, E. D. 1932. *The Mechanism of Nervous Action.* Univ. Pennsylvania Press, Philadelphia. [A personal summary of the experiments that form the basis of modern cellular neurophysiology.]

Cole, K. S. 1968. *Membranes, Ions, and Impulses: A Chapter of Classical Biophysics.* Univ. California Press, Berkeley. [A comprehensive theoretical and experimental account of the ionic basis of membrane electrical activity.]

Cooke, I. M., and M. Lipkin. 1972. *Cellular Neurophysiology: A Source Book.* Holt, Rinehart & Winston, New York. [An invaluable collection of reproduced classical original papers covering the nerve impulse, synaptic transmission, sensory neurons, and neuronal integration.]

Galvani, L. 1953. *Commentary on the Effect of Electricity on Muscular Motion* (trans. by R. M. Green). Elizabeth Licht, Cambridge, Mass. [English translation of Galvani's 1791 treatise on animal electricity. A detailed description of the experiments that led Galvani to conclude the existence of electric currents in animal tissues.]

Hodgkin, A. L. 1964. *Conduction of the Nervous Impulse.* Liverpool Univ. Press. [A highly readable, brief series of lectures on the experiments that gave rise to the ionic hypothesis of membrane electrophysiology.]

Kandel, E. R., ed. 1976. *Cellular Biology of Neurons* (*Handbook of Physiology. The Nervous System, vol. 1*). Williams and Wilkins, Baltimore. [A comprehensive collection of short monographs on all aspects of structure, chemistry, and function of neurons and glia. It includes synaptic transmission and neuronal integration.]

Nystrom, R. A. 1973. *Membrane Physiology.* Prentice-Hall Inc., Englewood Cliffs, New Jersey. [A view of the variety of current ideas regarding the structure, permeation mechanisms, electrical behavior, and chemistry of biological and artificial membrane systems. Extensive references to original literature.]

Plonsey, R. 1969. *Bioelectric Phenomena.* McGraw-Hill, New York. [A quantitative physical chemical approach to the electrical behavior of nerve and muscle membranes.]

5
TRANSMISSION AT NEURONAL JUNCTIONS

I. INTRODUCTION

The functional significance of the regions in which individual neurons make special contact with each other was appreciated, among others of his day, by Sherrington, who in 1897 introduced the term **synapse** (see Glossary) to describe these "surfaces of separation." He suggested that the existence of a physical separation between neurons would give to that region properties that might account for the differences between reflex-arc conduction and nerve-trunk conduction. Sherrington listed eleven properties of reflex arc conduction that differentiate it from nerve conduction, including irreversibility, dependence on oxygen, and weak correspondence between input and output. These properties were correctly attributed, in light of the cell theory, to intercellular transmission at synapses. Historically, two main classes of interactions between nerve cells at synapses have been considered: (a) the spread of ionic current from one cell to another, or electrical synaptic transmis-

sion, and (b) the secretion of a chemical transmitter from one cell to affect the permeability of another, or chemical synaptic transmission. A controversy raged between these two schools of thought—the spark school and the soup school—throughout the 1930's and 1940's. It is now recognized that both electrical and chemical interactions occur between neurons (Fig. 5.1). They are generally alternatives (i.e., in different sets of synapses); the chemical synapses appear to be by far the more common. At those loci where neurons influence effector cells—muscle, gland, or electric organ—transmission appears always to be mediated by chemicals.

The synapse is the prime locus in the integration of nerve signals—the chief business of the nervous system and the theme of this book. It is also a site where functional properties are altered as a result of previous activity and where a variety of chemical agents act to alter nervous system behavior. In the next

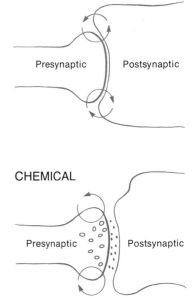

Figure 5.1
Two proposals for synaptic transmission, electrical and chemical.

ELECTRICAL

Presynaptic   Postsynaptic

CHEMICAL

Presynaptic   Postsynaptic

In electrical transmission, ionic currents from the presynaptic neuron cross the synaptic cleft to change the potential of the postsynaptic cell. In chemical transmission, the presynaptic neuron releases a specific chemical transmitter that diffuses across the synaptic cleft to alter the permeability pattern of the postsynaptic membrane.

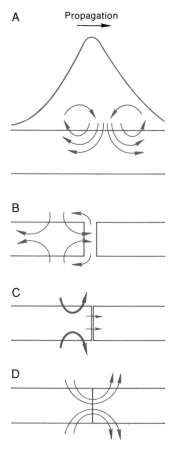

A Propagation

B

C

D

Figure 5.2
Electrical considerations in synaptic transmission. **A.** Diagram of current flow as a function of distance during a propagated action potential. **B.** At a normal synaptic cleft, when the impulse reaches the end of the axon, the current flows in the low-resistance extracellular fluid rather than crossing into the postsynaptic neuron. **C.** Where two neurons are in close apposition, most of the current leaves the axon through the single membrane rather than crossing a double membrane into the postsynaptic cell. **D.** If the membranes of the two cells are fused and the permeability of the fused membrane is low, current can readily cross from pre- to postsynaptic cell.

chapter other potential sites of integration, lability, and plasticity will be discussed.

II. ELECTRICAL SYNAPTIC TRANSMISSION

A number of synapses are now known in which ionic currents flow between neurons. Such synapses appear to be of advantage for very rapid transmission or for synchronizing the activity of a group of neurons. The main principles of electrical transmission are the same as those previously considered in electrical conduction along an axon. Inward current during the action potential in the presynaptic axon must flow outward in the postsynaptic axon to produce a depolarization. Such a synapse is known as an electrical excitatory synapse. The necessary features of such a synapse can be seen in the diagrams in Figure 5.2. An action potential propagating down the presynaptic axon is shown as a function of distance along the axon. We recall from the previous chapter that the steep rising phase of the action potential results from an inward movement of a small quantity of Na^+ ions. This decreases the membrane potential at that point and produces a depolarizing outward current in the region just ahead of the active membrane. When the action potential reaches the end of the axon, the outward current flows out of the axon into the intercellular cleft. The distribution of current after leaving the axon is determined by Ohm's law. That is, the current distribution depends on the resistance of the various barriers it encounters. It should be obvious that if the membrane resistance of the postsynaptic neurons is high compared to the resist-

ance of the extracellular fluid, as has been found in those neurons already examined, most of the current will flow in the extracellular fluid and little will enter the postsynaptic neuron. In fact, Katz has calculated that if the two neurons are 5 μm in diameter and separated by a cleft 150 Å wide (Fig. 5.2,B), a 100 mV change in the presynaptic axon would produce a 10^{-2} mV change in the postsynaptic neuron, far too little to influence activity in most neurons. Such a condition is fortunate in that over much of their surface, neurons in the central nervous system are separated from other neurons by clefts 100–200 Å wide; it is doubtful that the system could function if the action currents in one neuron affected activity in adjacent neurons at other than specific sites of apposition. Even if there were no gap, electrical transmission would be difficult. Suppose that the membranes of two neurons were in intimate contact with one another so that the extracellular space were either eliminated or greatly reduced and that the membranes retained their original electrical properties (Fig. 5.2,C). The resistance of this double partition at the terminal would be much higher than the resistance of the single membrane at the sides, and most of the current would flow out the single membrane into the extracellular cleft system. In order to have current flow readily from presynaptic to postsynaptic cell, it is necessary that the extracellular space be eliminated or greatly reduced and that the resistance of the two terminal membranes be greatly reduced. In addition, the neurons must be of comparable size or the postsynaptic process must be smaller, so that the current leaving the presynaptic axon is sufficient to change the potential of the postsynaptic structure.

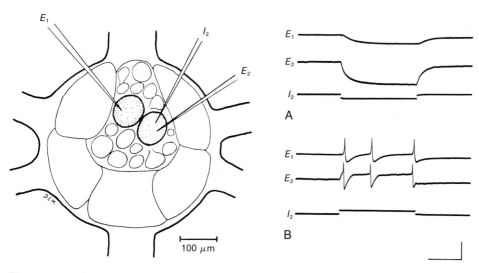

Figure 5.3
Left. Diagram of paired giant ganglion cells in a leech ganglion. The membrane potential is recorded from one cell (E_1) while current is passed (I_2), and membrane potential recorded from the second cell (E_2). **Right.** (A) Records of membrane potential from one cell (top trace) during passage of hyperpolarizing current (bottom trace) across the membrane of the second cell. The middle trace is the voltage change in the second cell. Note the larger voltage change in the cell containing the current-passing electrode. (B) Current is now depolarizing and produces synchronous action potentials in both cells. (Calibrations: time 0.1 sec; 20 mV, 10^{-8} A.) [Eckert, 1963.]

These conditions have been achieved by the formation of specialized junctions called **"gap junctions"** or "nexuses" between certain neurons. They are found in virtually all the major groups of animals, including coelenterates, annelids, arthropods, molluscs, lower and higher vertebrates. The operation of one such synapse in the leech central nervous system is illustrated in Figure 5.3. It can be seen that an action potential in one neuron produces an action potential in the other and, moreover, that even passive hyperpolarization of one of the neurons, by current injection, leads to a hyperpolarization in the second neuron. Thus the two neurons are electrically continuous. It should be emphasized that in this case such electrical synapses are specific between paired neurons with the same function on opposite sides of the animal. Electrical interactions are not found between some other neurons in the leech ganglion even though the cell bodies are closer together.

The above electrical synapse was found to transmit equally well in both directions. This is not always the case; the electrical synapse between giant motor fibers in crayfish exhibits properties of an **electrical rectifier,** allowing current to pass much more easily in one direction than in the other. An impulse in one axon produces an action potential in the second, but the converse does not occur.

Additional properties of electrical synapses are that there is essentially no delay between potential changes in pre- and postsynaptic cells and that transmission is relatively insensitive to blockage by lack of oxygen or changes in ionic or chemical environment that do not block the action potential. There is, in principle, less opportunity for control and plasticity than at chemical synapses because fewer steps are involved.

III. CHEMICAL SYNAPTIC TRANSMISSION

The hypothesis that neurons interact through chemical mediators is a complicated one. The general scheme is that an action potential in a presynaptic axon causes the release of a chemical that diffuses across the synaptic cleft to produce in the postsynaptic cell a change that leads either to excitation or inhibition. The process involves metabolic machinery for the synthesis and storage of the chemical in the presynaptic neuron, a release mechanism, specialized chemical sensitivity of the postsynaptic cell, and a mechanism whereby the chemical is inactivated. It now appears that this type of transmission is of general significance, and we want therefore to examine carefully the experimental basis for its general acceptance.

A. Historical Background

The basic idea of chemical transmission between neurons originated from studies of the mammalian autonomic nervous system done early in this century. Around 1900, J. N. Langley and his students noted the remarkable similarity between the effects of adrenalin, a naturally occurring substance isolated from the adrenal glands, and stimulation of neurons of the sympathetic nervous system. Both increase blood pressure and relax intestinal smooth muscle. One student, T. R. Elliott, went so far as to suggest in a preliminary note in the *Journal of Physiology* that "Adrenalin might then be the chemical stimulant liberated on each occasion when the impulse arrives at the periphery." The idea was not included in Elliott's full paper, published in 1905, possibly because he (or a cautious editor) felt the idea too revolutionary. Some years later H. H. Dale found that choline and its derivatives have effects similar to stimulation of parasympathetic nerves on peripheral effectors such as the heart, bladder, and salivary glands. In particular, the acetyl ester of choline, acetylcholine, was found to be the most potent. Dale raised the question of the possible physiological significance of the similarity between the actions of acetylcholine and adrenaline and the effects of stimulating the two parts of the autonomic nervous system, but hesitated to speculate further.

The most famous demonstration that stimulation of nerves leads to the release of an active chemical substance was made in a remarkably simple experiment by Otto Loewi in 1921. Loewi collected the fluid perfusing a frog heart before and after stimulation of the vagus nerve. When applied to a second heart, the perfusate collected before vagal stimulation had no effect, but that collected during stimulation inhibited the beat of the second heart in the same way as the addition of acetylcholine or vagal stimulation. The inhibition was blocked by the addition of atropine, a plant alkaloid known to inhibit the action of acetylcholine. He termed the substance released by vagal stimulation "Vagusstuff." Subsequent studies showed it to be acetylcholine.

In 1936, Dale and his colleagues collected acetylcholine after stimulation of motor nerves to skeletal muscle, thereby extending the hypothesis of chemical transmission to include the entire peripheral nervous system. These findings, however, did not unequivocally point to an intercellular role of the cholinergic compounds, for it remained to be shown (a) that electric currents in the axon are not sufficiently large to excite the muscle and (b) that sufficient quantities of the chemical are released from the presynaptic axon. The idea of chemical transmission was not generalized to the central nervous system or widely accepted until about 1952, just seven years before the first clear demonstration of an electrical synapse by Furshpan and Potter.

B. Evidence for Chemical Transmission

The vertebrate neuromuscular junction is the most completely understood of all junctions or synapses. As indicated above, Dale and his colleagues identified **acetylcholine (ACh) as the neurotransmitter** at neuromuscular junctions. They found that ACh could be collected from a muscle perfused with a physiological solution containing eserine (a compound that prevents the breakdown of ACh) when the motor nerves to the muscle were stimulated. When the sensory nerves were stimulated there was no such release; thus the ACh did not arise indiscriminately from nerve impulses. To rule out the possibility of ACh coming directly from the stimulated muscle, the nerves were cut in some experiments and

allowed to degenerate; when the muscle was stimulated to contract with electric shocks, no ACh could be recovered. As a further check, muscles were paralyzed with curare, a South American indian arrow poison that prevents the action of ACh on muscle. Stimulation of the nerves in curarized preparation did not lead to muscle contraction, but ACh was still released.

The most critical test for ACh being the chemical transmitter at the neuromuscular junction would require not only that the electric current in the presynaptic axon be insufficient to cause excitation but also that there be quantitative agreement between the amount of ACh released from the nerve terminal and the amount of ACh necessary to produce the observed excitatory effect on the muscle membrane. The best measurements currently available show that 10^{-17} M of ACh is liberated/end-plate/nerve impulse, whereas the smallest amount of ACh sufficient to obtain a muscle response is 10^{-16} M. Since the errors involved in collecting ACh from perfused muscle tend to decrease the measured value, and since the errors involved in applying ACh to an end plate with a microsyringe tend to increase the measured value, the agreement between the two values is actually very good. Figure 5.4 illustrates how the application of minute amounts of acetylcholine to a frog neuron can mimic the effect of nerve stimulation.

C. Properties of the Postsynaptic Potentials

At the vertebrate "twitch type" of neuromuscular junction there is normally sufficient transmitter released by a single

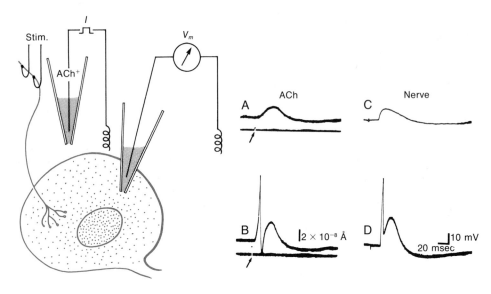

Figure 5.4

Left. Diagram of experimental arrangement for iontophoretic application of acetylcholine to the synaptic region of a ganglion cell in the frog heart and for stimulation of a presynaptic axon. One microelectrode records the membrane potential (V_m) while the second, filled with acetylcholine (ACh) is brought close to the surface of the cell. When the acetylcholine-filled pipette is close to a synaptic region, a brief current pulse ejects a small quantity of acetylcholine. **Right.** Comparison of the effect of acetylcholine and nerve stimulation on the ganglion cell. Current passed through acetylcholine-filled pipette is monitored (arrow) in the lower trace (in *A* and *B*). Records *A* and *C* show subthreshold responses; *B* and *D* show suprathreshold responses. The quantity of acetylcholine used in these experiments is estimated to be about 10^{-17} M. [Dennis et al., 1971.]

nerve impulse to depolarize the muscle fiber to threshold for the muscle action potential. This situation contrasts markedly with that at most central nervous system synapses, where simultaneous activity in a large number of presynaptic axons is necessary to produce an action potential in a postsynaptic neuron. Experimentally, one can partially reduce the effectiveness of ACh by chemically blocking the reactive sites on the muscle fiber with curare or decreasing the amount of ACh released from the nerve by reducing the external calcium concentration or increasing the external magnesium concentration. Under these conditions the synaptic potential is insufficient to reach threshold for the all-or-none muscle action potential, and the properties of the junctional potential (termed end-plate potential in vertebrate skeletal muscle) can be studied. Fatt and Katz studied the **end-plate potential (e.p.p.) distribution** by stimulating the nerve and recording the resulting potential at various points along the muscle fiber. The results are shown in Figure 5.5. Here it can be seen that the e.p.p. is largest and its duration briefest when the microelectrode is inserted just at the muscle end-plate. As the electrode is moved away, the potential becomes smaller and

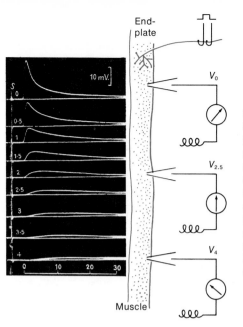

Figure 5.5
Changes in amplitude and form of the sub-threshold end-plate potential with increasing distance from the site of innervation. End-plate potentials recorded from frog muscle treated with curare. Numbers above each trace indicate distance from end-plate focus. S is time of stimulus artifact. The end-plate potential resulting from the action of the neurotransmitter on the muscle membrane decreases in amplitude and rate of rise with increasing distance from the end-plate because of the passive electrical properties of the muscle membrane. [Fatt and Katz, 1951.]

Box 5.1   Identification of Chemical Transmitters

The chemical identity of the neurotransmitter at many synapses is not known, particularly in the central nervous system. The list of possible neurotransmitters is large and includes the following substances: acetylcholine, norepinephrine, dopamine, γ-aminobutyric acid, glutamate, glycine, 5-hydroxytryptamine, histamine, and even adenosine triphosphate. There is evidence, in varying degree, that all of these compounds function as neurotransmitters (for a summary tabulation, see Altmann and Dittmer, 1974, Table 141). The best way to verify that a substance functions as a neurotransmitter is to collect a measured quantity of the substance after stimulation of the presynaptic neuron and to demonstrate that this quantity, when applied to the postsynaptic cell, is sufficient to account for the observed postsynaptic effects. This has not been done for any transmitter, although the studies with acetylcholine at vertebrate neuromuscular junctions come close to fulfilling this criterion. Other criteria that have proved useful in establishing whether a substance is a neurotransmitter include the following:

Anatomical. The radioactively labeled compound or its precursors should be shown by autoradiography to be present at presynaptic nerve terminals.

Biochemical. The substance itself, as well as enzymes for its synthesis, must be present in the presynaptic neuron. An enzymatic or other process that will remove or inactivate the transmitter should operate in the synaptic region.

Physiological. The postsynaptic effects of the substance should be identical to those of the neurally released neurotransmitter.

Pharmacological. Compounds that block synthesis, storage, or release of the substance should also block synaptic transmission. Agents blocking the inactivation enzymes or process should prolong transmitter action, and those altering the postsynaptic sensitivity to the substance should act similarly on the natural transmitter.

The value of these criteria is primarily that, if met, they can increase the probability of a substance being the natural transmitter. If not, they can rule out consideration of the compound.

Figure 5.6
Temporal summation.

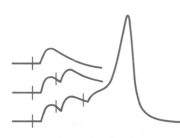

With repetitive stimulation, subthreshold end-plate potentials can sum and eventually reach threshold for a propagated action potential.

Figure 5.7
Intracellular recording of spontaneous miniature end-plate potentials from frog muscle.

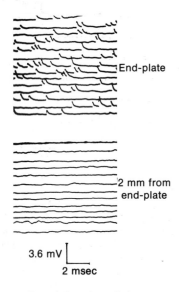

End-plate

2 mm from
end-plate

3.6 mV

2 msec

Top. *Recorded at the end-plate.* **Bottom.** *Same recording conditions but electrode 2 mm from end-plate.* [*Fatt and Katz, 1952.*]

its duration longer. These changes in e.p.p. are similar to those observed if one produces an electrotonic potential on the muscle fiber by passing current through an electrode. We can recall from the previous chapter that because of the cable properties of the fiber, a local potential change in an axon decreases in amplitude and increases in duration as an electrode is moved away. These properties are determined by the passive electrical behavior of the cell. Thus the end-plate potential is a local response of the muscle fiber produced at the end-plate, and its spread along the muscle fiber is passive; a few millimeters away from the end-plate, the potential is not recorded. This behavior contrasts with that of the action potential, which is of constant amplitude and duration all along an axon or muscle fiber.

A second important property of the synaptic potential, as indicated above, is that it may be **graded in amplitude** by varying the sensitivity of the muscle membrane with different amounts of curare or by varying the amount of ACh released from the nerve by changing the Ca^{++} and Mg^{++} concentrations. In addition there is no necessary refractory period to the synaptic potential. By repeatedly stimulating the nerve, one can cause the e.p.p.'s to add to one another (Fig. 5.6). If this process is continued the muscle will become depolarized to threshold, and an all-or-none action potential will be produced. This phenomenon of **temporal summation** is important because it permits a neuron in the central nervous system to add up all the excitatory influences on it from various nerve endings and eventually to produce an action potential.

D. Quantal Release of Synaptic Transmitters

It is now known that chemical synaptic transmitters are released into the synaptic cleft in "packets" containing many thousands of molecules. In the absence of action potentials in the presynaptic axon, the packets of transmitter "leak" out of the axon, producing spontaneous small postsynaptic potentials that occur at random intervals. An action potential invading the presynaptic terminal causes the synchronous liberation of a large number of packets. As the single multimolecular packet represents the minimum amount of transmitter to be released, the packets have been termed "quanta"; the number of quanta released by an action potential is the "quantum content." This general scheme of synaptic transmitter release has been found at all well-investigated chemical synapses, from invertebrate neuromuscular junctions to the brains of mammals. The mechanism operates with a variety of synaptic transmitters, including acetylcholine, norepinephrine, and possibly γ-aminobutyric acid and glutamate. The decisive experiments that led to the quantal hypothesis were first performed at the frog neuromuscular junction.

The initial observation of Fatt and Katz in 1952 was that **spontaneous, small random depolarizations** of the muscle fiber were recorded with an intracellular electrode in the region of innervation, the motor end-plate (Fig. 5.7). The potentials decreased in amplitude and increased in duration as the electrode was inserted at greater distances from the end-plate, and they were abolished by the addition of curare in a graded fashion. Thus these spontaneous potentials were found to have the spatial distribution and sensitivity to curare of the neurally evoked e.p.p. Because of their small size they are called minature e.p.p.'s.

It was considered possible that miniature e.p.p.'s might arise from the spontaneous leakage of single ACh molecules from the nerve terminal. This was ruled out, however, because curare was shown to decrease their amplitude in a graded fashion and because the addition of ACh to the preparation results in a continuously graded depolarization rather than an increase in frequency of miniature e.p.p.'s. Present evidence indicates that a miniature e.p.p. is the result of the **simultaneous release of 1000 to 10,000** molecules of ACh. A strong piece of evidence that miniature e.p.p.'s result from activity in the presynaptic nerve is that a small depolarization of the terminal portion of the axon increases their frequency of occurrence.

E. Transmitter Release by the Nerve Action Potential

The finding that small depolarizations of the nerve terminal increase the frequency of miniature e.p.p.'s suggests that a large depolarization, such as occurs during the presynaptic spike, might transiently produce an enormous increase in frequency and give rise to an end-plate potential. Using potassium to depolarize the nerve terminal, it has been found that depolarizing the nerve by 30 mV increases the frequency of miniature e.p.p.'s by a factor of 100. By extrapolation, a brief 100 mV depolarization of the nerve should produce an enormous

transient increase in the frequency of miniature e.p.p.'s. At the normal frog neuromuscular junction it has been estimated that the presynaptic action potential causes the **release of 100–300 quanta** of transmitter. Good evidence that the nerve impulse does in fact release ACh in the form of discrete quanta was obtained in experiments in which the release of ACh was reduced to a low level by lowering the Ca^{++} concentration and raising that of the Mg^{++} in the bathing fluid. Under these conditions it was found that the action potential in the nerve produces end-plate potentials whose amplitude fluctuates greatly from one impulse to the next. Such a result is shown in Figure 5.8. The important finding is that the amplitude of the postsynaptic potential is not continuously variable. Rather, it can be seen that sometimes the response is about 1 mV, sometimes 2 mV, and that often there is no response. By recording the amplitude of the spontaneous miniature e.p.p.'s, it became apparent that the smallest observable response to nerve stimulation was of the same amplitude as that of the miniature e.p.p. and the next smallest twice that amplitude. It is as if one had a box of marbles with a valve on the bottom. If the valve were opened for a brief period of time, one, two, or perhaps no marbles might fall out upon successive openings of the valve. One cannot obtain 0.3 or 1.6 marbles, and in the same way the axon puts out packets of ACh in whole units. Of course, there are small variations in the size of the spontaneous potentials, as one might expect in a biological system; these arise because of variations in the number of molecules of ACh per packet, differences in the sensi-

Figure 5.8
Evidence that the spontaneous miniature end-plate potential is the basic building block for the neurally evoked end-plate potential.

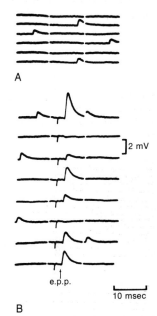

A
]2 mV

e.p.p.

10 msec

B

A. *Record of spontaneous miniature end-plate potentials recorded from rat diaphragm muscle in low calcium and high magnesium medium.* **B.** *The nerve was maximally stimulated once during each trace at the time indicated by a tick, causing an e.p.p. at the arrow. The postsynaptic response is not continuously graded; sometimes there is none. Those that do occur appear to be singles or multiples of the spontaneous potentials.* [Liley, 1956.]

tivity of different patches of the postsynaptic membrane, and for other reasons. By contrast, if one places a microsyringe with ACh near the end-plate, one can produce identical depolarizations with the identical doses and vary the amplitude of the postsynaptic response in a continuously variable manner by varying the amount injected. This result shows that the quantal nature of the end-plate response is a property of the way in which ACh is released from the nerve rather than the reactivity of the postsynaptic membrane. These results suggest that the action potential in the axon does lead to the release of packets of ACh identical to those released spontaneously in the absence of nerve impulses. Comparable results for other neurotransmitters have been obtained at crustacean neuromuscular junctions and mammalian spinal cord, in which neither a change of ionic environment nor the addition of blocking drugs is necessary to reveal the quantal process.

F. Statistical Test of Quantum Theory

In its original form, the quantum hypothesis assumed that there were a large number of packets of transmitter in the presynaptic terminal. At rest the probability of a packet being released is low; with an action potential in the presynaptic axon, the probability is greatly increased. If the probability of release in a series of trials is p and the total number of packets available for release is n, the average number of packets released in a series of trials, m, is given by $m = np$. The average number of quanta released can be determined in a **number of ways.** One method is to determine the mean amplitude of a

spontaneous miniature e.p.p. and that of the evoked e.p.p. The quantum content is given by

$$m = \frac{\text{mean amplitude of e.p.p.}}{\text{mean amplitude of miniature e.p.p.}}$$

Thus if the miniature e.p.p. is 1 mV and the average size of the e.p.p. 3 mV, then the quantum content is $3/1 = 3$. If p is small the distribution of e.p.p. amplitudes should follow the predictions of Poisson distribution. The assumptions in the Poisson distribution are that the number of units available, n, is large, the probability, p, of the release of any of the units is small, and the release of one unit is independent of the release of other units. The general equation for the number of responses containing x quanta is given by

$$n_x = \frac{Ne^{-m} \, mx}{x!}$$

where N is the number of trials and x the number of quanta in a given response. The number of trials giving zero response in $r_0 = Ne^{-m}$. Thus a **second method** of determining quantum content, m, is simply to count the total number of trials, N, and the number of times there was no response, n_0. Rearranging, we get $m = 1n \, N/n_0$. If our hypothesis is correct the quantum content calculated by the two methods should be identical. Figure 5.9 shows a comparison of the two methods of calculating m at a neuromuscular junction, and the agreement is quite satisfactory. A complete analysis of the amplitude distribution of e.p.p.'s and the theoretical predictions of the quantum hypothesis are shown in the example of Figure 5.10. All the results are in satisfactory agreement with the hypothesis that the synaptic transmitter is

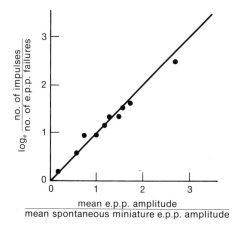

Figure 5.9
Comparison of two methods for calculating the quantum content of the end-plate potential. The line corresponds to complete agreement between methods. Note that the methods are independent. In one (data plotted on the ordinate), the number of stimuli and the number of failures are counted. In the other (data plotted on the abscissa), the amplitudes of the spontaneous and evoked end-plate potentials are measured. [Del Castillo and Katz, 1954.]

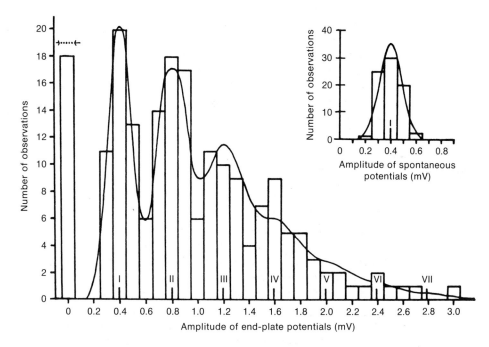

Figure 5.10
Amplitude distributions of spontaneous (*inset*) and evoked end-plate potentials in mammalian muscle, transmission partially blocked by increased magnesium. Peaks in the end-plate amplitude distribution occur at multiples (I, II, III, IV, etc.) of the spontaneous potential amplitude. The smooth curve is the expected distribution obtained by calculating the quantum content from the relation mean e.p.p./mean spontaneous e.p.p., with the assumption that the amplitudes of the individual quanta are distributed like those of the spontaneous potentials. The arrows above 0 amplitude indicate that 19 failures were expected; 18 were actually counted. [Boyd and Martin, 1956.]

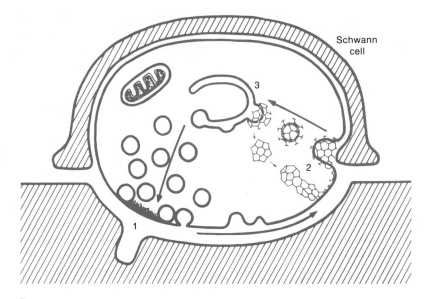

Figure 5.11
Summary diagram of hypothesis for transmitter release at the neuromuscular junction, based on electron-microscopic observations. *1*, vesicles coalesce with outer nerve terminal membrane and discharge their contents; *2*, equal amounts of membrane are pinched off in region of Schwann cell and appear as coated vesicles; *3*, coated vesicles then coalesce to form cisternae, which eventually reform new synaptic vesicles. [Heuser and Reese, 1973.]

released in the form of multimolecular packets. One or more of the above statistical tests have been applied at a variety of synapses, including those of the crayfish neuromuscular junction, frog sympathetic ganglion, sympathetic innervation of mammalian smooth muscle, and cat spinal motorneuron. The results in all cases have been consistent with the quantum hypothesis.

The evidence for quantal release of synaptic transmitters coincided with the finding by electron microscopists of synaptic vesicles in the terminals of presynaptic axons. The tentative hypothesis was suggested that the **vesicles are the structural basis** for quantal transmission.

That is, each vesicle contains an amount of synaptic transmitter that produces a quantal postsynaptic response. One hypothesis for release is that the vesicle membrane fuses with the external membrane, as shown in Figure 5.11, and the contents of the vesicle are emptied into the synaptic cleft—a process termed exocytosis. Electron micrographs that have been taken of neuromuscular junctions show vesicles opening into the extracellular space; similar pictures have been made by freeze-etching techniques at nerve terminals on fish electroplax. In addition, there is evidence that transmitter substances are stored in association with synaptic vesicles; differential centrifugation studies have found the concentration of transmitter to be highest in the vesicle fraction of nervous tissue.

G. Synaptic Delay at Chemical Synapses

Additional evidence for the interposition of a special series of processes at chemical synapses is an interval between depolarization of the presynaptic terminal and the earliest appearance of a postsynaptic response. This interval is called the synaptic delay (Fig. 5.12). At the frog end-plate the delay for the appearance of the postsynaptic response to a single packet of ACh is at least 0.4 msec. When ACh is directly applied to the end-plate through a micropipette, the delay between release and postsynaptic response is about 0.1 msec. As the external pipette is at least as far from the receptors as the nerve terminal, this value is the upper limit of the time necessary for ACh to diffuse across the synaptic gap and react with the postsynaptic cell. Thus most of the **synaptic delay is due to** a delay between the depolarization of the nerve terminal and the increase in probability of a packet of ACh being released.

The time for synaptic delay has been used in the study of reflexes in the central nervous system to **estimate the number of synapses** in a reflex pathway. To make this estimate one measures the total time for the reflex, subtracts the time for axonal conduction, and divides the remainder by 0.5 msec. For example, if total reflex time is 6 msec and conduction time in the axons is estimated to be 4.5 msec, the pathway is presumed to have three synapses. Such measurements are at best approximate because, not only does the synaptic delay vary, but

the time for conduction also varies with axon diameter, hence the exact conduction time in the fine branches of the axon terminals can only roughly be estimated.

H. Control of Transmitter Release

We will now consider how the presynaptic action potential leads to release of the chemical transmitter. In order to analyze these processes, it is an advantage to study a synapse in which one can control the membrane potential of the presynaptic axon and at the same time measure the amount of transmitter coming out of the terminal. Such a synapse has been found in the stellate ganglion of the squid, *Loligo,* and has been the subject of much study. At this synapse it is possible to place two microelectrodes in the presynaptic terminal, one to record the membrane potential and the second to pass current to set the membrane potential at a given level. A third intracellular electrode is inserted into the postsynaptic neuron to measure indirectly the amount of transmitter released by measuring the amount of depolarization produced. By bathing the ganglion in tetrodotoxin, action potentials are eliminated. This compound selectively prevents the increase in g_{Na^+} following depolarization, and so prevents the occurrence of all-or-none action potentials. It does not interfere with the release of the transmitter or the sensitivity of the postsynaptic cell to its actions. Under these conditions the presynaptic terminal can be progressively depolarized and the resultant postsynaptic potential recorded. Figure 5.13 shows the relation between presynaptic and postsynaptic membrane potentials. The **nonlinearity of the relation** is important because it

Figure 5.12

With a microelectrode positioned immediately over a synaptic region, one can record both the action potential current in the presynaptic axon and the postsynaptic current. The period between the two events is termed the synaptic delay. [Katz and Miledi, 1964.]

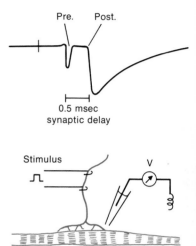

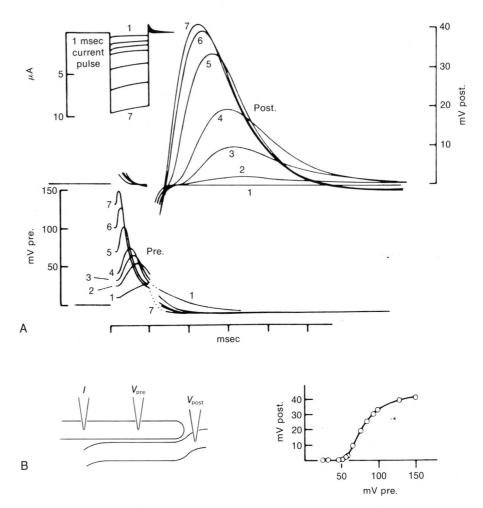

Figure 5.13
Relation between the presynaptic depolarization produced by a 1 msec depolarizing current pulse and the postsynaptic response at squid giant synapse. **A.** Seven superimposed tracings of each. **B.** Diagram of the position of the voltage-recording and current-passing electrodes and a graph of the relation between presynaptic and postsynaptic depolarization. [Katz and Miledi, 1967.]

indicates that small changes in presynaptic membrane potential can greatly affect transmitter release. Note that the relation is continuously graded. The result in the presence of tetrodotoxin is of particular interest in that as the increase in g_{Na^+} is blocked and transmitter is still

normally released, it would appear that transmitter release is not dependent on an influx of Na^+. Release of transmitter does depend on depolarization of the presynaptic terminal. Similar results have been obtained at the neuromuscular junction by producing graded depolarizations of the nerve terminal with an external current-passing electrode. At the neuromuscular junction, however, it is not possible to record the presynaptic membrane potential directly, as it is too small to be penetrated by a recording microelectrode.

I. Calcium Entry and Transmitter Release

Calcium ions have long been known to be essential for the release of synaptic transmitters. When the calcium concentration in the bathing medium is greatly reduced, transmission fails at all chemical synapses studied, including those of sympathetic ganglia, vertebrate and invertebrate neuromuscular junctions, and the squid giant synapse. When calcium is removed from the neuromuscular junction, the nerve impulse invades the terminal but fails to increase the probability for release. The action of calcium to release transmitter is shared, to some extent, with Sr^{++} and Ba^{++} and is antagonized by Mg^{++} and Mn^{++}.

It has been suggested that the entry of calcium into the nerve terminal increases the probability that a collision of a packet of transmitter with the inner surface of the membrane will lead to release. According to this hypothesis, depolarization of the membrane leads to the entrance of Ca^{++} in a way analogous to that which causes an increase in g_{Na^+} and subsequent entrance of Na^+. In fact, it has been found in squid axon that a small amount of calcium enters the axon with each action potential. The normal external Ca^{++} concentration is 10^{-3} to 10^{-2} M, and the internal concentration less than 10^{-5} M. The electrochemical gradient, therefore, favors the entrance

of calcium if $g_{Ca^{++}}$ is increased. If the membrane potential is clamped at E_{Na^+} in a squid axon, there is no net Na^+ entry. At the squid synapse it has been found that with very large depolarizations, transmitter release is suppressed during the depolarization, and occurs when the pulse is turned off. Such observations are in accord with the view that the inward movement of calcium is a part of the **coupling process between depolarization and transmitter release.** During depolarization to $E_{Ca^{++}}$, there is no net inward flux of calcium; when the pulse is turned off the calcium conductance remains elevated for a time, leading to transmitter release. As Figure 5.14 shows, injection of a small quantity of calcium into the presynaptic terminal is sufficient to release packets of transmitter.

J. Transmitter Effects on Postsynaptic Membranes

Following release from the presynaptic terminal, the synaptic transmitter diffuses across the synaptic cleft, some 200–500 Å in width, and reacts with specific molecules on the postsynaptic cell that are termed **receptors.** This process takes at most a fraction of a millisecond to occur. The nature of the receptive molecules is only partly understood. What is known from studies in which transmitters are applied to various portions of the postsynaptic cell is that the receptors are located in the vicinity of the nerve terminals and that the sensitivity to the transmitter is markedly reduced, by 1000 times or more, at nonsynaptic regions of the cell. We have seen in the previous chapter that potential changes across nerve membranes

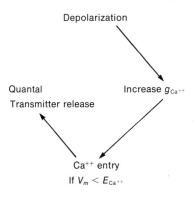

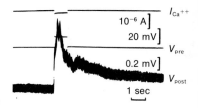

Figure 5.14

Postsynaptic depolarization (lower trace) resulting from the injection of a pulse of Ca^{++} (upper trace) into a presynaptic axon of squid. The preparation was bathed in 11 mM manganese solution, so that presynaptic depolarization (middle trace), per se, does not cause transmitter release. [Miledi, 1973.]

are the result of ionic movements resulting from permeability changes in the membrane. We can now ask what permeability changes are responsible for postsynaptic potentials. The most direct way to determine this is to measure the movement of radioactive tracer ions across the postsynaptic membrane in a manner analogous to the way in which Na^+ and K^+ movements were measured in axons during the action potential. However, as the area of membrane acti-

---

Box 5.2 Acetylcholine Receptors

Attempts to characterize, identify, localize, and isolate the acetylcholine receptor have been greatly facilitated by the discovery and purification of a neurotoxin from snake venom, α-bungarotoxin. This polypeptide (molecular weight 8000) binds selectively and irreversibly to acetylcholine receptors on vertebrate skeletal muscle, blocking neuromuscular transmission. The neurotoxin may be (a) labeled with radioactive iodine to permit autoradiographic localization or isolation of the receptor; (b) conjugated with fluorescent dyes to allow visualization of the receptor in living or fixed tissue in the light microscope; or (c) bound to peroxidase using immunological techniques to permit receptor localization in electron micrographs. The figure illustrates the way in which fluorescent-labeled α-bungarotoxin can be used to determine the distribution of acetylcholine receptors at the neuromuscular junction.

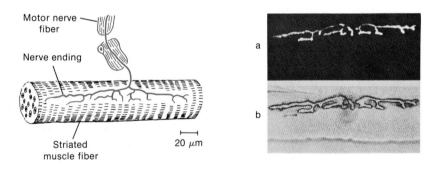

Distribution of acetylcholine receptors at the neuromuscular junction. **Left,** Diagram of an amphibian neuromuscular junction. **Right.** (*a*) Photograph with fluorescence optics to show the distribution of tetramethylrhodamine-labeled α bungarotoxin. (*b*) The same junction photographed using bright-field optics following a histochemical procedure to show the distribution of the enzyme cholinesterase (the enzyme that inactivates acetylcholine). Note that both the acetylcholine receptors and the enzyme cholinesterase are most concentrated at the region of nerve-muscle contact. The acetylcholine receptor and the enzyme cholinesterase are known from other studies to be separate molecules. [Anderson and Cohen, 1974.]

vated by the synaptic transmitter is very small, such measurements have not often been possible.

Instead, electrophysiological techniques have been used. The method involves **determining the equilibrium potential** for the permeability change produced by the reaction of the transmitter with the postsynaptic membrane. It should be recalled that the equilibrium potential for an ion is that membrane potential at which there is no net electrochemical force tending to make the ion move across the membrane. Thus, when the membrane potential equals the equilibrium potential for the ion, an increase in permeability to that ion will not result in current flow or a potential change. Figure 5.15 shows the results of an ex-

periment designed to determine the equilibrium potential for the action of a transmitter on a frog neuron. When the membrane potential of the neuron was increased by passing current through an intracellular electrode, the amplitude of the postsynaptic potential increased. Decreasing the membrane potential from the rest level decreases the amplitude of the postsynaptic response. The level at which the synaptic transmitter produces no response is called the equilibrium potential, or reversal potential, for the transmitter action. The latter term is appropriate because, if the membrane potential is further changed, the sign of the postsynaptic response reverses. In order to determine which ions carry the current responsible for the synaptic potential, it is desirable to measure the synaptic current directly, determine the equilibrium potential for the current, and vary the concentrations of different ion species to see if they influence the equilibrium potential.

Excitatory Synapses. At the frog neuromuscular junction it is possible to insert two microelectrodes, one to record the membrane potential and a second to pass current. With the experimental setup shown in Figure 5.16, one can voltage-clamp the membrane at a given level and measure the synaptic current resulting from the change in permeability produced by the transmitter. It is seen in Figure 5.16 that hyperpolarization of the cell increases the synaptic current, whereas depolarization decreases the synaptic current. An extrapolation of the data indicates that at a membrane potential of -15 mV the synaptic current would be zero. The linear relation between synaptic current and membrane potential and the occurrence of an equi-

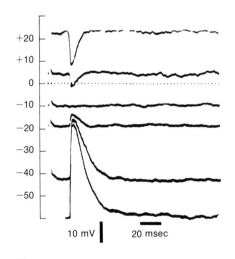

Figure 5.15
Determination of equilibrium potential for the action of the synaptic transmitter at an excitatory synapse on a frog parasympathetic ganglion cell. Two intracellular electrodes are used, one to pass current to set the membrane potential and the second to record the potential change produced as a result of nerve stimulation. The equilibrium potential for the synatpic transmitter in this case occurred at -12 mV. [Dennis et al., 1971.]

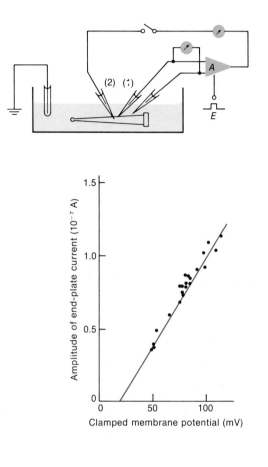

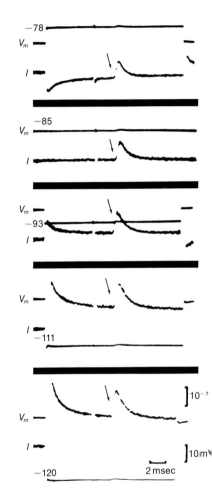

Figure 5.16

Upper left. Experimental arrangement for measuring the amplitude of the end-plate currents at various membrane potentials. The recording microelectrode (1) is connected to a differential feedback amplifier whose output is connected to the current electrode (2) in such a way that the current maintains the membrane potential at the command voltage (*E*). When the nerve is stimulated and the transmitter produces a conductance change in the postsynaptic membrane, the additional current that flows in the circuit to maintain the membrane potential at the command voltage is of the same amplitude as the synaptic current that would flow in the absence of the voltage clamp. **Right.** Records of end-plate currents (arrows) in response to nerve stimulation recorded at indicated membrane potentials. **Lower left.** Relation between clamped membrane potential and end-plate current amplitude. The equilibrium potential is obtained by extrapolating the line to the membrane potential at which the current is zero; in this case, about −15 mV. [Takeuchi and Takeuchi, 1960.]

librium potential at which there is no synaptic current are in marked contrast to the results of voltage-clamp experiments conducted on the action potential, discussed in Chapter 4. The **linear relation between current and voltage** indicates that the increase in membrane conductance produced by the transmitter does not vary with membrane potential. The existence of an equilibrium potential for the total current rules out the possibility that we are concerned with a sequential change in the conductance to more than one ion species, such as occurs for Na^+ and K^+ during the action potential.

The next question to be considered is the **nature of the permeability increase** during the transmitter action. To study this problem, the Takeuchi's varied the concentrations of the major ions in frog Ringer solution (Na^+, K^+, and Cl^-) and measured the effect of these changes on the equilibrium potential. Figure 5.17 shows the effect of varying the concentration of Na^+ or K^+. It can be seen that the equilibrium potential becomes more negative when the concentration of Na^+ or K^+ is reduced. Changes in the Cl^- equilibrium potential produced by varying the external Cl^- concentration had no effect. Suppose that the transmitter were to produce an identical conductance increase to both Na^+ and K^+. In this case the equilibrium potential would be halfway between the equilibrium potentials for Na^+ and K^+ (see p. 138). For frog muscle, $E_{Na^+} = +50$ and $E_{K^+} = -100$; $(+50 - 100)/2 = -25$ mV. As the equilibrium potential was found to be -15 mV, the best fit for the data suggests that the Na^+ conductance is increased 1.29 times as much as the K^+ conductance. Thus the transmitter appears to make the postsynaptic mem-

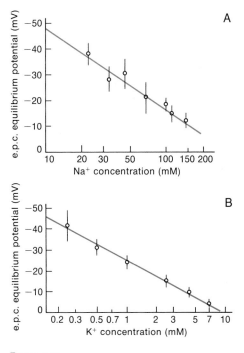

Figure 5.17

Relation between external concentrations of sodium **(A)** and potassium **(B)** and the equilibrium potential for the end-plate current determined as in Figure 5.16. The straight lines are drawn according to equations that assume the internal concentrations of the ions remain constant and that the increase in sodium conductance is 1.29 times the increase in potassium conductance during the action of the transmitter. [Takeuchi and Takeuchi, 1960.]

brane permeable to Na^+ and K^+ but not to Cl^-. Additional studies have shown that the Ca^{++} conductance is also increased by ACh. Normally the external Ca^{++} is low, so that the current contributed by calcium is small. It should be noted that the depolarizing effect of the transmitter would be more efficient if only the conductance to sodium were increased rather than that to both sodium and potassium; the potassium current only serves to reduce depolarization.

One can only speculate that the simultaneous increase evolved as part of the mechanism for opening and closing the ionic channel. In fact, at the frog neuromuscular junction, acetylcholine also increases the conductance of the postjunctional membrane to ammonium and its derivative hydrazinium, tetramethylammonium, and lithium. In other systems—for example, in some crayfish neuromuscular junctions—the increase in conductance induced by the excitatory transmitter appears to be much more selective for sodium ions.

Amplification at Chemical Synapses. One of the real advantages of chemical synaptic transmission is the possibility for amplification. At electrical synapses sufficient current must flow in the presynaptic terminal to depolarize the postsynaptic cell to the firing threshold. At the neuromuscular junction, for example, the fine twig of an axon could not possibly supply sufficient current to depolarize the large postsynaptic muscle fiber. To depolarize a frog skeletal muscle fiber to threshold requires a charge transfer of about 10^{-9} coulomb. This is accomplished by the net movement of about 10^{-14} mole of a univalent ion (10^{-9} coulomb/10^5 coulomb/mole). Because the area of the presynaptic axon in contact with the muscle fiber is small (less than 10^{-4} cm^2), the total movement of univalent ions across this membrane during an action potential is 10^{-12} M/cm^2 $\times$ 10^{-4} cm^2 = 10^{-16} M, which is 100 times less than that necessary to depolarize the muscle fiber. A small quantity of **acetylcholine opens channels** in the postsynaptic membrane to Na$^+$ and K$^+$, taking advantage of the energy stored across the postsynaptic cell in the form of ion concentration gradients. It is estimated that a single molecule of acetylcholine combining with a receptor opens a membrane channel for 1 msec and leads to a net charge transfer quivalent to about 50,000 univalent ions.

Inhibitory Synapses. Inhibitory synapses are probably just as common and important as excitatory synapses. To study the nature of the effects of inhibitory synaptic transmitters on the postsynaptic cell, one can isolate an inhibitory axon and stimulate it while recording the membrane potential from the postsynaptic cell. It is found that stimulation of inhibitory axons produces either a slight depolarization, a slight hyperpolarization, or, in some cases, no observable change in potential in different cells at rest. In all cases in which the cell is depolarized by stimulation of the excitatory axon, inhibitory stimulation produces a decrease in the depolarization.

The explanation for these results was obtained in an experiment illustrated in Figure 5.18. It is clear that the equilibrium potential for the action of the inhibitory transmitter is close to the resting potential. We know that the equilibrium potentials for both potassium and chloride are close to the resting potential, so that it is natural to ask if the permeability to either or both of these ions is affected by the inhibitory transmitter. By varying the concentration—and therefore the equilibrium potentials of each ion separately at crustacean neuromuscular junctions—it has been found that changes in potassium produce little if any change in the **inhibitory equilibrium level.** Changes in Cl$^-$ concentration produce changes in the inhibitory synaptic potential, however, which is consistent with the view that the inhibitory transmitter increases the permeability of the

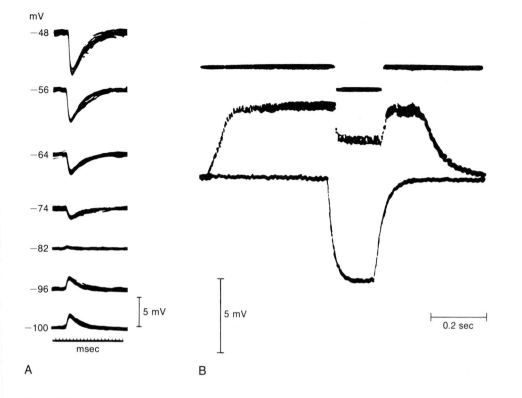

Figure 5.18
A. Inhibitory postsynaptic potentials recorded from a cat spinal motor neuron at various levels of membrane potential. The equilibrium potential for the conductance change produced by the transmitter was about −80mV; resting potential −74 mV. [Coombs et al., 1955.] **B.** Evidence for a conductance increase during inhibitory junctional potentials in crayfish. The lower trace indicates the resting potential with an electrotonic potential produced by a current pulse of 1.5×10^{-8} A monitored on the top trace. The middle trace shows the depolarization produced by stimulation of the inhibitory axon at 150 impulses/sec with an electrotonic potential produced by the same current pulse. The decrease in size of the electrotonic potential from 7 mV to 2 mV during inhibitory stimulation indicates a marked increase in membrane conductance. [Dudel and Kuffler, 1961.]

postsynaptic membrane to chloride. Identical permeability changes in crustacean muscle can be produced by the addition of γ-aminobutyric acid to the subsynaptic membrane. There is now good evidence that this substance is the inhibitory transmitter at crustacean neuromuscular junctions.

At first view one might be alarmed at the existence of depolarizing inhibitory synaptic potentials. The essential feature is that, for a transmitter to be inhibitory, it is only necessary that its equilibrium potential be more negative than the threshold for excitation.

A second system in which the inhibitory synaptic transmitter appears to increase the permeability of the postsynaptic cell, primarily to Cl⁻, is the mammalian spinal cord. In this system, it

specialized cells connected in specific ways. The neuron is like a miniature person—having a personality, having an array of unlike parts, having actions both spontaneous and upon stimulation. Its actions depend on the convergence of steady states, transient events, built-in weighting factors, and intrinsic tendencies; it speaks finally with one voice, which integrates all that went before.

II. THE NEURON AS RECEIVER AND FILTER: MECHANISMS OF EVALUATION

The neuron doctrine has had a quiet revolution, and it is stronger than ever in its new guise. As was set out in Chapter 2, the structures mediating the nervous functions are nerve cells and their processes. These are separated from other neurons by their own membranes and intercellular space, and come into orderly kinds of contiguity with them in a variety of specialized contacts. The neuron is a unit, embryologically, trophically, and anatomically, but the revolution led to an understanding that it need not act as a whole functionally—that its parts possess a degree of independence. The neuron is dynamically polarized in most cases, both by one-way synaptic transmission and by unequal spread in different directions within the cell; excitation is usually received by dendrites, and spreads to and down the axon. But an all-or-none impulse need not travel out the dendrites or even occur in some neurons at all. The doctrine permits crucial integrative roles to be played **not only by synapses** but by dendrites, somas, initial and terminal segments of axons, and branch points. The doctrine

does not exclude the possibility that important contributions to nervous integration are made by nonnervous cells, such as the neuroglia.

A. A Frame of Reference: Loci and Modes of Lability

Figure 6.1 shows most of the known forms of neuronal activity associated with information processing (over and above the resting metabolism, cytoplasmic movements, cellular syntheses, and the like). The majority are measured as membrane potential changes and are graded; only one, the nerve impulse, or spike, is all-or-none. Other forms of nerve cell activity are also important in integration. Neurosecretion is a major one. Movement, or change in dimensions, especially of axon endings or dendrites, is perhaps a major one, though not yet well known.

Each form of activity (i.e., each response) means, or manifests, a distinct form of excitability, a coupling function for that form of output (Fig. 6.2). Furthermore, each given form may, in different neurons or in different loci in a single neuron, differ in sensitivity or in responsiveness. "Sensitivity" refers to the stimulus intensity necessary for a given percent of available response, whether threshold, maximal or an intermediate. "Responsiveness" means the actual intensity of response for a given percent of maximal stimulus. We have met all these types of response potential before, but are now re-examining them from the point of view of their contribution to an adjustable, flexible, evaluative input-output relation.

Generator potentials (p. 168) may show a linear dependence on the logarithm of

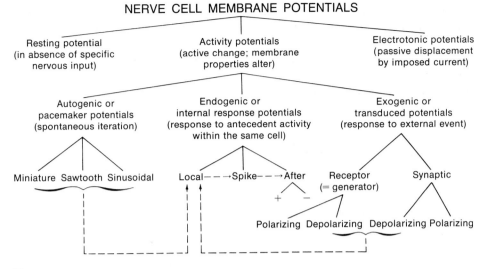

Figure 6.1
Diagram of the types of nerve-cell membrane potentials, with special reference to the potentials of activity. Arrows indicate that a sufficient level of one potential may cause the initiation of another. Plus and minus signify that there are those that increase the membrane potential and those that decrease the membrane potential, equivalent to polarizing (or hyperpolarizing) and depolarizing. [Bullock and Horridge, 1965.]

stimulus intensity over a certain range or they may exhibit some other dependence. There is also great diversity in the properties of the generator potential as a filter with respect to time; some cut off low stimulus frequencies and pass only high ones; others are low-pass, high-cutoff filters. **Synaptic potentials** (p. 181) are formally equivalent to generator potentials. Their excitability curve (presynaptic potential versus p.s.p.) can be extremely nonlinear (Fig. 5.13,B) and can be different for separate synapses on a neuron. **Pacemaker potentials** (p. 152) are truly spontaneous in that they do not depend on any external timer or trigger, but can tick like a watch by themselves. Nevertheless they are sensitive to steady conditions such as ionic milieu, osmotic and hormonal states, diffuse field potentials, maintained depolarization and temperature. When one of these is a relevant variable in a situation, we may call it a stimulus and to that extent what was spontaneous becomes a response.

A form of excitability at least as important as any other, but often neglected, is the slope of the dependence of **spike discharge rate** on the prevailing depolarization of a relevant part of the neuron, where spikes arise (Fig. 6.4); this slope varies among neurons and as a function of recent history. When spike discharge is repetitive, its frequency and pattern are not predictable from the single spike threshold but depend on multiple recovery processes and their changes with time. As we note elsewhere (p. 170) some neurons can fire only in brief bursts and are called phasic; others can sustain long

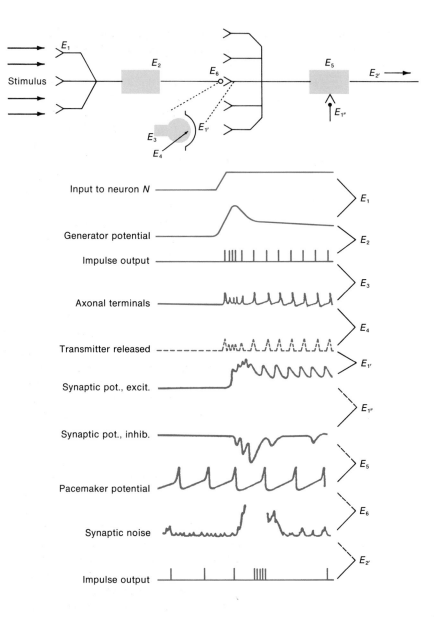

Figure 6.2
Schematic diagram and summary of events plotted against time (ca 0.1 sec shown), showing several successive stages in the transfer of information from one nerve cell to another. The E's represent the independent excitabilities, or coupling functions, and the sites of possible molecular participation. The boxes in the schematic represent two nerve cells of order N and $N + 1$; the synaptic contact between them is enlarged to show the E's involved. Release of transmitter is shown by a broken line, to indicate that the record is hypothetical. Broken leaders from some of the records indicate that the event just above that E is not the input for that E. [Bullock, 1968.]

trains of discharge and are called tonic. These properties strongly influence input-output relations to driving stimuli.

Local potentials and **afterpotentials** are graded labile functions of the antecedent activity. Afterpotentials vary greatly among neurons in sequence, amplitude, and duration. They influence integration by being the background on which successive inputs arrive.

Transmitter release contributes to integration by its nonlinear dependence on presynaptic depolarization and its time course. Moreover, these properties change with repetition and under the influence of presynaptic synapses. Different junctions, even on the same postsynaptic neuron, often differ markedly in these properties.

Each of these neuronal processes therefore has a coupling function—dependent to a greater or lesser degree on the present inputs and states, on any recent activity from which it is recovering, and on the relevant past (such as genetic and developmental background, instincts, learning, and plastic changes)—that has set the stage. Around these processes and their loci in the cell will center future research into molecular mechanisms for the effects of experience, drugs, hormones, changes in electrical fields, and the importance of intercellular spaces and neuroglia.

B. The Degrees of Freedom: Intracellular Permutations

Within this framework, we may review some of the parameters of integration within neurons. We ask at the intracellular level what variables are available to determine the output? These make the neuron a filter as well as a receiver.

Distribution on the Neuron of Types of Membrane. Integration depends on the weighting of inputs. One of the important factors in determining weighting is the area of synaptic contact; another is the proximity to the spike-initiating locus. The extensive **dendrites are the prime integrating structures,** the real seat of nervous function. Here electrotonic spread is the agent of mixing, smoothing, attenuating, delaying, and summing excitatory postsynaptic potentials (e.p.s.p.'s), inhibitory postsynaptic potentials (i.p.s.p.'s), and local and pacemaker potentials.

Consider such illustrations as Figures 2.2, 2.18, and 3.2 and the example in Figure 6.3. Different kinds of input gain access to the cell only through certain regions of it; that is, the spatial distribution of input is often not random but systematically restricted. Some imputs will enjoy a stronger influence than others because of more proximal access (closer to the soma), a higher density of junctions, or because the cell's dendritic branching pattern is more conducive to spatial summation. A **comparative microphysiology** correlated with microanatomy—barely possible technically today—should become a central subject in neurobiology. We know, for example, that in some cells spikes cannot invade the soma-dendrite region at all, whereas in other cells spikes invade some distance into the region; we know also that this is a labile property, subject to influence, for example, by axoplasmic transport-inhibiting drugs. Some types of synapse, such as those on retinal amacrine and bipolar cells are more plastic—that is, altered by experience—than others, such as those between retinal receptor and horizontal cells. Synapses on somas tend to be strategically placed

Figure 6.3
Schematic neuron, its functionally distinct regions (or loci), and its inputs. [*Alexandrowicz, 1932.*]

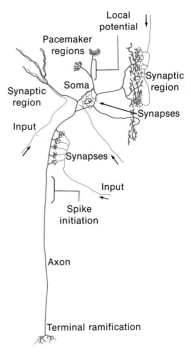

for inhibition. Patches of membrane of drastically different properties can be side by side—for example, patches of chemically and of electrically excitable membrane, or depolarizing and hyperpolarizing membrane. Responses from adjacent membranes—for example, e.p.s.p.'s and i.p.s.p.'s—can interact in different ways, at least partly dependent on the geometric relations. The wide variety of **characteristic geometries** of synapses is no doubt important not only for the weighting of different combinations of inputs, but in many ways we do not yet appreciate. The importance of cytoarchitecture is emphasized by many examples in Chapters 2, 3, and 9.

Forms of Excitation and Inhibition. As we have seen in earlier chapters, inhibition and excitation can result from any of several distinct causes. The following lists are not necessarily complete. **Inhibition** may be caused by (a) refractoriness following an impulse or train (sometimes called indirect inhibition); or (b) refractoriness following a subthreshold e.p.s.p. (like "Wedensky inhibition" in the axon); (c) specific transmitters that give rise to i.p.s.p.'s; or (d) forms of prolonged hyperpolarization distinct from i.p.s.p.'s and probably several in number; (e) any shunting conductance increase, even one causing a depolarization, if the synaptic potential is below threshold; (f) excess depolarization by e.p.s.p.'s; (g) diffuse, weak electric fields in the right orientation to depress pacemaker loci or hyperpolarize spike-generating loci; (h) certain specific electrotonic connections; (i) presynaptic inhibition.

Excitation also occurs in several ways not mutually equivalent and each potentially important in integrative properties (membrane mechanisms may overlap).

Among them are (a) depolarization to spike discharge threshold; (b) subthreshold depolarization that lowers the threshold for subsequent input; (c) suprathreshold maintained depolarization that increases the frequency of repetitive firing; (d) prolonged enhancement for up to several hours after a few minutes of excitatory input in certain pathways; (e) diffuse field potentials that influence other cells in the right polarity; (f) presynaptic excitation. It is important to realize that depolarization cannot be equated with excitation. In some places accommodation occurs (p. 151), and depolarization may cause little or no excitation. In some places facilitation occurs (see below), and excitation is accompanied by little or no visible change in depolarization. Furthermore, the tendency to iterate (fire repeatedly) is a property that varies widely (Fig. 6.4); some nerve cells characteristically fire in bursts or runs, others only a few times or singly, even in response to a prolonged depolarization. Not even the ability to develop regenerative action and fire an impulse is universal. The breadth of this repertoire bespeaks a range of dynamic properties that can determine output as some function of input in various ways in different neurons.

What happens when excitation or inhibition or both arrive from two different sources at the same time? **The interaction of overlapping synaptic potentials** from different sources can depart from the simple sum. Two subthreshold monosynaptic excitatory postsynaptic potentials in the motor neuron of the cat spinal cord add almost linearly, but inhibitory postsynaptic potentials do not; the inhibitory equilibrium potential is too close to the resting potential. Two polysynaptic e.p.s.p.'s may sum with a large deviation

Figure 6.4
Diverse input-output functions.

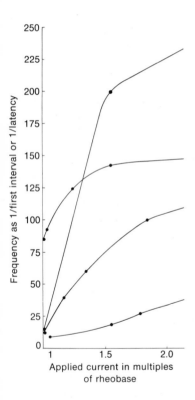

Four types of neurons differing in initial frequency of discharge to imposed direct current. [*Hodgkin, 1948.*]

Box 6.1 Differentiation Among Axons: Velocity Varies More than 2000-fold

In Chapter 4 we saw some properties that could influence the velocity of propagation of an impulse along an axon, but did not pause to consider the actual range of velocities among axons or the factors that may cause such differences.

True spike conduction can probably be slower than 0.1 m/sec in some axons and faster than 200 m/sec in others—both at 20°C. Each axon conducts at a consistent velocity, if rested. The spectrum of velocities shows peaks due to classes of fibers, and the relative abundance of the classes, usually measured as a fiber-size spectrum, is characteristic for each nerve and central tract. There are two nomenclatures for vertebrate nerves; tabulated below is one of them, with values for mammalian axons. (For more detail, see Altman and Dittmer, 1974, Table 138, pp. 1143–1149.)

Fiber type	Velocity (m/sec)	Diameter, incl. sheath (μm)
Aα*	70–120	12–20
Aβ	30–70	5–12
Aγ	15–30	3–6
Aδ	12–30	2–5
B	3–15	<3
C	0.6–2.2	<0.3–1.4

* Aα fibers are often called group I fibers, from another nomenclature that is little used otherwise.

Among the factors determining velocity, fiber diameter is an important one, since the larger the axon, the lower its internal resistance, and this resistance acts on velocity in the same way as external resistance, as we saw already in Chapter 4. But if we plot velocity against diameter for a wide array of invertebrate and vertebrate axons, we obtain at least half a dozen curves with positive slopes, displaced but roughly parallel to each other. Other factors must also play a role. Myelin sheaths have an important influence, as we saw in Chapter 4, but it is not readily evaluated numerically. Other factors may be lumped under the term "intrinsic," and could include the dependence of sodium conductance upon membrane potential, the time constant thereof, and the other membrane parameters that determine threshold and the rate of reaching it. The largest giant fibers (ca. 1.8 mm) are 1/3 to 1/7 as fast as some crustacean giants (up to 210 m/sec at 20°C). The lower limit of true spike propagation is not clear. (For more detail, see Bullock and Horridge, 1965, Table 3.4, p. 150.)

Axon differentiation affects not only velocity but other functionally significant properties. Spike duration approximately equals the absolute refractory period; afterpotentials even more roughly equal the relative refractory period, though this relation is actually complex. Accommodation (rising threshold membrane potential with slowly rising stimulus) is insignificant in well-studied axons, but may be important at some spike-initiating loci and axon terminals. Certain axons readily fire repetitively to a long maintained depolarizing stimulus, others only a few times or only once. The safety factor (transmembrane spike height/threshold depolarization from resting potential) is commonly 10 or 20 but may be low, especially locally—for example, near axon terminals, at some branch points, or early in the relative refractory period. When it falls below 1.0 the spike becomes a local potential and decrements with distance.

Figure 6.5
Diagram of two alternative consequences of repetition of synaptic input at a suitable frequency.

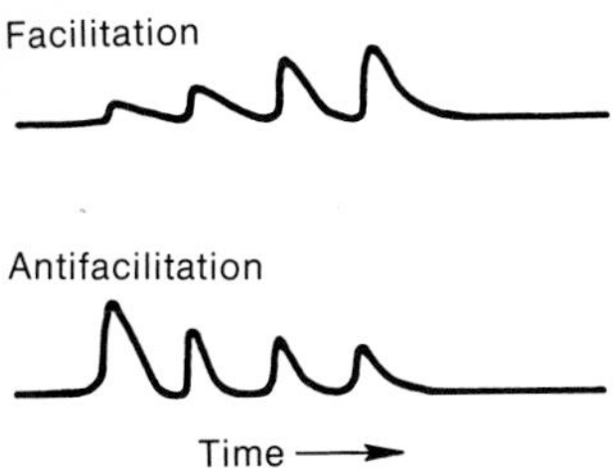

Facilitation

Antifacilitation

Time ⟶

Successive synaptic potentials may facilitate or antifacilitate.

from linearity that differs according to their sources and times of arrival. A monosynaptic e.p.s.p. that occurs during an inhibitory hyperpolarization may show no change in size, as seen by an electrode in the soma, though the spike that results may be delayed. The adequacy of shunting and familiar membrane potential dependent mechanisms to account for these properties has not been fully checked out, but their importance for us is the variety of integrative possibilities permitted by permutations of the cellular events.

The same dose of arriving excitation or inhibition—for example, the same stimulation of an input pathway to a postsynaptic cell—can **interact additively or multiplicatively** according to the amount of background activity. Additive interaction means that a constant *number* of impulses per second are added or subtracted from the background when the same dose of input is delivered on different levels of ongoing activity. Multiplicative interaction means that a constant *proportional* change from the background activity is caused. There are other relationships in other cases. For instance, we may consider the input-output curve for certain spinal motor neurons; the curve is plotted by adding to a constant, stretch-receptor input varying amounts of depolarizing current injected directly into the cell. The cell fires at a rate that increases with current in two linear ranges, a lower-frequency, "primary" range with a small slope abruptly changing to a higher-frequency "secondary" range with a steeper slope. Algebraic summation of inputs obtains in the primary but not in the secondary range. Facilitatory and inhibitory inputs to such cells have not only the control of firing rate but of the range and operating slope.

Facilitation and Antifacilitation. Another important integrative property, which differs greatly from junction to junction, is brought to light by repeating stimuli at various intervals (Fig. 6.5). In some junctions successive impulses arriving via a single presynaptic pathway cause increasing synaptic potentials, if the intervals are not too short or too long. This increase is said to be due to facilitation. On the basis of the curve of dependence on interval, we can characterize the process of facilitation as growing to a maximum following each impulse and then decaying. If the response—for example, a p.s.p.—lasts longer than the interval between impulses, so that the second response rises from a residue of the first, we say there is **temporal summation.** Only if the response increment due to the second impulse is larger than the preceding can we say there is **facilitation.**

Facilitation is found in quite diverse degrees, forms and durations. Its duration may be in the range of tens of milliseconds, or hundreds or thousands or even tens of seconds. Its rate of growth differs greatly also, and is not necessarily correlated with its rate of decay. It may saturate with a very few successive intervals or only with many. One consequence (referred to on p. 223) is sensitivity of some junctions to temporal pattern in impulse trains at the same mean interval. The same junction may have early- and late-developing facilitation.

Facilitation at long intervals may grade from forms of sensitization to learning. For example, in experimental studies on animal models of epilepsy, if a one-second stimulus train of 30 shocks/sec is delivered to a part of the limbic system (see Chapter 10) at a suitable, moderate intensity, there will follow only a slight electrical afterdischarge at first. If the

one-second stimulus trial is repeated once per day, it gradually elicits more afterdischarge ("**kindling**"), and on about the tenth trial (tenth day) an overt, convulsive seizure. The interval between trials may be even longer than a day, but if it is too short (e.g., 15 minutes) kindling does not occur.

A special case is **posttetanic potentiation,** or p.t.p. This name is given to a form of enhancement brought on by a long period (e.g., 10 sec) of high-frequency stimulation (e.g., 200/sec.). During the stimulation no enhancement may be seen, but afterwards test stimuli at various intervals reveal a marked increase in response, up to manyfold more than that before the "tetanic" stimulus.

When input converges upon one postsynaptic cell via two different presynaptic pathways—a situation discussed in the preceding section—it may sometimes be found that input *A* enhances the response to input *B* in a certain range of intervals. This is called **heterosynaptic facilitation.** It is typically irreciprocal (i.e., *B* does not facilitate *A*). It must be distinguished from spatial summation, which merely means that the responses to separate inputs add whether facilitated or not.

The mechanisms for facilitation are only partly known. It is likely that there are several. A probable explanation for certain cases, in terms of changes in the presynaptic terminals, in given in Chapter 5.

In some junctions successive impulses cause decreasing synaptic potentials; this is called antifacilitation. Its mechanism is even less well understood. There may be facilitation of some junctions and antifacilitation of others on the same neuron. Still others may show neither change.

In the cases permitting examination of the same junction in different specimens, the kind, degree, and time course of these enhancements and depressions are found to be dependable characteristics. It is likely, however, that this dynamic property is plastic in some situations (i.e., can be altered more or less enduringly).

In the frog spinal cord a monosynaptic response to controlled presynaptic stimulation shows p.t.p. to high-frequency input and classical **habituation** to low-frequency input. The synapses responsible are at least overlapping if not completely identical sets. The habituation is not due to presynaptic changes in this case, as it is in more familiar cases in crayfish and *Aplysia;* nor is it due to postsynaptic membrane potential or resistance or recurrent collaterals. It is believed to be due to changes intrinsic to the synapses themselves.

Aftereffects. Here we refer to changes in activity or state following the cessation of a period of arriving input, whether there is a correlated afterpotential or not. In some places and circumstances there may be no noticeable aftereffects. In others, the effect of the input, whether excitatory or inhibitory, may continue for some time, gradually wearing off (Fig. 6.6). In still others **rebound** in the opposite direction is released at the moment of cessation. These tendencies can be quite strong. They can follow each other in sequence. They can last for many seconds and even minutes. Obviously they will contribute markedly to determining output. Post tetanic potentiation belongs in this category; it is treated just above.

Presumably **accommodation** (p. 151) is one of the prime factors involved, but we cannot assume a 1:1 relation between

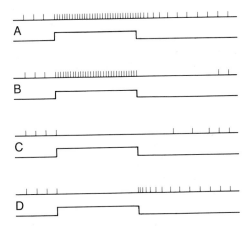

Figure 6.6
Aftereffects and rebound. An artificial scheme showing positive aftereffects (**A, C**) and negative aftereffects (**B, D**; rebound) following excitatory and inhibitory input. Duration of input marked by the lower trace. [Bullock and Horridge, 1965.]

accommodation, afterpotentials, and afteractivity. Possibly other factors, such as intercellular accumulations, countertendencies for rebound, iterative properties, or diffuse field effects, may be involved in different cases. It is the formidable range of degrees of freedom available in different neuronal systems, which result from combinations of cellular mechanisms partly known and largely unknown, that is important in providing for diversity in integrative properties.

Effects of Milieu, Hormones, and Substrate Acting on Nerve Cells. A wide variety of states and agents can influence the activity of nerve cells. Some are probably useful factors in the functioning system; others are probably sources of unwanted perturbation or noise. It is difficult to choose between these alternatives with an incomplete knowledge of the functional system involved. Among the conditions and agents known to be effective in determining neuronal output, at least in certain places, are oxygen tension, CO_2 tension, hypo- and hypertonic media, temperature, light, mechanical distortion, and certain ions and organic substances.

Some of the neurons influenced by these factors are recognized as receptors because their role in the system appears to be specialized for the detection of one or more such conditions and agents as stimuli. Some of these receptors are within the brain, others in the periphery. It is more than likely that many central neurons not regarded as specialized receptors are under such influences as part of their normal integrative operation. Examples are temperature in poikilotherms, generalized osmotic pressure in species that are imperfect regulators, and pH and K^+ concentration in the microenvironment of any cell. Under the influence of activity some neurons have been found to **release substances** other than transmitters, including amino acids, nucleosides, enzymes, and glycoproteins; some of these are involved in transneuronal transfer and may have metabolic and/or neural signaling roles.

A major class of effects is represented by **hormones.** Neuroendocrinology has brought to light many examples in which a change in the level of a circulating homone has altered behavior of a neuron. Often this effect is part of a feedback, since the nervous system also controls the release of hormones into the blood. At the neuronal level little is known about the mechanisms of excitation and inhibition by these agents, but it is clearly an area in which important advances will be made (see Box 6.2, p. 216).

Movements of Parts of Neurons. Outstandingly important, if it occurs, is change of position, dimensions, and relations by movement or growth of axonal terminals, dendrites, and somas. Such movement, though well known in embryonic development and in tissue cultures, is little documented in the adult nervous system. It is suggested by many fragments of evidence. For example, in the cortex changes are reported at the light-microscope level in dendrite spines, in branching, and in protein incorporation with stimulation. Motor nerve terminals and end-plates have been shown to increase in size and complexity with increased use of the synapse. Cinefilms are said to show movements of processes around parasympathetic postganglionic cells in the wall of the toad bladder and alterations with preganglionic stimulation. Transfer functions should be influenced strongly and probably enduringly by movement or growth.

The degrees of freedom available even at this low level can provide for an almost **unlimited degree of complexity.** These integrative parameters of single nerve cells, processes, and junctions are well developed even in the invertebrates. There is a wider spectrum of development of integrative properties among the neurons of a given species than we can recognize between major groups. So great are the possibilities for lability, variability, and complexity of transfer functions, even at the neuronal level, that the problem of extrapolating upward has changed its complexion from that of a few years ago when one asked, "Can we get really complex behavior from large numbers of all-or-none units?" to quite a different one. Now one asks, "Which of the many possible mechanisms are ac-

tively operative? What constraints on the possible permutations are there, and how is consistent performance ensured in the face of the possibilities for inconsistency?"

III. THE NEURON AS AN ENCODER: MECHANISMS OF SIGNALING

We turn now from the factors that influence input weighting to the output variables. How does each neuron or array of neurons encode and signal its messages to a receiving neuron or array? How is information—the coin and business of the nervous system—represented in the neurons? What coding principles are employed? We turn necessarily toward more organized groups of cells and try to use the criterion for significance that the cells receiving the coded message show an ability to react differently according to the output variables in question. Some of these variables, while carrying information discernible to an investigator, may only be epiphenomena, like the sound of an automobile, and must be treated as no more than candidate codes.

The problem may be subdivided into (a) coding by identity or label on the line that specifies by foreknowledge the dimensions of information variables within the activity range of each line, (b) coding by spike train parameters, and (c) nonspike and pooled activity that may carry information in units and assemblies. One of the important mechanisms of nonspike and pooled activity as a signal is neurosecretion. This and mechanisms intermediate between it and ordinary synaptic transmission are treated in Box 6.2 (pp. 216–217).

A. Labeled Lines: Modality and the Meaning of Messages

The most important way in which axons carry information is by their "labels"— that is, their identity, as defined by central connections. Whatever is encoded in a stream of impulses, the main message is, in effect, "something going on in line so and so." The relevant receiving part of the central nervous system, sometimes called the analyzer, distinguishes the quality or meaning of the messages as though it knows the source of impulses. One line has the meaning "optic stimuli," another one "acoustic stimuli," whatever the actual cause of impulses in those lines. Thus a strong blow to the head can produce vivid visual sensations, in the absence of light. It was just such a blow, claimed by a victim to have allowed him to identify his assailant by the light it produced!, that led Johannes Müller, more than a century ago, to formulate the "law of specific nerve energies" for sense organs. This states that each receptor normally carries information about one type of stimulus; if excited in some other way, it still "tells" the brain what it is built to say.

Modalities and Submodalities; Spatial Representation. The old list of five senses (=modalities)—sight, hearing, smell, taste, and touch—left out many important afferent systems. Classification by sensation was replaced by one based on the source (exteroceptors, interoceptors, proprioceptors) or, still better, on the nature of the stimulus. Photoreceptors, thermoreceptors, chemoreceptors, elec-

troreceptors, nociceptors, mechanoreceptors (subdivided into phonoreceptors and tactile, stretch, acceleration, position, and pressure receptors) are the usually recognized types. Each is really a class that takes different forms in different groups of animals.

Evidence from anatomy, physiology, and the clinic indicate that most of the major distinguishable sensations and most of the major forms of stimulus **have their own systems** of labeled lines. But such sensations as tickle, wetness, some forms of pain (p. 221), and perhaps others, may be based on combinations. Moreover, some forms of stimulus (e.g., temperature) may normally affect more than one afferent system (see Box 10.2, p. 478).

Within modalities and central afferent systems, labeled lines play a large role in defining **subsystems for distinct aspects** of each modality—for example, pitch, color, specific tastes, or the exact locus of a mechanical stimulus on the skin. Here we come to the constellation code of the next section. Unexpected or unfamiliar submodalities or stimulus specificities— that is, labeled lines—are likely to continue to come to light. Examples are the type 1 ganglion cells of the frog retina (pp. 251–255), which respond best to small, dark, moving objects in the absence of surrounding movements, and the distinct part of the porpoise auditory system specialized for the echolocating type of sound. It may commonly be difficult to determine the proper label in terms of meaning to the organism.

Lines for different modalities can be followed in almost all of them to at least third order, and in some (vision, hearing) to fourth-, fifth-, or sixth-order neurons before losing their general modality labels. Within a modality, spatial mapping ("local sign") can be seen to persist almost as far. Nevertheless, at most of the relays between orders of neurons, integrative processes take place, so that well before a line reaches a level at which it is integrated with other modalities or at which output is formulated, its exact label changes; that is, its meaning or **representation is altered.** An example in visual processing is given on p. 383. Some important problems related to the representation and meaning of messages concern ambiguity, reliability, redundancy and recognition. These are considered below (p. 233 et seq.).

Another problem for many modalities is **dynamic range.** We naturally approach the issue from the standpoint of a design engineer and ask what are the ways of gaining sufficient intensity discrimination over a sufficiently wide range of stimulus intensity. In some modalities the success is impressive—vision and hearing, for example. Our explanations in terms of known physiology are still inadequate. Here we can only mention some of the principles employed. One is **range fractionation;** unit receptors may operate over a limited fraction of the total range. The intensity continuum can in this way be treated like the topographic continuum of points on the body surface. The constellation code (p. 223), by specifying the set of lines and the degree of activity in each, would define intensity.

Another principle that takes diverse forms among the modalities is **central control of sensitivity,** discussed also on pages 170 and 255. The middle ear muscle can contract and condition the tympanum for loud sounds. There is also a bundle of efferent axons in the cochlear nerve that inhibit auditory receptors right at the periphery, under brain control. Additional inhibitory actions descend from higher levels to influence the input at early central stations. (We discuss gamma efferents to muscle stretch receptors on p. 267.) Yet it is not clear that these mechanisms are primarily useful as extenders of dynamic range; their roles may be more complex.

Spontaneity of receptor unit discharge (see also p. 152) enhances sensitivity and extends the range downward. The logarithmic stimulus-response relation commonly found (p. 204) permits a limited output range (e.g., of receptor potential depolarization and nerve impulse frequency) to represent a wider range of intensity. Still, units rarely cover an amplitude range much more than 100:1, and the range of hearing and vision of more than 100 db (100,000:1) is not fully explained.

There are other, more qualitative **forms of differentiation.** Within the same system of lines having a common broad label—for example, in the lateral line nerve of fish, in inflation receptors in the lung, in touch units around cat's whiskers, and elsewhere—there are found distinct populations of thick fibers and thin fibers; the two apparently carry signals of different meaning for the organism. Thin fibers come from receptor units with lower threshold, lower discrimination of intensities, lower rate of adaptation, and greater tendency to continuous background discharge than the thick fibers. Though it is not clear just how, there is probably a difference in role in reflex control.

Many systems employ reciprocal push-pull submodalities. Vision, temperature, joint position, and head movement are often signaled by two sets of neurons, with opposite best stimuli.

Another form of differentiation exists in the respective weight or value of

Box 6.2 A Spectrum of Neurochemical Communication Mechanisms, Conventional and Neurosecretory

Besides conventional synaptic communication, we have referred repeatedly to another form of signalling, the phenomenon of neurosecretion. Certain neurons in the hypothalamus of vertebrates and of specific parts of the ganglia or plexuses in all invertebrates with nervous systems engage in secretory activity much more pronounced than that involved in the synthesis of neurotransmitters.

Typically, the conspicuous neurosecretory granules are released not at intimate or private junctions with individual target cells, but into the blood, usually via storage depots called **neurohaemal organs,** whence the general circulation distributes these chemical messengers, called **neurohormones,** to remote target cells.

A remarkable part of the pathway is the bulk transport of granules of secretion down the axons, from sites of synthesis in the soma to sites of release at the terminals—for example, from hypothalamic nuclei to the posterior lobe or pars nervosa of the pituitary gland (neurohypophysis) (see figure). The substances so transported and stored are the posterior pituitary hormones vasopressin and oxytocin, each from its own neurosecretory cells. Their release into the blood stream is controlled by impulses down the respective axons and perhaps also down parallel, nonneurosecretory axons.

Another group of neurohormones, called releasing factors or **hypophysiotropins,** likewise derived from hypothalamic neurosecretory neurons, are transported down their axons to the median eminence of the hypothalamus, which acts as a neurohaemal organ. There they are released into capillaries that, by their confluence, form portal veins that pass from the hypothalamus to the anterior lobe, here dividing again into capillaries that deliver the releasing factors to the pituitary gland cells, controlling the release of their hormones. Thus the control of the pituitary, the "master gland," and all its six distinctive hormones is by neurohormones from the hypothalamus. This in turn is under the influence of afferent (including feedback) signals,

sensory input from receptors, and spontaneous changes of central state.

The same general principles operate also in the invertebrates. They show how neurons can adapt their

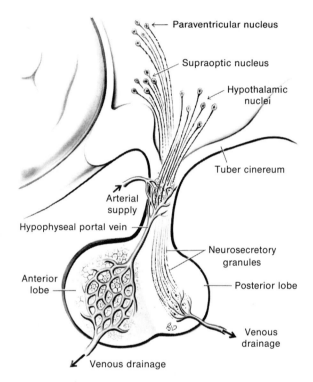

The hypothalamohypophyseal tract and the hypophyseal portal vein. Neurosecretions are transported via nerve fibers of the hypothalamohypophyseal tract from the supraoptic and paraventricular nuclei to the posterior lobe of the hypophysis; they are conveyed from this lobe via the bloodstream. Other neurosecretions, elaborated in other hypothalamic nuclei, are conveyed via the hypophyseal portal vein to the anterior lobe of the hypophysis. [Noback, 1967.]

Box 6.2 (*continued*)

communication mechanism to the needs and properties of different effectors. The endocrine effectors are amenable to activation relatively slowly, simultaneously, and not so privately as are other neurons or striated muscle. In general, throughout the animal kingdom, neurosecretory systems seem to be fundamental to control of endocrine glands, which in turn control major aspects of physiology, including growth and reproduction.

Neurohormonal—that is, blood-borne chemical messengers—are at the other end of a spectrum from conventional neurotransmitters at ordinary synapses, working extremely locally, at close range, for short times, with small amounts, rapid release, and rapid inactivation. The neurohormones work on widely distributed targets (millimeters to centimeters) at long range (millimeters to centimeters) for long periods (seconds or more), probably in much larger quantities, slowly released, and slowly metabolized. The substances seem to be, for the most part, polypeptides, hence the respective neurosecretory neurons are called peptidergic. The importance of this mechanism may be stated in the words of B. Scharrer (1974): "The effectiveness of the integrative functions of the body depend on appropriate mechanisms for the exchange of information between the neural and the endocrine control systems. The most central, and therefore crucial, step in this chain of events is that by which the nervous system addresses itself to the first way-station in the endocrine hierarchy. It is the 'final common path' which handles the integrated commands resulting from a multitude of afferent signals intended for endocrine centers, or for delivery directly to nonendocrine 'terminal target cells'."

The **continuity between these extremes** is suggested by intermediate cases (a) In the first place, endocrine glands are not exclusively controlled by neurohormones; several, including the adenohypophysis, receive some standard chemical, probably aminergic synapses, though certain of their ultrastructural characteristics tend to be missing, and there may be some specialization for the properties of endocrine effectors. In these respects such innervation resembles that of exocrine gland cells. (b) In some teleosts there are peptidergic fibers that make direct contact with anterior lobe gland cells, instead of communicating only via the portal circulation. Some insect striated muscle also receives peptidergic nerve endings. (c) In other teleosts there is only a layer of extracellular stroma (connective tissue) separating neurosecretory terminals from anterior lobe cells. Certain striated and, especially, smooth muscles likewise receive nerve endings, not in intimate contact with every cell but via an intervening stromal compartment. This presumably provides a graduated degree of localization or privacy of activation. "At any rate this mode of communication is neither neurohormonal nor neurosecretomotor, but somewhere in between. It seems to operate with neural signals of relatively long latency and extended duration" (B. Scharrer, 1974). (d) A variant is neurosecretory fibers, in which the mediator is aminergic instead of peptidergic. The mediators are blood-borne and seem to be brought into action for relatively short-term responses. (e) A nonvascular route for the extracellular pathway is the cerebrospinal fluid. This is believed to serve some chemical mediators, although normal receptor sites are not known. (f) Even conventional neurons may be subject to the influence of neurochemical signals other than those of the usual synapses. For example, some insect brain neurons are affected by neurohormones from the corpus cardiacum or other neurosecretory centers. A wide array of nerve cells in the mammalian brain is under the tonic influence of the circulating neurohormones. (g) "A class of neurochemical messengers, quite apart from those discussed so far, is exemplified by a diffusible **'neurotrophic substance'** which is released from sensory as well as motor fibers and under certain conditions, even from some nonneural tissues. Such trophic substances regulate growth and maintenance" (B. Scharrer, 1974).

different modalities in competition for central influence. Different modalities are not merely parallel avenues of inflow of news of the world, with equal access to centers of command. Pain input is more compelling than postural. Olfactory input has a peculiarly overriding effect in man. When vision and proprioception disagree, the former dominates in man. The relative values are difficult to measure or make precise, and depend on the criterion of weight. They probably differ importantly between species and may be altered to a significant degree by experience and motivation.

Evolution of Modalities. In our present stage of knowledge it is difficult to generalize about the evolution of the modalities and their differentiation. The direction has been from few to many, from general to subdivided, and probably from mixed to pure. The principal classes of receptors appear to be approximately the same in higher and lower animals. Even in the coelenterates there are several clearly distinct types, and we cannot place an upper limit on the number, owing to the difficulty of recognizing functionally differentiated but not anatomically distinct receptors. This is all the more important, for even though one might at first glance expect that the evolutionary level would also be indicated by total number or density of receptors, the more recent histological and physiological findings—for example, in coelenterates and annelids—show very large populations of primary sensory neurons. In man the total number of afferent nerve fibers entering the central nervous system has been estimated to be about five times that of the efferent fibers leaving, and in some annelids and arthropods the ratio is higher by at least a factor of ten.

This probably means not a relative decrease in the number of afferents in man, but an increase in the number of efferents. Multiplication of sense cells appears to be simple and to occur in lower forms, but establishing new kinds of labeled lines is more difficult and perhaps emerges more slowly in evolution.

B. Representation of Information in Spike Trains; Neural Codes

What are the parameters of impulse firing that could be relevant to communication? In single fibers they must derive from the only basic variables in a train of events that do not themselves vary: number and interval (Fig. 6.7). With respect to all-or-none spikes, the nervous system is a **pulse-coded analog device,** since the intervals are continuously graded. It is sometimes mistakenly called a digital and binary system, but it is not, because the intervals are not quantized. If it were a digital system, time would be divided into arbitrary quantal periods, agreed upon between sender and receiver; if it were a binary system, the state of the input would be represented by one of two states of the output for each such period. Let us look at some of the parameters inherent in a stream of events and then ask which are candidate neural codes.

In a stream of discrete events, such as nerve impulses, each of which can be pinpointed in time, there is a **distribution of interval lengths** between the individual events. In any sample or epoch this can be expressed by the shape of the interval histogram (Fig. 6.8). The distribution differs significantly between examples. Motor units in voluntary muscle in man, for instance, fall into two popu-

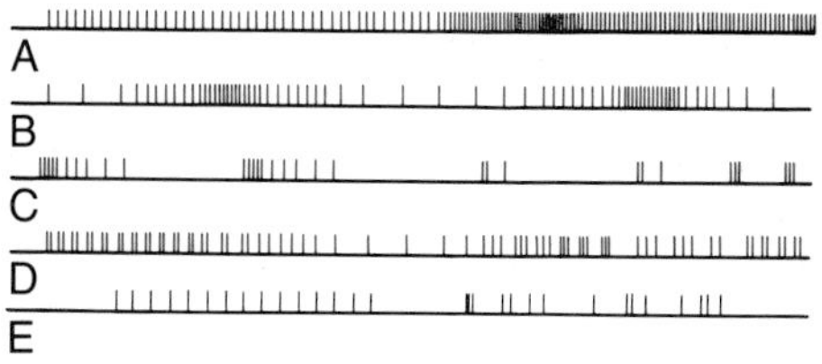

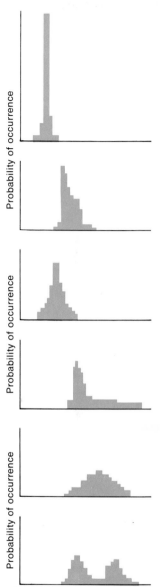

Figure 6.7

Temporal patterns. Diagrams of types of sequences available for pattern coding of impulses in single lines. **A,** Long-maintained, constant, or slowly changing frequencies can signal steady or slowly changing states, not divisible into messages. The first diagram is drawn to emphasize that it grades into **B,** in which slowly changing frequency signals events rather than states and hence can be regarded as divided into messages. Diagram **B** grades into **C,** in which the messages are still more discrete and the code may utilize both the number of impulses in a message and the general form of the frequency change during a message. Diagram **C** grades into **D,** in which the microstructure is considered to be significant. Unequal but regular and meaningful intervals, such as doublets or triplets of spikes, may carry information different from the same average frequency with a different ratio of intervals. Diagram **D** is drawn to show different ratios and the gradation into regular spacing. **E,** Not a distinct category, but a feature of importance in each category, is the tolerance of irregularity or random fluctuations of interval or the significance of regularity of rhythm as such; extremes drawn here are like those commonly seen. The random spacing at right suggests that only average frequency, integrated over some time, is utilized; if true, the high regularity at left has no significance. [Bullock and Horridge, 1965.]

lations that differ in this respect; one population shows more variability, the other more regularity of the interval between action potentials, measured at a given mean frequency. The biological meaning remains to be determined. Nerve cells in the motor cortex of monkeys have a standard deviation of interval that is higher when the animal is awake than when it is asleep (at the same

mean frequency). First-order auditory units in the cochlear nerve have broad, simple, Poisson-like histograms, and differ from second-order units in the cochlear nuclei, which have more complex, often Gaussian or bimodal shapes. The coefficient of variation (standard deviation/mean) in some neurons is larger at low mean frequencies than at high, whereas in others it is virtually constant. Most of these features are typical of the class of neurons under given conditions, but apart from mean frequency we know little about the possibility of physiological conditions that may systematically alter statistical features.

Another set of properties depends on the **sequence of longer and shorter intervals** than the mean. Some neurons show no serial dependence, some a negative and some a positive correlation of successive intervals. Thus one neuron may fire with alternately long and short intervals, giving pairs of spikes and a high negative correlation for successive intervals. Another may fire with a faster or slower waxing and waning of frequency, giving bursts of high frequency separated by periods of lower than mean frequency, therefore a high positive correlation for successive intervals. It has yet to be discovered what kinds of cells or situations produce each type of sequence.

When the impulse discharge is in response to a stimulus that can be identified in time, other methods are in use, including the post-stimulus time histogram, latency histogram, and phase histogram (Figs. 6.9; 6.10). These utilize the moment of stimulation as a point of reference and average many responses to reveal certain consistent features of temporal pattern. More sophisticated ways are available for examining interval dis-

Figure 6.8
Diversity of interval histograms.

Gaussian or normal distributions may be narrower or wider (upper graph in each pair). Trains of spikes may approach Poisson, gamma, or more complex distributions (lower graph in each pair). Potentially meaningful measures may be the mean, standard deviation, variance, coefficient of variation, and higher moments.

Probability of occurrence

Length of interval between spikes

Figure 6.9
Various ways of plotting the consistency of intervals in response to stimuli over many cycles of stimulation.

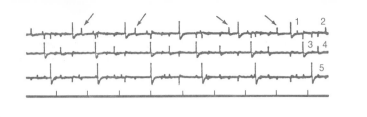

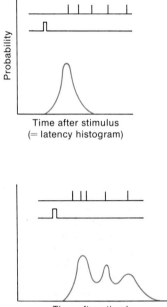

Probability

Time after stimulus
(= latency histogram)

Time after stimulus
(= post-stimulus time histogram)

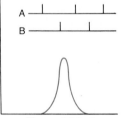

A

B

Phase of B in A
(= phase histogram)

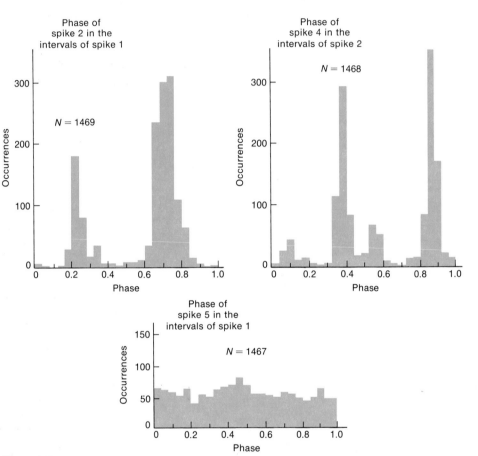

Phase of
spike 2 in the
intervals of spike 1

N = 1469

Phase of
spike 4 in the
intervals of spike 2

N = 1468

Phase of
spike 5 in the
intervals of spike 1

N = 1467

Figure 6.10
Multistable firing patterns in flight motor neurons of the fly. In the electromyogram, the upper two traces are from two electrodes in the same muscle. Several different motor units can be discerned in each trace. The four numbered units in the top traces can be identified throughout a long record. Phase relationships between selected units are illustrated in the top two histograms. The phases of unit 2 in the interval of unit 1 are bimodally distributed (*upper left histogram*). (Interval is the time between two impulses in the same channel. Phase is defined as the amount of time from an impulse in one channel to an impulse in a second channel, divided by the time from the same impulse in the first channel to the next impulse in that same channel—i.e., its interval.) In the short section of record in the upper trace, the two quasistable patterns are shown with a sudden transition between them. The phases of unit 4 in unit-2 intervals have four modes (upper right histogram). Unit 5 (third trace), which is in another muscle, shows no preferred phase with respect to unit 1 (lower histogram) or any other unit in that muscle. (Lower trace is a time marker; intervals are 0.1 sec.) [Wyman, 1966.]

tribution and sequence during adaptation.

Various theoretical problems can be attacked with these and related concepts, but they are beyond the scope of this book. Examples are the consequences of different modes of spike initiation, the origin of noise in interval length, and the problem of what set of operations by model neurons would yield the most sensitive detector of weak signals in the presence of noise while preserving a reasonable temporal resolution.

With the several properties of impulse trains in mind we may now ask, "What can be said about coding information in single channels (nerve fibers)?"

Nerve impulse coding in single channels has classically been disposed of by the inference that **average frequency** is *the* variable that represents the relevant gradations of the input. Certainly the

---

Box 6.3   The Problem of Pain

In this context of comparative and evolutionary biology the problem of pain deserves special mention. The sensation in man is remarkable among the senses in its strong negative affect or unpleasant quality and its demand for attention. These are based on its association with consciousness and with a separable experience that may be called suffering. The adaptive withdrawal from nociceptive, potentially damaging stimuli is, however, not dependent on consciousness, and occurs even when consciousness is reduced greatly by several types of drugs, light gas anesthesia, clinical lesions, surgical intervention, and congenital and hypnotic conditions. Pain cannot be judged, therefore, by reactions to stimuli, even if empathy and common feelings are assumed, unless a complex of signs is read and the behavior of the species is well known. Even in fully conscious man the relation of felt pain to suffering is not obligate; in several types of conditions pain is reported but tolerated without distress. Plainly, we cannot reliably "measure" distress or suffering by observing behavior or by comparing the effect the same stimuli would have on ourselves. This conclusion applies to man and, therefore, presumably to other animals.

Although the judgement is not reliable and suffering may be falsely imputed, humane treatment of other species consists in assuming within reason that they have the same feelings as ourselves. How far down our hypothetical scale of animals is this assumption intellectually respectable? Although this is virtually unanswerable in a precise or definitive way, the propositions are defensible on neurophysiological, anatomical, and ethological grounds that (a) pain and even more, suffering, cannot be equally present in corals, flatworms, and man, and (b) something like pain at least cannot be assumed to be a unique attribute of man. Therefore it is reasonable to believe that pain has evolved and is felt in different degrees, from trivial to significant, in different animal groups. In animals other than ourselves, sensation, when felt, is probably substantially less than ours, meaning less cognitive, affective, esthetic, or rich, if not less vivid or real. Even more clearly, **pain cannot be estimated by withdrawal** responses alone.

Given this conclusion, we must decide in practical cases—in the operation of zoos, farms, abattoirs, and laboratories—just what degree of conscious sensation and potential suffering to impute to insects, squids, fish, frogs, turtles, birds, rats, sheep, cats, monkeys, and other animals. Although the rational basis for decision is meager and crude and we are not free of a social milieu with traditions, we can strive, on the basis of individual responsibility for our actions, to apply both moral and intellectual criteria, in any form of imposition of our will, depending on the procedures, purposes, and animal subject.

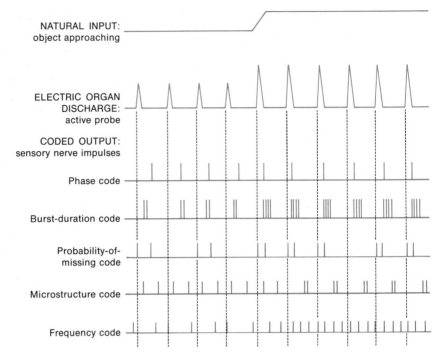

Figure 6.11
Summary diagram of several events plotted against time (in the range of 0.1–1.0 sec for the whole sweep), showing main types of nerve impulse codes, all of which probably occur in one place or another. Since several are best seen in electroreceptor nerve fibers of weakly electric fish, the chart includes a diagrammatic record of the fish's own electric organ discharge as seen by an electro-receptor on the skin, after the stimulus (approach of an object of different conductivity from that of the water) has modified its amplitude locally. [Bullock, 1968.]

discoveries of Adrian, Zotterman, Matthews, Hartline, and others—that increasing the adequate stimulus to sense organs causes an increase in mean frequency in afferent axons—were epochal in understanding the nervous system. The inference is more than reasonable; mean frequency is probably the code in most cases, though little attention has been given to defining it in terms of the receiver—for example, averaging time, weighting function, and forgetting function. Characteristic input-output functions have been found

(Fig. 4.28). But newer evidence suggests that there are in fact several distinct coding principles, in addition to mean frequency, normally employed in different places in the nervous system, or even in the same lines as read by different receivers (Fig. 6.11).

Each theoretically possible code becomes a candidate code as soon as there is evidence from some living preparation that it is either available in or readable by the system. That is, if normal input elicits a change in the parameter under consideration, that code can be said to be available. If we artificially impose changes in that parameter alone, upon trains of presynaptic impulses, and elicit changes in the response of the postsynaptic cell, that code can be said to be readable. But only if both are found in the same preparation and no other parameter is concomitantly available and readable can the candidate code be said to be the actual code employed. These criteria have not been met for most candidate codes, although each property has been found separately for several of them. Let us list the principal ones.

(a) **Time of firing,** like the ring of a doorbell, contains the essential information—for example, "something happened just now," or "now jump," clearly useable in many systems. (b) **Variance** of interspike intervals, as reflected in the interval histogram, may signal some information independently of that carried in the mean frequency. An example is the difference in state between the dark- and light-adapted eye of *Limulus.* This code may use variance but ignore the precise sequence. (b) **Temporal pattern** or microstructure in sequences of intervals, such as alternately short and long intervals of different ratios, at the same mean interval, has been shown to be readable by certain neuromuscular junctions in crustacea and by neurons in *Aplysia,* lobsters, and others. The greater effectiveness of some patterns is presumably a consequence of certain rates of rise and decay of facilitation in pattern-sensitive junctions. Whether such pattern, fine-grained and maintained, is normally available is more difficult to say. It is found here and there, though it is not clear that these cases represent a consistent encoding of some meaningful information. The term "pattern" is

often loosely used for differences in the structure of phasic bursts that would be better described in terms of profiles of frequency modulation. (d) **Number of spikes or duration** of burst can encode stimulus intensity in some electric fish receptors without involving change of spike intervals.

Two types of multichannel coding are treated here; others are considered in Section C, below. (e) Spike delay or **phase,** relative to some reference in another neuron, is the principle used in certain electric fish electroreceptors that fire at a high, fixed frequency. Some units shift in latency with changes in the adequate stimulus (position of an object in the water near the fish), while others do not, providing the central nervous system with the basis for a code that can function at high sampling rates. Arrays of parallel mechanoreceptor fibers from the cat foot pad code the locus of a touch by a "flying wedge" of spikes; fibers innervating the center of the zone of a touch have short latencies, adjacent fibers long latencies (Fig. 6.12). Spatial information is transformed into a spatio-temporal pattern. (f) The chief classical candidate for a multichannel code is based on the identity of the neurons activated, rather than temporal relationships (i.e., on a **"constellation code"**). This is actually a multichannel application of the labeled line principle discussed above. Receptive fields (the part of the stimulus spectrum or extent that exerts an effect on a given receiving unit) typically overlap extensively. On the body surface and in the eye receptive fields are topographic. In the mammalian ear the receptive fields of the afferent units are bands of sound frequency. In taste and smell the range of substances that excite or inhibit each unit define

Code in TIME of 2 simultaneous events   Code in SPACE of 2 successive events

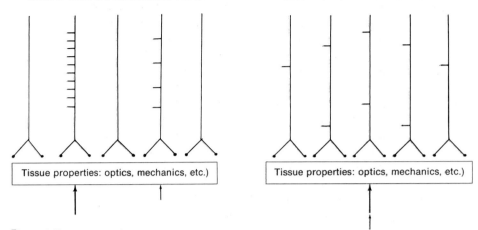

Figure 6.12
Coding intensity and locus of a stimulus. Two types of "coding": frequency (left) and spatial (right) representations. Impulses are shown traveling upward from receptor ends to central nervous system. [Courtesy of J. A. B. Gray.]

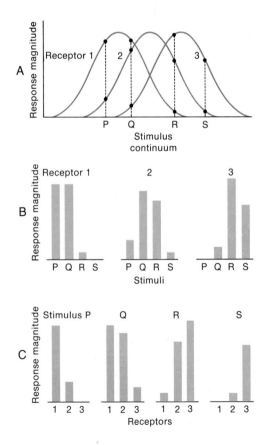

Figure 6.13
Parallel fiber coding with overlapping receptive fields (a hypothetical case). **A.** Receptive fields of three units (afferent fibers or central neurons). Curves 1, 2, and 3 represent the response areas of three hypothetical fiber types along the relevant stimulus continuum, which may be topographic distance, wavelength of light, frequency of sound, etc. *P, Q, R,* and *S* represent four stimuli along this stimulus continuum. The responsiveness of a fiber type to one of these stimuli is indicated by the intersection of the response curve and the ordinate erected at the stimulus. **B.** Responsiveness of the three fiber types to the four stimuli in part A. In each of the bar graphs is shown the responsiveness of one of the fiber types to each of the stimuli in A. If recordings were obtained from one of the fiber types shown in A using these stimuli, one of these three "response profiles" would be obtained, depending upon which fiber type was being sampled. There would be as many "response profiles" as fiber types. **C.** Across-fiber patterns. In these bar graphs are shown the patterns of activity across the three fiber types produced by the four stimuli in A. Each stimulus produces a characteristic pattern across the three fiber types. There would be as many across-fiber patterns as stimuli. [Erickson, 1963.]

equivalent receptive "fields," and they overlap (Fig. 6.13). That is, any given stimulus will excite a certain constellation of receptors, and is therefore encoded in the particular set that is activated. The unit receptors do not have to be highly specific to permit encoding a large number of distinct stimuli. The same may be true of modalities like pain, temperature, pressure, stretch, and others. If the brain can take note of which combination of fibers is firing, it can achieve a higher resolution than by using an equivalent system with non-overlapping units.

These and other forms of representing information are all quite likely operating, each in certain subsystems. Each requires the subsequent stages to "know" a priori the coding principle and the calibration.

C. Nonspike Signaling: Electric Fields of Units and Assemblies

Intracellular recording and extracellular microelectrode recording (ca. 1–10 μm uninsulated tip diameter) are not the only means of seeing electrical activity in nervous tissue. Macroelectrodes (ca. 50–1000 μm) placed on or in nervous masses also pick up fluctuating potentials. A potential difference may be recorded between two electrodes in the same part of the brain ("bipolar recording") or between one in an active area and another in an inactive area ("monopolar recording").

It is commonly assumed that the fluctuating potentials seen by macroelectrodes are no more than the sum of potentials arising in cells. Probably there are also steady or slowly changing potentials due

to (a) gradients of concentration in intercellular fluids, (b) the flow of blood, (c) movements of such structures as the eyes, and (d) other extracellular causes. It is still debated how the various kinds of unit (cellular) activity are summed, spread, and filtered. This means there is disagreement also on **how to interpret** records in terms of the synchronization, locus, character, and sequence of unit activity. The uncertainty is in part due to the geometric problem posed by several kinds of activity that spread through the ramifying branches of diverse cell types, in part to the technical difficulty of recording intracellularly from all the types of cells and their dendrites, and in part to the complications of having to record from enough of these simultaneously.

Until it is feasible to record in this way, the gross electrode record can give less detailed but still useful access to patterns of population behavior, especially in response to external stimuli (evoked potentials). Empirically, a range of diverse forms of activity is found in different kinds of nervous masses and functional states. Most of our knowledge is confined to fluctuations above about 0.3 Hz. Steady and very slowly changing potentials are found as well, however, and these exhibit some characteristics and correlates that, as we will see, are consistent and therefore perhaps important.

Extracellular micro- and semimicro-electrodes are likely, in most central nervous structures, to "see" a number of units in such a way that spikes are recorded superimposed on slow waves (mainly from 1–20 Hz) of greater or lesser amplitude. Finer electrodes generally see fewer units, but, in addition, unknown factors seem to be important in detecting single units, hence the prepara-

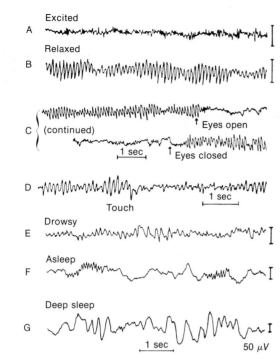

A Excited

B Relaxed

C (continued) ↑ Eyes open
↑ Eyes closed
⊢ 1 sec ⊣

D
⊢ 1 sec ⊣
Touch

E Drowsy

F Asleep

G Deep sleep
⊢ 1 sec ⊣ 50 μV

Figure 6.14
The electroencephalogram (EEG), or "brain waves," recorded from the scalp of a human subject. **A.** In the alert subject, the EEG has a low-amplitude, fast-wave pattern. **B.** In the relaxed state, alpha waves appear. **C.** Opening of the subject's eyes (arrow) blocks the alpha rhythm, but it returns soon after the eyes are closed again. **D.** A blocking occurs upon touching the subject. **E.** In drowsiness, a transition to the next stage is seen. **F.** During sleep, runs of 14-per-sec rhythms are found superimposed on the slow waves; they are termed "sleep spindles." **G.** In deep sleep, large-amplitude slow waves replace all fast activity. This figure does not show the last stage—fast-wave, or rapid-eye-movement (REM) sleep (see Box 10.5, p. 492). [Jasper, 1941.]

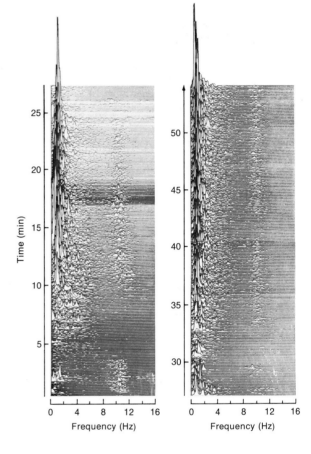

Figure 6.15
Use of the power spectrum to compress an hour's EEG. Recording from human scalp before and while entering sleep. [Courtesy of R. G. Bickford.]

tion of "good" electrodes is highly empirical.

Ongoing Activity in Assemblies; Brain Waves. Macroelectrodes see only smoother, slow, sinusoidal waves in midbrain and higher levels of vertebrates; the power spectrum is broad but shows little energy above 50 Hz. This activity ("brainwaves," electroencephalogram = EEG) can be recorded through the skull and scalp by simple electrodes on the skin. This fact, together with the complex form and changes in form, amplitude, and distribution in such states as sleep, wakefulness, inattentiveness and alertness, problem-solving, and learning, and in hyperventilation, epilepsy, and other clinical conditions, has attracted a great deal of work and continual refinement of techniques.

Figures 6.14, 6.15, and 6.16 illustrate the general character of brain waves and some types of alteration that make them

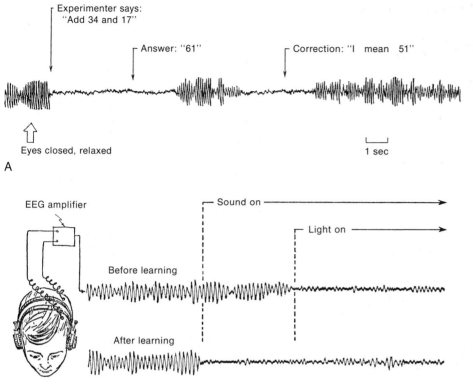

Figure 6.16
Changes in EEG with thinking and learning. **A.** The 10-per-sec alpha waves of the awake individual disappear during mental effort on an arithmetic problem. **B.** A weak sound usually does not flatten the alpha waves, but a certain light always does. When the subject realizes that light always follows sound, the 10-per-sec rhythm disappears as soon as the sound is heard. [Galambos, 1962.]

interesting, both theoretically, in terms of cellular interpretation, and practically, in terms of localization, assessment, and diagnosis. The parameters that can vary include the frequency components, the amplitude of each, phase relations among them, changes with time in any of these, like waxing and waning, and differences between regions of the brain.

The degree of **synchronization** of generating neural elements should be one of the decisive parameters in determining the amplitude of particular frequency components. An extreme case is the class of large brief waves in a subject experiencing convulsive seizures, when apparently a massive discharge of neurons combines to produce waves that perhaps themselves help to pull cell firing into step. More normal are the large slow (0.3–3/sec) waves and the waxing and waning medium-frequency (7–11/sec) waves, called spindles, which are probably due in part to greater degrees of synchronization than obtain during the low-voltage fast activity in alert subjects as well as in paradoxical sleep. Yet the true degree of synchronization is actually difficult to assess with a small number of microelectrodes. Imagine trying to measure, with a few microphones, the degree of synchrony of shouting at a political rally where a small but significant number of well-organized individuals coordinate their outbursts, though scattered through the crowd.

Either because of synchronization or differences in intrinsic activity of the generators, there are two basic differences in the macroelectrode recorded activity of **vertebrates compared to invertebrates** of many groups. So far we have spoken of vertebrate brain waves. It is significant that these are extremely similar from fish to man. Their basic character, as shown in a power spectrum, does not depend on brain size, complexity, or the presence of a cerebral cortex. This is true in spite of subtle differences clearly visible in a given species in states like sleep and epilepsy, and also between waves recorded from different regions of the brain. Invertebrates—at least earthworm, crayfish, Limulus, cockroach, and snail (but not octopus, in which waves are more like those of vertebrates)—are alike in showing relatively little slow and much fast activity in their central ganglia. The two remarkable puzzles, not yet explained, are: Why does an insect or a lobster show so little slow (<25Hz) activity and a goldfish or man so little fast (>50Hz) activity with macroelectrodes? It seems likely that a significant correlation and possible cause lies in differences between the roles of dendrites of vertebrates and the analogous processes of invertebrate neurons, together with the organization of intercellular relations

such as slow potential synchrony in the neuropile (see Fig. 10.58, p. 438).

The **origin and cellular basis** of brain waves are still not understood. Most authors believe large numbers of cells must be relatively synchronized, the more so the larger the waves. Opinion has swung away from assigning a major role to summed spike activity. Many would now assign the main role to synaptic potentials, whereas others would add a significant contribution from pacemaker potentials and afterpotentials, and from changes in neuroglial cell polarization that result from neuronal activity. Generally the dendrites are considered the central structures, and some speculation about means of synchronization would have them interacting weakly without the necessary intervention of impulses. Some slow waves travel along the cortex at millimeters per second or slower, and may even be synchronized across complete cuts, suggesting that diffuse fields can mediate the synchrony without requiring a common synaptic drive. Triggering loci in the thalamus appear to pace some rhythms synaptically in widespread areas of cortex. Spikes in single cells may show no relation to waves in the region or may tend to fire at preferred phases, once or several times per wave. Similarly, intracellular slow potentials in single neurons may show no relation or a good correlation with the macroscopic field potential.

The possible **significance of brain waves** in brain function is problematical. Some treat them as epiphenomena, the result of brain activity but without causal influence, like the noise of a car. Others suggest the fields of current may exert some effects on the probability of firing or on the synchrony or other aspects of neuron activity. Only if the last speculation is true would brain waves and diffuse fields around cells represent a form of code. At the least, we can point to some intriguing correlations between EEG changes and behavioral states (Figs. 6.17; 6.18).

Evoked Potentials; Time-locked Activity. This term is used for responses, generally of a population of units, that are time-locked to a distinct stimulus. It is therefore a wave complex superimposed on the ongoing brain-wave background (Fig. 6.19). If it is small relative to the brain waves, it may be brought out by averaging the activity after each of many repetitions of the stimulus.

The "direct cortical response" is such a potential, evoked in the cortex by electric shock to nearby cortex. Electric shocks applied to subcortical structures and evoked potentials in other parts of the brain have also been much studied. Although such synchronous and anatomically defined stimulation is quite abnormal, it permits analysis of organization by experimenting with consistent

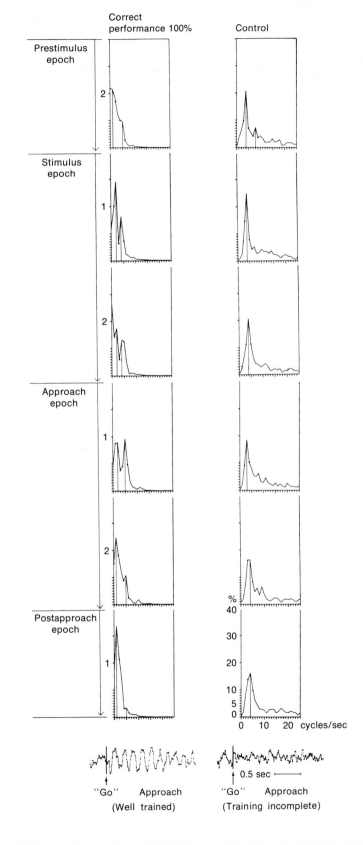

Correct performance 100%   Control

Prestimulus epoch

Stimulus epoch

Approach epoch

Postapproach epoch

0   10   20   cycles/sec

"Go"   Approach
(Well trained)

"Go"   Approach
(Training incomplete)

0.5 sec

Figure 6.17
EEG while making a choice in a light-dark discrimination. **Above.** Computed spectral density (relative power at each component frequency) of EEG's from dorsal hippocampus of cat. *Left:* record made during choice by well-trained cat. *Right:* no-stimulus controls. Notice the complex shifts of peaks. [Elazar and Adey, 1967.] **Below.** Two samples of records like those analyzed above. At the arrow, the cat receives the signal to make its choice between a correct or an incorrect signal. The rest of the sample is taken during its approach to the choice point. Each trace is the average of twenty runs. [Adey and Walter, 1963.]

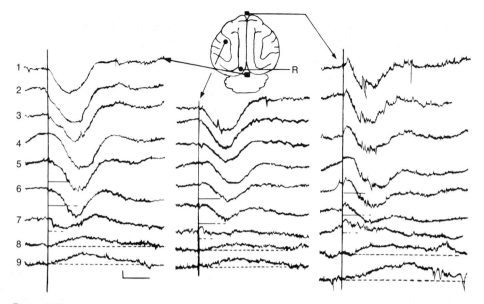

Figure 6.18
Slow wave changes with lapping of milk and gradual satiety in a cat. Records from three brain loci during nine serial offerings of liquid food to a cat deprived of food and water for 24 hours. Horizontal solid lines show the period of lapping (shown on trial 5 for trials 1 to 5). Trials 6 to 7, partial satiety with intermittent lapping; trials 8 to 9, sated animal looking at food but not eating. Large, 3-minute, negative shift reverses to a positive shift with reduction of the drive. R, reference electrode. Calibrations: 100 μV, 2 min. [Rowland, 1967.]

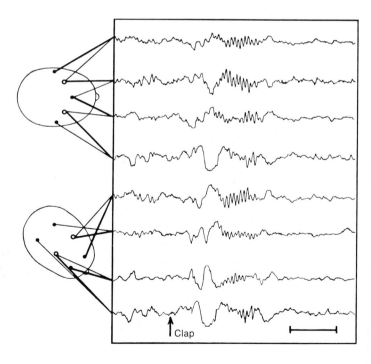

Figure 6.19
The evoked potential. Effect of a sensory stimulus (clap) on the EEG; human subject, asleep. A complex sequence of slow waves, superimposed on the ongoing brain waves, includes a "spindle" of faster waves. Calibration: 1 sec. [Brazier, 1968.]

properties not otherwise observable. Potentials evoked by "physiological" or "adequate" stimuli (i.e., relatively normal sensory input) are also used extensively, wherever the onset of the stimulus can be timed, to allow demonstration of its causal influence, training of the sensory pathway (Fig. 6.20), analysis of the integration of converging influences, assessing possible damage to parts of the brain in clinical diagnosis, and other kinds of studies.

The **interpretation of evoked potentials** in terms of cellular events is still uncertain, despite a vast amount of work including unit recording that reveals close correlation of some neurons to certain phases. The phase can change with subtle alteration of stimuli. The evoked potential probably includes components due to arriving action potentials in long afferent axons, postsynaptic potentials, local spikes, and afterpotentials. There is time for recurrent activity and for repetitive and complex interrelations among

layers of the cortex, or subcortical loops. Since the possible cellular basis is complex, not only in underlying processes but in the roles of subsets of the population and the sequences of the relevant events, it is usually necessary to deal with these potentials phenomenologically, as though a new form of graded

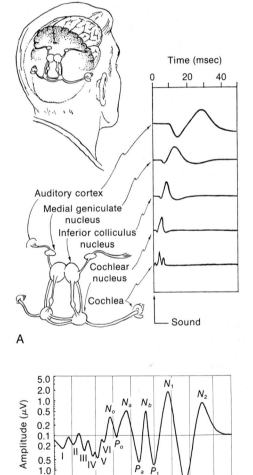

Figure 6.20
Evoked potentials at various stages in the sensory pathway. **A.** Sound activates the cochlea and then the successive auditory centers, diagrammatically shown. The evoked responses, as recorded directly from each center, are also shown diagrammatically. [Galambos, 1962.] **B.** An actual record of responses to clicks (led off by an electrode on the skin on top of the head) referred to an "indifferent" electrode on the mastoid area. The complex sequence of responses is exhibited on a logarithmic time scale to accentuate the small early peaks. This requires averaging many hundreds of trials. The early peaks are attributable to brainstem centers, the late waves to higher centers. The waves are quite consistent in healthy brains, and are assigned standard symbols, as shown. [Picton et al., 1973.]

responsivity had been found and required characterizing. Recovery of evoked potential responsiveness after a stimulus can be complex, with both facilitation and depression of different component waves. Recovery to 50% amplitude can require more than 200 msec or less than one msec, in different cases —providing an evidence of specialization in different structures or species.

It should be emphasized that the macroelectrode-recorded evoked potential cannot tell us what kinds of unit responses the microelectrode might find in the same mass of cells. Nor can even a large sample of single cell records (in general, selecting larger cells and not representative of the majority) tell us with any confidence what the evoked potential properties will be in the usual brain center, especially in the absence of depressing drugs used for anesthesia. Each form of recording is a useful window, pending the technical achievement of many-channeled, simultaneous, unit-recording from a significant sample of the cells. It is a bit like recording from the crowd at a football stadium, using a single remote microphone or a few microphones close to some individuals; they are not redundant, and **neither one nor both together tell the whole story.**

The question of **possible causal significance** is like that mentioned above for brain waves. Perhaps much or all of the summed activity of the assembly, even to normal stimuli, is merely a sign of the pattern of action, useful only to an instrumented observer. But arguments can be marshalled that suggest at least some role for the field potentials in influencing the probability or timing of neuron activity in some situations. Mainly the case rests on evidence that cells can be sensitive to external fields of

this strength. Cells not specialized for electrical reception (e.g., crayfish stretch receptor neurons) have been reported to be influenced by potential gradients of a few microvolts across the cell ($1 \mu V/\mu m$), perhaps less. Cells specialized for electroreception in cutaneous sense organs of certain fish are normally responsive to even feebler fields. ($<0.1 \mu V/cm$). The same may be true of cochlear hair cells. In no case, however, do we know the actual gradient at the detecting locus, and measurements of sensitivity of cells in the brain are still unsatisfactory. Indirect evidence that weak fields may be influential comes from behavioral effects of low voltages imposed between plates positioned in the air on either side of the head. When the resulting gradient in the brain is said to be microvolts or fractions of a microvolt per centimeter, people may show changes in circadian rhythms or in estimating times.

Whatever the causal significance or lack of it, the evoked potential provides an easy way to study the overall activity of a population of cells, organized and dynamic, in respect to either natural or artificial stimulus parameters. For example, we may ask the question, "What stimuli and states can we, as investigators, discriminate using evoked potentials as our measure?" Since the measure is crude and blurs or even cancels responses of equal and opposite polarity, we may assume the **system must actually discriminate** much more than we can by looking at averaged evoked potentials. But they may be our first direct physiological sign that the organism can detect or discriminate some suspected stimulus. Thus when sound gives a good evoked potential in the midbrain tectum of snakes, we can really say for the first time that the brain of a snake is not deaf.

The same method lends itself to determining the range of hearing in sea lions versus porpoises or in normal versus deaf human infants. The organism must hear at least as well as indicated by the evoked potential, and probably substantially better.

Besides the threshold of an evoked potential occurrence, the wave form can sometimes reveal a difference in **response to rather subtle differences** in stimuli or background states of the organism (Fig. 6.21). Thus when a flash of light illuminates a word, (carrying associations) printed on a card, the form of the response in man differs from that elicited by the same letters in nonsense order. In another example, if a subject is asked to indicate which of two sounds occurred, "ba" or "da," both of the same low, initial fundamental frequency, the evoked potential over the left (speech) hemisphere is different from that recorded when the task is to indicate whether the sound was "ba" at low frequency (104 Hz) or at a slightly higher initial frequency (140 Hz; the syllables are synthesized but sound as though sung). Over the right hemisphere there is no difference. Probably the difference over the left hemisphere requires that these sounds be discriminated in the language of the individual. It is not astonishing that the brain can make such a discrimination; but perhaps it is surprising to be able to record a sign of them; cells that can do this might be hard to find!

D. Reliability, Redundancy, and Recognition

In an information machine of some complexity, noise and the reliability of components become important problems. It

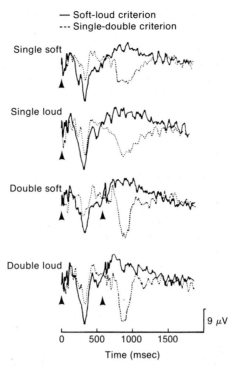

— Soft-loud criterion
--- Single-double criterion

Single soft

Single loud

Double soft

Double loud

9 μV

0   500   1000   1500

Time (msec)

Figure 6.21
Evoked potential sensitivity to context, with the same physical stimulus. Average wave forms obtained in response to four types of clicks. Solid lines represent wave forms obtained when the subject is guessing "soft" versus "loud." Broken lines represent wave forms obtained when the subject is guessing "single" versus "double." Triangles indicate points at which clicks were delivered. Labels at the left specify the physical characteristics of the stimuli. [Sutton et al., 1967.]

is often said that the nervous system, with its $>10^{11}$ cells, operates primarily probabilistically, with noisy and unreliable nerve cells, and that it must average over large numbers of equivalent or redundant cells to escape a serious, intrinsic uncertainty. Since there is bound to be some noise, the question is one of degree. Attempting to estimate degree also compels the question of distinguishing noise.

Box 6.4 Recapitulation of Some of the Variables Available for Integration in Neurons and Networks of Them

In earlier chapters we emphasized the common denominators of nervous elements—nerve cells, axons, synapses, impulses, and p.s.p's—as a first approximation to understanding the neural machine. It is justifiable simplification to treat neurons as all alike, up to a point.

Here we restate some of the reasons, brought out in this chapter and elsewhere, that a more sophisticated view of how the nervous system works must emphasize the differences among nervous elements. For brevity, the following list groups alternatives that vary on the same dimension. Even within one of the alternatives mentioned, neurons can differ importantly in degree. Items are arranged in an approximate order from lower to higher level.

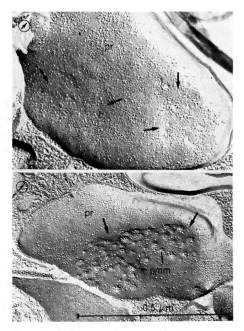

Freeze-etch electron microscopy. Synaptic regions in anesthetized and alert animals. **Above.** Presynaptic terminals of anesthetized rat seen by viewing the inner leaflet after fracturing the outer leaflet away. Shallow indentations are from nonpenetrating vesicles. **Below.** Same from alert rat; more signs of deep vesicle indentation. Lower center shows fractured section of cytoplasm with vesicles. [Akert and Livingston, 1973.]

1. Spike threshold can be high or low (i.e., depolarization from resting level necessary to fire).
2. Recovery cycle through the relatively refractory period can be slow or fast.
3. Subthreshold local potential can be lower or higher in the steepness of its nonlinear response function.
4. Safety factor can be high or low, especially at axon branch points.
5. Accommodation can be large or small; fast or slow.
6. Time constant of membrane can be large or small, especially in dendrites and terminals.
7. Space constant of membrane can be large or small, especially in dendrites and terminals.
8. Afterpotentials can be large or small in the magnitude and duration of each successive phase.
9. Iterativeness can be tonic or phasic or both; fast or slow or both in succession.
10. Authorhythmicity can be large or small; the neuron spontaneously active or driven or both.
11. Pacemaker units can be invaded and reset by input or not.
12. Influence by hormones, CO_2, temperature, and other agents can vary in direction and degree.
13. Quantal miniature junction potentials can be great or few in number; large or small in amplitude.
14. Synaptic transmission can be chemical or electrical.
15. Synaptic transmission can be excitatory or inhibitory.
16. Synaptic transmission can be polarized or unpolarized.
17. Synaptic transmission can be high gain or low gain.

Box 6.4 (*continued*)

18. Postsynaptic potential can be monophasic or biphasic.
19. Postsynaptic response can be slow or fast.
20. Postsynaptic decay can be solely passive or partly active.
21. Postsynaptic response can depend mainly on potential change or on conductance change.
22. Postsynaptic inhibition can occur with increased or with decreased conductance.
23. Postsynaptic inhibition can occur with discrete i.p.s.p.'s or i.l.d. (inhibition of long duration, up to many minutes).
24. Postsynaptic activity may or may not exert influence back on presynaptic ending.
25. Postsynaptic response can be facilitating or antifacilitating or neither.
26. Postsynaptic response after the end of input can continue or rebound or neither.
27. Facilitation can be fast or slow or both in succession.
28. Input sequence and interval can be critical (e.g., A *then* B is critical = heterosynaptic facilitation) or not.
29. Firing rate/depolarization function can be steep or shallow.
30. Firing rate can be more or less subject to autoinhibition or some form of saturation.
31. Firing rate can be regular or patterned or bursting or irregular.
32. Proportion of signal to noise in neuronal activity can be higher or lower.
33. Electrotonic connection between cells can be low or high in resistance.
34. Electrotonic connection between cells can be low or high in capacitance.
35. Synchrony in subthreshold potentials between units can be strong or weak or at chance level.
36. Synchrony of spike firing can be strong or weak or at chance level.
37. Recognition units can have lower or higher complexity of input requirements.

38. Command units can be of greater or lesser effectiveness (i.e., one or several units command a large musculature); sustained or triggering; high or low in responsibility.
39. Redundancy of neurons can be high or low.
40. Central determinants or sensory determinants of the time course of activity can be relatively more important.
41. Set point of a control circuit can be adjustable or relatively fixed.
42. Potentiation can be present or absent.
43. Recruiting response can be present or absent.
44. Augmenting response can be present or absent.
45. Kindling can be present or absent.
46. Evoked slow waves can be large or small, early or late, simple or complex.
47. DC and slowly fluctuating potentials across cells and masses (the EEG and related states) can form fields of larger or smaller equivalent dipoles, simple or complex form, widely different power spectrum, and varying coherence with other areas.
48. Plasticity as a result of environmental influence or experience can be high or low.

In order to appreciate, even in such a condensed overview, the true repertoire of variables available for integration, it is important to add to this incomplete physiological list at least a reference to the many chemical and anatomical parameters. These are basically the underlying mechanisms for the functional properties listed. The richness of variety in both domains, but especially in the anatomical, with its permutations of connectivity, offers a virtually limitless substrate for physiological operations. The microstructural parameters should not be viewed simply as various forms of fixed wiring, but are themselves dynamic, changing to some degree with some states. The figure exemplifies one of the forms of alteration at junctions; others include the size and number of synaptic contacts, the size and number of dendritic spines (see also pp. 41, 213), and the resistance of electrotonic junctions.

The conclusion often voiced—that a large degree of uncertainty prevails and requires the brain to operate probabilistically—is attractive and plausible, but evidence that is more than compatible is meager. The obvious arguments based on the large number of neurons available, the seemingly high redundancy, and the irregular background activity of the brain waves are more impressionistic than exclusive. The facts do not speak against the alternative view that the brain is very compex, the cells fairly reliable, and the redundancy for typical cells incomplete. The principal hard data for unreliable neurons are the variable responses of single units and evoked potentials to repeated, physically identical stimuli.

On the one hand, the limitation of the last argument is that **variability in response cannot be assumed to be slop,** or noise (see Glossary for definition of "noise"). The nervous system may have changed even though the apparent stimulus has not, by shift of mood, attention, or the like. The measure of response we use may be irrelevant, whereas variation in terms of the relevant code may be much less. Apparent noise or jitter may in some cases be advantageous and functional. In short, the probabilistic assertion, as stated above, is virtually untestable.

On the other hand, it is demonstrable that **neurons can act quite reliably** in a growing list of cases. Interval fluctuation in a train is commonly only 2 or 3% (e.g., "clock" neurons in insect CNS, vestibular neurons in vertebrates) and can be as small as 0.01% (standard deviation/mean interval, in electric fish pacemakers)! Some unit responses to repeated stimuli are astonishingly regular. For example, a lateral line electroreceptor afferent of a fish responded to single, 0.5-msec stimulus pulses with 14 ± 1 spikes, and all the intervals but the last were constant within less than 1% in 192 trials. Reliability is also indicated by well-specified connectivity (Chapter 3). Sensory recognition units, such as small, dark, moving object detectors (treated in Chapters 7 and 9) have such complex stimulus requirements to make them fire that a relatively high specificity of connections and transfer functions must exist. The same is true of command cells (Chapter 7) that trigger complex behavioral acts. In Chapters 2, 3, 7, and 10 we cite the impressive array of recently established "identifiable neurons"— specific in connections, spontaneous rhythm, pharmacology, and coupling functions. To be sure, these are still chiefly known only in invertebrates. But the specificity of connections in the vertebrates is also becoming greater with each advance in neuroanatomy, and the uncertainty, slop, and potential redundancy correspondingly less. To deny the existence of neuronal individuality is as unjustifiable as it would be for a visitor to a strange land to deny individuality in its inhabitants before he learns to recognize it. The demand on genetic specification does not seriously limit individuality, for a few instructions in terms of gradients of influence, permuted parameters, sequenced milieu, critical periods, neighbors, and input could generate great complexity.

There are certainly sources of meaningless fluctuation, including threshold fluctuations, the convergence of independent rhythms, spontaneous activity, and other events with good causes but no signal value to the system. This is true **noise,** and it may be large or small in proportion to signals; highly structured, or quasirandom. Since it is defined not by its causes or structure, but its lack of meaning or value, it is difficult to identify in any given case. We rarely know the system well enough to be confident of what has no value. Heuristically it is better not to label unexplained activity or fluctuation as noise or uncertainty. Calling it something like "unexplained variation" may encourage the search for both causes and possible significance.

The same holds for structural **redundancy** (see Glossary); there is no doubt that it exists in the nervous system, but the problem is one of degree and type. The term is used here not in the sense of "superfluous" or "exceeding what is natural," but in the sense of "equivalence," suggesting the existence of many neurons with either complete or partial overlaps of their input and output fields. Overlap is especially conspicuous in sensory pathways. Large numbers of receptors with overlapping receptive fields and nearly the same threshold are characteristic of verterbrate sensory systems, and even in invertebrates, in which economy in cell number is common, sensory fibers constitute a large proportion of the total nervous system.

It is doubtful, however, that these units are entirely equivalent; a certain degree of nonoverlap in either receptive field or output influence is usually found, and may be critical to sensory function. To that extent the units are not redundant. It is better to regard the design feature we have been considering as the extensive use of partial overlap, so that neither the redundant nor the nonredundant fraction will be overlooked.

Several thousand cells in the same cortical column for a particular visual field and preferred orientation of light-dark boundaries (Fig. 3.12) may respond to similar stimuli, but the equivalence of the cells may be reduced by quantitative differences in threshold, in preference for contrast, image size, disparity of left and right eye axes, ocular dominance, efferent connections, and other properties.

Redundancy does not necessarily mean unspecified or random connectivity. Nor is it necessarily wasteful, superfluous, or primarily a protection against possible injury or failure of components. Some **possible values of redundant units** are easy to see. It can permit greater sensitivity and resolution, improve signal-to-noise ratio, and reduce the demand for precision in ontogenesis and for stability of performance. Most of these benefits derive from the proposition that a significant part of the physiological activity of each unit is not governed by the desired signal but represents apparent noise, either as baseline spontaneity or irregularity in the coding of generator potentials into spike potentials. Assuming there is irrelevant activity, such as coding noise, it is likely to be independent in parallel neurons; thus any measure of the activity of a population would be sensitive to lower stimulus intensities, smaller increments, briefer events, higher frequencies, and wider dynamic ranges than a single neuron, and less disturbed by independent drift, or instability, or loss.

Spontaneous activity has the value of providing a background level of activity against which the organism can detect both increases and decreases in signal strength. By using receptors with a wide range of different thresholds, but the same stimulus specificity and receptive field, a wide dynamic range can be achieved. The vertebrate eye, for example, functions over an intensity range of 10^{10} to 10^{12}, whereas any given receptor has a dynamic range of only 10^2 to 10^3. Note that, to the extent spontaneous activity is useful and relevant to the function of the subsystem, it is not properly called noise. Its independence in parallel units is what makes the redundancy valuable.

We know that there are many identifiable—that is, unique cells (pp. 54, 99)—but we cannot as yet place any upper limit on their numbers. It seems likely that many degrees of equivalence will be found—from a very few identical cells to at least hundreds, in different places and species. Still more numerous, no doubt, are nonidentical cells with varying degrees of overlap of function.

Integration of large numbers of converging inputs to give a single output is of course normal for most neurons, as it is for muscles and for joints and higher entities. This can, in a sense, be thought of as averaging and as a **probabilistic operation.** Probabilistic methods are certainly necessary and powerful tools to be used in studying what we can, in practice, measure; they are surely used by the brain too, at least to some degree, for some transactions.

This consideration also makes cogent the question, "Where are the crucial convergences of behavioral interest?" For example, where are the averages taken or the integrations accomplished that determine either/or decisions such as the onset of flight, of song or attack, or, without immediate motor consequence, of recognition? As Kornhuber has pointed out, our awareness combines into a unit the various sensory inputs from a brown, long-legged and long-haired, moving, barking, odorous dog. Such "decisions"

may be at a high level in some terms and beyond the scope of neuronal integration, but we cannot be sure, hence it is as well to introduce the problem here.

Decisions of all levels of complexity or importance depend on the inputs, or intrinsic states, achieving a preset criterion—a threshold—defined by quality and quantity. If the integration is performed in the nervous system, then there must be, before the command enters the effectors, a final **determination by some competent functional unit** whether the criterion has been met. All relevant input and central predispositions must be read and finally evaluated by a single integrator after all the partial evaluations. Whatever its composition, it is a unit, because it is necessary and sufficient to perform this act.

If averaging of many penultimate integrators is involved, something must perform the averaging and deliver an unequivocal answer, or else the action is not centrally determined. No doubt some actions are integrated finally only by mechanical averaging of the activity of motor units. But it seems likely there are many that are integrated centrally, such as the examples mentioned above. For each of these, or more precisely for each pair or set of alternative events, there must be a single final functional unit (or a number of equivalent units) upon each of which all preceding lines converge.

Since the final functional unit is, like a military general, competent to render an answer, it may be called a "decision" unit, or recognition-of-criterion unit, or high-level trigger unit. The term "decision" may suggest a lack of determination by input. We assume determination for all actions and use "decision" precisely because it is those actions that have been said to be "decided" that we wish to cover—that is, to include with all other actions determined by input and existing biases and patterns. Not only are there many such units—one for each pair or alternative set of recognitions that the organism has acquired by birth and training—but there is physiological evidence to support the theoretical expectation that many **successive labels of recognitions** to subcriteria occur, which gradually converge on the narrowest bottleneck in the stimulus-response sequence. The analogy to an army suggests the hierarchy of convergences and integrations of lines of incoming information, the encoding, decoding, and re-encoding, the predetermined differences in the value and the meaning of the same message depending on what line it comes in on, the processing of data from lower levels in order to make interpretation possible, and other similarities. All these are features of our present picture of the nervous system. A more extended discussion of the characteristics of decision-making in the nervous system can be found in Bullock and Horridge (1965, pp. 282 *et seq.*).

What are these decision units, which, like miniature sentient beings, receive all that goes before and act adaptively to determine an output? They must be numerous, but they need not be everywhere the same, either in composition or mechanism. Several possibilities seem compatible with known neurophysiology, and not mutually exclusive.

1. Single neurons can certainly be decision units. Every neuron is a decision-making element when it changes from one state of activity to a different one. They differ in the level of complexity of their inputs and the responsibility or consequences of

their outputs. Some well-known single neurons, like Mauthner's and other giant cells, demonstrate the suitability of this type of unit for complex criteria and life or death responsibility. Seemingly even more subtle criteria can be recognized by cells in the mammalian cortex that strongly prefer certain visual or acoustic patterns. We have no reason to place an upper limit on the complexity of recognitions achievable by sequences of neurons.

2. A theoretical possibility is that a mass of randomly connected neurons constitutes a trigger unit. Activity with a sharp threshold can spread through such a mass, which provides redundancy of a different kind than the duplicate, specified neurons plausible for the preceding case. The input and output connection of each mass would still have to be as specified as for any other alternative.

3. A third class of possibilities, perhaps closer to the usual view, is a multiple-input, metastable feedback loop. Here a specified meshwork of mutually interacting neurons has a threshold and diffuses the responsibility. While requiring a high degree of specificity of connections, into, within, and out of the circuit, redundancy can be added, as in each of the foregoing alternatives. One neuron can be part of several such unit circuits.

4. Some people respond to the question, "What is the unit that decides?" by pointing either to the whole organism or to the whole set of neurons activated by the adequate inputs for that decision. The former begs the question by overlooking the large parts of the organism not essential. The second alternative does the same to a lesser degree. It also begs the question how this set has a sharp threshold (for the either/or class of decisions we have chosen to discuss). It implies that many sets can have the same meaning, without any necessary convergence between them—for example, those that lead to the decision, "There's grandmother," via voice, footsteps, or snapshots. Nothing tangible has to connect the different sets with the same meaning; and nothing tangible is asserted to distinguish the large number of

sets that have this kind of sharp meaning (all of our relatively unequivocal recognitions) from the virtually infinite number of potential sets without such discrete meaning. Nevertheless, it is a real alternative and belongs in our list.

The four cases mentioned may be merely extremes on continua, in different dimensions. Theoretically any of them can provide the performance, stability, tolerance of disturbance, and modifiability we should like to impute. But the differences are large, and the problems that remain are major. In respect to theory we can hope for precise formulation and formal development of alternative propositions. The common appeal, for example, to a population-processing or probabilistic analysis of input to achieve recognition and decision is untestable in its vagueness, but might be formally stated in a model. In respect to experiment, we can look forward to learning the "natural history" of the types of neurons and dependencies, rules, and regularities there are, and to performing careful circuit analysis, leading to explanations of measured transfer functions in terms of neuronal properties and connectivity.

SUGGESTED READINGS

Brazier, M. A. B. 1960. *The Electrical Activity of the Nervous System* (2nd ed.). Macmillan, New York. [A small, general text dealing with this aspect of activity, from membranes to brain waves.]

Perkel, D. H. and T. H. Bullock, 1968. Neural coding. *Neurosciences Research Program Bulletin,* **6:**221–348. [The first systematization of spike and nonspike, single-unit and multi-channel codes and candidate codes in the nervous system.]

Quarton, G., T. Melnechuk, and F. O. Schmitt. 1967. *The Neurosciences: A Study Program.* Rockefeller Univ. Press, New York. [In each of three carefully planned volumes (see Schmitt, below) there are sequences of chapters that deal with neuronal integration.]

Regan, D. 1972. *Evoked Potentials in Psychology, Sensory Physiology and Clinical Medicine.* Chapman Hall Ltd., London. [A systematic treatment in textbook form, including a substantial section on methods.]

Schmitt, F. O., ed. 1970. *The Neurosciences: Second Study Program.* The Rockefeller University Press, New York.

Schmitt, F. O., and F. G. Worden, eds. 1974. *The Neurosciences: Third Study Program.* M.I.T. Press, Cambridge, Massachusetts.

Shepherd, G. M. 1974. *The Synaptic Organization of the Brain.* Oxford Univ. Press, New York. [An organized introduction used in a seminar course, relevant to many of our chapters. It has more detail on many aspects than the present book.]

Uttal, W. R. 1972. *Sensory Coding. Selected Readings.* Little, Brown & Company, Boston. [A convenient assemblage of original papers.]

Uttal, W. R. 1973. *The Psychobiology of Sensory Coding.* Harper & Row, New York. [A text on "the central problem."]

7

INTEGRATION AT THE INTERMEDIATE LEVELS

I. INTRODUCTION: DOMAINS IN THE INTERMEDIATE LEVELS

We are concerned in this chapter with the transactions that occur in organized subsystems of receptors, neurons, and effectors. Their mechanisms will be central to the issues of how the nervous *system* works as a communications machine that recognizes, decides, and commands. Of the various domains of operations that are available for our scrutiny, we include here five, and these form the headings of sections II to VI.

This is the first encounter, between these covers, with the physiology of useful arrays of neurons. It will help to introduce two classes of function that emerge from and pervade the activities of such arrays.

Each part of the nervous system—but especially the receiving side, from the battery of receptors to the networks of higher-order neurons of the afferent systems—can be thought of as a **filter.** The sense organs send a patterned stream of impulses, in space and time, to the afferent centers, and these represent to the afferent centers a coded form of the sense organs' selection, or filtrate, of the actual stimuli. The afferent networks—meaning the interconnected second-, third- and higher-order neurons that receive their input primarily from the receptors, though also from other sources—do not merely pass on this information. They use convergence of separate channels (comparison), divergence of each channel (parallel processing), lateral inhibition (enhancing contrast), and other processes to modify the signal. Special attributes of the original input are passed on (recognition), and much information is discarded. The structure and coupling functions of the network determine what gets through.

Comparable networks exist on the output side. They are also filters, but since they formulate and send to the effectors commands that are crucially patterned in space and time, they are often thought of as **pattern generators.** They convert triggering input or steering input from receptors or their own spontaneous discharge or a mixture of these

Figure 7.1
Recruitment in a muscle with three motor units.

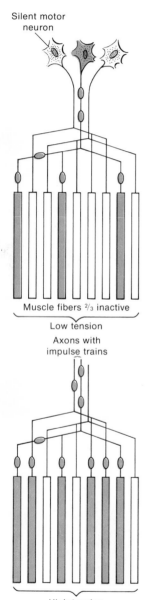

Silent motor
neuron

Muscle fibers ²/₃ inactive
Low tension
Axons with
impulse trains

High tension

into adaptively patterned ("coordinated") streams of impulses in the output channels. Again the connectivity and dynamic properties of the network determine its output, but not merely by passive filtering. Often there are intrinsic rhythms of spontaneous impulse bursts, and these result in the generation of specified constellations and sequences of activity in populations of units and finally in the effectors. Let us look first at the final link.

II. NERVOUS CONTROL IN EFFECTORS: DIVERSITY OF PERIPHERAL INTEGRATION

The best known vertebrate skeletal muscles consist of muscle fibers whose cell membrane is capable of producing propagated all-or-none impulses like those of nerve fibers. Each muscle fiber is innervated, in general, by only one axon with one terminal at the specialized end-plate. A motor nerve impulse arriving there gives rise to an end-plate or junctional potential, like an e.p.s.p., which is usually suprathreshold for the initiation of a propagated muscle action potential. Each motor neuron innervates many muscle fibers. The motor neuron and its muscle fibers comprise a **motor unit.** Each muscle contains from one to several thousands of motor units. An impulse in one unit, or a synchronous volley in many, causes a brief contraction to summate if they overlap, or to fuse into a smooth contraction called a **tetanus** if the frequency is high. When there are many motor units per muscle, a means of grading the strength of contraction of the muscle is to vary the number of active units. This method of control is called **recruitment** (Fig. 7.1). A

second means is to vary the frequency of impulses in each unit, since either the average tension in a series of twitches or the tetanic tension is a function of frequency. You can observe frequency and recruitment control phenomena by placing a stethoscope over your eyelid and listening to the twitches of its muscle while you control them willfully. Recruitment may be important in vertebrate skeletal muscle only at low tensions.

Since there are many motor units in most vertebrate skeletal muscles, operating these units out of phase with each other can give a smooth overall contraction even when the frequency in each is so low that it contracts in a series of twitches. Another possible utility of having many units is that there could be a rotation of activity during low or medium work loads, so that units could rest. But this old notion is apparently not confirmed in the best studied materials.

Other muscle fibers in vertebrates are incapable of producing propagated action potentials. They are innervated not by a single end-plate, but by numerous spatially distributed motor nerve terminals. Such **multiterminally innervated** muscle fibers (Fig. 7.2) respond electrically with junctional potentials only, but since these occur at many sites, electrotonically spreading depolarization can activate the whole fiber. Muscle fibers of this type may be mixed with others, as in some frog muscles. They are usually innervated by small diameter motor axons.

Since individual vertebrate muscles are excited by many axons, we think of them as being driven by pools of motor neurons. There is a diversity of size within the pool of motor neurons. Generally, **the smaller ones have lower**

Figure 7.2
Types of striated muscle innervation.

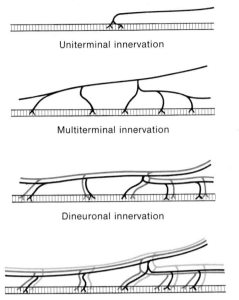

Uniterminal innervation

Multiterminal innervation

Dineuronal innervation

Polyneuronal innervation

thresholds for natural stimulation and are more tonic in their discharge. They innervate relatively fewer muscle fibers and produce weaker contractions, but are probably principals in the ordinary load of muscle work. The larger phasic fibers are called forth by stronger stimulation and produce vigorous action. This relation among threshold, fiber size, and tonic versus phasic mode of discharge is probably of general significance in neurophysiology. It is easily demonstrated in sensory systems as well.

Many whole arthropod muscles, and some in annelids, are innervated by only one or a very few motor axons. In muscles innervated by only one axon, recruitment cannot be like that in vertebrates, which add motor units by central enlistment of motor neurons. But a single axon can recruit increasing numbers of the muscle fibers it supplies because differences in the facilitation properties of the synapses allow transmission to be effective at different frequencies of arriving nerve impulses in different fibers. There are also differences in excitation-contraction coupling threshold in some muscles.

Innervation of all these muscles is of the multiterminal type. At normal frequencies of arriving nerve impulses, the junctions commonly exhibit the time-dependent properties of facilitation or antifacilitation. Contraction strength can be controlled not only by average frequency of motor axon impulses, but, in some muscles, also by detailed temporal structure of the impulse train. For example, the opener muscle of the crayfish claw is innervated by only one excitatory neuron, and the myoneural junction shows strong facilitation. A steady rhythmic train of impulses causes a contraction of a given strength. But if the same total number of impulses in a given time is delivered grouped in pairs, the contraction is stronger. The junction is considered to be **pattern sensitive** in that it responds to details of innervating temporal pattern, not just average frequency (Fig. 7.3,A). Recordings from intact moving animals show that the CNS sometimes issues motor commands in impulse doublets, but it is not yet certain how this potential code is employed, whether independently of or correlated with mean frequency.

Muscles may be **polyneuronally innervated** in that single fibers can receive input from more than one motor axon (Fig. 7.2). Each innervating axon differs in the properties of its endings on the same muscle fiber. One of the axons is usually a so-called "fast" or phasic axon that normally carries short bursts of

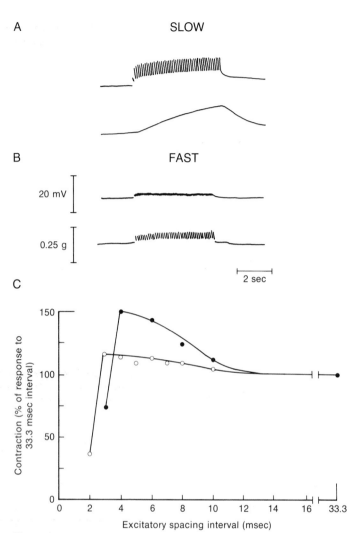

A SLOW

B FAST

20 mV

0.25 g

2 sec

C

Contraction (% of response to 33.3 msec interval)

Excitatory spacing interval (msec)

Figure 7.3
Differences among muscles in response to stimulus interval. **A.** A muscle fiber in the claw-closer muscle of a crab responds to repetitive stimulation of a "slow" axon with large action potentials (upper trace) having slow facilitation, and with gradually summating smooth tetanic contraction (lower trace shows tension). **B.** The same muscle fiber excited via a "fast" axon shows small, quickly plateauing action potentials and small, clonic (= nonfused) contractions. [Atwood, 1967.] **C.** Pattern-sensitive (filled circles) and pattern-insensitive (open circles) muscle of crayfish. At a constant *mean* frequency of 30 shocks/sec (= 33.3 msec intervals), trains of shocks are delivered either at uniform intervals (extreme right) or at alternately short (abscissa values) and long intervals; response plotted as a percentage of the evenly spaced contraction. Refractoriness reduces responses at the shortest intervals. [Ripley and Wiersma, 1953.]

impulses at high frequency for quick movements and produces large junctional potentials in the muscle fiber (Fig. 7.3,B), and these elicit spike-shaped local response potentials and rapid contractions. Another axon is a "slow," or tonic, nerve fiber that normally carries long trains of impulses at low frequency; these impulses decrement severely as they approach the nerve ending, and produce smaller and more facilitating junctional potentials that cause slowly developing contractions. Commonly there is also an inhibitory axon. Each innervating axon may end upon a large fraction of all the fibers of the muscle, the fast axon preferentially reaching the fast (short sarcomere) and intermediate muscle fibers; the slow axon, the slow (long sarcomere) and intermediate muscle fibers (Fig. 7.4). The presynaptic endings of the same axon upon muscle fibers of different sarcomere length and hence contraction properties differ in effectiveness, in arousing postjunctional potentials, and in degree of facilitation. An extraordinary heterogeneity of muscles and actions and their smooth gradation can now be understood, even in such animals as arthropods, in which only a few axons supply a whole muscle. This heterogeneity is the result of the different combinations of phasic, tonic, and inhibitory axons, the different degrees of polyneuronal innervation, the diversity of type of terminal of each axon upon different muscle fibers, the diversity of type of muscle fiber, and the diversity of rise and fall times of facilitation.

Inhibitor axons exert two distinct effects, presynaptic and postsynaptic (Fig. 7.5). **Postsynaptic inhibition** causes hyperpolarization and increased conductance of the muscle fiber membrane;

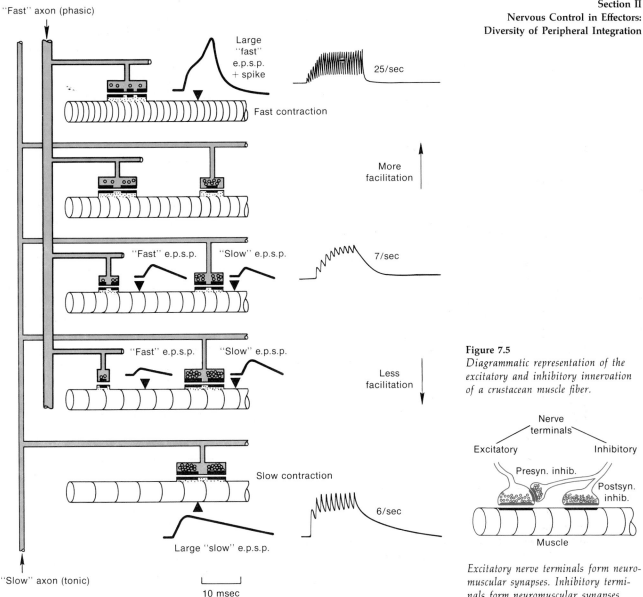

"Fast" axon (phasic)

Large "fast" e.p.s.p. + spike

Fast contraction

25/sec

More facilitation

"Fast" e.p.s.p.   "Slow" e.p.s.p.

7/sec

"Fast" e.p.s.p.   "Slow" e.p.s.p.

Less facilitation

Slow contraction

6/sec

Large "slow" e.p.s.p.

"Slow" axon (tonic)

10 msec

Figure 7.4
Crustacean neuromuscular innervation. The diagram illustrates the matching of phasic ("fast axon") and tonic ("slow axon") innervation with different categories of muscle fiber (indicated by sarcomere length) in a crab muscle. [Atwood, 1973.]

Figure 7.5
Diagrammatic representation of the excitatory and inhibitory innervation of a crustacean muscle fiber.

Nerve terminals
Excitatory   Inhibitory
Presyn. inhib.
Postsyn. inhib.
Muscle

Excitatory nerve terminals form neuro-muscular synapses. Inhibitory termi-nals form neuromuscular synapses (postsynaptic inhibition) and axo-axo-nal synapses upon excitatory terminals (presynaptic inhibition). [Lang and Atwood, 1973.]

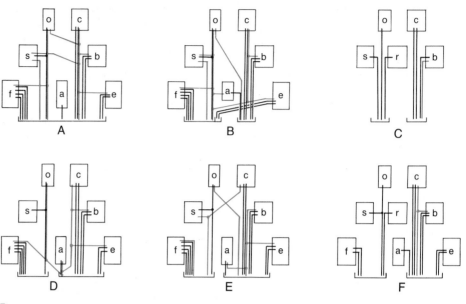

Figure 7.6
The anatomical distribution of axons to the distal thoracic limb muscles of several groups of
Malacostraca. Black lines represent motor excitor axons; colored lines represent inhibitory axons.
Brackets indicate the distribution of axons between nerve bundles. **A.** Brachyura. **B.** Anomura. **C.**
Stomatopoda. **D.** Palinura. **E.** Astacura. **F.** Stenopodidea (Natantia). The boxes represent the muscles
in the three distal segments of the legs, as indicated by their common names: *a*, accessory flexor; *b*,
bender; *c*, closer; *e*, extensor; *f*, flexor; *o*, opener; *r*, rotator; *s*, stretcher. [Wiersma and Ripley, 1952.]

presynaptic inhibition reduces miniature
e.p.s.p.'s and the amount of excitatory
transmitter released per impulse. Both
depend on the arrival time of the inhibitory impulse relative to the excitatory,
but in different ways. They are widely
different in importance. Inhibition in
some muscles is mainly postsynaptic; in
others, presynaptic. In some muscles the
response to a slow excitor is inhibited
presynaptically, and the response to a
fast excitor in the same muscle is less
inhibited. In the crayfish claw opener
muscle, inhibition caused by 10 impulses
per second is 90% presynaptic, but at 40
impulses per second it is only 50% presynaptic. Inhibitory transmitter is re-

leased with facilitation, which varies
among junctions.

An additional degree of complexity of
muscle innervation in arthropods can be
appreciated if one examines the innervation of the muscle of a whole limb, such
as the walking leg of a crayfish. Several
muscles may be innervated by a single
motor axon (Fig. 7.6). The overlapping
but nonidentical innervation patterns
still allow independent action. For
example, the opener and stretcher muscles are innervated by the same excitor
axon but by different inhibitors. Nevertheless, the whole limb is innervated by
only 12 excitatory and 3 inhibitory motor
fibers, and this fact of small numbers

makes it seem possible that we will be able to understand fully at the level of single nervous units the normal control of movement in arthropods.

The important lesson from the examples chosen is that even at the last link, the nervous control of effectors, there is diversity in the principles employed, some cases exhibiting a substantial degree of integration in the periphery. Wags have said crabs can think (evaluate) in their legs. We have not exhausted the diversity: fast insect flight muscles, a variety of smooth muscles (Box 7.1), electric organs, glands, cilia, chromatophores, and other effectors manifest additional principles of series and parallel control (Fig. 7.7).

III. ANALYSIS OF SENSORY INPUT: PARALLEL AND SERIES PROCESSING

The general principles of sensory reception are common to systems as different as vision and taste and to inputs that rarely reach consciousness, such as those from pressure receptors and chemoreceptors in the great arteries, as well as those that do. Foremost is the **principle of parallel channels:** many receptors independently sample the stimulus world and send their impulse-coded reports to the central nervous system in parallel afferent fibers. The basic independence may in some organs be abridged by superimposed influence from the central nervous system (centrifugal or efferent influence) or from neighboring receptors (lateral inhibition).

Intimately related is the second principle, that of overlapping receptive fields: neighboring receptors are gener-

ally stimulated by a fraction of the impinging world that is large but circumscribed (the excitatory receptive field, ERF) and which overlaps substantially with that of contiguous channels. This applies in some systems to topographic overlap and fields, such as areas of the skin, retina, and basilar membrane (which distributes sound frequencies to auditory afferents). In some systems the overlap involves ambiguity in modality (Chapter 6, p. 214).

The third principle relevant to the present section states that sensory input typically exhibits divergence and convergence in the central nervous system, the former distributing information to analyzers concerned with different aspects of the stimulus world, the latter permitting resolution of ambiguities, sharpening of fields, and abstraction of special features. Generally, **divergence predominates in sensory pathways,** so that even though each neuron still receives synapses from many different presynaptic cells, there are increasing numbers of cells at each higher neural level enroute to the sensory cortex. One characteristic example is the first- to second-order neuron stage in the auditory system, already mentioned (p. 108). The value of such wide divergence is that different postsynaptic cells, receiving input from subtly different, overlapping, presynaptic populations, can perform different types of information-processing operations in parallel, more fully extracting all available information. Thus a visual signal can be analyzed by different cell populations for brightness, shape, color, distance, and movement; an auditory signal, for frequency, rate and direction of change of frequency, duration, intensity, presence of overtones and temporal patterns, localization of the

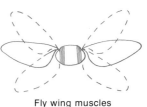

Figure 7.7
Various types of effectors.

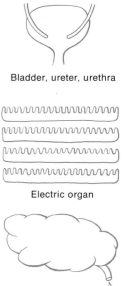

Fly wing muscles

Snail heart

Bladder, ureter, urethra

Electric organ

Salivary gland

Ciliated band

Chromatophore

Box 7.1   The Nervous Control of Smooth Muscle

We will briefly review the state of knowledge with respect to vertebrates. Prosser (1973), from whom we take much of this account, divides smooth muscles of vertebrates into two kinds, unitary and multiunitary, with some intermediates. Unitary muscles include those of the major viscera—uterus, ureter, and gastrointestinal tract. These show spontaneous rhythmicity, and distribution of excitation is electrical from muscle fiber to fiber. They can be stimulated by stretch and modulated by nerves. Multiunitary muscles include nictitating membrane and pilomotor, ciliary, and iris muscles. These are normally activated not spontaneously or by stretch, but by nerves or hormones; distribution of excitation within the muscle is normally by nerves. There may be facilitation of p.s.p.'s; several nerve fibers may influence one muscle fiber.

Unitary muscles and some others are typically arranged in bundles of fibers connected to each other by "nexuses," areas of close apposition and electrical low resistance. The motor nerve terminals release transmitter from large numbers of varicosities (swellings). Only some muscle cells receive direct innervation (see figure). Others are excited by their direct electrotonic coupling to these. Still others are excited indirectly, perhaps by a combination of electronic coupling and diffuse transmitter. In different viscera the smooth muscles vary widely in the proportions of these three types, as well as in cable properties, ion dependence, the effects of transmitters and drugs, sympathetic and parasympathetic innervation, and spontaneous activity pattern.

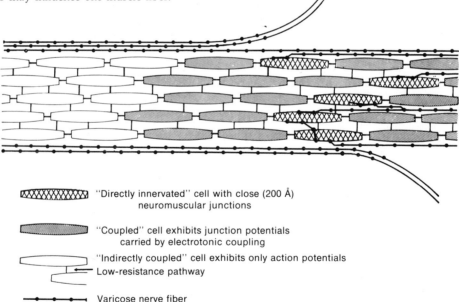

☒☒☒☒☒ "Directly innervated" cell with close (200 Å) neuromuscular junctions

▨▨▨▨ "Coupled" cell exhibits junction potentials carried by electrotonic coupling

▭ "Indirectly coupled" cell exhibits only action potentials
— Low-resistance pathway

•—•—•—•— Varicose nerve fiber

Schematic representation of the types of autonomic innervation of smooth muscle. All the smooth-muscle fibers are interconnected by low-resistance electrotonic junctions, shown here as bridges. Some receive direct nerve endings (dark shading); others (light shading) do not, but are close enough to the foregoing to show junction potentials that have spread electrotonically. The most remote (unshaded) show no junction potentials; they may be excited by electrotonic spread of action potentials from the preceding class and by transmitter released from the varicose nerve fibers at some distance. [Burnstock and Iwayama, 1971.]

source in space, etc. After building up sharply specific response requirements, these elements can again converge to combine their specificities.

Animals abstract from their total sensory input special qualities of that input before formulating a motor command. A frog jumps and snaps at any small, dark, moving object within range if it is in the mood. A male stickleback fish attempts to court with any oval, silvery, red-bottomed object of suitable size, whether it be a female stickleback or a crude model. A few characters of the whole constellation of visual inputs associated with the female stickleback seem to be the only relevant ones to the male (see Chapter 8). How might filtering of this quality (recognition) go on in the nervous system?

We could discuss filtering networks for any sensory modality, but space does not permit a survey. Instead we will concentrate on **visual filtering.** Since notions of modality, submodality, labeled lines, and temporal coding have been dealt with in Chapter 6, we can go directly to a consideration of relatively complex network functions.

The histological structure, cell types, and connectivity of the vertebrate retina have already been described (pp. 121–126), together with some of the electrophysiological properties expressing the coupling functions of the connections.

At least thirteen functionally distinct types of optic nerve fibers carry information to the brain from the eye in the cat. Table 7.1 gives a recent classification. Note that 92% have excitatory receptive fields (ERF's) that are concentrically organized. These are either brisk or sluggish, referring to the promptness and vigor of their responses, and either transient or sustained, referring to their

mainly phasic or mainly tonic character. For each of the four combinations of these properties there are two types, distinguished by the sign of their concentric organization. There are ON-center-OFF-surround units and the converse, OFF-center-ON-surround units. Units of the former sort are excited by the ON of a small spot of light in the central zone of the ERF or by the OFF of an annular illumination of the surrounding zone. Units of the latter sort are converse in character. The table lists all types of units in the order of their axon diameter, the largest first. The largest and fastest units, those in the brisk-transient group, are also called Y-neurons; the next largest units, those in the brisk-sustained group, are also called X-neurons. (All the rest, both concentric sluggish and nonconcentric, are collectively called W-neurons, but this term embraces too heterogeneous a set to be useful.)

For many years the cat was described as having essentially only two types of optic nerve fibers, ON-center and OFF-center. In contrast, the rabbit, ground squirrel, and gray squirrel were described as more frog-like in having several common nonconcentric and more complex types—for example, those preferring movement, some specific for direction of movement, some even for orientation. It now appears that the cat has about the same variety of nonconcentric units, including these complex types, but in much smaller proportion, 8% compared to 34% in the rabbit. In the cat, "local edge detectors" are five times as common as "direction selective" units; in the rabbit the latter predominate, except in the visual streak (the equivalent of the area centralis, for high resolution).

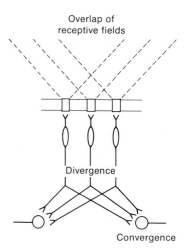

Overlap of receptive fields

Divergence

Convergence

All of these optic nerve fiber types bespeak transactions that process the information encoded by rods and cones. Lateral interactions, convergence, and highly specified connectivity (see Chapter 3) among receptor, horizontal, bipolar, amacrine, and ganglion cells are all indicated. This is usually referred to as early extraction of features to distinguish it from further changes in the meaning of cell discharge at later stages in the visual pathways. It also evidences parallel processing for central destinations of quite different function. The central targets of these optic nerve fiber types are known only in part. In the cat, X- and Y-fibers go mainly to the dorsal lateral geniculate, each to its private class of geniculate neurons; these in turn project to the visual cortex, X-fibers to the so called simple cells and Y-fibers to the complex cells. W-fibers go to the midbrain, largely to the tectum. It should be pointed out here that there are at least six central targets of optic nerve fibers, of which the dorsal lateral geniculate and the tectum are only two. Evidence suggests that they have different functions.

The retinas of frogs, lizards, and pigeons have even fewer concentrically

Table 7.1
Receptive field types of 960 cat retinal ganglion cells. Latencies are the ranges of antidromic conduction times from the optic tract stimulus site, and can be taken as proportional to the reciprocal of axon diameter.

Types	Number	Percentage	Latency (msec)	
Concentrically organized	887	92		
Brisk	774	80		
Transient	243	25	1.0–2.4	Y-cells
ON-center	115			
OFF-center	128			
Sustained	531	55	2.5–5.9	X-cells
ON-center	271			
OFF-center	260			
Sluggish	113	12		
Sustained*	44		4.6–24.0	
ON-center	22			
OFF-center	22			
Transient*	27		6.1–18.7	
ON-center	13			
OFF-center	14			W-cells
Nonconcentrically organized	73	8		
Local edge detector	45	5	6.6–15.9	
Direction-selective	11	1	6.1–12.4	
Color-coded	6	<1	3.8–14.2	
Uniformity detector	5	<1	8.7–13.9	
Edge-inhibitory OFF-center	3	<1	3.9–6.6	
Unclassified†	3	<1		

Source: Cleland and Levick, 1974.
* In a further 42, there were insufficient observations to distinguish whether sustained or transient (21 ON-center, 22 OFF-center).
† Insufficient observations to reach a conclusion.

organized ganglion cells and many more movement-specific cells. This is probably not a phylogenetic trend; it may have some relation to habit of life and the roles of vision. Frogs are known best from the pioneering work by Lettvin, Maturana, McCulloch, and Pitts, who used stimuli of more natural kinds than flashes of diffuse fields or focused beams, and from the quantitative work of the Grüssers (Fig. 7.8). In contrast to the cat, in which all optic nerve fibers are myelinated, the vast majority in frogs are very fine and unmyelinated. The five types of optic nerve afferents may be summarized in the following way. (a) Type 1 fibers, called "sustained-edge detectors," respond to a sharply focused edge of an object, light or dark, moving or recently having moved into the 1°–3° ERF. (b) Type 2 fibers, called "convex-edge detectors" (Fig. 7.9) respond only to a small object, darker than the background, that moves into or has recently moved into the field; they must detect not only the change of position in the ca. 3° ERF but also that there is little or no change of position in the ca. 15° surrounding inhibitory receptive field (IRF). (c) Types 3 fibers, called "changing contrast detectors," respond to any edges in motion, large or small, dark-on-light, or light-on-dark, if the contrast and rate are adequate and the edge is not too fuzzy. These fibers give a weak response to nonmoving temporal changes in light; they are classical ON-OFF units, but much prefer motion. (d) Type 4 fibers, called "dimming detectors" respond to any dimming or darkening, whether caused by motion or not; these are OFF fibers but not like the concentric OFF-center units of the cat. (e) Type 5 fibers respond to any brightening; these are ON fibers, but not concentrically organized,

with an OFF-sensitive surrounding excitatory field, as in the cat; they are more sensitive to blue light than to white or other colors.

Types 1 and 2 are fine, unmyelinated axons in the optic nerve and by far the most numerous. Types 3, 4, and 5 are myelinated. Types 1 to 4 go to the optic tectum of the mesencephalon, type 5 to the dorsal lateral geniculate of the diencephalon. Histological types of ganglion cells with distinctive dendrite branching patterns are probably associated with these fiber types (see Fig. 3.4).

Types 1 and 2 do not respond to general illumination, and their responses to movement are not influenced by the level of illumination over a range of intensity of more than a hundredfold. Given their requirements they discharge at a rate that encodes contrast of the moving object (but not light level) and rate of motion of the object minus that of background objects in the IRF (but not relative motion). The discharge is ambiguous for certain severely constrained combinations of object size, contrast, speed, and amplitude of movement.

Change of position is as good a stimulus as visible movement; thus a good response is elicited by briefly illuminating a static scene (no response) and then, during the dark period between flashes, moving an object within the ERF. The "memory" of the position of objects during the first flash lasts through at least one second of darkness (Fig. 7.10). An additional and remarkable property of type 2 units is **erasability.** When such a unit fires upon motion of a sufficiently convex edge toward the ERF center, it continues to fire at a lower rate for some seconds after the motion stops; but it

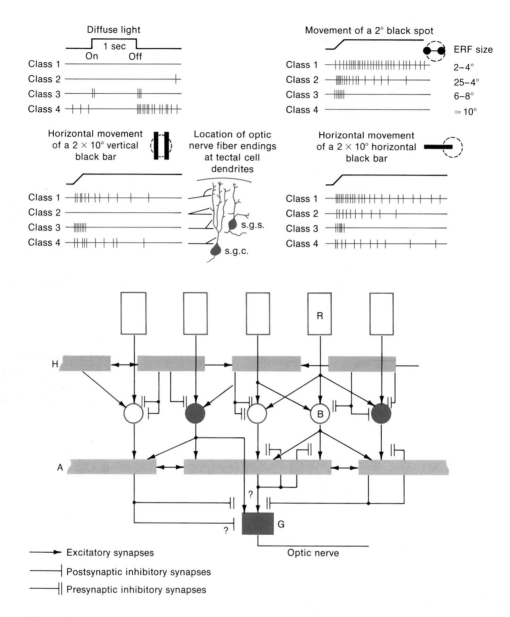

Figure 7.8

Types of ganglion cells and their optic nerve fibers in the frog. **Above.** The four types found in the optic tectum, distinguished by their responses to four kinds of tests. *ERF,* excitatory receptive field; *s.g.c.,* stratum griseum centrale of the tectum; *s.g.s.,* stratum griseum superficiale. **Below.** Diagram of the connections between the elements of the retina converging on a ganglion cell; presumably differences in the details of these connections and their transfer functions account for the types. *A,* amacrine cells; *B,* bipolar cells; *G,* ganglion cells; *H,* horizontal cells; *R,* receptor cells. [Grüsser and Güsser-Cornehls, 1972.]

253

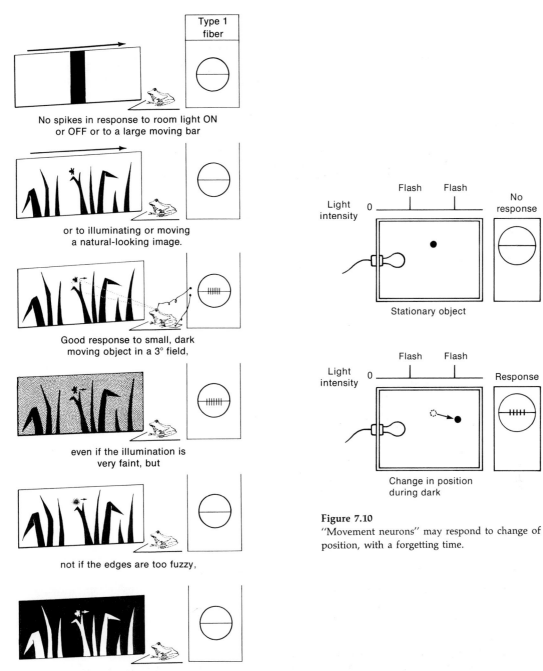

No spikes in response to room light ON
or OFF or to a large moving bar

or to illuminating or moving
a natural-looking image.

Good response to small, dark
moving object in a 3° field,

even if the illumination is
very faint, but

not if the edges are too fuzzy,

or to reversed contrast.

Figure 7.9
Abstraction early in an afferent pathway. Impulses recorded from a frog optic nerve fiber in the roof of the midbrain. [Based on data of Maturana et al., 1960.]

Stationary object

Change in position
during dark

Figure 7.10
"Movement neurons" may respond to change of position, with a forgetting time.

promptly ceases if the light is turned off briefly, and does not resume after the light is back on.

Notice that units of type 2 respond to any small, dark, moving object, especially if the motion is jerky. A hungry frog will jump at any such object. Normally any object having these characteristics in the environment of a frog will be a bug, an edible object. Since there are many detectors of type 2 spread over the visual field, the activity of one or more fibers of that type can both signal the presence of a bug and give its spatial coordinates, hence both activate and steer a jump and a tongue flick. Tadpoles may lack some of the types more important for the metamorphosed, hunting frog. Type 4 units in the frog's optic nerve signal general dimming, perhaps the approach of a predator or any large object.

The known visually oriented behavior of frogs in a natural environment includes finding food and avoiding obstacles and large, threatening objects. The filtering processes that occur in the retina in frogs can abstract the relevant aspects of the whole visual input signal for those behaviors. Studies by Ingle and by Ewert, using single-unit recording, ablation, and stimulation suggest that central processing builds on the retinal filtering by separately analyzing optic input for **different behavioral meanings in different structures.** The frog's attraction to blue depends on the dorsal geniculate; its avoidance of large obstacles in jumping and its retreat from large threatening objects may depend on distinct, overlapping parts of the pretectum, and its approach to food upon the tectum. We may speculate that the phase locking of its circadian rhythm with the environmental photoperiod depends on a hypothalamic center, the suprachiasmatic nucleus, as has been shown in rats. Each of these four central structures receives its own optic input, consisting largely of a distinct mix of the ganglion cell types.

As activity proceeds through first-, second-, third-, and nth-order neurons in a sensory pathway, the meaning of the impulse activity changes. We have been considering differences in meaning in parallel neuronal pathways as a result of divergence; we should note the differences in meaning in successive neurons as a result of convergence. When the criteria for firing include a significant fraction of the complex features of a stimulus that releases normal behavior, we may speak of **recognition cells;** this is a matter of degree. Higher-order cells may add dimensions to the criteria, such as novelty or familiarity. A novelty unit deep in the frog tectum may fire in response to a small, dark, moving object anywhere in a large (30°) field, but will soon cease, only to resume if the same or another object wiggles in a fresh part of the field. A familiarity unit fires in response to a similar object, but, instead of ceasing, continues, even maintaining a low rate of discharge ("muttering") for many seconds if the object stops moving. It now ignores fresh, moving objects within its large (30°) field. It will flare up if "its" object slowly moves about within the field, but will lose it and go silent if the object jumps too far—to a fresh part of the field!

In the auditory sphere too, we know of

units of a wide range of complexity of criteria. In bats, for instance, a series is found leading to cells that do not fire in response to any pure tone at any intensity but only to frequency-modulated tones with a certain range, rate, and direction of modulation, like the bat's echo-ranging cry. In squirrel monkeys, cells are found that strongly prefer a certain one out of some 20 tape-recorded sounds chosen from the 35 or so natural vocalizations in the species' repertoire.

Some workers distinguish between recognition of natural stimuli and feature extraction; the former is potentially more complex and may result from convergence of cells of the latter type. The term "feature" in this context means limited aspects of a complete natural stimulus, such as duration or frequency modulation. Our information is still too limited to decide whether some subsystems in some animals work differently from others in a fundamental sense. It is often supposed, for example, that some subsystems funnel successive feature detectors down to a single recognition unit, like a "bug detector," whereas others never quite converge the set of relevant feature detectors. We discussed the same problem from another direction on pp. 238–240. By whatever means, complex natural stimulus recognition must occur widely in nervous systems—sometimes early in the pathways, sometimes late.

Modulation of input by central influence via centrifugal (efferent) fibers is a potentially important part of the active filtering in many sense organs, as was noted on pages 170 and 225. Inhibitory

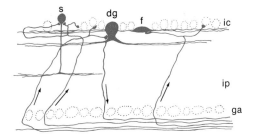

Figure 7.11
Efferent fibers to a sense organ. Axons from the brain to the retina in the pigeon. Golgi preparation. *dg*, displaced ganglion cell; *f*, flat amacrine cell; *ga*, ganglion cell layer; *ic*, inner nuclear layer; *ip*, inner plexiform layer; *s*, small parasol amacrine cell. [Maturana and Frenk, 1965.]

fibers to muscle receptor organs in crayfish are illustrated in Figure 2.75. The γ-efferents to muscle spindles are treated below (p. 267 *et seq.*). The mammalian cochlea (Fig. 2.80,C) and the avian retina (Fig. 7.11) are also well-known examples, but the functional significance of the efferents is not yet adequately understood.

IV. ELEMENTARY NEURONAL NETWORKS: EMERGENT PROPERTIES OF CIRCUITRY

In Chapter 3 we introduced some well-studied examples of connectivity and certain general principles of circuits of neuron-like units. Here we extend the discussion to emphasize the physiological consequences. Three kinds of arrays will be chosen; these may be called "networks," following the usage in the literature on neural modeling, meaning any assemblage of connected neurons.

A. Mutually Excitatory or Positive-feedback Networks

If two or more neurons are capable of exciting each other, then input that exceeds the threshold of one will likewise excite the others (Fig. 7.12A). Should the others exceed threshold as a result, they feed excitation back to the first, and a runaway process ensues. The whole network may come into a state of maximum activity and, without limiting processes, might stay in that condition. Most neurons have relatively long-term self-inhibitory processes, such as adaptation, accommodation, or fatigue. As the neurons in the positive-feedback network begin to fatigue, one or more of them will decrease in frequency. They then excite the others less, and hence

also receive less excitatory feedback. Just as the whole network was able to run away to maximum activity, it now runs away negatively to a minimum state. Once the adaptation or fatigue has worn away, a new cycle can begin. Networks of cells, each of which may not be capable of rhythmic bursts of activity, can produce bursts of more-or-less synchronous activity in all members.

Networks of this kind have been demonstrated in several cases and may be widespread. Inspiratory interneurons in the medulla are probably so connected, perhaps leading to their rhythmic bursting. Certain cells in the brain in several gastropod molluscs have been found to be positively coupled. They produce synchronous rhythmic bursts that act as triggers in the control of feeding and other activities. Decapod crustacean hearts are controlled by ganglia containing only nine cells. Ex-between some of these is known at least to aid in the buildup of the heart-beat.

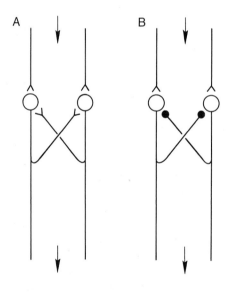

Figure 7.12
Simple networks with **(A)** mutual excitation and **(B)** mutual inhibition. Input comes from presynaptic axons. Excitatory synapses shown by forked axon terminals, inhibitory by terminal balls.

B. Mutually Inhibitory or Negative-feedback Networks

Two cells, or two clusters of cells that inhibit each other, may produce alternating single impulses or rhythmic bursts. This arrangement, called reciprocal inhibition, is a common feature of the control of antagonistic effectors or actions. The activation of motor neurons of one group of muscles is often coupled, both by feedback and feed forward, to inhibition of the motor neurons controlling antagonistic muscles (see Section V, p. 266). This kind of reciprocal inhibition is diagrammatically simple in

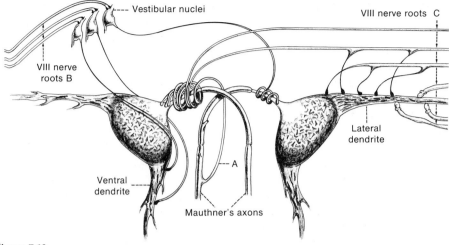

Figure 7.13
A reciprocally inhibiting pair of neurons. The giant cells of Mauthner in the medulla of fish, schematically shown, with the inhibitory collateral (*A*) indicated only on the left, the indirect VIII nerve afferents (*B*) only on the left, and the direct VIII nerve afferents (*C*) coming only from the right, although all these components are really bilateral. [Retzlaff and Fontaine, 1960.]

the case of the paired giant **Mauthner's fibers** of teleost fish and tailed amphibia (Fig. 7.13). The cell bodies and large dendrites of these neurons are situated in the medulla, where they collect input, including particularly that from the eighth cranial nerve (see also Fig. 2.17). The axons cross over before descending the spinal cord to synapse on the motor neurons of the longitudinal musculature. Each axon has a branch that ends in an inhibitory synapse on the contralateral Mauthner's cell axon hillock, near the spike-initiating site. (This is an electrical synapse.) An impulse in one cell prevents a simultaneous one in the other. Each descending impulse activates, nearly synchronously, an extensive longitudinal body wall musculature on one side, causing a twitch-like curvature of the posterior body region or tail. This startle reaction begins the familiar, sudden "jump" of some fish when the aquarium glass is struck. The vibrations excite the sensory endings in the vestibular apparatus of the ear. Although these cells have the largest axons in the body, and although each has thousands of input terminals and must be often bombarded with very many impulses per second, they produce only the occasional output necessary for startle reactions.

Another system of one-to-one alternation is found in the neurons driving the tymbal muscles in some cicadas. A pacemaker interneuron firing about 200 impulses per second drives two motor neurons, which each fire at half this rate in exact alternation and precisely phased with respect to the pacemaker. The alternating clicks of the two tymbals during song double the sound frequency possible if they were synchronous.

Reciprocal inhibition can lead to

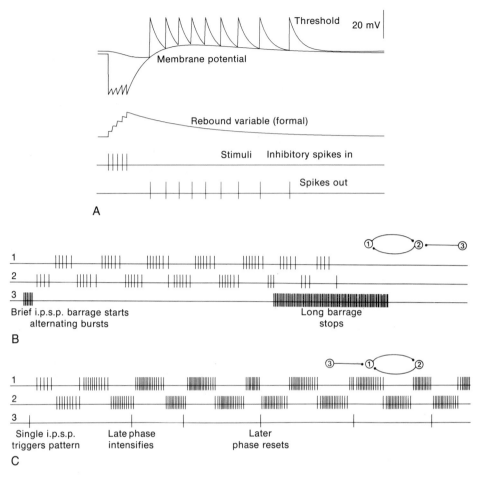

Figure 7.14
Reciprocal inhibition and postinhibitory rebound provide flexible mechanisms for generating bursts.
A. Postinhibitory rebound in a model neuron. A neuron that is not spontaneously active receives an inhibitory input and produces spike output by rebound. **B.** Two such neurons, reciprocally inhibitory, can give a long series of alternating bursts to a brief input barrage from a third neuron. **C.** A single input impulse has different effects according to the phase of the alternation when it arrives. [Perkel and Mulloney, 1974.]

alternating bursts of activity in otherwise nonbursting cells. The networks shown in Figure 7.14,B and C consist of only two interconnected cells, but they could represent two populations. Input to the network is inhibitory in this example, and reaches only one of the cells. The common neuronal property of **post-inhibitory** rebound (Fig. 7.14,A) is invoked in this model, and causes bursts that suppress the other cell. As the rebound burst in the first cell slows down, the second is disinhibited. Released and rebounding in its turn, the second fires and inhibits the first. As the second cell slows, the first recovers, and the whole cycle repeats.

Whether dependent on rebound or not, some such simple mechanism for **alternating burst activity,** though difficult to demonstrate, seems to operate in favorable materials, like the lobster stomatogastric ganglion. We believe it to be an important mechanism for many kinds of rhythmic and alternating behavior. Inspiratory and expiratory interneurons in the mammalian medulla probably inhibit each other. Locomotory systems in many animals may involve reciprocal inhibition between pacemakers for antagonistic muscle sets.

Reciprocally inhibiting networks can perform a variety of functions, according to the particular transfer functions of the synapses and the input and output connections. The following examples are theoretical and qualitative; proof that reciprocal inhibition is the mechanism that operates in living animals is incomplete. Given certain properties, reciprocally inhibiting networks can act as **gates,** switching rapidly from control of the output by one input line to another. This can recur at a steady rate, for a steady-state input, providing a **pacemaker** in which no single cell is the essential element, as discussed above. With certain dynamic properties the same simple circuit can act as an intensity-to-time converter that may be useful in comparing the strength of two inputs by their relative duration of control of some downstream system. Slight changes could provide **sensory scanning** by periodically or irregularly sampling each of a number of input lines and giving them control, in turn, of some later elements. In a cell with a large number of input lines converging on it, a drastic increase of activity in one line would ordinarily be drowned in the background activity of the others; but reciprocal inhibition in the same cell might allow one line to dominate if its activity were to rise above some level, and thus serve to direct attention or to switch control. Still another possible use of such circuits, with only slight changes in the coupling functions, might be as an **alarm system.** Time- and load-sharing delay lines, null detection, and filtering are other theoretically available consequences of such networks.

An array of inhibitory cross-connections can be thought of as an arena in which there is competition between parallel streams of impulses. Both the nonlinearity of dependence on activity levels and the critical inflections in the input-output functions would allow one stream to win control. A spectrum of properties is possible from democratic to oligarchical to dictatorial.

The reliability of performance of networks is considered on p. 233.

C. Lateral Inhibition Networks

This term differs from the preceding heading in directing attention more to layers or arrays than to alternate cells or groups. Many neural tissues consist of layers of similar neurons that inhibit each other either directly or indirectly. The inhibitory connections spread from any particular cell to make contact with neighboring and more distant members of the layer, but with decreasing density, as we saw in Chapter 3 (p. 111). Hence the effectiveness of inhibition decreases with distance.

Some effects of lateral inhibition are illustrated schematically in Figure 7.15.

sensory systems in both vertebrates and invertebrates and may operate at the earliest stage in the pathway and/or at later stages, even in the cortex (Fig. 7.16).

We have seen in Chapter 3 how lateral inhibition might result from recurrent collaterals of the axon ending on neighboring cells (Fig. 7.17). We saw how such inhibition can be complicated in the cerebellar cortex by the inhibitory influence of recurrent collaterals of Purkinje cell axons, not only on neighboring Purkinje cells but on basket cells, thereby disinhibiting Purkinjes in a certain geometric pattern. Renshaw cell lateral inhibition via motor neuron recurrent collaterals is treated on page 269 (Fig. 7.24).

D. Mixed Networks

If both excitatory and inhibitory connections exist inhomogeneously in a set of cells, the variety of possible outputs expands to the degree that it is useless to make *a priori* or generalized, unconstrained models. A recently studied real example is, however, worth describing (Fig. 7.18). The **stomatogastric ganglion** of crustaceans controls the stomach, one part of which contains the gastric mill used to grind food. The ganglion contains 30 cells, almost all of which are identifiable and constant in connections and influence. Ten are motor neurons to the gastric mill part and 14 to the pyloric part of the stomach, but both groups also have direct influence upon each other. Furthermore, two interneurons are present with connections to both sets of motor neurons. There are 123 known inhibitory connections and only 6 excitatory junctions. Twenty-nine of the junctions are electrotonic, the rest un-

known and presumably mainly chemical. All the functionally established connections can be anatomically justified by fibers visualized by injection of Procion yellow. How much spontaneity there is cannot be stated, but it must be considerable. Among these junctions, a variety of integrative input-output properties are found. It seems likely that if we could unravel the web of influences, there would be both excitatory and inhibitory reciprocity, lateral effects, and mixtures of spontaneous rhythmicity with imposed burst-shaping effects. The normal activity is largely a rhythmic series of bursts of repeatable but labile spike pattern; the known connections explain much of the detail of the bursts. Although many of the connections of this network are known (nearly all; hence far more in proportion than for any other known system of some complexity), it is difficult to assign causes to the bursting phenomenon itself. Does the positive feedback, by itself, cause one group of cells to burst, or is the reciprocal inhibition relationship with another group necessary, or even sufficient? In theory, either mechanism could produce bursting, but physiological evidence suggests that both kinds of connections exist. Perhaps both mechanisms operate synergistically to make the whole system more stable.

E. Connections Ensuring Synchrony of Activity

Although exactly synchronous activity of neurons is not known to be required in many instances, it is important in a few, and these are interesting in showing the flexibility of design with which cells can be connected, among themselves and to sensory systems in both vertebrates and invertebrates and may operate at the earliest stage in the pathway and/or at later stages, even in the cortex (Fig. 7.16).

We have seen in Chapter 3 how lateral inhibition might result from recurrent collaterals of the axon ending on neighboring cells (Fig. 7.17). We saw how such inhibition can be complicated in the cerebellar cortex by the inhibitory influence of recurrent collaterals of Purkinje cell axons, not only on neighboring Purkinje cells but on basket cells, thereby disinhibiting Purkinjes in a certain geometric pattern. Renshaw cell lateral inhibition via motor neuron recurrent collaterals is treated on page 269 (Fig. 7.24).

D. Mixed Networks

If both excitatory and inhibitory connections exist inhomogeneously in a set of cells, the variety of possible outputs expands to the degree that it is useless to make *a priori* or generalized, unconstrained models. A recently studied real example is, however, worth describing (Fig. 7.18). The **stomatogastric ganglion** of crustaceans controls the stomach, one part of which contains the gastric mill used to grind food. The ganglion contains 30 cells, almost all of which are identifiable and constant in connections and influence. Ten are motor neurons to the gastric mill part and 14 to the pyloric part of the stomach, but both groups also have direct influence upon each other. Furthermore, two interneurons are present with connections to both sets of motor neurons. There are 123 known inhibitory connections and only 6 excitatory junctions. Twenty-nine of the junctions are electrotonic, the rest un-

known and presumably mainly chemical. All the functionally established connections can be anatomically justified by fibers visualized by injection of Procion yellow. How much spontaneity there is cannot be stated, but it must be considerable. Among these junctions, a variety of integrative input-output properties are found. It seems likely that if we could unravel the web of influences, there would be both excitatory and inhibitory reciprocity, lateral effects, and mixtures of spontaneous rhythmicity with imposed burst-shaping effects. The normal activity is largely a rhythmic series of bursts of repeatable but labile spike pattern; the known connections explain much of the detail of the bursts. Although many of the connections of this network are known (nearly all; hence far more in proportion than for any other known system of some complexity), it is difficult to assign causes to the bursting phenomenon itself. Does the positive feedback, by itself, cause one group of cells to burst, or is the reciprocal inhibition relationship with another group necessary, or even sufficient? In theory, either mechanism could produce bursting, but physiological evidence suggests that both kinds of connections exist. Perhaps both mechanisms operate synergistically to make the whole system more stable.

E. Connections Ensuring Synchrony of Activity

Although exactly synchronous activity of neurons is not known to be required in many instances, it is important in a few, and these are interesting in showing the flexibility of design with which cells can be connected, among themselves and to

Figure 7.15 (*facing page*)
Lateral inhibition. **A.** A pattern of uniformly gray areas with sharp edges, representing a visual field seen by an eye. (Mach bands, an illusion in which apparently darker and lighter bands are seen around each edge, are not evident to us in this configuration.) **B.** The stimulus plotted as intensity (horizontal) against spatial extent (vertical), emphasizing the uniformity of the physical intensity within each area. **C.** A network of receptors and second-order neurons with reciprocal inhibitory connections via interneurons (broken lines). Spontaneous activity in the receptors (*S*) is augmented by excitation (*E*) due to light. The network causes the second-order neurons to show *S* activity augmented by *E* and/or reduced by inhibition (*I*) in single or double (*2I*) dose. **D.** The output, equivalent to our sensation, plotted as darker or lighter than the background due to *S*.

First consider the behavior of one spontaneously active cell in the network while a stimulus is moved about its input field. As the stimulus approaches, it first excites neighbors of the recorded cell. Since they inhibit that cell, it responds by a decrease in firing frequency. If the stimulus passes directly over the recorded cell, it is excited to fire above its normal rate, but as the stimulus moves on it is again depressed by its neighbors. Any cell in the network may be characterized as having a receptive field that has an excitatory center and an annular inhibitory surround. This is comparable to the ON-center ganglion cells of the cat retina.

If, instead of looking at the response made by a single cell in the spatial array of Figure 7.15 to a moving stimulus, we examine the output of a whole line of cells in the array while one half of the array is stimulated more than the other half, we see an abstracting function of the network. Cells in either uniformly stimulated half of the field all inhibit each other symmetrically, but those at the edge do not. Cells on the strongly stimulated side of the edge are inhibited weakly by their neighbors across the edge while they strongly inhibit those same neighbors. The result is especially high and low firing frequencies at the stimulus edge. In terms of the firing frequencies of the cells in the network, the stimulus edge has been enhanced; there has been a **spatial differentiation** of the input signal. A second layer of neurons could be so constructed that it would detect the edge only. Lateral inhibition probably explains our psychophysical illusion known as "Mach bands" (Fig. 7.15,D), and perhaps operates widely to enhance the sensitivity to contrasts.

Since the lateral spread and extra synapse take time, there is a delay in the inhibition. This confers a temporal property of **frequency selection** or filtering that acts to attenuate rapid changes as a function of distance from the center and to prolong the exaggeration of the primary response at the center, again favoring low frequencies.

It can also be seen that the inhibition exerted on a given cell by those surrounding it can be reduced by stimulation of a slightly more distant population. The distant population reduces the level of activity of the nearer one, which **"disinhibits"** the center.

These phenomena were first described by Hartline and his associates in a series of elegant experiments on the compound eye of *Limulus*. The same lateral interaction exist, however, in virtually all

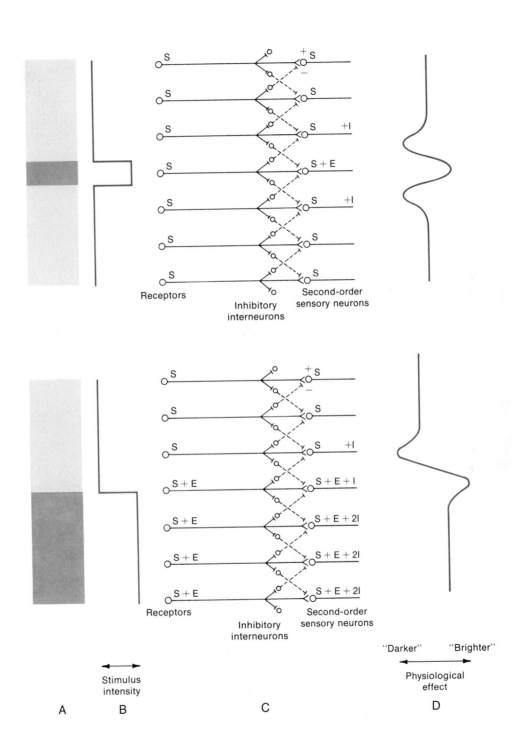

Receptors

Inhibitory interneurons

Second-order sensory neurons

Stimulus intensity

"Darker"   "Brighter"

Physiological effect

A   B   C   D

and inhibits the first. As the second cell slows, the first recovers, and the whole cycle repeats.

Whether dependent on rebound or not, some such simple mechanism for **alternating burst activity,** though difficult to demonstrate, seems to operate in favorable materials, like the lobster stomatogastric ganglion. We believe it to be an important mechanism for many kinds of rhythmic and alternating behavior. Inspiratory and expiratory interneurons in the mammalian medulla probably inhibit each other. Locomotory systems in many animals may involve reciprocal inhibition between pacemakers for antagonistic muscle sets.

Reciprocally inhibiting networks can perform a variety of functions, according to the particular transfer functions of the synapses and the input and output connections. The following examples are theoretical and qualitative; proof that reciprocal inhibition is the mechanism that operates in living animals is incomplete. Given certain properties, reciprocally inhibiting networks can act as **gates,** switching rapidly from control of the output by one input line to another. This can recur at a steady rate, for a steady-state input, providing a **pacemaker** in which no single cell is the essential element, as discussed above. With certain dynamic properties the same simple circuit can act as an intensity-to-time converter that may be useful in comparing the strength of two inputs by their relative duration of control of some downstream system. Slight changes could provide **sensory scanning** by periodically or irregularly sampling each of a number of input lines and giving them control, in turn, of some later elements. In a cell with a large number of input lines converging on it, a

drastic increase of activity in one line would ordinarily be drowned in the background activity of the others; but reciprocal inhibition in the same cell might allow one line to dominate if its activity were to rise above some level, and thus serve to direct attention or to switch control. Still another possible use of such circuits, with only slight changes in the coupling functions, might be as an **alarm system.** Time- and load-sharing delay lines, null detection, and filtering are other theoretically available consequences of such networks.

An array of inhibitory cross-connections can be thought of as an arena in which there is competition between parallel streams of impulses. Both the nonlinearity of dependence on activity levels and the critical inflections in the input-output functions would allow one stream to win control. A spectrum of properties is possible from democratic to oligarchical to dictatorial.

The reliability of performance of networks is considered on p. 233.

C. Lateral Inhibition Networks

This term differs from the preceding heading in directing attention more to layers or arrays than to alternate cells or groups. Many neural tissues consist of layers of similar neurons that inhibit each other either directly or indirectly. The inhibitory connections spread from any particular cell to make contact with neighboring and more distant members of the layer, but with decreasing density, as we saw in Chapter 3 (p. 111). Hence the effectiveness of inhibition decreases with distance.

Some effects of lateral inhibition are illustrated schematically in Figure 7.15.

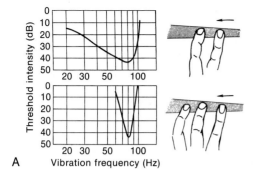

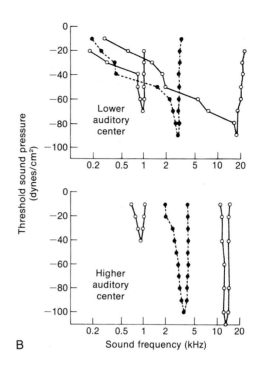

Figure 7.16
Sharpening by suppressing sensitivity on each side of the best frequency, with converging input. **A.** Thresholds for sensation as the endpoint; vibration felt by 2 or 3 fingers. **B.** Thresholds for single neuron firing as the endpoint; unit responses to sound at lower and higher auditory centers. It should be noted that such narrow curves at higher centers are not common; units with many types of curves are found. [Von Bekesy, 1967.]

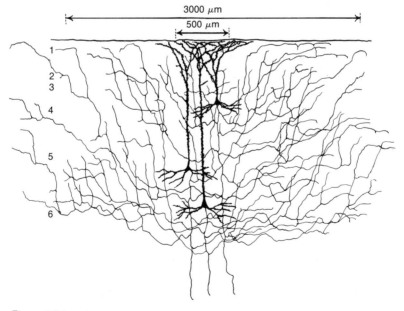

Figure 7.17
Recurrent collaterals. The fine fibers are the array of collaterals of three pyramidal cells in the cortex of the kitten. [Scheibel and Scheibel, 1970a.]

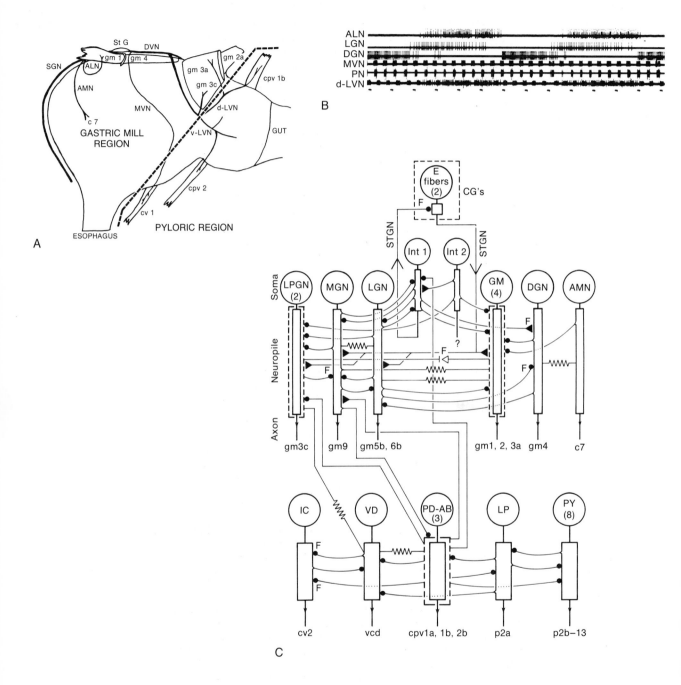

the periphery, to accomplish precise timing. These can be regarded as special cases of the general problem of assuring precise timing of sequences of activity. Among the best examples of systems requiring nearly exact synchrony are the **electric organ discharges** of electric fish. (Other examples are the oculomotor neurons and the neurons innervating sound-producing muscles and wings in certain animals.) The electric organ cells (electrocytes) are in most cases composed of modified muscle cells that are oriented in series so that their depolarizations can be summed. It is necessary in these organs that all the electrocytes be activated nearly simultaneously, and in some species within ≪0.1 msec. The command or pacemaker neurons are usually located in the medulla or midbrain and act through interneurons and motor neurons on the electrocytes. In all cases, these neurons are electrically coupled to each other via gap junctions (see p. 333). The electrotonic connections may be between cell bodies, or dendrites, or via presynaptic fibers that form electrical synapses on many of the cells. These synapses, by equalizing the level of depolarization between different cells, are both excitatory and inhibitory, but ensure synchronous activity of all coupled cells. The simplest example of such coupling is in the electric catfish, *Malapterurus,* in which there are two electromotor neurons, one on either side

Figure 7.18 (*facing page*)
A simple system of some thirty neurons; the stomatogastric ganglion of a lobster. **A.** Side view of lobster stomach. The two principal functional divisions, the gastric mill region and the pyloric region, are separated by a broken line. Part of the stomatogastric nervous system is shown together with a few of the stomach muscles that it innervates. The stomatogastric ganglion (*StG*) can be seen on the dorsal surface of the stomach just above gastric mill muscle 1 (*gm*1). Other lettering identifies muscles and nerves. **B.** The two basic rhythms produced by the isolated ganglion can be seen in the extracellular recordings from nerves supplying the two different regions of the stomach. The top three nerves (*ALN, LGN, DGN*), not all shown in part A, supply muscles that operate the gastric mill. Note that the bursts of activity bear a particular phase relationship with each other and that the duration of each burst is several seconds. The three lower traces contain axons of motor neurons (*MVN, PN, d-LVN*) supplying pyloric muscles. The bursts are much shorter in duration, and the overall frequency is about seven times that of the gastric mill. Note also that the bursts of activity maintain a particular phase relationship. The *d-LVN* trace contains axons innervating both regions (see above) and the long bursts seen in this trace are from axons to muscle *gm 3a.* [Parts A and B, Selverston and Mulloney, 1974.] **C.** Neuronal connectivity diagram for the lobster stomatogastric ganglion. All the cells except interneurons 1 and 2 are motor neurons. The top ten neurons control the gastric mill cycle, and the bottom fourteen cells control the pyloric rhythm. The axonal pathways, as well as the muscles innervated by the cells, are known. Soma, neuropile, and axonal parts of the cells are indicated on the left. Broken lines around some of the neuropile areas indicate that cells of that group are electrotonically connected and can be considered together. Known connections with the central nervous system are shown at the top. Round dots represent chemical inhibitory synapses; triangles represent chemical excitatory synapses and resistors (= electrotonic junctions); *F*, functional synapse with strong effect but no clear unitary postsynaptic potential; *E*, excitatory fiber input from commissural ganglia. *LPGN*, lateral posterior gastric neuron; *MGN*, median gastric neuron; *LGN*, lateral gastric neuron; *Int 1* and *2*, interneuron neuron *1* and *2*; *GM*, gastric mill neuron; *DGN*, dorsal gastric neuron; *AMN*, anterior median neuron; *IC*, inferior cardiac; *VD*, ventricular dilator; *PD*, pyloric dilator; *AB*, anterior burster; *LP*, lateral pyloric; *PY*, pyloric; *CG*, commissural ganglia; *STGN*, stomatogastric nerve. [Courtesy of A. Selverston.]

of the first spinal segment, tightly coupled to each other. In other fish, there may be as many as 20–50 pacemaker cells coupled to each other, all of which electrically excite internuncial relay cells that are also electrotonically coupled.

Given a synchronized command, how are electrocytes at different distances from the brain activated synchronously? This is accomplished, in most cases, either by systematically shortening the length of branches to the progressively more distant electrocytes, or by altering their diameter so that the slower conduction velocities of axons innervating the nearer electrocytes compensate for the shorter distance. In several species the graded delay is built into the electrocytes. Distant ones are innervated near their principal surface, nearer ones at the end of long, slow-conducting stalks of the electrocyte.

The existence of these mechanisms for building compensating delays into neural circuits opens the possibility that similar refinements of structure may be involved in other systems that are sensitive to the precise timing of inputs, such as the parts of the auditory system responsible for sound localization or the motor systems controlling speech, eye movements, middle ear muscles, and the like.

V. STIMULUS-TRIGGERED REACTIONS: THE ORGANIZATION OF REFLEXES

With some of the principles of effector control, of sensory input, and of elementary circuits in mind, we can now turn to the lowest levels of motor response to natural stimuli. What can we say is the simplest nervously mediated response?

The **vertebrate stretch reflex,** while specialized for simplicity and speed rather than being primitive, is a good starting point because it is monosynaptic—that is, no interneurons are interpolated between afferent fibers and motor neurons (Fig. 7.19). Skeletal muscles contain numerous sense organs, called muscle spindles, that are sensitive to stretch in the axis of the muscle fibers (Fig. 7.20). Passive stretch, as by the action of gravity or of other muscles, causes a train of impulses to arise in the terminals of a sensory axon in a muscle spindle, and these are conducted via the dorsal root to the spinal cord, there to be distributed in axon collaterals to the dorsal and ventral horns of the same segment on both sides of the cord, to nearby segments up and down the cord, and to the dorsal columns ending in nuclei in the medulla. Of these destinations one is the large alpha (α) motor neurons of the same muscle, which are excited. This excitation is distributed by the axon branches of the motor neuron to a group of muscle fibers, usually between 100 and 1000, called a motor unit. Contraction of the motor unit tends to cancel the stretch. This proprioceptive reflex acts as a tonic muscle-length servo (like a gyrocompass), tending to maintain the length against any change in load either way. Gravity is an important normal stimulus; this reflex arc is the principal antigravity circuit.

At the same time, collaterals of the spindle afferents fire interneurons that in turn excite synergistic motor neurons and inhibit motor neurons of antagonistic muscles on the same side while

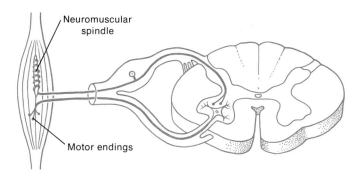

Figure 7.19

The monosynaptic stretch reflex pathway in mammals, simplified by omitting the other destinations of the same afferent neuron, the other inputs to the same motor neuron, its output recurrent collaterals, and the motor control of the spindle. [Gardner, 1968.]

A sense organ excited by passive stretch, or activation of its own intrinsic muscle fibers by gamma motor axons.

motor neurons of the homologous muscle and its synergists on the contralateral side are inhibited and antagonists excited. The phenomenon of reciprocal innervation contributes to coordination by preventing antagonists from working against each other.

Given this length-maintaining reflex, how can the organism walk or voluntarily change muscle length? If higher centers were simply to command α-motor neurons to greater or less activity, the stretch reflex would quickly cancel the effect. Somehow, the set point (Sollwert) must be changed. This can be done by adjusting tension in the muscle fibers within the spindle, called intrafusal fibers.

There is another set of motor neurons in the spinal cord that also send their efferents to the muscle. These are the **gamma (γ) efferents,** emanating from small motor neurons (Fig. 7.21). They innervate the muscle fibers in the muscle spindle and alter its sensitivity. In the central region of the spindle fibers is a noncontractile enlargement filled with nuclei, and it is to this zone that the stretch receptor fibers come (Fig. 7.20). The end regions of the spindle fibers are contractile and are innervated by the γ-efferents. When the γ-efferents increase their activity, the spindle fibers shorten, the central region is stretched, and the stretch receptor fibers are excited. It is as if the muscle had been passively stretched. The opposite effects obtain if the efferents have reduced activity or the main muscle has increased activity.

The **set point** of the postural reflex can be affected by tension changes in the spindle muscle fibers. When these contract they stretch the muscle receptors but do not change the length of the muscle. The resulting increase in stretch

Figure 7.20 267

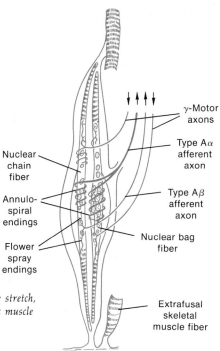

The mammalian muscle spindle.

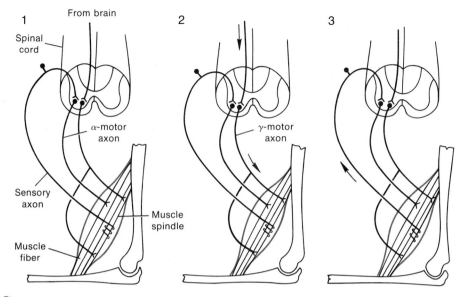

Figure 7.21
The γ-loop servomechanism. Commands from the brain may operate by contracting the muscle fibers of the spindle via the γ-efferents in the ventral root (2). This excites the stretch reflex (3) and hence the main muscle. There are also direct connections (not shown) from the brain to the main motor neurons. [Merton, 1972.]

receptor firing rate excites α-efferents, which elicit greater tension and hence muscle shortening. Shortening continues until stretch receptor firing rate is reduced again to normal values. The central nervous system can command a long-lasting new length by varying frequency in the γ-efferents and thus changing the sensitivity of the stretch receptors with respect to muscle length. The setting of muscle length by way of the γ-efferent is called γ-loop activation.

The motor neurons (both γ and α) for one muscle are loosely grouped in the ventral horn of the spinal cord. They receive many inputs, usually in parallel. The stretch receptor afferents excite only the α-efferents. If they excited γ-efferents the reflex would comprise a positive feedback loop, which would run away to

maximum or minimum tension. Descending excitation or inhibition from the brain impinges on both γ- and α-fibers. We have already encountered the general rule—that small units are more tonic and have lower thresholds for natural stimulation. The γ-neurons are smaller and more sensitive than the α's. A weak command from the brain excites the γ's, which in turn change the set point of the muscle-receptor–α-efferent reflex, and muscle length changes to compensate for this. Strong descending-movement commands excite both γ's and α's. The faster-conducting α's initiate a movement that would be later cancelled by reflex function were it not for the more slowly developing effect of γ-excitation, which changes the set point of the reflex. Thus voluntary and other brain control

Figure 7.23
Spindle versus tendon receptors.
[Patton, 1965.]

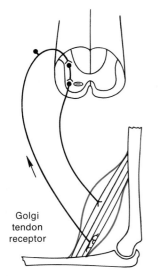

Figure 7.22
The tendon receptor for stretch. Another sense organ, besides the spindle, is in the tendon, which differs by being stretched (excited) when the muscle contracts as well as when it is loaded. The sign of its influence is such as to prevent overcontraction.

function, restraining the motor neuron from causing the muscle fibers to contract too violently.

One more automatic subsystem is important in helping to determine motor neuron activity locally; this is the **Renshaw cell** negative-feedback loop, acting on motor neurons in the spinal cord (Fig. 7.24). Situated in the ventral roots there are branches of α-motor axons called recurrent collaterals because they turn back and reenter the ventral horn to synapse with small interneurons, named for Renshaw, who discovered them physiologically. These cells fire at high frequency when excited by motor neuron output and have the effect of inhibiting the same and neighboring motor neurons. The roles of this inhibition may include preventing motor neurons from excessive activity, focusing activity upon certain cells and perhaps suppressing phasic responses more than tonic ones.

Many reflexes are elicited by extramuscular stimuli. For example, painful cutaneous stimulation often gives rise to flexion of a limb (Fig. 7.25). The flexor reflex to **nociceptive stimuli** is not sharply localized; it may affect all the muscles of a limb. Both strength of response and degree of spread of response are related to stimulus strength. Input fibers do not impinge directly upon motor neurons, but on interneurons; the reflex pathway is polysynaptic. The input excites flexor action and at the same time inhibits extensor motor neurons. If the stimulus is quite strong its effects may spread even to the contralateral limb, where they are opposite in sign. The crossed extension reflex prepares one limb to bear the extra weight shifted to it when the painfully stimulated one flexes. Similar reflex con-

of movement operates largely via the γ-loop balanced in various degrees with coactivation of α's.

Another set of receptors is on the muscle tendons, and these respond to stretch of the tendon (Fig. 7.22). In contrast to spindles, therefore, they respond in the same direction to imposed load and to contraction of the muscle (Fig. 7.23). These receptors (Golgi **tendon organ receptors**) have no monosynaptic endings, but, via interneurons, they send (a) inhibitory input to motor neurons of the muscle they innervate and to its synergists, (b) excitatory input to motor neurons of antagonist muscles, and (c) opposite inputs to motor neurons of the corresponding muscles on the opposite side of the body. These inputs might be said to perform a tension-regulating

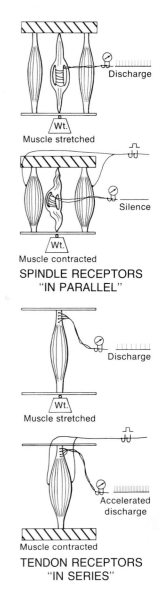

**SPINDLE RECEPTORS
"IN PARALLEL"**

**TENDON RECEPTORS
"IN SERIES"**

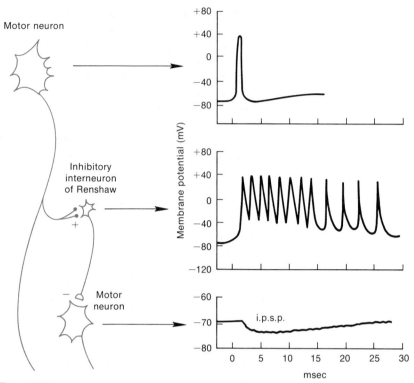

Figure 7.24
The Renshaw type of inhibitory interneuron. Axon collaterals from motor neurons activate Renshaw cells to high frequency discharge, which sets up summating inhibitory postsynaptic potentials in neighboring motor neurons of the same pool. [Eccles, 1964.]

nections are seen for cutaneous touch, pressure, and temperature receptors.

We have seen now some quite general contrasts between flexion and extension reflexes, nociceptive and proprioceptive, pain and postural, cutaneous and muscle-afferent, phasic and tonic reflexes. These are overlapping but not synonymous dichotomies. Another useful one, referring to their roles in causing normal behavior is the distinction between elemental and tuning reflexes; the former cause the basic sequences, the latter adjust to momentary conditions.

Sherrington stated the **concept of the reflex,** one of the truly great and fruitful abstractions in biology, with these words: *"The unit reaction in nervous integration is the reflex,* because every reflex is an integrative reaction and no nervous action short of a reflex is a complete act of integration"* (1906, emphasis his). Many authors have pointed out the limitations of the concept or criticized it as artificial, and Sherrington, as clearly as anyone else, emphasized the non-existence of a discrete circuit insulated from others. Nevertheless, the funda-

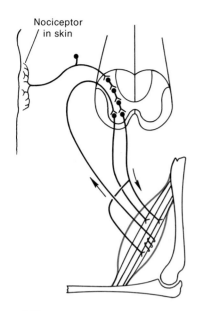

Figure 7.25
Nociceptors and the polysynaptic pathway. Another set of inputs impinging on the motor neuron come from pain and similar receptors. The sign of their action is generally excitatory to the ipsilateral flexor and contralateral extensor muscles and inhibitory to the ipsilateral extensors and contralateral flexors.

mental usefulness of recognizing this category of responses has been amply proved by the insight that experiments based upon it have provided. The reflex is a useful abstraction, but we qualify the broad statement that it is the unit of all nervous integration: (a) the arousal functions of the reticular activating system in mammals cannot be resolved into reflexes, nor can (b) mere sensing, (c) autochthonous action (arising from within), or (d) many instincts.

Familiar examples of the phenomena under consideration are the stretch reflex, the flexion, crossed extension, shake and scratch (dog) reflexes, the reflexes of micturition and defecation,

the pinna, swallowing, stepping, sneezing, salivation, blink, accommodation, and tonic neck reflexes, placing, hopping, and righting reflexes. They are ready-made, unlearned, adaptive movements, prompt and coordinated. The **coordination** is not observed just within each reflex, which we might explain as a fixed pattern, but is observed even when conflicting reflexes are stimulated simultaneously, for there is almost invariably a resolution of the potentially maladaptive conflict or intermediate action in favor of adaptive selection among them. Cooperative interplay is also marked between simple segmental reflexes, long spinal intersegmental reflexes of forelimbs and hind limbs, and coordination of body, limbs, neck, head, and eye reflexes, as in visually guided walking.

Though the present account of reflexology necessarily draws mainly from mammalian literature because lower forms have been less studied in this respect, we believe the principles enunciated are probably general. Implicit in the concept of reflexes—since there is almost invariably more than one reflex utilizing a given muscle, and hence more than one central mechanism converging on the same motor neurons—is the concept of the **final common path.** The very expression emphasizes the integrative function.

The properties of reflexes (see Box 7.2) and the rules by which they are used make the best case for their reality and importance. We have already learned many of these rules in the examples detailed above. Let us look further at the ways they combine.

Separate reflexes may be either compatible or incompatible. The former combine adaptively into compound reflexes. For example, a gravity reflex that keeps a fish upright adds to a light reflex that keeps the dorsal side toward the light, so that if light comes from the side, certain fish tilt to a degree graded according to the light intensity, the strength of gravity, and a central evaluation that multiplies each input by a weighting factor. The weighting depends on time of day, temperature, hunger, and other inputs, such as chemical signals associated with food, mechanical inputs that indicate a substratum, and visual input sufficiently imaged and "understood" to represent a substratum.

Incompatible reflexes are those that cannot be accomplished at the same time. If the hind leg reflex to scratch the back is elicited on one side in a dog and at the same time the stimulus for an extensor thrust of the same side is given,

the nervous system does not release them both. One or the other is suppressed, at each moment. Incompatible reflexes do not add algebraically or combine linearly; instead, switches or patterned control insure normally adaptive interaction. This may show either inhibition or facilitation. It is as though the system were preorganized for useful movements.

The tonic neck reflexes and a number of related postural reflexes due to vestibular and proprioceptive input interact with each other and with phasic movements as though adding a bias or "tuning" the response to the conditions of the moment. For example, if a load is lifted by wrist flexion, more work can be done (stretch reflexes facilitated) with the head bent down or turned away from

---

Box 7.2 Properties of Reflexes

What are the **properties of reflexes** that manifest integration? (a) The threshold stimulus is very much dependent on conditions. (b) Above the threshold, gradation of response does not closely correspond with gradation of stimulus. (c) If the stimulus is repetitive, there is usually a poor correspondence between its rhythm and that of the reflex response. (d) Single afferent impulses are usually not adequate; temporal summation is usually necessary to elicit a response. (e) A depressed excitability typically follows a reflex and is often quite long. (f) Afterdischarge, or the prolongation of the motor neuron activity after the cessation of the stimulus, is a prominent feature of many reflexes, as though the mechanism were organized to complete a certain movement in a controlled way. (g) Spatially and temporally **patterned control** of several muscles is probably involved in all reflexes.

The timing of inhibition is coordinated with that of excitation, as is clearly seen in the alternating reflexes, shaking, stepping, and scratching. Even simple flexion and crossed extension reflexes show characteristic temporal patterning. (h) **Irradiation** with increasing intensity of stimulus occurs in some reflexes—for example, the protective flexion reflex. At threshold a response may involve a limited part of a synergic muscle group across one joint, but with irradiation it may spread to other joints of the same appendage, to other appendages and segmental levels, to the head and neck. The spread is generally saltatory and is confined strictly to certain lines or muscle groups. However, apparently uninvolved muscles may in fact be involved as objects of inhibition. The possible movements are thus circumscribed in a characteristic pattern. These properties help to define reflexes, to bring out their integrative nature, and to emphasize the central determination of details of form and timing.

that arm, because of the tonic neck reflexes; if the load is met by wrist extension, the opposite head movements enhance output.

Reflexes are normally woven into an integrated fabric, without sharp lines. They are graded in amplitude by influences descending from higher centers and combined under the rules of a hierarchy. We cannot help wondering whether the same circuits that are reflexly triggered from the periphery might be centrally triggered in patterned sequences.

Having erected this edifice of plausible assumptions and conclusions from studies of reflexes, we should now raise the question as to what evidence there is for the central origin of patterned impulse discharge.

VI. CENTRALLY SCORED BEHAVIOR: PATTERNING IN SPACE AND TIME

One way to state the function of the nervous system is that it formulates appropriately patterned messages to drive the effectors. A core question in the study of intermediate level integration is: "How is this done?" The patterning in time may be treated as a more serious issue than that in space. Theoretically it could arise in either or both of two ways: by following (a) timing cues from peripheral sense organs or (b) timing cues from central pacemakers or pattern generators. We may call the second mechanism a central score to emphasize its potential complexity of detail, its modifiability from outside on any given occasion, and its reality as a stored program apart from external inputs. A score has timing built in, though subject to external influence; a program might have the same, but the term applies also to cases that merely call for reactions, leaving the timing to effectors, transducers, loops, and largely peripheral events.

Instead of a simple dichotomy we may distinguish several possible mechanisms (Fig. 7.26). In A, we have a simple reflex, such as an eye blink, a swallow, or a cough, triggered by a stimulus. Startle responses mediated by giant fibers belong in this category (see Box 7.3). Even here the temporal pattern of messages to various muscles is determined by central pathways and integrative junctions. In B, there is sensory feedback from proprioceptors early enough to determine a rhythm of recurrence; a chain reflex accounts for frequency, phasing, and amplitude. The extreme case, in E, is purely centrally timed, without any immediate feedback. C and D are combined mechanisms in which the central spontaneity can determine the basic rhythm but feedback may alter either the rhythm (C) or only the details of the expression of the rhythm (D).

Probably all five mechanisms are common, though usually there has not been enough analysis to be sure which class an activity belongs to. In this section we are concerned particularly with some examples of C, D and E. In each example, there is permissive or essential input from sense organs that may start or stop the whole pattern or influence the overall "central excitatory state," to use Sherrington's phrase. This is what we mean by spontaneity, not that there is independence of the environment for permissive conditions, but only for triggering the succession of actions. Spontaneous rhythms can be influenced in

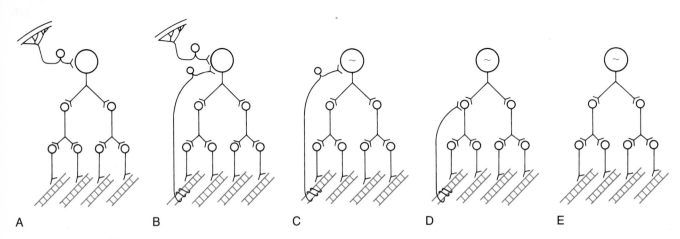

A   B   C   D   E

Figure 7.26
Five mechanisms of pattern formulation. The three levels of neurons are understood to represent
branching chains in whose junctions integrative properties may alter the actual impulses and distribute them spatially as well as temporally to the effectors (*bottom*). **A** and **B** are shown with receptors;
C, D, and **E,** with spontaneous pacemakers giving simple or group discharges. **B** and **C** have proprioceptive feedback acting on the trigger neuron; **D,** only on the shaping of the pattern. [Bullock, 1961a.]

Box 7.3 Giant Fibers and Startle Responses

Striking among the behaviors that are merely triggered
by environmental stimuli, but not further guided by
either environmental or proprioceptive inputs, are the
responses mediated by giant fibers. Giant fibers are
found in many invertebrate animals and fish (see
Chapter 10). Their phyletic distribution is scattered and
their structures are diverse; hence they are likely
correlated only in function, not through evolutionary or
developmental relationships. Because of this diversity,
the following generalizations may have exceptions.
Because of their size they conduct impulses rapidly.
Perhaps related to size is the fact that they can have a
large divergence ratio—that is, one giant fiber can excite
many other neurons or muscle fibers. Moreover, the
current available for electrical transmission of the
action potential to downstream neurons is large; electrical transmission may be common in these systems. In
all of the adequately studied cases giant fiber action

results in rapid, nearly synchronous, widespread muscle activity. Behaviorally, giant fibers ordinarily fire
only with rather special input requirements; these often
have the characteristics we call startle.

The **giant axons of squid** are the best known of all
nerve fibers. We may briefly review their functional
anatomy (see also Chapter 10, p. 434) and behavioral
role. The muscles of the mantle of the squid are doubly
innervated. Many small motor axons from the stellate
ganglion excite relatively few muscle fibers each, and
these small neurons control the slower movements
involved in respiration and ordinary swimming. The
single giant fiber in each of the 6 or 8 stellar nerves on
each side together innervate most or all of the muscle
fibers of the mantle. Each giant fiber has many cell
bodies in a lobe of the ganglion and is therefore a fused
syncytium of many motor neurons. "The" giant fiber of
the squid is the last, longest, and largest of the 6 or 8 on

Box 7.3 (*continued*)

each side; these fibers are properly the third-order giants. Their input is from two second-order giant fibers that arise in the visceral lobe of the brain, enter the stellate ganglion, and synapse on each third-order giant fiber in that ganglion. The giant synapses are one-to-one relays; every input impulse causes an all-or-none twitch throughout the mantle, nearly simultaneously, and a vigorous ejection of water through the funnel. When a squid is startled, it activates the giant fiber system probably via the single bilaterally fused first-order giant unit in the pedal lobe of the brain. This unit is a command unit, as defined on page 279. There is probably no immediate feedback influencing the operation of the giant system.

The **giant fiber system of earthworms** consists of two parallel chains of electrically coupled segmental command units: a median fiber and a lateral electrotonically coupled pair of fibers (see p. 405). These premotor intereurons control largely overlapping musculature, the longitudinal or shortening muscles, plus separate muscles for the setae of anterior and posterior segments. The inputs have some labile overlap. The median giant is activated by startling stimuli, such as a vibration or a tap anywhere on the anterior third of the worm, and causes anchoring by protrusion of setae in the tail, plus shortening, which therefore retracts the head. The lateral giant pair is activated by mechanical stimuli in the posterior two-thirds, and causes anchoring of the head end, hence pulling up the tail. Each giant is therefore a unique, consistent chain of neurons acting as a decision unit. The adequate input is from many receptors, and reaches threshold only when some subtle criterion is met that involves rate of rise, recent history, spatial pattern, and a central excitatory state.

The crayfish giant system has been more completely studied, and its circuitry is diagrammed in the sketch.

Mauthner's fibers in teleosts and aquatic amphibia are likewise premotor command units receiving a large input from many sources, especially vibration receptors (see pp. 30, 106, 257, 440). They normally fire only once or twice, first one side and then the other, causing the

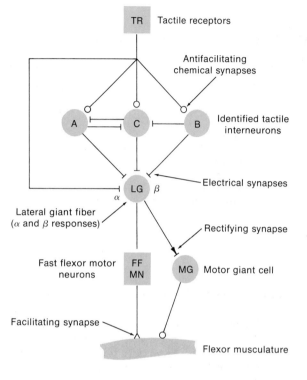

Zucker's cicuit for the rapid tail flexion of the crayfish. Schematic diagram of the known elements and connections for phasic mechanical stimuli to abdomen. [Zucker, 1972.]

initial body bend of a startle response. The wealth and variety of synapses known—on the soma, axon hillock, and large dendrites—suggest tens of thousands of impulses may arrive per second, during the reaction time of a single afferent impulse, representing a high degree (but perhaps not atypical for neurons) of integrative filtering, recognizing, deciding, and commanding.

occurrence and frequency both by tonic input and by higher central levels (e.g., by changes of mood).

A. Central Rhythms and Reflex Modulation

The peripheral or reflex hypothesis fails to account adequately for respiratory control of vertebrates. A crucial test, total deafferentation, leaves a functioning central system. Similar tests have shown that many other rhythmic control systems have built-in central scores. The motor neuron discharge sequence, which withdraws the mantle and closes the valves in a clam, *Mya*, occurs after deafferentation. The copulatory movements of a praying mantis, the flight rhythm of insects, the walking patterns of insects and amphibia, respiratory movements in insects, beating of the swimmerets in decapod crustacea, stridulatory singing in crickets and grasshoppers, side-to-side alternation of longitudinal muscle activity in sharks, heartbeat and gastric mill control in crustacea, and the swimming beat of jellyfish have been shown with varying degrees of rigor to be centrally controlled.

A few of these examples deserve more detailed discussion. But first we should point out that proprioceptive feedback functions have also been demonstrated in nearly all of them, and in the discussion of centrally patterned motor output we should attend to the role of reflexes as well. In studying cases of oscillatory behavior, as in all considerations of oscillatory phenomena, three measures always merit notice: **frequency** (or the reciprocal of frequency, the period) of the oscillation, **amplitude,** and

phase of one element of the oscillatory pattern relative to another. In several of the known centrally driven behaviors, we can relate the functions of peripheral feedback to one or more of these particular parameters.

Locust Flight. One of the most decisive studies on central rhythms is that of Wilson on wingbeat control in grasshoppers. These animals will sometimes fly when the ganglia of the head and abdomen are removed, so that the pattern generator must be present within the thoracic segments. The normal motor score has been analyzed in such detail that the temporal sequence and phasing of every motor neuron impulse during flight is known. This pattern of motor neuron discharge can be recognized in the central stumps of the thoracic nerves, even after all those nerves have been severed. Isolated thoracic nerve cord preparations are not spontaneously active in the flight rhythm, but random electrical stimulation of the nerve cord can elicit nearly the same pattern as is found in intact animals during flight. The output pattern of deafferented preparations is deficient in one major respect; it is low in frequency. The decreased frequency can be ascribed to lack of input from four stretch receptor cells, one in the hinge of each wing. These stretch receptors discharge when the wing is elevated. During flight they fire one to a few impulses toward the end of the upstroke in each wingbeat cycle (Fig. 7.27). The number of impulses in the burst is correlated with wingbeat amplitude, the timing of the burst with wingbeat phase, and the burst repetition rate with wingbeat frequency—all the measures of an oscillation. This information,

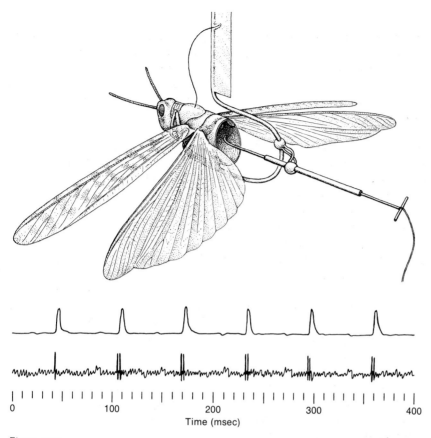

Figure 7.27
Proprioceptive feedback in a flying insect. Sensory discharges in nerves from the wing and wing hinge in a locust, recorded with wires manipulated into the largely eviscerated thoracic cavity of a locust. The top record is of downstroke muscle potentials, which are repeating at the wing-beat frequency. The bottom record is of a sensory (stretch) receptor from one wing, firing one or two times per wing beat. [Wilson, 1968.]

upon entering the CNS, would be adequate to trigger and control the events of the next cycle. Apparently, however, it does not even significantly affect the next cycle. In spite of this rich detail of information about wing position, there is little input/output correlation except a relatively long-termed one relating average input frequency in the stretch receptors to average wingbeat frequency and amplitude. The ganglion integrates and smooths the input over several wingbeat cycles, thereby almost but not quite losing the phasic information. The filtering process is analogous to the smoothing and integration in a resistance-capacitance electrical network. In sum, locust flight reafference primarily plays a role in controlling the average excitation of the motor pattern generator, thus affecting wingbeat frequency and power. It has a very weak effect on wingbeat phase.

An important lesson to be learned from this example is that, even though one can demonstrate that proprioceptive input in a rhythmic system fits the requirements for a reflex feedback model, that input may not be necessary for the normal pattern, and significant information parameters for the peripheral hypothesis may not even be used in the normal operation of intact animals. They may be discarded in a filtering process.

Swimming in Sharks. When many fishes swim, the longitudinal musculature of the two sides contracts in alternate metachronal waves. If a shark is curarized to the point of total paralysis, motor output in the segmental nerves may still result when the animal is stimulated, but since no movement occurs there can be no correlated proprioceptive feedback. In this circumstance, bursts of impulses still issue alternately through contralateral nerves, but at unusually low frequency. Proprioceptive input may, as in locusts, be necessary for tonic excitation of central state. But in sharks, comparison of the outputs on the same side in different segments shows them all to be synchronous. The metachronicity is lost. On the basis of presently available evidence, it appears that proprioceptive feedback is necessary for the phasing of outputs in proper segmental sequence as well as for the maintenance of normal frequency (see Fig. 10.63).

Crayfish and Lobster Swimmeret Beat. The abdominal appendages of decapod crustaceans also have a metachronal rhythmic beat. The completely isolated abdominal nerve cord can be stimulated to produce the swimmeret beat command, and the output of a deafferented or isolated preparation is normal in both frequency and segmental phasing (Fig. 7.29). The known proprioceptive reflexes cannot even modulate these parameters. They do, however, modulate the force of each stroke, affecting the velocity and amplitude by influencing the number and repetition rate of motor impulses during each cycle—in other words, the magnitude of the motor discharge.

B. Command Cells

A superficially quite different category of centrally scored pattern is that called up by command cells, actually just the end of a spectrum of mechanisms that occur in all degrees. Command units, first discovered in crayfish, are now known in many arthropods, annelids, molluscs, and vertebrates (see Box 7.3, p. 274). Command units, or redundant clusters of similar units, may turn out to be quite general. The concept and the term come from the observation that certain single units, upon stimulation, are capable of causing actions resembling major pieces of normal behavior. Some cause static posture, others phasic sequences. In higher invertebrates, in which they are probably all potentially identifiable, nonidentical subsets of command cells may act together to determine the form of the behavior. They are presumed to trigger or release the action, not to instruct the motor neurons in the temporal pattern of their firing; that pattern is already preformulated by other neurons.

The best-known command fibers in crayfish are not of extraordinary size, are not motor neurons themselves, and in general do not even synapse directly on motor neurons. They are higher-order interneurons that run through several ganglia or even the whole length of the neuraxis. They excite whole motor systems, small or large, controlling posture or locomotion (Fig. 7.28). They are found repeatedly in different animals. Each is uniquely characterized by position in the nerve cord, axon diameter, output function, and other properties. The pattern of output does not depend much upon command fiber frequency or pattern, and the output may continue well after the stimulation of the command fiber has ceased.

The command fibers are obviously labeled lines. The temporal pattern of activity in each is relatively unimportant. What is important is which command fibers are active, since they drive diverse behaviors. Each command fiber controlling posture of the crayfish abdomen seems to have unique but overlapping output fields. Perhaps the several fibers commanding swimmeret beat are also unique and produce somewhat different outputs (Fig. 7.29). Whether they do or not, swimmeret movements can be differentially controlled by combining the action of fibers driving the oscillatory mechanism with those affecting posture. Command fibers may turn motor systems on or off or bias them for purposes of steering or orientation.

Some giant fibers (see Box 7.3) are command units—namely, those that are not motor neurons but interneurons receiving from many small cells. They are specialized for prompt, synchronous, and brief actions that do not continue after the giant fiber stops firing. Local sites in the hypothalamus of mammals, where stimulation rather reliably triggers characteristic behavior, may represent a cluster of cells acting as a kind of command center (pp. 316–319).

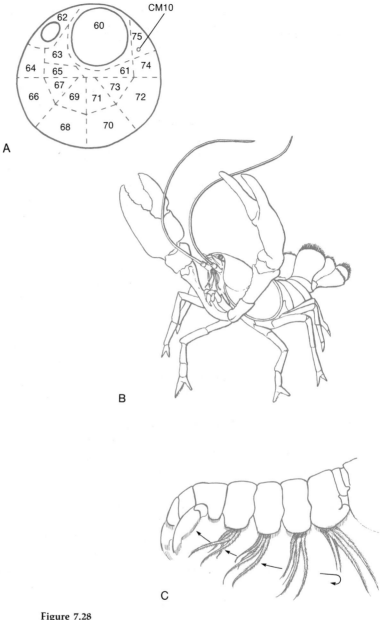

Figure 7.28
A. A command fiber for posture. This fiber, called CM10, travels in area 75 of the circumesophageal connective of the crayfish. When stimulated at a minimum of 20 shocks per sec, it releases the "defensive posture" shown in **B,** involving stereotyped positions of all appendages. This is one of the static, postural responses triggered by any one of a small number (3–5) of specific command neurons; others are dynamic, rhythmic movements, such as swimmeret beating (**C**). [Part A, Wiersma, 1958.]

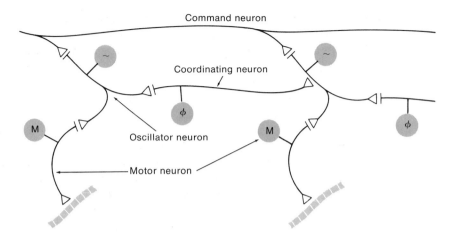

Figure 7.29
Circuit for swimmeret beating in crayfish. The segmentally repeated circuit is connected both by the multisegmental command neuron and by the intersegmental coordinating neuron that controls the actual time relations of the wave of beating. [Stein, 1971.]

What sets off a command cell? In general, we do not know, but an informed guess might be that it usually requires a number of background conditions plus some triggering input. This means it is highly integrative, acts as a recognition unit (p. 236) to detect these criteria, and is therefore a decision unit (p. 238) in a significant sense. The input it requires may come from sense organs, but in some cases it seems likely that centrally arising changes in mood or readiness might do the same thing (see vacuum activity, p. 315, central spontaneity, p. 325).

C. Hierarchical Structuring of Motor Systems

If rhythmic motor output can be driven by nonoscillatory impulse trains in the command fibers, where does the oscillation arise? Could it be due to interactions between the motor neurons? Several sorts of motor neuron interaction are known. Mutual excitation and reciprocal inhibition between motor neurons in the stomatogastric ganglion of crabs and lobsters are responsible, at least in part, for the way in which the output pattern of that ganglion reinforces and adjusts a spontaneous rhythm or command fiber response (p. 279). Some motor neurons exciting the same muscle in insects are electrotonically coupled and tend to fire together. The same is true for contralaterally paired inhibitory motor neurons innervating the abdominal postural muscles in crayfish. Motor neurons innervating the same flight muscle in certain flies inhibit each other. Synergistic motor neurons in vertebrate spinal cords are also inhibitorily linked, through known short-axon interneurons, the Renshaw cells (p. 269).

If interaction between motor neurons were generally responsible for their rhythmic activity, one might expect that antidromic stimulation of sets of motor neurons could reset or modify the output of the whole network. In general, this is not true. At a more detailed level of analysis one would expect to find that intracellular stimulation of one motor neuron would give rise to synaptic potentials in functionally related ones, but again this is not generally true, or else the effects are quite weak. Current thought on the matter is that motor neuron interactions are usually inadequate to account for the patterns of output in motor systems.

This conclusion leaves us looking for a process, or even a structure, that mediates between the commands and the motor discharge itself. The search is on for interneuronal **pacemaker cells or networks** that produce rhythmicity (Fig. 7.29). Consistent with the notion that there is a hierarchy in motor systems, with input commands driving oscillators that drive motor neurons, is the fact that the same set of motor neurons can be used in more than one behavioral pattern. Frogs swim or jump with synchronous output to homologous contralateral muscles, but alternate them during walking. Insects use some of the same muscles to move the wings and legs, but they do so according to different synergistic/antagonistic relationships, depending upon whether the command says "walk," "jump," "sing," or "fly." We are pushed into thinking that even in the lower ganglia or spinal cord there is a multiplicity of pattern generators or oscillators that can each be turned on or off or be modulated in frequency or amplitude by command input, and that these pattern generators each converge upon identical or overlapping pools of motor neurons. Thus the motor neurons are, to use Sherrington's phrase, the "final common paths" transmitting to the muscles signal patterns that are produced at a higher level.

If we turn from a preoccupation with the genesis of rhythm to the question of how **alternative motor patterns** that involve higher level switching are selected and programmed, we have to deal mainly with conceptual models that seem compatible with principles of physiological organization.

The rather widely accepted notion today is that actions commanded by higher centers are not specified in detail by those centers, nor primarily determined by peripheral stimuli, but triggered in a preprogrammed language calling up combinations of elementary acts in a hierarchy of levels. The highest command, as well as the lower-level instructions to still lower levels, may simply specify the sequence, strength, and duration of the next lower components, finally eliciting movements as though the adequate peripheral stimuli had occurred that can reflexly elicit them. Analysis of six gaits of horses shows that they *could* be produced simply by calling up components equal to certain local spinal and long spinal reflexes in formulated sequences and durations, given a few fixed rules (Easton, 1972). Whether the horse actually works this way, we cannot tell, but the rules are simple, and the model might work. Of course, superimposed on whatever "calling up" the brain initiates, proprioceptive input as well as visual and somesthetic reafference (sensory input caused by one's own actions) will be important in shaping and correcting the centrally patterned program, by

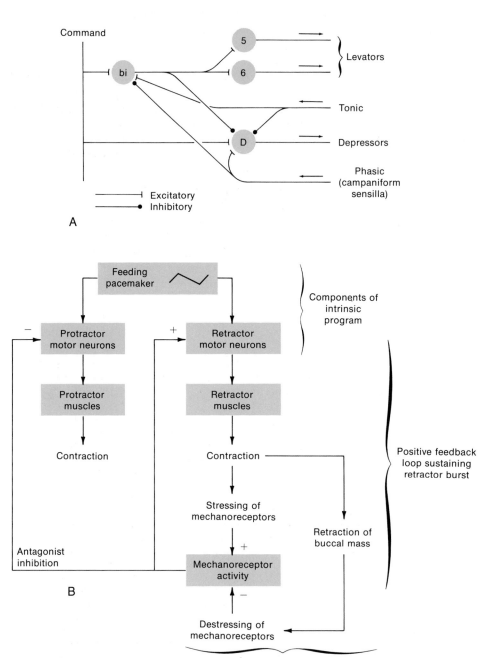

Command

5
6
} Levators

bi

Tonic

D

Depressors

Phasic
(campaniform
sensilla)

⊢ Excitatory
● Inhibitory

A

Feeding
pacemaker

} Components of
intrinsic
program

−
Protractor
motor neurons

+
Retractor
motor neurons

Protractor
muscles

Retractor
muscles

Contraction

Contraction

Stressing of
mechanoreceptors

Retraction of
buccal mass

+
Antagonist
inhibition

Mechanoreceptor
activity

Positive feedback
loop sustaining
retractor burst

B

−
Destressing of
mechanoreceptors

Delayed negative feedback loop
ending mechanoreceptor burst

simple adjustments of strength and phase of particular components.

Motor patterning systems have been so incompletely studied that we know very little of their actual mechanisms. Partial models with some supporting evidence can be made in many cases. (Examples are shown in Figure 7.30 and in the figure in Box 7.3, p. 275; see also p. 333.) Since it has been so difficult to make an analysis of even relatively simple motor systems in terms of individual neuron activities, perhaps we should expect to find eventually that presently unknown concepts of neural function are involved.

D. Centrally Scored Pattern by Sensory Tape

Another possible mechanism of central programming, besides the motor score, has been suggested by Hoyle. He calls it a sensory tape, or to use the phrase of the ethologists, a sensory template (see p. 309). Sensory tapes or templates have not been demonstrated, though strongly inferred for the control of some bird song. The idea is worth more discussion. Suppose the CNS contained an instruction that said, "Produce a motor output that results in a specified feedback from

Figure 7.30 (*facing page*)
Circuits for insect walking (A) and snail feeding (B). **A.** Hypothetical scheme for the observed discharge patterns of levator (5 and 6) and depressor motor axons (D). A bursting interneuron (*bi*) is excited by the command neuron and in turn excites the levators while inhibiting depressor motor neurons. The command fiber is believed to excite the depressor, so that an increase in command input decreases interburst interval while producing a less marked decrease in burst duration. Certain sensory input tonically facilitates the *bi* and inhibits *D*; other input phasically excites *D* and inhibits *bi*. [Pearson and Iles, 1973.] **B.** Diagram showing the functional interconnections that give rise to the temporal relations of activity in the retractor and protractor elements of the 25 pairs of muscles responsible for feeding in the snail *Helisoma*. The neurons of the pacemaker generate an autonomous rhythmic output that appears with different amounts of phase shift in the various motor neurons. The rhythm is symbolized by the sawtooth wave and the two representative populations of motor neurons at the top of the diagram (upper brace, right-hand side of diagram). The retractor neurons drive the retractor muscles. Contraction in these muscles excites mechanoreceptors, and their output is fed back positively via excitatory synapses on the retractor motor neurons. This positive-feedback loop (lower brace, right-hand side) sustains the retractor burst. Contraction of the retractor muscles produces, after a delay (due to excitation-contraction coupling and the viscoelastic and inertial properties of the system), a retraction of the buccal mass. When it is retracted, contraction of the muscles no longer stresses the mechanoreceptors, and these shut off; this series of events constitutes a negative-feedback loop with delay, and it limits the retractor burst by opening the positive-feedback loop after retraction is complete (low brace). The protractor motor neurons and muscles are excited at the opposite phase of the cycle from that of retractors. They fire until inhibited by input from mechanoreceptors, which are stretched by the contracting retractor muscles. This accomplishes a unidirectional antagonist inhibition with delay, which allows the protraction phase to be sustained until retractor tension is developed. Antagonist inhibition in the reverse direction (i.e., protractors inhibiting retractors) is absent, and this allows the overlapping activity in the two groups of muscles at the beginning of the retraction phase. [Kater and Rowell, 1973.]

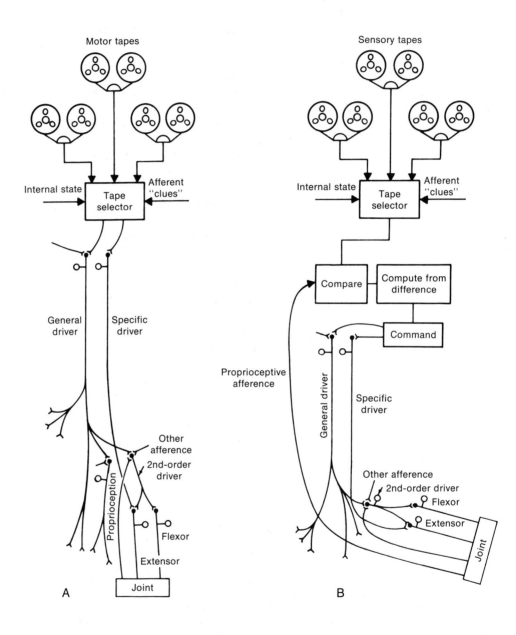

Motor tapes

Internal state

Tape selector

Afferent "clues"

General driver

Specific driver

Other afference

2nd-order driver

Proprioception

Flexor

Extensor

Joint

A

Sensory tapes

Internal state

Tape selector

Afferent "clues"

Compare

Compute from difference

Command

Proprioceptive afference

General driver

Specific driver

Other afference

2nd-order driver

Flexor

Extensor

Joint

B

proprioceptors or other sense organs." This instruction would not lead inevitably to a single stereotyped motor output, as from a motor score generator, but it could guide motor output to achieve a goal (Fig. 7.31). Output might at first give incorrect results, but feedback could modulate the output on successive cycles of loop operation. Consistent with this notion of central programming by comparison of sensory feedback with a centrally stored goal pattern is the fact that diverse motor outputs may be associated with apparently identical leg movements during walking in insects. In some cases the flexors and extensors alternate. In others, one muscle contracts tonically while the other oscillates. The resulting movement is the same.

The only reasonably strong case of a sensory template is found in bird song control (see Fig. 8.18). In the invertebrate cases in which a sensory principle may be operative, it seems to be superimposed upon a motor score type of central generator. Insect flight is basically programmed by a motor score, but exteroceptive as well as proprioceptive inputs can modify that score for purposes of stability, steering, or compensation for inherent error or damage to body parts. Perhaps in these systems the notion of a sensory template really reduces to reflex modulation of a motor score.

E. Coordinated Movement to Gross Stimulation of the Brain

A clear progression is evident if one compares the responses to crude electrical stimulation of structures at successive neural levels. Ventral roots give a segmental, local contraction closely related to the duration and strength of stimulation. Lower motor centers in the cord or brain stem give little more, though the distribution may be more functional (e.g. flexion of certain joints). The responses relevant to this section are the quite normal actions involving sequences of movements, such as can be elicited from the hypothalamus of mammals (p. 471). These are so natural as to suggest a central pattern, since the stimuli are like lightning bolts. A pocket mouse may stuff invisible seeds into its cheek pouches at a high rate, a cat may arch its back, hiss, unsheath its claws, erect its hair. The value for our purposes is the same whether we assume that the stimuli trigger motor patterns or sensory "hallucinations": the patterns are central and need only an adequate trigger.

Figure 7.31 (*facing page*)
Two types of control by centrally determined sequences. **A.** System driven by sequences that determine motor output directly. Driving commands can be general or specific to any degree of detail, and at a lower level they can be modified by proprioceptors. **B.** System driven by tapes of sensory feedback that must be expected. A comparator makes the actual commands on the basis of the differences between the proprioceptive afference and the instruction from the tape. There can also be proprioceptive control lower down, as before. This system is more adaptable, but requires much more circuitry in addition to a comparator and a command center, which themselves must produce highly complex adaptive sequences of impulses. Most invertebrate responses investigated use an "inline" system, as in **A.** [Hoyle, 1964.]

Box 7.4 Chronology and Background of Ideas on the Physiology of the Nervous System

We present here a selection of highlights in the history of ideas, from the earliest times up to 1929. The interpretation of the brain in terms of cells is highlighted in Box 3.1, p. 102. The roots of brain chemistry, membrane biophysics, pharmacology, sensory and psychophysiology, and behavioral analysis are not attempted here.

The reader is urged to look further, for more balance and adequate representation. Some useful, more-or-less condensed accounts are by Nordenskiöld (1935), Dampier (1948), Singer (1959), Brazier (1961), Sirks and Zirkle (1964), Gardner (1965), Clarke and O'Malley (1968), and McHenry (1969). Special aspects are dealt with in Fearing (1970), Brazier (1961), and Swazey and Worden (1976).

1700 B.C. An Egyptian document, translated centuries later and published as The Edwin Smith Surgical Papyrus, includes 13 case descriptions of head injuries. Aphasia, paralysis, and seizures were described, and suggested the functions of the brain. Nevertheless, disease continued to be generally attributed to supranatural influences and whims of the gods.

800 B.C. Homer's works and the flowering of Greek art and intellectual life led slowly and incompletely to the idea that the world is knowable.

500 B.C. Alcmaeon performed the first recorded dissection of a human body. He paid some attention to the brain and discovered the optic nerves. His teacher, Pythagoras, taught that the brain is concerned with reasoning. In the next century more dissections and similar speculations were made by others.

400 B.C. Hippocrates of Cos countered the mystics and the entrenched supranaturalists to introduce rational medicine. This required the systematic accumulation of clinical experience, and his description of epilepsy went unsurpassed until the work of Hughlings Jackson. However, Hippocratic teaching was dominated by the idea that function derives from the combination of four humors: blood, phlegm, and black and yellow bile. The brain is the organ of intelligence and dreams, but it also secretes phlegm and cools the blood. Even in the Golden Age the Greeks did not easily follow his example. The lack of autopsies delayed progress.

340 B.C. Aristotle systematically pursued comparative anatomy and in his 19 books set a high water mark of natural knowledge that lasted until the Renaissance, but he did little to change ideas on the brain.

300 B.C. Herophilus and the great school of Alexandria in Egypt dissected many cadavers and really founded anatomy. He distinguished sensory and motor nerves and showed that they connect from spinal cord to periphery.

250 B.C. Erisistratus postulated a mechanism of the brain function: blood and two kinds of air are carried in the veins, arteries, and nerves; air is changed to vital spirits in the heart and these to animal spirits in the brain ventricles, whence they go via the nerves to distend and shorten the muscles.

200 B.C. Galen culminated the classic period, writing more than 400 works that were definitive for a millenium. By now much of the naked-eye anatomy of the nervous system had been discovered, including most of the cranial nerves. Among the few advances in the understanding of function were the descriptions of symptoms following section and hemisection of the spinal cord in lower mammals.

400–800 The Dark Ages lasted more than 12 generations. Greek knowledge was forgotten in Europe. Men did not ask to understand themselves or nature but to be told the supranatural or religious meanings of things.

600–1200 Islam spread from Asia minor to Spain, carrying Greek knowledge unknown in the Christian world. Jewish traders introduced Arabic translations to Europe. Long-lost Latin versions of Greek writings were rediscovered in monastery storerooms. Men did not yet ask about nature but about their heritage.

1200–1300 This interest in book learning and the new wave of ideas induced scholasticism, which in turn

Box 7.4 (*continued*)

brought on humanism as a reaction. Universities sprang up widely during the last period of the crusades.

1400–1500 The invention of movable type stimulated printing. Voyages of exploration expressed the new attitude toward discovery.

1478 Mondino represents the height of classical, authoritarian (Galenic) anatomy. His manual, illustrated crudely and ascribing functions such as fantasy to the anterior part of the lateral ventricle, was little influenced by actual dissection. Nevertheless, it was used for 200 years.

1500 Leonardo da Vinci manifested the new curiosity by making his own dissections and drawing more accurately than anyone before; but, failing to publish, he had little influence on the progress of anatomy.

1543 Vesalius broke with the tradition of Aristotelian and Galenic authority, inaugurating the modern idea of the authority of original observations. His landmark work, *De Humani Corporis Fabrica*, contains many plates of brain dissections showing numerous features for the first time.

1608 Harvey was the first to reason that the blood circulates. He multiplied the capacity of the heart (with its one-way valves) by the heart rate, both long-known quantities. The method of inductive reasoning, lost since the Greeks, was re-established.

1662 Descartes, the leader of 17th-century physiological thought, crudely conceived the idea of reflex action powered by a Galenic mechanism. He broke new ground also in the mind-body problem, placing the seat of the soul in the pineal.

1664 Thomas Willis published one of the first separate works on the brain, the most complete and accurate so far, introducing several of our current terms. He suggested that the cerebrum presides over voluntary motions and the cerebellum over involuntary movements; he manipulated the cerebellum in a living mammal and noted that the heart stopped.

1691 Robert Boyle pointed to the existence of a motor cortex. A knight suffering a depressed skull fracture showed a sustained paralysis of arm and leg; the paralysis disappeared promptly after surgical removal of a spicule of bone.

1730 Stephen Hales opened the door to reflex physiology by noting that the legs of a decapitated frog would withdraw upon pinching but that such "reactions" disappeared when the spinal cord was destroyed. The terms "stimulus," "response," "reflex," "afferent," and "efferent" came into use by the 1770's. Robert Whytt (1751) played an important role in the drama.

1740 Swedenborg considered the basal ganglia the seat of primary sensibility of body and soul and the route of "all determinations of the will." He distinguished upper and lower motor centers and correctly subdivided the motor cortex.

1791 Galvani started electrophysiology by inadvertently stimulating the muscles of dissected frog legs when they completed a circuit with two dissimilar metals. Volta used the discovery of a source of electric potential to develop the battery and voltaic pile. Galvani, mistakenly believing that the potential came from the tissue, went on to discover bioelectricity by observing that a nerve is excited when it completes a circuit between an injured and an uninjured tissue. The use of a nerve-muscle preparation as a biological detector, amplifier, and indicator was an ingenious physiological technique that permitted the discovery of millivolt level bioelectricity many years before the galvanometer was invented. Electricity came just in time to fill the gap as improved anatomy excluded the hydraulic model of nerves and the new physics and chemistry raised doubts about "animal spirits." The stage was set for a rational, mechanistic physiology.

1809 Rolando removed the cerebellum in fish, reptiles, and mammals and saw disturbances in voluntary movements without influencing sensation.

1822 Francois Magendie made firm an earlier claim by Charles Bell that dorsal roots are sensory and also showed unequivocally that ventral roots are motor. Like all such experiments in these pre-anesthesia days, his work was based on vivisection.

(*Continued on next page.*)

Box 7.4 (*continued*)

1823 Pierre Flourens showed that vision depends on the cortex; ablation on one side in pigeons, rabbits, and dogs was found to cause contralateral blindness.

1826 Johannes Müller, wide-ranging German natural philosopher, sensory physiologist, and comparative anatomist, enunciated the "law of specific nerve energies," which states that each sensory nerve gives rise to its own characteristic sensation, however it is stimulated. For example, electrical, mechanical, or chemical stimulation of the optic nerve causes a sensation of light.

1833 Marshall Hall recognized segmental, intersegmental, and suprasegmental reflexes. "The spinal cord is a chain of segments whose functional units are separate reflex arcs which interact with one another and with the higher centres of the nervous system to secure coordinated movement." He recognized the temporary depression of reflexes below a spinal transection and called it spinal shock.

1848 Du Bois Reymond showed that activity in a nerve is invariably accompanied by an electrical change ("negative variation"). He described the properties of neuromuscular transmission and opposed the prevailing doctrine of vitalism.

1850 Helmholtz measured the velocity of conduction in nerve tissue and began what would become a continuing effort to improve the instruments of electrophysiology.

1851 Claude Bernard developed the landmark concept that the body maintains a constant internal environment for the cells by means of the extracellular fluids. Among other things related to sympathetic function, he described the vasomotor nerves, which play a key part in this regulation, later called homeostasis. An enthusiastic experimentalist, he wrote influentially on the experimental method, advocating rigor, controls, and formulation of testable predictions.

1861–1898 Hughlings Jackson, British neurologist, developed concepts on the underlying principles of brain function from clinical observations. One was the concept of "release" to account for various signs of injury to higher parts of the brain, such as spasticity, that are more positive than negative; the resulting overactivity of the surviving lower centers bespeaks a normal restraint imposed by the higher. A second concept was that although evolution has been a process of increased differentiation and heterogeneity, with integration keeping pace, disease reverses this, such that higher parts go first and the lower take control. A third concept, growing out of intense study of patients with speech disorders, those with sensory, motor, or psychic epilepsy and hemiplegias, was that the cortex has many localized functions.

1863 Sechenov studied "reflexes of the brain," meaning cerebral activity that arises from sensory stimulation and mediates psychic experience and causes voluntary action, subject to modulation by other brain centers, including inhibition by the midbrain. This he obtained by placing salt on the optic lobes. He recognized temporal summation of subthreshold stimuli; also muscle sense, later called proprioception. He emphasized the physicochemical analysis of metabolism and excitation. Trained with Claude Bernard in Paris and Du Bois Reymond in Berlin, he is regarded as the father of Russian physiology.

1865 Pflüger systematically investigated inhibition, mainly via autonomic nerves. Searching for inhibitory nerves to skeletal muscle, Pavlov (1885) found them in the fresh-water mussel, *Anodonta,* a bivalve mollusc; Biedermann (1887) found them in the crayfish. They are still unknown in vertebrates.

1870 Gudden's finding that specific thalamic nuclei degenerate when certain areas of the cerebral cortex are destroyed was a milestone in experimental anatomy, as well as calling attention to retrograde degeneration and opening the modern period of study of the thalamus.

1874 Bartholow, in the U.S.A., stimulated and mapped the motor cortex in man. He found, incidentally, that the brain itself is insensitive to manipulating and cutting.

1875 Richard Caton, in England, observed electrical waves from the exposed brains of rabbits and monkeys;

Box 7.4 (*continued*)

his finding was overlooked, but the waves were rediscovered later in Russia (1877), Poland (1890), and Austria (1890). Caton was looking for action potentials in the brain, inspired by Du Bois Reymond's in nerve, hoping they would provide a method for localizing sensory areas. In this he succeeded, discovering evoked potentials and, incidentally, DC shifts with activity, as well as the ongoing EEG.

1898 Langley introduced the term "autonomic" and 7 years later, "sympathetic" and "parasympathetic."

1902 Pavlov, investigating the physiology of digestion, saw clearly the road he would follow for 34 years, analyzing the psychological properties of conditioned reflexes. His influence on neurophysiology was "almost nil" (Fulton, 1949) until recent years.

1903 Brodmann, Vogt, and Campbell each made their first communications on the architectonics of the cerebral cortex, mapping the distribution of different types of cortical structure. It was some years, however, before the 6-layered stratification was fully exploited; then its embryological origin was emphasized. By 1929 Ariëns Kappers and others had added an evolutionary origin from a primitive 3-layered mantle. Enthusiasm for the 6-layered structure long delayed a concern for the neuronal organization and connections.

1906 Sherrington, in England, published his landmark treatise, *The Integrative Action of the Nervous System,* in which he systematically analyzed how the nervous system works, by close examination of simple and compound reflexes. Primarily he used mechanical recording of contraction of individual muscles. Most of the ideas on pp. 271–272 are his.

1909 Karplus and Kreidl began the first experimental study of the hypothalamus, their results appearing in a long series of papers. Many others joined in, including the celebrated surgeon Harvey Cushing, who in 1912 discovered that removal of the pituitary or merely transecting its stalk causes an adiposogenital dystrophy. Nevertheless, the modern period of research, in which the hypothalamus is related to autonomic, emotional, and instinctive functions and the

control of the pituitary, did not really begin until the 1930's.

1911 Henry Head and Gordon Holmes, using psychological concepts and testing, studied sensory deficits after clinical lesions. They were more concerned with the nature rather than the locus of cortical sensory processes in man. They showed that the cortex is especially involved in discriminative and higher aspects of perception. Head is remembered also for severing a nerve in his own arm to study the loss and return of sensation with regeneration.

1917 Keith Lucas firmly established the all-or-none law and quantitative relations in excitation, such as the minimal slope of a slowly rising current necessary for excitation.

1924 Kato, in Japan, settled a controversy by showing nondecremental conduction in nerve.

1924 Gasser and Erlanger, in the U.S.A., used the cathode ray oscilloscope to describe the components of the compound action potential of the whole nerve.

1924 Rudolph Magnus published his landmark monograph on posture. Starting his investigations 16 years before with Sherrington, he early realized the importance of twisting the head on the neck, which reduces the tonus of postural muscle and their reflexes on the side of the forward ear in the human or the lower ear in the dog.

1926 Adrian and Zotterman recorded from single sensory nerve fibers and found that, as in motor fibers, the repetition rate of impulses is graded. Stronger stimuli result in a greater frequency of spikes.

1929 Denny-Brown recorded from a single motor neuron activated by a normal reflex stimulus; he used the stretch reflex in a decerebrate cat.

1929 Berger reported that brain waves could be recorded through the skull in humans. After a general scepticism, this was confirmed by Adrian in England and taken up by Davis and others in the U.S.A.

SUGGESTED READINGS

Autrum, H., R. Jung, W. R. Loewenstein, D. M. MacKay, and H. L. Teuber. 1971. *Handbook of Sensory Physiology*. Springer-Verlag, New York. [Projected to make up eight volumes, some in parts, this will be a comprehensive treatise.]

Bach-y-Rita, P., C. C. Collins, and J. E. Hyde. 1971. *The Control of Eye Movements*. Academic Press, New York. [A symposium on a system especially favorable for revealing principles. Other more recent symposia and reviews can be found on this system.]

Granit, R. 1970. *The Basis of Motor Control*. Academic Press, New York. [A systematic analysis from sensory receptors to higher cerebral levels controlling motor neurons.]

Grodins, F. S. 1963. *Control Theory and Biological Systems*. Columbia Univ. Press, New York. [A compact textbook, useful both as an introduction to this approach and to some insightful examples, which are partly neural.]

Horridge, G. A. 1968. *Interneurons*. W. H. Freeman and Company, San Francisco. [A collection of diverse examples of neural mechanisms that involve interneurons at several levels of analysis.]

Reiss, R. F. 1964. *Neural Theory and Modeling* (Proc. 1962 Ojai Symposium). Stanford Univ. Press, Stanford. [A symposium combining experimental and theoretical treatments of examples of sensory and motor behavior.]

Schmitt, F. O., and F. G. Worden. 1974. *The Neurosciences: Third Study Program*. [See Suggested Readings in Chapter 8 for comments on this and two preceding volumes in the series.]

Stein, R. B., K. G. Pearson, R. S. Smith, and J. B. Redford. 1973. *Control of Posture and Locomotion*. Plenum Press, New York. [A symposium including invertebrate and vertebrate sensory, central, reflex, and rhythmic control.]

Stark, L. 1968. *Neurological Control Systems: Studies in Bioengineering*. Plenum Press, New York. [Exemplary analyses from an engineering approach of some subsystems controlling seeing and manipulating.]

Wiener, N., and J. P. Schadé. 1965. *Cybernetics of the Nervous System* (Progress in the Brain Research, vol. 17). Elsevier Publishing Company, Amsterdam. [Although many chapters are dated or mainly of historical interest, the collection exemplifies how diverse are the problems and stages of progress under this rubric. Many modern symposia and articles are readily found, in addition to the journals whose titles carry Wiener's term.]

Wiersma, C. A. G. 1967. *Invertebrate Nervous Systems: Their Significance for Mammalian Neurophysiology*. Univ. Chicago Press, Chicago. [A symposium dealing at many levels with results relevant to mammalian neurophysiology.]

8

INSIGHTS INTO NERVOUS INTEGRATION
FROM BEHAVIORAL PHYSIOLOGY

I. INTRODUCTION

In this chapter we deal with neural integration at a higher level. Whereas Chapter 6 deals with neuronal integration and Chapter 7 with the reflex level, the subject area of this chapter may best be compared to the German category "Verhaltensphysiologie," or behavioral physiology. In contrast to the approach in the textbooks on animal behavior and on physiological psychology, which treat behavior comprehensively and apply some neurophysiological principles where possible, we will select only a few examples of behavior and attempt to draw inferences about what goes on in the brain.

The school or approach called **Verhaltensphysiologie** owes much to the late Erich von Holst, and we will draw several cases from his work. It is an approach aimed at making formal statements about the nature of relationships, without immediate concern for underlying componentry. It led naturally to the

use of the methods and terminology of control-systems analysis and cybernetics.

Another school of animal behavior study, **ethology,** was developed in Germany, Holland, and England under the leadership of Lorenz, Tinbergen, Thorpe, and others. Ethology may be thought of as the comparative study of species-characteristic behavior. Its points of departure are the findings that behavior patterns may be attributes of taxa larger than the species, and that behavioral differences between species are correlated with other indications of species relatedness. Ethologists propose that whole behavior patterns are genetically specified—that is to say, that much of the determination of behavior patterns is based on information genetically stored within the central nervous system. This is a view strongly opposed to an older one held by many physiologists and psychologists—that behavior consists only of reflexes and learned activities. We will not discuss complex behavior extensively in this book, but will attempt to relate some aspects of ethology and psychology to neural mechanisms through an examination of the effects of stimulation or extirpation of specific parts of the nervous system.

To augment and clarify much of this chapter the student is urged to consult texts on animal behavior, such as those by Marler and Hamilton (1966), Brown (1975), and Alcock (1975), and also works on physiological psychology, such as those by Thompson (1975), McGaugh (1971), and Greenfield and Sternbach (1972).

One thing we shall emphasize in this chapter is another way of viewing nervous system function, in which the main facts and ideas spring from an analysis of behavior or input-output relationships of whole animals or large systems. This is a level of analysis in which few generalities are known and even the terminology is not agreed upon. It is the level at which we wish to examine the operations performed by whole organs or systems, not just the properties of their parts. Though we may have faith that, given enough information, we could predict organ or system properties from sufficient knowledge about the parts, this approach by itself is quite impractical. A serious limitation on the state of neurophysiology at higher levels of brain function is probably that the relevant categories of function are unfamiliar or hard to recognize.

An analogy may be appropriate to clarify the significance of saying that unfamiliar functions may be important in a complex system. A corporation or a university has many units—people—and these have quite similar basic properties. They all talk, listen, move about, read, manipulate, think, and remember. Al-

though they may use these properties to different degrees, the generality of such abilities allows us to recognize and study them easily and to call them Level I operations. At an arbitrarily defined Level II in the organization, operations are performed by people who use the basic properties in specialized ways or contexts: phoning, typing, filing, programming, lecturing, and grading. These are a bit harder to understand, if we are observers investigating the system, unless we recognize telephones, typewriters, files, programs, classes, and examinations. At Level III might be listed operations that would require still more insight into the web of relationships for an observer to understand, such as those of the bookkeeper, librarian, supervisor, analyst, scholar, physician, or safety officer. These functions in turn might exist at Level IV within any one of several subsystems, such as the university departments or schools. Here the difficulty for an outsider to understand is much more serious. In certain cases, this level may be the most conspicuously circumscribed geographically and yet the defining function of some units can be cryptic or unexpected (counseling service, planning division, extension division, ethnomusicology institute, privilege and tenure committee, endowments office), although others are more obvious, at least superficially (museum of anthropology, architecture office, archeology department, publication division). Analysis of a complex system by mapping connections, recording input-output relations, stimulating or ablating may be severely limited by our ability to guess or imagine the real function that defines a subsystem.

In the nervous system the neurons have Level I properties like impulse production, accommodation, encoding, transmitter release, facilitation, pace-making, inhibition. Level II operations might be filtering, pattern-generating, storing, switching, modulating, contrasting, comparing, erasing, feature-extracting, and synchronizing. Level III might be represented by widely distributed special groups of neurons that use common abilities and particular connections to accomplish a higher-level function like recognizing, expecting, arousing, altering set-point (see Glossary), orienting, focusing attention, controlling input, commanding output. The subsystems at Level IV might be like departments responsible for locomotion, respiration, appetite and eating, reproduction, temperature-regulation, hearing, vision, speech, body image, and the like. Or they might be quite unfamiliar, subtle, or unexpected, and therefore hard to demonstrate or accept. Some of these functions have readily localizable central and outlying offices—for example, centers for breathing, eating, hearing; others may not. Some functions—perhaps memory and motivation—may occur at both lower and higher levels, suggesting the need to distinguish types and hierarchies. The value of such an analogy is heuristic (suggests new research); it may "loosen" our thinking about possible domains of function for the structures of the brain, or modes of organization, and encourage rigor in designing and interpreting experiments.

II. ELEMENTARY FIXED ACTION PATTERNS AND INSTINCTIVE BEHAVIOR

The simplest elements of behavior are not called reflexes by writers in ethology, although some of them—for example, a cough, a swallow, a tail flick, or a scratching movement—would be by physiologists. With some reservations, ethologists recognize a category called fixed action patterns as the lowest-level behavioral acts. Examples are biting, drinking, cleaning, jumping, defecating, coition, threatening gestures, and specific vocalizations (Fig. 8.1). They are components of both instinctive and learned behavior.

What is significant for us is that fixed action patterns are (a) typically genetically specified, (b) little modified by the environment, (c) triggered only by specific stimuli, and (d) as characteristic of the species or larger taxon as are anatomical features. We may confidently infer that the nervous system contains genetically specified circuits (connectivity, see Chapter 3) and, what is equally important, physiological properties of integration (Chapters 6 and 7), such as timing, thresholds, frequencies, after-effects—in short, "coupling functions" representing the complete act in coded form, including specific sensitivities to both triggers and background conditions. Female canaries use a special weaving movement to push strands of material into place in their nests, and they perform the same movement even if given no material or nest site. Fixed action patterns cannot be broken down into a succession of components at the behavioral level, each with its external stimulus. They may reflect a real limitation of the nervous system, since they are amazingly resistant to learning. No bird has ever been known to develop a movement different from its characteristic one for retrieving eggs that roll out of the nest. "Skill" is largely the aiming and timing or the sequencing of action

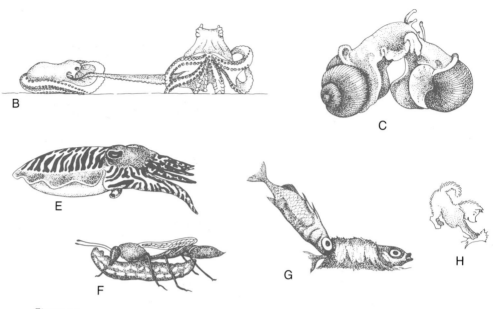

Figure 8.1
Elementary fixed action patterns. Examples of the species-characteristic components of higher behavioral sequences. Learning may modify the use of these actions, but very little their actual pattern.
A. Fiddler crab, *Uca pugilator,* waving its claw. [Tinbergen, 1951.] **B.** Mating in *Octopus.* [Buddenbrock, 1956.] **C.** Mating in *Helix pomatia.* [Tinbergen, 1951.] **D.** European wildcat striking with its paw. [Lindemann, 1955.] **E.** Male *Sepia officinalis* in sexual display. [Tinbergen, 1951.] **F.** The digger wasp *Ammophila campestris* carrying prey. [Tinbergen, 1951.] **G.** Male three-spine stickleback stimulating the female to spawn by quivering. [Tinbergen, 1951.] **H.** The mouse jump in the domestic dog. [Lorenz, 1954.]

patterns that are self-maturing in normal environments.

We are not here concerned with environment and genome as sources of the information in development, or with generalizations about all fixed action patterns. The insight or principle that is offered by certain kinds of behavior is that we can infer both its structural basis and its dynamic coupling functions to be built in during the development of the integrative organization of the brain. If this seems trivial or tautological, then (a) we have come a long way and do not know it (i.e., we take for granted a causal relation between behavior and neurons that is historically recent) and (b) we should re-examine what we take for

granted to see if it is adequate today (i.e., we may need to think about its consequences for our general view of how the central nervous system works).

Pursuing the latter goal, it is important to place the fixed action pattern in perspective. Each animal has a large repertoire of such acts, and they occur in restricted combinations and sequences. As Tinbergen points out, the **organization is hierarchical** (Fig. 8.2). Many writers view the higher categories, or alternative spheres, as **the major instincts**—reproduction, feeding, and the like. Lower categories like fighting, building, and mating include numerous elementary acts, and these are not independent but occur in sequences. There

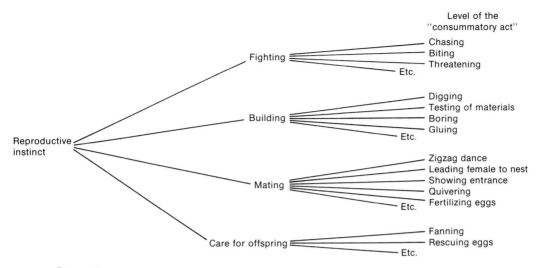

Figure 8.2
The principle of hierarchical organization. The reproductive instinct of the male three-spine stickleback. [Tinbergen, 1951.]

is often an initial "appetitive" behavior, which may appear to be aimless searching, followed by stages of narrower movements, each ending when a certain stimulus configuration is encountered, constituting a releaser for the next. Think of a hawk circling, approaching, diving, grasping, killing, tearing, and feeding. For neurobiology it is significant that typically it seems not to be the filling of a basic biological need (e.g., nourishment ingested) but simply the accomplishment of the consummatory act or the sensory input normally resulting from it that diminishes the "drive" (see below, p. 313) behind that activity, as measured by its probability of recurrence.

A key difference between the higher and the lower categories is the greater dependence of the higher upon the state of priming or readiness for the particular sphere of action. We must expect that a neural correlate for this readiness and for the switch that shifts spheres will be found.

The adaptiveness of the whole web of interconnections, including its malleability in certain respects and its steering by certain environmental features (see Section IV), adds up to a staggering degree of specification built into the circuits during development. This is "merely" the message of complexity and is only news if we have come to think of behavior in terms of rats in a forced choice situation or of the brain as a stimulus-response switchboard.

III. KINESES AND EXPLORATORY BEHAVIOR

A simple kind of behavior, which should be amenable to neural explanation, is exploratory locomotion with random turning and some aspect of activity graded with the intensity of a nondirectional stimulus, such that animals accumulate or spend more time where the intensity is maximal (or minimal). This is a kinesis. Sow bugs (*Porcellio*) aggregate

Figure 8.3
Klinotaxis.

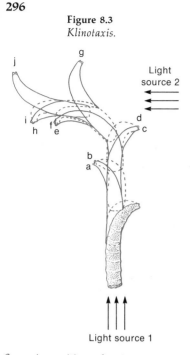

Successive positions of a maggot, reacting at first to light source 1 only. At position d, *light source 1 was switched off and 2 switched on.* [Mast, 1911.]

Figure 8.4
The dorsal light response.

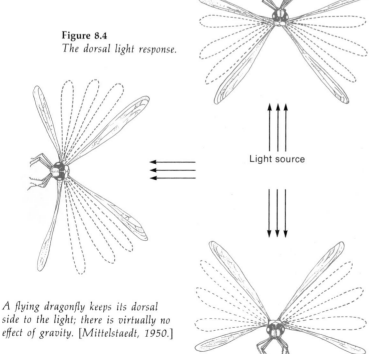

A flying dragonfly keeps its dorsal side to the light; there is virtually no effect of gravity. [Mittelstaedt, 1950.]

in damp places, flatworms (*Dendrocoelum*) in dark places, in this way. Kineses are of neuroethological interest partly because of their simplicity, but, surprisingly, no example has yet been studied physiologically. We may pose questions about the inferences that have been proposed. In *Porcellio* only the speed or time spent in locomotion is supposed to be graded by the stimulus (orthokinesis). In *Dendrocoelum* the rate of turning (degrees per minute) is graded (klinokinesis). Will we find these parameters among the neurons, including a suitable encoding of the relevant intensity, and a graded coefficient of speed, time spent, or turning rate? Is adaptation required to explain accumulation, and can its locus and dynamics be confirmed physiologically? It is of basic neurobiological significance that here is one of the instances of a

useful consequence of a random sequence of neural events—the impulses determining turning direction. Random sequences should not automatically be called "noise," which in our usage means useless events or irrelevant fluctuation. In exploratory behavior it is quite adaptive to search randomly, even though more sophisticated strategies may be developed.

IV. TAXES AND ORIENTING BEHAVIOR

A. Taxic Behavior

In taxic behavior the body is oriented relative to a directional stimulus. If the animal then moves, it will go directly toward or away from (or at a fixed angle to) the stimulus source. If a large sample of positively phototactic animals is placed in a uniformly lighted environment, they may distribute themselves randomly, but adding illumination at one side will cause them to crowd up against that side. It is common to find the small organisms of a pond or aquarium concentrated at one edge or corner, often toward the sun. A dozen maggots (house fly larvae) of a certain stage will march like a troop away from a light source; this is a negative phototaxis. They do this by swinging the head back and forth, comparing the light intensity they receive at successive points in time, and controlling as a result the appropriate compensatory movements to orient away; the temporal comparison makes this a klinotaxis (Fig. 8.3).

Taxic behavior appears compulsive, stereotyped, and often nonadaptive when studied in the isolated laboratory situation (Fig. 8.4). But taxes are not

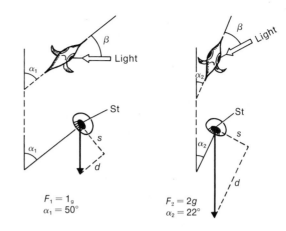

$$F_1 = 1_g$$
$$\alpha_1 = 50°$$

$$F_2 = 2g$$
$$\alpha_2 = 22°$$

Figure 8.5

Vector summation in the control of behavior. A fish swimming into a current of water is depicted head on. It assumes a tilt as a function of the direction and intensity of gravity and light. When the apparent gravitational force is doubled by means of a centrifuge, but the intensity and angle of light remain the same with respect to the fish, the tilt is much less. The vector diagrams show the position of a utricular statolith, St, in the inner ear; s and d are the shearing and pressure vectors, respectively, of the force F exerted by the statolith. [Mittelstaedt, 1964a.]

altogether simple. A phototaxic organism may be attracted to dim light and repelled from bright light, and the threshold for this crossover of sign may vary with time of day. Geotaxis may change sign with the diurnal rhythm too. Both these taxic sign reversals aid in the vertical daily migrations of planktonic organisms. Two or more taxic influences can act at the same time, adding so that the animal takes the direction of the vector sum. Phototaxis and geotaxis in fish usually work together to keep the dorsal side up, but a light from the side causes a tilt of the dorsoventral axis to a compromise angle between the two stimuli.

Certain tall, thin fishes, such as the fresh-water angel fish (*Pterophyllum*), are convenient to use in studies of gravity orientation; their dorsal and ventral fins are excellent pointers. Fish have gravity receptors or otolith organs as part of their inner ears, as do other vertebrates. In the dark they orient vertically, unless these organs are destroyed, in which case they do not orient at all. In a dark, centrifuged aquarium they orient along the line of the vector sum of the earth's gravitational force and the centrifugal force. Even with the otolith organs destroyed, angel fish can orient if the environment is lighted, as long as their eyes are intact and the lighting is not uniform. They turn their dorsal fin toward the region of greatest average light intensity, in a reflex called the **dorsal light response.** Using a single source of light, such as a lamp, angel fish whose otolith organs have been destroyed can be induced to assume any spatial orientation. If the otolith organs are intact and a lamp is moved away from the vertical, then the fish tilts toward the lamp, but not all the way (Fig. 8.5). Its deviation from vertical depends upon the position of the lamp and its intensity. If, on the contrary, the lamp position and intensity are held constant and the gravitational direction and force varied by placing the aquarium in a centrifuge, the fish orients with a deviation from the light source, which is a function of the direction and strength of the gravitational field and measures, in effect, **the value the CNS places upon input** from the two sense modalities.

A further outcome of this type of work answers the question, "Are these orienting taxes turned off during voluntary movement?" (See also p. 306.) Fishes sometimes tilt "voluntarily." Given background information on the average magnitude of these voluntary tilts, one can show that under increased gravitational force the angle of the average tilt is smaller. The reflex is therefore operative; the voluntary command is superimposed.

An example from man will emphasize the power of this approach in predicting specific connections in the nervous system. If seated in a rotatable chair and surrounded by rotatable walls with visible contours, most subjects will refer any relative motion to themselves—that is, they feel that the walls are stationary and the chair rotating, even when the opposite is true. The sensation has no affective quality (good or bad feeling) and is nothing more than that experienced in a train station when the next train starts to move. However, if the head is tilted to rest on one shoulder during this experience caused by rotation of the walls, a distressing sick-to-the-stomach feeling like sea sickness is added to the sensation of self-rotation, even though there had not in fact been any rotation of the inner ears. Visual dominance makes the apparent motion so real that it interacts with head tilt to elicit the unpleasant sensation in the same way as does real rotation. Whether there is a **descending message from visual to vestibular centers** is an experimental question of great interest. Recent evidence suggests that there is such a message. Also from human experience, it is noteworthy that damage to one inner ear or to the vestibular nerve on one side, even if complete, causes only a temporary period of imbalance. People soon recover from this loss, thereby displaying a prominent property of the nervous system—**the ability to compensate** or regulate after certain kinds of damage. This must imply an alteration in the function of

undamaged centers. Microelectrodes in the vestibular nuclei of guinea pigs show changes in the sustained activity level after injury and again after compensation. A labyrinthectomy on the left causes loss of input to, and decreased activity in, the left vestibular nuclei; the asymmetry causes profound imbalance and distortion of posture. Compensation is accompanied by gradual return of sustained activity on the left (from whence is little known) and decline of activity on the right, toward the same levels; posture and equilibrium then recover. This example deserves further study as one of the outstanding cases of central compensation in mammals.

The classes of taxes in which orientation is directly toward or away from the source of stimulus are believed to depend on a comparison of the intensity of afferent input from the relevant sense organs. The comparison may be made at successive moments—for example, as the head of a maggot waves and alternately exposes left and right sides to the source (Fig. 8.3)—or it may occur simultaneously and not require side-to-side swings. In either case, the locomotion directly toward or away from the source implies a matching or equalizing of the left and right receptors and effectors and, in the nervous system, a built-in balanced sensitivity of the transducers, encoders, filters, and commands on the two sides. Moreover, since most eyes have many receptors of differing, overlapping receptive fields, the organism's goal in keeping a light source straight ahead (or behind) implies a certain excitation formula or defined distribution of excited receptors that is equal on the two sides. This is the kind of physiological prediction from observed behavior that is sought in the neuroethological approach.

Another class of taxes permits us to infer the maintenance of a fixed, unequal ratio between the two sides, since the animal orients at a fixed angle to the source, as in **sun compass navigation.** The angle is set by some recent experience and can be called a **"set point,"** or "Sollwert" ("should-be value" in German); it must have a neural correlate.

An interesting illustration is provided by the behavior of some ant species that orient relative to sun position. If their view of the sun, wherever it is relative to their trail, is blocked by a shield, they may be redirected by the light reflected from a mirror, which gives them a false view of the sun's angle. Nocturnal insects sometimes orient with respect to the moon. A moth flying at constant angle relative to the moon may encounter other, brighter light sources, such as street lamps, and change its flight path. Orientation reactions to point stimulus sources can have an either-or characteristic (Fig. 8.6). Night-flying insects that enter the sphere of influence of a noncelestial light source will behave in one of two ways. If their original path was at an angle greater than 90° relative to the orienting light, then they will move away from it. If the path angle was less than 90°, they will spiral inward toward the light. This result of the compass reaction probably accounts at least partially for the tendency of insects to accumulate around night lights or to fly into candle flames.

In compass reactions it is as if the animal orients by keeping the visual source focused on a particular spot on its retina. Actually, two additional observations show that the mechanism is much more complex than that. In bees, and probably many other invertebrates, solar orientation is possible even when the sun is hidden by clouds. They can still orient because they are able to detect the plane of polarized light. The atmosphere has a polarizing effect upon the sun's rays. We cannot normally see the effect, but to a bee it must appear as a radiating or concentric pattern around the sun in the polarization-sensitive visual submodality. If only a portion of this regular geometric pattern is visible, its center can be predicted, so that a compass reaction is still possible.

Another complication in celestial navigation is that the celestial objects move relative to the earth. Most of these movements are cyclic and have periods of one day or one year. (The lunar cycle may not be accounted for in navigation in these animals.) Birds, bees, and ants can maintain constant course over periods of time that ought to result in deviations in path if their angle of orientation were truly constant. Hence, they must make use of mechanisms that correspond to both of the classical tools used by human navigators, an angle-measuring device (sextant) and a clock. **Biological clocks** that represent the time of day (as distinct from stop watches that measure time differences) are little known, but we will introduce some ideas on such mechanisms on page 326. Here we can prove their necessity by illustrations from ant behavior. A group of ants following a celestially oriented course can be interrupted by keeping them enclosed within a box so that they cannot continue on their way or see the sun. If, after one

Figure 8.6
Telotaxis.

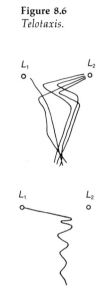

Paths of hermit crabs (above) and isopods (below) alternately crawling toward one or the other of two lights (L_1, L_2). [Above, von Buddenbrock, 1922; below, Fraenkel, 1931.]

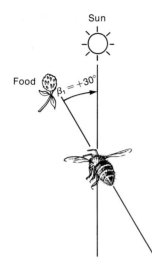
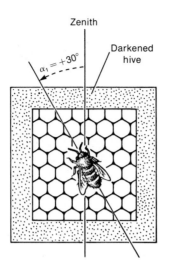
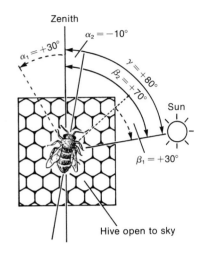

Figure 8.7

Angle transformation and summation. Experiments of Frisch (1948, 1962) with bees, indicating the deviation angle of a food source from the sun (β_1, left picture) by "dancing" on the vertical honeycomb in a dark or homogeneously lit room at the same angle to the zenith (α_1, middle picture). In the critical experiment (right picture), the vertical honeycomb is exposed to the sky, so that both light and gravity act on the bee. Under these conditions, the dancing direction (α_2) exactly bisects the angle between the broken lines. Recent work shows that intermediate angles can occur if light intensity or amount of gravity vector is varied. [Mittelstaedt, 1964a.]

or two hours, they are released, the sun will have moved 15–30°. Some proportion of that amount, dependent upon vertical angle of the sun, will appear as a change of the sun's azimuth angle relative to their former path. Nevertheless, they set out along the former compass path. They must be able to predict the new sun position in order to change their angle to it by the right amount for the time of day.

Some birds orient consistently in a planetarium with artificial stars rotating around an empty "polar" part of the sky; they maintain a certain angle to the extrapolated "north star."

Another sign of **complex encoding** of orientation angle in the brain is shown by bees forced to travel an indirect route to a food source that lies behind an obstacle. When they dance after returning to the hive, they communicate the correct angle for the "bee-line," which is the vector sum of the indirect path. The horizontal azimuthal angle to the sun is transposed in the nervous system to the same angle relative to gravity on the vertical comb surface in the dark hive. This is not done by an unequivocal switch of modality, since if a view of the sky is permitted in the hive, bees may show vector summation by dancing at a

Figure 8.8
Fixed calibration reflex steering. This is the first of a series of diagrams of examples of behavior graded by the complexity of the organization of the control system. (See also Figures 8.14, 8.15, and 8.19.) To represent the main functional features of a system, some kind of conventions are necessary. We use a simplified form in which boxes stand for a mechanism or operator, such as a visual system or postural system, and arrows indicate input and output signals. Here the firefly exemplifies the open chain class, in which there is no time for feedback.

vertical angle that is intermediate between the transposed azimuthal angle and the directly observed vertical inclination of the sun (Fig. 8.7).

All those properties of taxes, like those of the simpler forms of behavior treated earlier and the more complex forms to be discussed below, give insight into nervous organization, because we now appreciate the wealth of both qualitative and quantitative connectivity and coupling functions that must be specified.

B. Analysis of Behavior as a Control System

All these examples could have been described with the language and conventions used by control systems engineers. It will be worthwhile to consider a brief series of examples using a simplified systems analysis to illustrate this powerful approach. Even without entering into quantitative analyses of the notions of control systems, we can still gain some insights about behavioral systems. Indeed, in earlier chapters we have already encountered some of the principal notions involved (pp. 247, 255). Given information about input-output relationships across a whole behavioral system, one can make deductions that might suggest underlying mechanisms. More accurately, one can say from such relationships that some model has the same behavior as that characteristic of the system under study. This approach has revealed phenomena that might never have been discovered otherwise, and at the least provides a kind of intuition about the phenomena themselves, which helps by suggesting implications at a lower level of analysis.

Straight-chain Command Without Feedback. Two examples worthy of mention have similar control features: in both of them, once an act is initiated it cannot be further steered by a feedback servomechanism. Orientation or steering takes place before the motor action takes place, as in the firing of an unguided missile. Our first example of such a mechanism is the orientation of a male firefly toward the female's light flash (Fig. 8.8). The system may be thought of as a case of fixed-calibration reflex steering. The male turns toward a flashing

Figure 8.9
*Motor pattern resistant to
environment.*

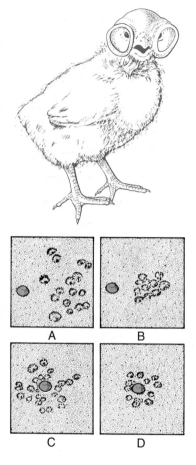

(A). *A day-old chick fitted with plastic prisms that throw the apparent image 7° to the right misses a brass nail by 7°.* (B). *With experience, a 4-day-old chick clusters its strikes, but the average score is still 7° to the right. A control, fitted with flat windows* (C), *strikes close to the target from the start and* (D) *also becomes more accurate with experience.* [Hess, 1956.]

light source, but no feedback correction is possible because the female's light flash is over before the male has executed his orienting movements. The input signal is so brief that there is no opportunity for the animal's turning to be corrected by renewed comparison with the stimulus; nor does the output (male's motion) influence the input (angle between axis of male and direction of light source); the output begins after the input ends. The adaptive quality of the output depends entirely upon the accuracy of the built-in performance characteristics, or calibrations, of the sensory, integrative, and motor systems. Perturbations or defects in these cannot be compensated for during the act.

The second example in this category is complementary in the sense that the limitation on time available for correction is imposed by the duration of the motor act itself and not the persistence of the input signal. When chickens peck at grains of corn the original aim is visually directed, but the strike is abrupt and twitch-like. Neither proprioceptive nor visual feedback pathways can be traversed during the strike itself. Hence, if there is statistical variation or miscalibration in the aim, the chicken may miss the grain. Miscalibration may be induced experimentally in an otherwise normal chicken by fitting it with prism glasses. Simple prisms offset the visual world by some angular deviation along one axis (Fig. 8.9). Chickens outfitted with prisms that give a deviation of only a few degrees can never achieve the goal of the peck—seizing the corn. They continue indefinitely to err by an angle equal to that of the prismatic aberration. This result demonstrates two important things. One is that the system cannot be recalibrated through learning under

these conditions. The second, more relevant here, is that there is no (or at least insufficient) visual feedback correction during the strike, even though the strike is visually guided. If there were continual correction the linear error would become indefinitely small (even with a large constant angular error) as the target was approached.

Feedback Control and Superimposed High Command. Figure 8.10 shows the difference between two forms of control. Feedback ("loop") control waits for an error signal to occur; in contrast, open-chain, or feed-forward ("mesh"), control has quite different dynamics and can anticipate. Experimentally we can test whether a given example of behavior depends on feedback by "opening" the presumptive loop—that is, preventing the movement from altering the input. Figure 8.11 shows a way to do this with a fly. (Figs. 8.12, 8.14, and 8.16 show modifications for other animals.)

Many animals, perhaps all of those with form vision, show a tendency to move with a moving visual field in such a way that their retinas keep a constant angular relationship to the visual field. This tendency is called **the optomotor response;** it is most conspicuous in the absence of conflicting factors (e.g., locomotory movements or orientation with respect to more than one input modality). Hence the optomotor response is often studied in a special situation that eliminates to a large extent the effects due to other inputs or drives. In the arrangement for this purpose, shown in Figure 8.11A, the animal is fixed in space and allowed to "walk" on a lightweight paper disc that it holds. As the disc turns it gives a measure of the optomotor response without visual feedback. The

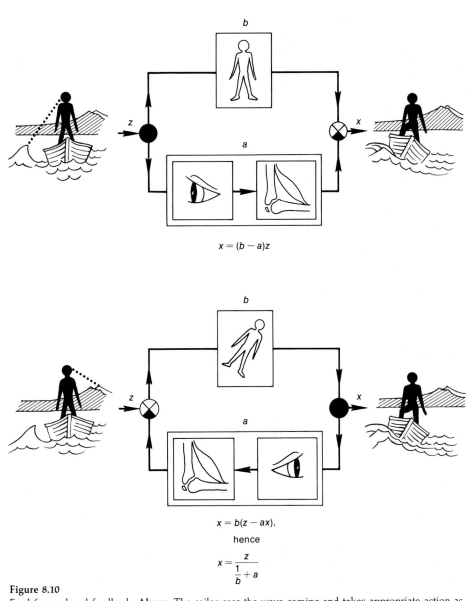

$$x = (b - a)z$$

$$x = b(z - ax),$$

hence

$$x = \frac{z}{\frac{1}{b} + a}$$

Figure 8.10
Feed-forward and feedback. **Above.** The sailor sees the wave coming and takes appropriate action as it arrives. The feed-forward direction of the parallel pathways of signal flow is sometimes called a mesh, to distinguish it from the next type. **Below.** The sailor sees the change in his orientation by sighting a reference object and takes action after some tilting has occurred. This is feedback, sometimes called a "loop." The symbols z and x refer to the magnitudes of signals, controlling and controlled, respectively; a and b refer to the frequency characteristics or filter properties of the corresponding components. Boxes represent amplifiers (in a broad sense) in which an input magnitude is acted upon—for example, with mechanical inertia (upper pathway) or sensing and muscle contraction (lower pathway). Circles are signal-summing junctions; black sectors signify that the signal is subtracted. [Mittelstaedt, 1961.]

Figure 8.11
*Method for measuring the optomotor response of a fly under "open-loop" (**A**) and "closed-loop" (**B**) conditions.*

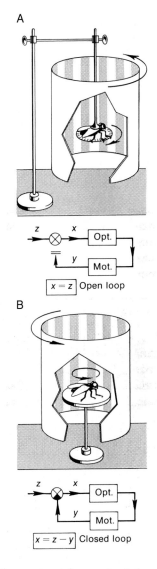

A

$x = z$ Open loop

B

$x = z - y$ Closed loop

*The experimental apparatus (schematized) is shown above; diagrams of the control patterns, below. **A.** The fly is fastened at the thorax, and the velocity of a paper disk held by the fly is measured. **B.** The fly moves freely on a standing platform, and its velocity relative to the ground is measured. [Mittelstaedt, 1964b.]*

reaction can be compared quantitatively with various parameters of the stimulus, such as light intensity, velocity of visual field movement, and stimulus pattern. Other forms of apparatus may be employed to measure the turning tendency as a series of left-right choices or as isometric rotational force or as output activity in motor units. One may also record other relevant nervous activity through electrodes placed in sensory nerves or the CNS. Such experimental situations have been used to test several aspects of the processing of visual information (p. 308), and provide a powerful tool even when no electrophysiological correlates are available. As one example, it has been shown that some arthropods can respond to visual fields rotating at less than the angular velocity of the sun relative to the earth. Another is that discriminability between light and dark stripes disappears only as the intensity difference becomes so small that the difference in the number of photons entering sets of receptors from separate stripes becomes a statistically unreliable estimate of the intensity difference in the short period of time available for averaging.

In the above situation the animal is unable to move in space; therefore, the turning tendency, or force, does not result in actual following of the stimulus. If it were free to move, the movement would reduce the relative magnitude of the stimulus movement. The two conditions—of spatial fixation and freedom—are, respectively, the open-loop and closed-loop conditions of a negative-feedback system. The motion of the visual field relative to the animal may be considered as an error signal; the degree to which it is reduced by the animal's own movement is a measure of the efficacy of the feedback loop.

To illustrate the first step in quantitation, Figure 8.12 shows how the gain of the loop can be measured in a crab. The loop is opened by fixing the seeing eye rigidly, as was done with the fly, above. But instead of using the legs to measure the optomotor response in the crab, we observe the movements of the other eye, which is carefully blindfolded with paint or shielded from seeing any contours. The seeing eye drives the unseeing eye via the brain. In the closed loop, with the seeing eye free, both eyes move with a velocity just a little less than that of the stripes; the slip, or error signal, is small. With the seeing eye fixed, the unopposed error signal is large, and because of the feedback amplifier the driven eye moves at a velocity much higher than the stripes. From such numbers one can learn a good deal about the formal properties, dynamics, and couplings in the system. For example, (a) the ratio of eye velocity to stripe velocity in the open loop gives the feedback gain; (b) the stripe speeds effective under these conditions extend over a thousandfold range and down to extremely slow motion; (c) the gain is large at low stripe speeds and falls gradually over most of the range; (d) the gain is most variable in the same low velocity range in which the range of variation of eye speed is least.

One additional feature that points to an important neural property of the crab in this situation is its ability to remember the angle between the optic axis and a stationary stripe (not relative to other visible landmarks) for a period in the dark. After it has viewed a stationary drum, the light is extinguished, and during the dark period the drum is turned slightly to a new stationary position, with no clue to the crab. When the drum is reilluminated the eye turns toward but not quite to the new position; the longer the dark period, the greater the discrepancy, but the memory of stripe position can be accurate to 0.1° and can last at least 8 min, even though the eye may have moved during the dark period.

What is the utility of the optomotor response in the normal life of an animal? Consider by analogy a machine, such as an automobile or a rocket, which is set into motion along some path toward a goal or a target. In general, such machines are imperfect and environmental conditions along the path vary in unpredicted ways, hence the vehicle deviates to some degree from the intended path unless continuous or periodic corrections are made. In an automobile, guidance is accomplished by a human driver who senses deviations and makes compensatory turns. In a locomoting animal, the optomotor response can have the same function, given that the visual world is for the most part stationary relative to

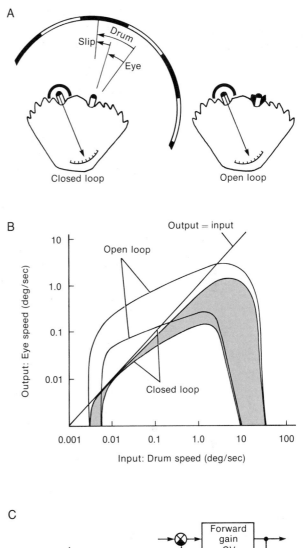

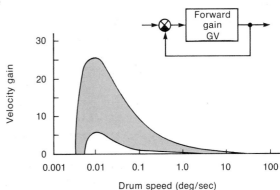

Figure 8.12

Comparison of performance in the "closed-loop" (normal feedback condition) and "open-loop" (interrupted feedback condition). **A.** The optomotor response of a crab whose carapace is held immobile and whose left eye is covered and unseeing but free to move and carrying a pointer. The right eye is either free to move (closed loop) or immobilized (open loop). When free, it follows the striped drum motion, but never reduces the velocity error to zero. **B.** Performance in the two conditions is plotted as eye speed versus drum speed over a wide range; variance is indicated by the width of the bands. This permits the calculation of the amplification, or "gain," shown in **C** at different drum speeds. Where the gain is highest, it is also most variable. [Horridge, 1966.]

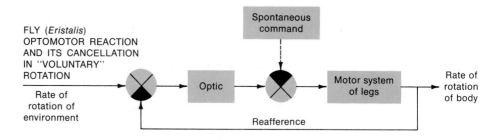

FLY (*Eristalis*)
OPTOMOTOR REACTION
AND ITS CANCELLATION
IN "VOLUNTARY"
ROTATION

Figure 8.14
Feedback control and higher command. Unless there is a higher command, the fly follows a rotating drum with a slight slip or error, exemplifying feedback control. Commands from higher levels in the brain, as in "voluntary" walking, can override without shutting off the feedback.

Figure 8.13
Experimental reversal of the sign of reafferent feedback.

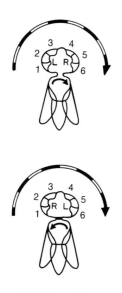

A. Normal fly follows the striped drum (Fig. 8.11). B. Fly with head twisted 180° turns against the drum. [*Mittelstaedt, 1961.*]

the intended path. If the animal makes an unintended deviation, say to the right, there is an equal relative motion of the visual world to the left. The latter motion induces a turn to the left, which at least partially corrects the error.

In accord with this view of its role, the reaction should operate during voluntary locomotion. But then, we may ask, "What about the role of the optomotor response in normal behavior: in view of this strong reflex, how is it that an animal can turn voluntarily in a stationary environment? Is the reflex turned off during voluntary rotations? To be sure, the reflex is commonly weaker to stimuli such as normally occur in locomotion than to the opposite stimuli; thus a frog reacts more to stripes moving from back to front than to stripes moving front to back. One test of whether the optomotor reaction does function during locomotion is provided by the possibility of artificially changing the feedback signal. A sign reversal may be accomplished by outfitting the animal with glasses that invert the visual image. In some vertebrates the eyeballs are rotated surgically, and in some insects the whole head is

rotated 180° and fixed in that position (Fig. 8.13). Now a movement of the visual world to the right is perceived as a movement to the left. The sign of the feedback loop is positive, and errors are not corrected, they are enhanced. With the first deviation from the intended direction, the animal begins to circle compulsively. This shows the optomotor response is not turned off, but still operates, merely overridden by the locomotor command (Fig. 8.14).

Analysis of optomotor responses in monkeys, and of other kinds of eye movements as well, yields the interesting conclusion that proprioceptive feedback from eye muscles plays no role in these movements. The system does very well with permanent calibrations and visual feedback.

Combinations: Open Chains with Feedback Subsystems. A system is whatever piece of nature we choose to isolate for consideration. Most behavioral systems are more complex than the foregoing examples. There may be more inputs and more loops or chains. In Figure 8.15 an example is summarized from Mittel-

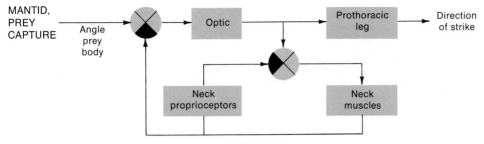

Figure 8.15
Open chain with feedback subsystems. One of the range of possible combinations is exemplified by the prey-catching strike of a praying mantis. Although the strike is so rapid that it is not controlled by feedback, once initiated (open chain), it is aimed by a visual feedback acting upon the head angle to the prey, which brings into play also a subloop of proprioceptively measured head angle to the body.

staedt's analysis of the praying mantis and its strike at a fly. The eye is on the head, with a movable neck between it and the leg-bearing thorax. There is a negative-feedback loop (proprioceptors keeping track of neck position) within a larger negative feedback loop (neck angle influencing the image of the prey on the eye), and these work together to steer the strike. But the final strike is so fast that there is essentially no opportunity for feedback from it in time to correct the movement. Hence the overall system is an open chain.

V. EFFECTIVE STIMULI AND INPUT FILTERS

A. Releasers and Recognition Mechanisms

This area is one in which neurobiology should be in the closest touch with ethologists, whose approach to animal behavior has enjoyed special success in experimentally isolating the effective features of stimuli that release acts characteristic for the species. By com-

bining the optimal dimension in each feature ethologists can even make artificial stimuli more effective than the natural ("supernormal"). Since this filtering is probably only partly peripheral (i.e., in the receptors), and mainly in the CNS, we may hope to find some neural correlates. It is difficult to know what to expect to find by our methods of probing with microelectrodes. In Chapters 3, 6, 7, and 9 we show that neurons in many animals are found to have some degree of sophisticated feature extraction. A few have been reported that approach specificity to a single natural sign stimulus or releaser. The type 2 units in the frog optic nerve have already been described (p. 251). Some cells in the temporal cortex of squirrel monkeys respond specifically to only one type of call—for example, the "isolation peep"—when stimulated with tape recordings of 13 of the 35 or so known natural vocalizations of the species. But in this case the essential features of the call have not been determined, either for these units or for the behavioral recognition.

An example from the behavioral side is the effort to specify what equivalent

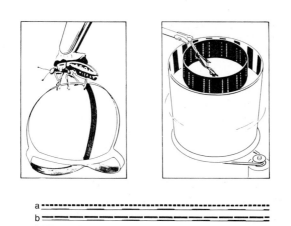

Figure 8.16

Experimental situation for analysis of the mechanism by which movement is detected in the visual field. A beetle (*Chlorophanus*) is fixed to a support, but carries a light grass "globe," which turns as it walks. At the intersection it chooses right or left according to the impinging visual stimuli. These are delivered in various patterns by rotating stripes seen through stationary slots. The four examples (*a* to *d*) show some possible sequences. In each, the upper, interrupted lines represent the stationary slots; the lower lines represent the moving stripes. Seen through the slots, *a* appears as a steady, forward-moving pattern; *b* as a fluctuating, backward-moving pattern; *c* as a fluctuating, forward-moving pattern; and *d* as a steady, backward-moving pattern. [Hassenstein, 1951.]

Figure 8.17
Model of a motion detector.

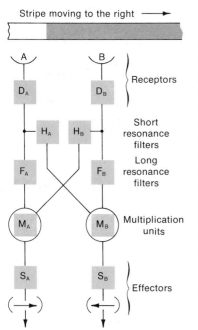

Example of a formal or equivalent model based on the optomotor system of Chlorophanus. *Specified in terms of operations, not neurons, it permits precise prediction, but is not the only model that will fit the data. [For detailed explanation, see Hassenstein and Reichardt, 1959.]*

operations must be performed in the nervous system in detecting motion and its direction. Hassenstein and Reichardt used horizontal motion of striped cylinders around a beetle (*Chlorophanus*), which while fixed to a support, will "walk" for hours by turning the grassblade globe it carries, making hundreds of right/left choices (Fig. 8.16). After testing with ingenious permutations of stripe width, stationary window spacing (Fig. 8.16, inner cylinder), and consequent apparent motion, two classes of conclusions were drawn. (a) The beetle adds linearly the signals of dark-to-light and light-to-dark transitions from neighboring facets of the compound eye (ommatidia) and from neighbors-but-one without employing comparisons of more separated ommatidia. (b) The magnitudes and the signs of the signals (positive making the beetle turn in the same direction as the striped cylinder and negative making it turn in the opposite direction) are processed as though filtered, multiplied, and compared in an equivalent circuit the authors have specified (Fig. 8.17). This circuit or model is not the only one that is equivalent to the observed relations and does not, of course, propose the anatomical substrates of the functions represented. Its main value for deeper mechanistic search is heuristic—for example, in suggesting the search for physiological signs of multiplication.

B. Sensory Templates

At least apparently related to the expectation principles on page 310 is a behavioral control mechanism based on the comparison of feedback input to an expected pattern called the sensory template (p. 283). This has been reasonably well shown only in the learning of songs by birds. The development of bird song deserves a brief account here in order to explain the principle of the sensory template.

The songs of many passerine birds,

like chaffinches and white-crowned sparrows (unlike swallows, orioles, wrens and others), are learned—that is, their development depends upon the bird having heard a suitable model, normally its own father, during a critical few weeks of its first months of life. When the young male begins singing, many months later, models are unnecessary. The song is produced crudely at first, but is gradually perfected. If the young of these species are hatched and reared in isolation, so that individuals cannot hear their own kind, they develop somewhat abnormal song (Fig. 8.18). If a young bird hears the normal song or a closely related one during the critical time and is then isolated, it develops that song. If it is allowed to hear a suitable model, but is then deafened before learning to sing, it cannot do so. This set of results is interpreted as follows: the young bird hears its species call and stores an image of it, a sensory template. Later in life it begins to sing, but the motor output does not produce proper song. The auditory feedback is compared to the template, and a mismatch gives rise to changes in output. The learning of song consists of serial comparisons of feedback to the template until the match is complete.

Although the mechanisms of the sensory template and of efference copy or corollary discharge (see next section) are similar in that feedback in each is compared to a central model, the two differ importantly. The sensory template, once learned, is fixed, whereas corollary discharge gives rise to a continually updated expectation. Furthermore, the latter appears to be primarily a sensory mechanism, whereas the sensory template is more fundamentally an output-control mechanism.

VI. REAFFERENCE AND EXPECTED INPUT

Another kind of behavioral control system is implied by questions like this, raised by Helmholtz: "Why does the world not seem to move when we voluntarily move our eyes?" Try the following. Move your eyes slowly left and right, or move your whole head or body without fixating your vision. The usual report is that the visual world does not appear to move. But the visual image of the world has moved across the retina. Now cover one eye and move the other eyeball by gently pushing it from one side with a finger. The visual world *will* appear to move! What is the difference between these two cases? In the former the efferent control of eye position is through ordinary channels; in the other, it is not. Von Holst and Mittelstaedt suggested a hypothesis to explain the difference in perception in the two cases (Fig. 8.19). This hypothesis distinguishes between ordinary sensory input from external causes, called **afference,** and input that may be identical but is due to the individual's own movements, called **reafference.** It states that for certain commands the CNS stores a "copy" of its own output command, which these authors called an efference copy, actually a **specific central expectation** of input. The CNS uses this expectation to predict the change in retinal input, the reafference, that results from the commanded movement. It is assumed that this reafference expectation carries specific information to another integrative center in the brain where the calculations of visual change are made. The expectation is supposedly compared to the input and a difference interpreted as motion of the surroundings. When the eyeball is

Figure 8.18
Sensory template.

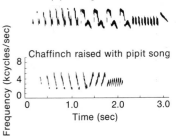

Chaffinch raised in isolation

Chaffinch raised normally

Tree pipit song

Chaffinch raised with pipit song

A male chaffinch, Fringilla coelebs, raised in isolation, develops a simpler song pattern than one that is allowed to hear other chaffinches singing. The song of a tree pipit, Anthus trivialis, resembles chaffinch song: If it is played to a chaffinch raised in isolation, the song of the chaffinch may have some characteristics of that of a tree pipit. [Thorpe, 1961.]

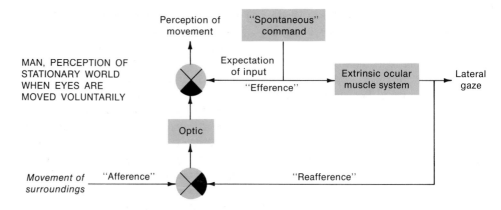

MAN, PERCEPTION OF
STATIONARY WORLD
WHEN EYES ARE
MOVED VOLUNTARILY

Figure 8.19
Expectation of input due to voluntary movement. Analysis of certain eye movements (slow lateral gaze) shows the existence of a connection (yet to be found physiologically) that is here called expectation of input, and must be quantitatively precise.

moved by means of the finger, there is no such expectation, and motion of the world is perceived. When the eyes are moved voluntarily by the lateral and medial rectus muscles at a moderate speed the expectation and visual report match, and no motion is perceived. Not all eye movements are covered by this analysis. When the eye is stationary and the real world does move, motion perception results, since the input is different from the expectation "no change."

Another combination of expectation and input is possible: by paralyzing the extraocular muscle, one can arrange that the eye will not execute willed movements. In this case the hypothesis predicts an expectation of change in the input; but the input does not change. A perception of motion should therefore occur, and it should be in a direction opposite to the willed eye movement. According to von Holst, who had his own eye muscles paralyzed by a temporary nerve block, this result does occur. The same result had been noted in

patients with lateral gaze paralysis. The relevance to this chapter is that a specific neural connection and message is hypothesized, and neuronal signs of this should be found.

There are now appearing reports of changes in neuronal discharge with intention to move—for example, in cells in the mesencephalon just before an eye flick (saccade). Questions now arise, therefore, such as how to recognize the expectation signal physiologically. How is a message that is suitably coded for initiating movement quantitatively recoded into a form suitable for "expectation"? How and where is input compared with expectation? What kinds of movements do or do not create an expectation? Is the apparent rolling of the dock experienced by a sailor just off a ship a related phenomenon? Special terms are used by some authors for signals related to some form of expectation—namely, **"corollary discharge"** and **"efference copy."** These are not synonyms, but the differences have not been clarified (see

Glossary); the concepts and terminology will surely change as physiological analysis advances.

In a few cases we now know the neuronal mechanism of expectation. When a crayfish brain initiates the command for the tail flip used in escape swimming, there is an adaptive or "protective" change in the receptors of the tail, which are due to be strongly stimulated by the bending. They are inhibited just before the command takes effect, as you might shut your eyes before switching on a bright light. In this case, however, the stretch receptors are inhibited in the peripheral sense organ, and the tactile receptors are inhibited at the central axonal terminal. In other cases we know there is an effect, though the neurons and pathways are not fully traced out. Just before bats vocalize they protect the auditory system by several means, including a middle ear muscle contraction that stiffens the tympanum and a midbrain neuronal response attenuation that reduces sensitivity another 5-fold. We most probably do the same everytime we speak even in a conversational tone of voice.

Another example is the suppression of our vision before and during eye movements. Psychophysically this is a graded suppression, not a switch-off, and it is partly due to the command, partly a reflex result of the visual input. The strong input from motion over a good part of the retina can catch up with some visual stimuli that occurred tens of milliseconds prior, and suppress the sensation before it reaches consciousness (backward masking). In what must be still incomplete catalogs and characterizations of relevant neurons in cats and monkeys, many types are found. Some that respond to moving bars imaged on the stationary retina no longer respond to the same relative motion when the eye flicks across a stationary bar; others do. If the eye flicks across a moving bar, the suppression is incomplete or absent. Some units are "space constant" and respond to targets in a certain part of the real world, even when the head is tilted and the image shifted on the retina; the unit or its input must compensate for the tilt. As knowledge of the diversity of specific neuronal types increases, it is likely that the old dichotomies, issues, and formulations of *the* question will alter beyond recognition.

VII. CHANGES IN PROBABILITY OF SPECIFIC BEHAVIOR: DRIVES AND MOTIVATION

Let us turn now to the evidence of significant lability in central excitatory state with respect to moderately complex normal behavior.

A. Adjustment of Set Point

A simple type of change in probability of specific actions is that due to a shift in the set point. Like readjusting the thermostat, this alters the threshold for the given stimuli to trigger certain movements. Adjustments can be short or long term.

Our skin temperature to which we behaviorally regulate clothing and environs changes with acclimation. Von Holst's angel fish, which tilt to a certain angle representing the brain's relative weighting of the amount of horizontal light received and the vertical pull of gravity (p. 297), will change the angle of

Figure 8.20
Measurement of changes in weighting function.

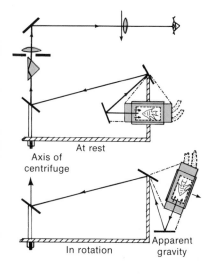

Axis of
centrifuge

At rest

In rotation

Apparent
gravity

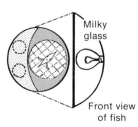

Milky
glass

Front view
of fish

Apparatus for measuring the tilt of angel fish as a function of gravity, light, and food juice. The centrifuge permits illuminating and viewing the fish during rotation. [Von Holst, 1950b.]

tilt if food juice is dropped in the water (Fig. 8.20). Since it is a species that hunts food by sight, this increases the weight placed upon visual stimuli, and it tilts further toward the horizontal—incidentally providing a test for the effectiveness of the chemical. The particular angle that a bee wants to maintain with respect to the sun or to the polarized light pattern of the sky—the Sollwert, or "should-be value" (see p. 300)—is of course adjusted not only by the time of day but with each switch to a new food source.

It would seem that such quantities must be direct reflections of some quite local neurophysiological variable, but it remains an intriguing open question as to just what and where.

B. Brain Stimulation and Drives

The preceding comments may suggest that complex motor control systems, as well as memory traces or complex sensations, might be activated by focal excitation. Such is the case; and, because motor acts are directly observable, such results are obtainable with animals as well as with conscious humans. Many examples are known. Hess pioneered in this field, and described, among many other things, how a cat stimulated in the amygdala (a primitive part of the cerebrum) will exhibit all the signs of rage or aggression or bizarre sexual behavior, depending upon exact electrode placement. A pocket mouse with electrodes implanted in one part of the diencephalon will compulsively stuff imaginary seeds into its cheeks with a high-frequency rhythmic movement. A cricket with electrodes placed appropriately in the brain can be made to sing. A nudibranch mollusc will perform prolonged swimming movements when a single impulse is initiated electrically in a single neuron. The list is long.

It is usually difficult to determine precisely what structures are affected directly by the electrical current. In particular, it can not be told whether applied currents affect only local cells and dendrites or whether they also affect axons that are passing through. It is not certain how many cells are affected, yet a high degree of local specificity is indicated by the fact that the electrode need be moved only a small distance, such as a millimeter, to erase the effect or give rise to a different one.

We would like to know more of the nature of these behavioral control centers. We can be reasonably certain that they do not contain the motor output pattern generators. Cricket song can be coordinated in a headless animal. The brain center that calls forth singing appears to turn on lower centers that are themselves the pattern generators. The higher centers are clearly not simply sensory processing stations either. Let us suggest as a prelude to the next section that these centers are the sites at which decisions of some sort are made—decisions that determine from moment to moment the course of action of the whole animal, and that they do this by weighing input from many sources and then commanding into action motor subroutines inherent in other nervous structures. When one stimulates such a center, it is as if he has made a decision about the required course of action for an animal. Of course, decisions come in higher and lower levels of influence; we are not attributing to one locus a choice among all the alternatives in an animal's repertoire. Higher-level decisions may occur elsewhere, and by their nature be less readily counterfeited.

We have been using the word **center** without definition; it has no rigorous limitation in the literature, but clearly does not imply either that the region in question has only to do with the one function or is the only locus dealing with it. A center may be quite diffuse, like the respiratory centers of the brain stem; it may topographically overlap, include, or be included in other functional cell groups. Centers for the regulation of motivated behavior in mammals lie mostly in the anterior brain stem, especially the hypothalamus. By "drive" or "motivation" we mean some force that elicits or propels some complex behavioral function, like eating, drinking, or mating. The drive seems to come from within. External circumstances alone do not provide the observer with sufficient information to predict the behavior, although they do influence the behavior.

The measurement of drive and some other aspects of the control of driven behavior can be illustrated most simply

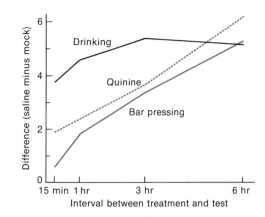

Figure 8.21
Three methods of assessing "thirst." Rats were given 5 ml of 2 M saline by stomach tube. The resulting "thirst" was assessed at different times after the treatment by measuring (a) the volume of water drunk to satiation, (b) the concentration of quinine the rats would tolerate, and (c) the number of times the rat pressed a bar, and was reinforced with a small amount of water, on a variable interval schedule. The ordinate indicates the amount by which the experimental test exceeded the control test. Actual values for milliliters of water drunk to satiation are twice those shown; for the number of bar presses during 9 minutes, ten times those shown. [Miller, 1956.]

by observations on drinking in rats. If a rat is deprived of water for a considerable time and then allowed free access to water, it will at first drink steadily for some seconds. After this first bout of drinking it will pause and perhaps pursue some other activity. It will then return to drinking repeatedly, but the bouts will gradually decrease in duration and repetition rate. Curves like that in Figure 8.21 are accepted as measures of thirst or motivation to drink. The figure shows the change in thirst by these measures as time passes after the injection of saline. Differences in the picture shown by the different measures are taken to mean that the simplest hypothesis—assuming a single intervening variable—is excluded; thirst must be more complex than that.

The hypothalamus contains a drinking center. Electrical stimulation there induces drinking even in a water-satiated rat. Ablation of that center by electrocoagulation results in the cessation of drinking even in a deprived rat, and this may continue to the point of fatal dehydration of the body. Another means of stimulating the drinking center is to inject salt solution directly into it, in a quantity much too small to affect the water balance of the whole animal. Hypertonic solutions cause drinking; hypotonic ones, suppression of drinking, even when the behavior is not consistent with bodily need.

The drinking behavior at any given state of motivation is a function of the quality of the stimulus. If provided with salty water that is too highly concentrated, the rat will only taste it; if the solution is physiologically tolerable, the rat will drink amounts that correspond to the salinity. This effect is mediated by the peripheral senses, not by direct effects upon the drinking center. The taste of concentrated solutions inhibits drinking before a change in blood concentration occurs.

The drinking center cannot be defined with anatomical precision, since electrodes and syringes are relatively large, crude devices. Nevertheless, we feel that it is not a clean-cut anatomical entity. Probably the drive control centers are like the respiratory control centers of the lower brain stem, in that cells involved in diverse functions may be interspersed. It seems likely that drive centers overlap or interdigitate and have ill-defined boundaries, and that this lack of discreteness is real, not just an artifact of technique.

There is only one known drinking center on each side of the brain, and it is positive in the sense that excitation there promotes drinking behavior. Other drive centers are double, in that, of the two on each side, one is excitatory, the other inhibitory. This more general situation is exemplified by the eating centers. Ablation of one of the hypothalamic eating centers leads to overeating and obesity, whereas stimulating it electrically inhibits eating. These manipulations of the other center give the opposite effects. The centers are affected by parameters of the internal state, such as blood glucose level, and by exteroceptive inputs, such as sight, smell, or taste of food. It is thought that the two centers reciprocally inhibit each other. Experimentally induced increases or decreases in the activity of one are correlated with opposite effects in the other. There are also paired centers for the motivation of sexual behavior, and in addition to the features described above, these centers are directly responsive to hormones. Implantation of a tiny pellet of sex hormone into one of the centers changes the electrical activity of that center and affects behavior accordingly. The pellet is too small to have its effect by entering the bloodstream and reaching distant target organs in adequate concentration.

The interaction of drives could lead to contradictory or conflicting states of intended motor output. For the most part, an animal cannot exercise simultaneously two kinds of driven behavior—for example, courting and caring for young or drinking and copulating. This difficulty could be partially resolved by building into the nervous system a hierarchy of importance or priority of the several drive systems. Such a priority system is evident in the case of respiration, which can override any other behavior, although it can be interrupted briefly, as during swallowing, speaking, or swimming underwater. But there is no absolute measure of the intensity of a drive, and therefore comparisons of the strengths of different drives are always relative to the very moment at which the measure is made (Fig. 8.22). How, then, can we understand why an animal performs one kind of driven behavior at any moment to the exclusion of others, even though each drive system provides some motivation and even though stimuli appropriate to the execution of several behaviors may be present? This is essentially the question, "How does the animal decide what to do?"

A digression will be useful, in order to give some background about the older ethologists' ideas and observations on drives. They postulated that for any driven behavior, like eating, there is a steadily increasing force—in the example of eating, a force related to the state of excitation of the positive eating center, which is itself related inversely to the blood glucose level. As the state of motivation increases, weaker and weaker stimuli will release eating behavior, as when starving humans chew on pebbles. In the extreme case, increasing drive might lead to **vacuum activity,** overt behavior related to the drive in the absence of any appropriate stimulus. (The notion of vacuum activity is not in vogue now, but our knowledge of spontaneous activity in neurons and large nervous structures seems to permit it on neurophysiological grounds.) Motor output (behavior) that is caused by the drive decreases the drive; that is, it is consummatory. Eating, drinking, copulation—all decrease the immediate tendency to do more of the same. The consummatory

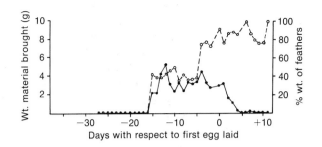

Figure 8.22
Nervous bias. Certain states, which have been called moods or sets, are manifested by posture, but include a complex of changes in readiness and receptivity, and therefore involving integrative functions in the nervous system. The graph shows an example of a change in the response to nesting material provided continuously to a naive female canary. Until two weeks before the first egg was laid, no nest-building activity was evident; after starting abruptly, it changed in character, so that the percentage of feathers used increased greatly even after the first egg, when the tendency to gather material (measured by weight) decreased. The nests built were removed daily. [Hinde, 1958.]

effect of output behavior on the degree of motivation might be a direct nervous effect, but it seems equally likely that it is a feedback effect; eating stimulates the oral and nasal senses, distends the gut, and increases blood glucose level. An animal deprived of the normal triggering stimuli for some time becomes easier to trigger and more likely to act, but only with respect to certain drives, such as eating; others, such as intraspecific aggression, become less likely and harder to trigger, at least in some well studied species.

Behaviorists have studied **situations of conflicting inputs** and drives—for example, simultaneous attraction to a mate and fear of the same animal. Another example is the familiar conflict between fear and aggression. A common finding is that when two drives conflict, the

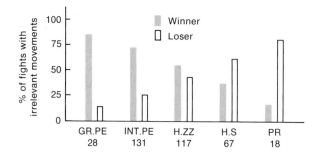

Figure 8.23

Displacement activity. Several types of movements apparently irrelevant to the situation, made during fighting by Burmese red jungle fowl. The incidence of the different movements varies reciprocally depending on whether the individual was an eventual winner or loser. The numbers below the columns indicate the number of fights, out of 250, in which each movement occurred; several may occur in one fight. It seems likely that physiological correlates of these probabilities could be found in the brain. *GR.PE*, ground pecking; *INT.PE*, intention ground pecking; *H.ZZ*, head zigzagging; *H.S*, head shaking; *PR*, preening. [Kruijt, 1964.]

output behavior is not intermediate, but is clearly that appropriate to one drive or the other. A mechanism exists for determining decisive action, but the decision may be short-lived; an animal in conflict may alternate abruptly from one behavior to another and back again. In other cases conflicting drives may result in a third behavior not obviously related to either of those resulting from the single drives. Again, the third behavior is not of intermediate type. It is what ethologists call a **displacement activity** (Fig. 8.23). Thus conflicting drive or input does not usually result in intermediate or confused behavior, but instead in unitary action. At least for each short span of time the whole animal behaves as if it were of single purpose.

Some of these points are illustrated in a series of experiments with chickens.

Implanting electrodes in the brain stem, von Holst found that virtually every aspect of chicken behavior could be elicited by local stimulation. Single motions, like turning the head, simple sequences like vocalizing or walking, or complex behaviors like courtship could be elicited repeatedly at separate foci. The foci were not sharply defined. Instead, as an electrode was advanced through a region triggering any particular act or sequence, a threshold profile could be plotted. The same behavior might result along one to three millimeters of one electrode track, but there was a peak sensitivity region for which the smallest voltages were required (Fig. 8.24). At any one point, say the most sensitive one, the time to onset of the behavior after onset of the stimulation —the latency—decreases with stimulus strength. The shortest latencies were only tenths of seconds, but with low stimulating voltage the latency could be several minutes.

Von Holst found that **simple behaviors,** like sitting, standing, preening, grooming, stretching, and calling, did not show a mutual antagonism of the kind discussed above. As long as the two behaviors were not mechanically incompatible they might be combined when both were stimulated together. For example, one chicken with two implanted electrodes could be made to sit down, to look to one side, or to sit down and look to one side. In the case of mutual exclusion, simple acts may combine by cancellation. Simultaneous, equal stimulation of command sites for looking left and looking right produced no effect at all.

Foci eliciting **complex behavioral sequences** gave results suggesting that these were drive centers in the sense of

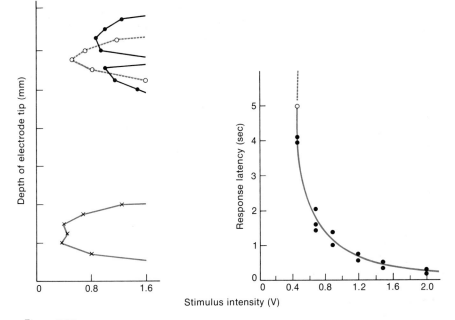

Figure 8.24

Quantitation of behavioral response to brain stimulation. **Left.** An indwelling electrode in the brain stem (hypothalamus?) of an unanesthetized chicken elicits different responses according to its position (shown here as depth in mm from an arbitrary zero) and the intensity of stimulation (shown here as threshold in volts). Filled circles indicate looking up; open circles, turning to the left; x, flight. **Right.** Above threshold the latency of a behavioral response may systematically decrease with increased stimulus voltage. The response here is a warning call (cluck). The open circle is the first stimulus, which failed to elicit overt reaction. [Von Holst and St. Paul, 1960.]

the above discussion (Fig. 8.25). Each of the individual acts in the sequence could be stimulated by electrodes in some other brain site. The response was ordered, leading to a goal, but not compulsive. The voltage required to elicit complex behavior depended upon whether the mood or set of the animal was correct, what other drives were active, and the sensory environment. The goal-oriented aspect of the control mechanism is demonstrated by the fact that different combinations or sequences of individual acts might occur, depending upon prior state of activity or input, but all se-

quences led to the same end state. A rooster that was eating when stimulation of his "sleep center" began, stopped eating, looked around, walked, fluttered his eyelids, yawned, sat down, fluffed his feathers, retracted its head, and closed his eyes. Had he already been sitting down, the sequence would have been shorter. Had he been flying instead of eating, the first part of the sequence would have been different.

In the case of **simultaneous activation of two drive centers** for the more complex behaviors, simple addition does not occur. Sometimes both behaviors appear

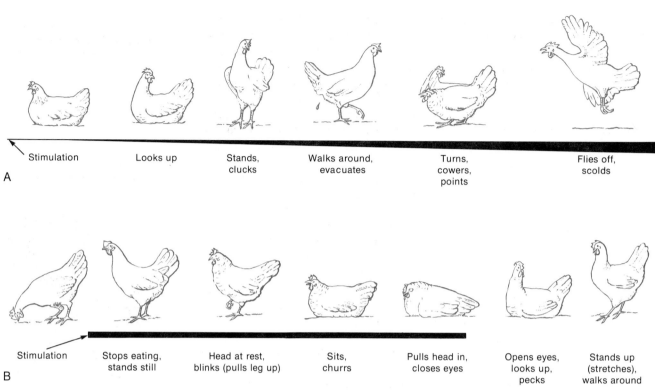

Figure 8.25
Sequence of behavioral response during slowly increasing or maintained brain stimulation. **A.** An electrode in the brain stem of a chicken carries a slowly increasing voltage, shown by the gradually thickening line. **B.** From another site stimulation elicits sleep if maintained and if external conditions are conducive. Cessation of stimulation causes arousal. [Von Holst and St. Paul, 1960.]

at full strength, but alternately. A chicken stimulated to eat and to reconnoiter will make a few pecking movements, raise its head abruptly to look about, peck, and look about again. Only rarely will simultaneous stimulation of two drives elicit a third, not obviously related behavior. Von Holst found one example—a chicken in which two electrodes were implanted could be excited either to aggressive pecking or to fleeing behavior. With balanced stimulation of both foci the hen would emit piercing cries and run frantically back and forth with its wings raised. While not simply related to the expected results, the behavior in this conflict situation may be adaptive in nature. A hen incubating eggs will not readily flee even in the face of a superior attacker, but at the last moment will exhibit this frantic, perhaps distracting behavior.

That different **drives mutually suppress each other** is shown by several of von Holst's experiments. One drive can suppress another even though more voltage is applied to the stimulating site for the latter. In the case of total suppression of overt activity from a weak drive, it can be shown that the drive center was in fact active. In some cases activity related to the weaker drive will occur after stimulation has ceased. Suppression of only a part of a whole sequence may occur. Fleeing behavior consists of attentive alertness, getting up, walking about, freezing, and finally jumping away. If a sitting hen is stimulated to flee and at the same time to sit, at first nothing happens, but at last the chicken suddenly jumps away. The suppressed fleeing behavior erupts violently rather than developing through the normal sequence of steps.

We have been speaking of the major drives and how they are turned on and off, but there are other states that influence the weighting of sensory signals; they may be subsumed under the common expressions **mood or set,** broadly defined as synonyms. The integrative mechanisms of the nervous system determine not only quantities but qualities, as distinct as pleasure is from pain. The set or the pleasure/pain balance—the **affect**—decisively colors the interpretation of input and the formulation of output. At first glance this may seem anthropomorphic. But there is abundant objective evidence that lower animals (octopuses, insects, even medusae and sea anemones) have lasting phases of responsiveness different from those in other periods. One cannot avoid thinking of moods; either integrative sensory input or spontaneous changes can set up a disposition to react in a certain limited direction and to place more weight upon certain types of stimulation than otherwise. A "sulking" octopus that has been punished in a learning trial (Fig. 8.26), a wasp in the act of stuffing food into a hole in the ground, a polychaete worm that has been removed from its tube, or a satiated sea anemone—all place an altered value upon many forms of sensory influx. We assume they do so because a definite pattern of nervous activity has so predisposed them. These patterns are important parts of both innate and learned behavior. The central circuits are so constructed that numerous alternative pathways of discharge are available to each integrative center. Without any alteration in its wiring plan, fleeting shifts in central state can link the same input signal to any of various responses and simultaneously inhibit each of the alternatives. Normally the nervous system does not get confused by activating inappropriate mixtures. There is an adaptive unity in decisions of either-or character, an integrative phenomenon par excellence.

The final **unity of action** is remarkable, and is one of the outstanding achievements of the nervous system, as well as one of the least understood. Sherrington put it this way: "The resultant singleness of action from moment to moment is a keystone in the construction of the individual whose unity it is the specific office of the nervous system to perfect. The interference of unlike reflexes and the alliance of like reflexes in their action upon their common paths seem to lie at the very root of the great psychic process of 'attention.'" Not only is an antagonist reflex or fixed action pattern inhibited, but also each one of a long list of alternative courses of action extending to the highest levels of behavior.

We can summarize this section on driven behavior as follows. There are centers in the brain that, when suffi-

Figure 8.26
Two behavioral states in octopus.

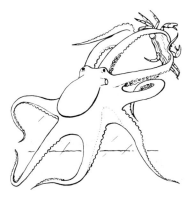

Swift attack on prey.

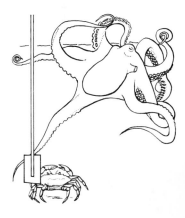

Cautious investigation, hesitation, and retreat after several crabs had been associated with negative reinforcement. [*Wells, 1962.*]

Nonaggressive

Aggressive

ciently excited, command the whole complexes of behavior that result in consummation of the drive. They do this not by a detailed commanding of each muscle, but instead by setting into action lower motor centers that are themselves the pattern generators for specific muscle action sequences. Different drive centers may call up in similar or different orders some of the same motor programs. For example, running may be involved either in escape or in capturing prey. Each drive center is active in proportion to some internal state, such as glucose titer or hormone level, and each center receives excitation or inhibition from sense organs. When the internal state is strongly motivating, weak stimuli may suffice to trigger output behavior. Many classes of behavior often show true spontaneity, reflecting an internal state so strongly motivating that no triggering stimuli are needed but only permissive or mood stimuli. If the output leads directly to a reduction of the internal signal, the driving "energy" is "consumed" (i.e., the readiness state declines). If it does not, the output may at least lead to a new sensory situation, and hence release another round of output. Motor output, which is related to a drive, but which does not immediately reduce the drive, falls under the ethologist's category of exploratory or appetitive behavior. The different drive centers somehow inhibit each other so that at any moment one dominates. There are probably different sorts of drive centers, but as used here they are distinguishable from sensory centers and attention mechanisms, which are not drive-specific, and from specific command centers, which are at the lower level of alternative acts within a drive domain.

From this point of view, behavioral sequences cannot be considered as simply serial responses to external stimuli. The response to any stimulus set depends upon the state of each of the drive centers and on the interactions between them.

C. Switching Between Alternate States

We have repeatedly encountered either-or decision-making, from the level of switching between incompatible reflexes to that of choosing between drives. It behooves the neurobiologist to consider the neural mechanisms available to account for this important behavioral integration (see Chapters 6 and 7). Wherever the switching is all-or-none, we may refer to the switch as a **decision unit,** however complex it may be and however many converging lines of input are compared—lines representing the permissive conditions (e.g., time of day and year), suitable internal states (e.g., temperature), and appropriate trigger conditions (e.g., small moving object with suitable contrast, direction, and background). Imagine a toad about to jump toward a tempting object or away from a threatening object that approaches; before a command triggers the motor-pattern centers, some competent functional unit must determine whether the predetermined criterion for all the converging inputs has been met. Perhaps there is usually a sequential and hierarchical series of more and more significant subdecisions. Like the grades of feature extractors, these may be successive filters and convergences for each of the relevant inputs, including sensory afference and central states. Identifying such decision units is partly a matter of definition and criteria. The existence of a wide range of models—from single neurons through specified circuits to unspecified masses of cells with specified inputs and outputs (see p. 238)—is compatible with our present knowledge of neurophysiology. We know that there exist intermediate-level recognition units that are single neurons, more or less redundant with others of similar properties but each encoding a complex stimulus situation. We also have examples, especially in invertebrates, of single command neurons that can trigger complex normal behavioral sequences. We need to know to what extent such command cells and highly specified recognition units are characteristic of higher levels in higher animals, and whether they explain the decision-making in complex behavior. Such a question receives very different response from equally competent neurophysiologists today, since it is so much a matter of willingness to extrapolate, and of "betting on horses." At least some authors are betting that there will be more evidence to show that vertebrates have specified units that account for some relatively complex recognition-switching, and that there are widely varying degrees of redundancy of units. In lower vertebrates, perhaps the midbrain contains decision centers for some moderately high-level behavioral choices, though others are certainly in the diencephalon or telencephalon. In mammals it seems that more of the behavioral decisions require the forebrain.

D. Nonspecific Changes in State: Sleep and Arousal

At least in higher vertebrates there is another class of alternative states of behavioral importance—the several stages or types of sleep (p. 494) and arousal, grading into alertness and directed attention. We know something about the neural mechanisms involved; anatomically, the main one is the reticular formation of the brain.

The **reticular formation** comprises the central core of the brain stem (Fig. 8.27). It is named from certain regions that consist of neuron somas and fibers mixed in criss-crossed array—gray matter cut up by bundles of fibers. As implied above, it receives inputs from all the afferent paths, sometimes from collateral branches of fibers running in the specific sensory pathways, and from other parts of the brain, especially descending fibers from above. Single neurons in the reticular formation may receive input from several sensory modalities. Similarly, it has output to widespread areas of the brain and spinal cord.

If the spinal cord of a mammal is severed, the region below the cut can of course not be controlled by the specific descending pathways, such as the cortico-spinal tracts. But in addition to this expected result, the isolated spinal cord shows another behavioral malfunction; it loses even the local reflexes. For a period of time the animal exhibits spinal shock. Sensory stimulation, which would normally result in reflexive action, is without effect. The reflex motor systems of the spinal cord must be kept in a state of tonic readiness ("central excitatory state") by nonspecific influence from the brain. Because of its tonic excitatory effect, a portion of the reticular formation is known as the reticular activating system.

With respect to spinal function, higher regions of the brain may inhibit the reticular activating system. A cut low in the brain stem gives rise to spinal areflexia, whereas one higher up causes

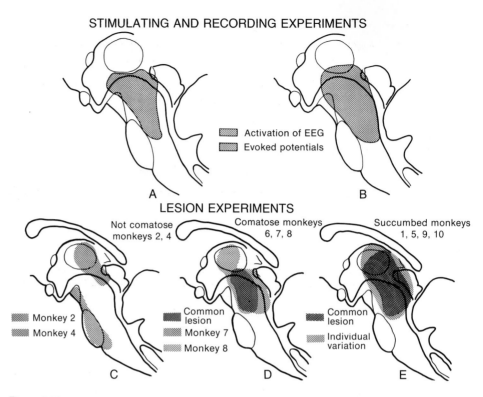

STIMULATING AND RECORDING EXPERIMENTS

Activation of EEG
Evoked potentials

A   B

LESION EXPERIMENTS

Not comatose   Comatose monkeys   Succumbed monkeys
monkeys 2, 4   6, 7, 8   1, 5, 9, 10

Monkey 2   Common   Common
Monkey 4   lesion   lesion
                  Monkey 7   Individual
                  Monkey 8   variation

C   D   E

Figure 8.27
The reticular activating system. Midsagittal plane of a monkey's brain stem; shading indicates the area whose stimulation evoked EEG arousal **(A)** and from which potentials are evoked by sensory stimulation **(B)**. Shown in **C, D,** and **E** are the distribution of experimental lesions made in different cases; those in D and E resulted in coma. [French and Magoun, 1952.]

hyperreflexia (exaggerated reflexes). Hypertonia (high level of tonic muscle contraction) is especially conspicuous, because of the stretch reflexes of the antigravity muscles. A decerebrate cat exhibits extensor rigidity in all four limbs due to removal of the higher-level inhibition on brain stem reticular facilitation of postural stretch reflexes. Bipedal primates have hypertonic extensor reflexes for the hind limbs and overactive flexor reflexes for the arms, and

tree sloths show flexor rigidity in all four limbs.

In the ascending direction the reticular formation output also has activating function. Its tonic effect upon the cortex and other structures is required for wakefulness, consciousness, and alertness. When sizable portions of the reticular formation are destroyed, a subject becomes comatose, exhibiting brain waves characteristic of the sleeping state. Sensory stimulation, which normally

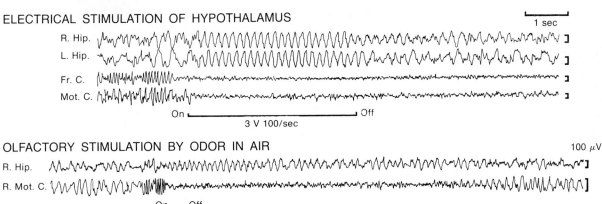

ELECTRICAL STIMULATION OF HYPOTHALAMUS
1 sec

R. Hip.
L. Hip.
Fr. C.
Mot. C.

On _____ Off
3 V 100/sec

OLFACTORY STIMULATION BY ODOR IN AIR   100 μV

R. Hip.
R. Mot. C.

On   Off
Olf. _____

Figure 8.28
EEG arousal. Electrical activity of the hippocampus (R. *Hip.*; L. *Hip.* = right and left hippocampus) and cerebral cortex (*Fr.C.* = frontal cortex, *Mot. C.* = motor cortex, R. *Mot. C.* = right motor cortex) before, during, and after stimulation in two ways. Note that arousal is accompanied by low-voltage, fast ("desynchronized") activity in the cerebral cortex and high-voltage, 6-per-sec waves in the hippocampus; the former outlasts the latter. [Green and Arduini, 1954.]

would awaken the subject, causes at most a transient alerting, even though the specific sensory pathways are operative. On the other hand, sensory stimulation can result in full arousal and concomitant EEG changes in an animal with intact reticular formation, even though its upper specific sensory pathways have been cut. No matter which sensory modality is excited, EEG changes occur all over the cortex. An animal with stimulating electrodes implanted in any of a wide range of sites in the reticular formation can be awakened by applying current. Electrodes in a few restricted sites can induce sleep.

There is a remarkable difference in the order of magnitude of the behavioral response to local electrical stimulation of different brain regions (Fig. 8.28). In the motor cortex, electrical stimulation causes localized, uncoordinated muscle twitching. In certain hypothalamic nuclei, stimulation orients the whole animal toward one mode of behavior or another. Stimulation of the reticular formation affects the overall state of activity of the entire nervous system, and also the synaptic transmission in each of the sensory relay nuclei (see Chapter 10).

We have stressed the nonspecific function of the reticular activating system. It is, however, not entirely unselective in its output effects. Some mechanisms, which are not yet understood, allow it to gate a particular modality of sensory input or activate especially only one cerebral system. These mechanisms give rise to the phenomenon of selective **attention.** The organism can be alerted to a single modality. By a complementary maneuver the nervous system can at various levels "detune" one monotonous input or reduce its effectiveness. This

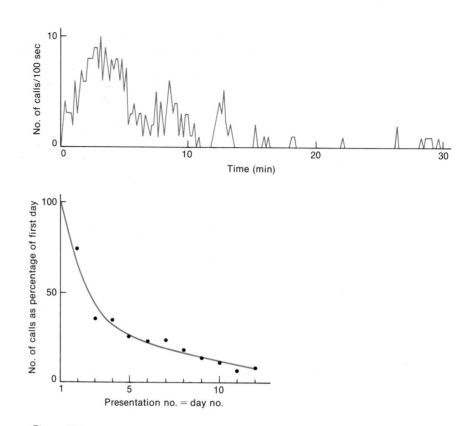

Figure 8.29
Habituation. The rate of calling by chaffinches while mobbing owls. **Above.** Frequency of
calls made by three individuals as a function of time after a given presentation of a stuffed
owl. **Below.** Response to daily 20-min presentations of a live owl. Numbers of calls are
group means. [Hinde, 1954.]

process, called **habituation** (Fig. 8.29), leads to the ignoring of irrelevant input by the specific systems; it is not the same as adaptation, accommodation, or fatigue (see Glossary). Habituation is generally regarded as one of the simplest forms of learning and has been found at the neuronal level—in certain cases (e.g., *Aplysia* ganglion), even localized to the presynaptic terminals.

Are there nonspecific systems in lower animals? Although this question is quite unanswerable at present, it seems worthy of some comment. Knowledge on this score would give some understanding of the evolution of nervous function that could not be obtained through other routes. A corollary question, "Is the reticular activating system evolving toward more importance in higher groups, or is it relatively more important in lower forms?" is also difficult to answer. Two lines of data and argument give conflicting answers. First, the reticular formation of mammals is held to be homologous to the major descending spinal tract of fishes (the medial longitudinal fasciculus); hence it is anatomically primitive. On the other hand, the major influence of the reticular activating system on the spinal cord—maintenance of a tonic excitatory state—is more highly developed in higher forms. A measure of this is the duration of spinal shock after spinal transection. In all animals the isolated lower part of the spinal cord sooner or later recovers to a more or less normal state of reflexive function. In man this recovery takes months; in lower mammals, days or weeks; in frogs, a minute or two; in fishes, shock is hardly recognizable. Perhaps as the higher centers of the cerebrum become elaborated, some of the functions of the hindbrain regions were transferred upstream and the lower centers acquired new roles and more dependence on the descending influence. The reticular system is both primitive and more elaborated in higher groups; we cannot measure its relative importance, but it has more diverse importance in mammals than in elasmobranchs.

Comparable "nonspecific" systems are practically unknown in invertebrates. One example that may be analogous involves a grooming reflex in grasshoppers. One or the other front leg will occasionally wipe a skeletal protuberance on the ventral side of the first thoracic segment. The movement can be triggered by stimulating sensory hairs on the skeletal structure, but not every time they are touched. If distant parts of the nervous system are ablated, the reflex is more easily elicited. The efficacy of the stimulation increases with each ablation, being maximal when the first thoracic ganglion is entirely isolated from the rest of the central nervous system. This particular reflex, which is mediated by the prothoracic ganglion alone, is inhibited by "nonspecific" input from the rest of the nervous system, and the degree of inhibition is a function of how much of the whole nervous system is in connection with prothoracic ganglion. A second, seemingly relevant observation on insect behavior is the effect of brain ablation in the praying mantis. The normal animal locomotes part of the time, but is mostly quiescent, waiting for prey. If the brain is destroyed, the animal locomotes continuously. This again appears to be a case of tonic inhibition of automatous lower centers by higher ones, quite a common finding with brain ablation in various invertebrate phyla. The female mantis may decapitate the male and thereby release copulatory behavior, which is adequately patterned and commanded by lower centers.

VIII. SPONTANEITY AND RHYTHMICITY

Though the man in the street knew it all along and would be surprised by the fuss, one of the great revolutions in our views on the physiology of behavior stemmed from the "discovery" of spontaneous activity. One of the great contrasts between views is that between the older behaviorist school, which emphasized peripheral causation, and those who hold a more recent and ascendant view emphasizing central (internal) causation. To those of the earlier school, behavior in general could be treated as stimulus-response ("S–R") sequences; behavior was even defined as response. Complex movements like locomotion, coition, and breathing were regarded as chain reflexes.

A. Nonrhythmic Spontaneity

Somehow it did not impress behaviorists that the fly on the table sometimes takes off without any apparent stimulus. They believed that causes like hunger arise in the peripheral nervous system (e.g., may be initiated by low blood sugar). Today it is rather generally believed that many starts or switches are centrally initiated —that is, they are not independent of input, but, in the presence of permissive and necessary, tonic, steady-state input, are triggered by some event in the central nervous system.

Historically, however, it is the ongoing, rhythmic activity that has attracted the most attention. The appreciation of the EEG in the early 1930's was one of the influential factors (see pp. 226, 289); renewed interest in circadian rhythms and other rhythmic activity in the 1940's and 1950's was another.

B. Biological Clocks and the Nervous System

The spectrum of biological clocks extends from neurons firing more than 1000 times per second to an individual's full life span (one complete cycle per individual). It is not known how many of these clocks are in the nervous system, but probably the answer is not, "Only those with short periods, milliseconds to minutes." Since this chapter is on behavior, we will deal here briefly with long-period, approximately 24-hour (= circadian) rhythms. Locomotor rhythms were treated in Chapter 7 and neuronal spontaneity in Chapter 6.

Many, if not all, animals show daily cycles of activity—in metabolism, courtship behavior, or other physiological processes. In the absence of obvious cues as to the time of day, such as light/dark or temperature cycles, these rhythms persist, commonly being free running and accurate to within minutes in a 24-hour period. That such rhythms have an endogenous basis is suggested by the fact that organisms isolated from timing cues may cycle at periods slightly different from 24 hours. Circadian activity rhythms (*circa dia* = about a day) are known in whole organisms, in bits of tissue, and in single cells. A noteworthy example of circadian activity at the cell level is the free-running rhythm of slow potentials and spikes seen in a number of neurons in the isolated parietovisceral ganglion of *Aplysia* (Fig. 8.30). The ganglion can be cultured for several weeks outside the body, and many cells maintain such rhythms, each with its own period. The intact *Aplysia* exhibits a circadian rhythm of rest and active locomotion. The various independently rhythmic cells, and the rest of the body, are believed to be synchronized normally by the circadian release of hormones from cells that have been localized to the eye.

Recent evidence from lesion studies on rats points to the somewhat surprising conclusion that the circadian clock may be highly localized in a mammal. Destruction of the tiny **suprachiasmatic nucleus** of the hypothalamus abolishes the signs of a circadian rhythm, whereas lesions in many other places, nearby or far away, either cause no effect at all or merely alter some readout.

Given that even at the single neuron level there are activity cycles with periods as long as 24 hours, two analytic problems remain. First, what normally keeps the approximately 24-hour endogenous rhythm in phase with the astro-

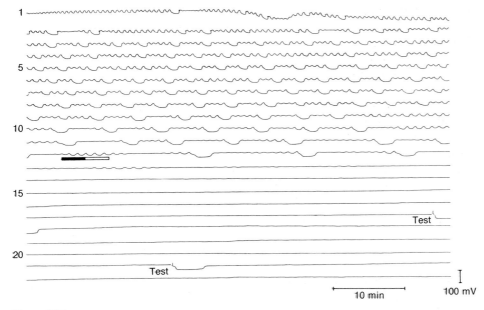

Figure 8.30
Probable circadian rhythm in a neuron. Long-term intracellular recording of spontaneous, subthreshold fluctuations in membrane potential in the parabolic burster neuron (R15) of *Aplysia*. The parietovisceral ganglion was isolated from an *Aplysia* previously conditioned to light/dark cycles whose dawn would have occurred at the time shown by a black-to-white symbol on line 12; it was maintained in culture at 14°C in continuous darkness. Nerve impulses were blocked by tetrodotoxin to eliminate synaptic input to the cell. Each line of the record represents one hour. At "test" (lines 17 and 21) about 3×10^{-9} A of depolarizing current was passed across the cell membrane for about 15 sec, resulting, *after* current flow, in a long-lasting hyperpolarizing response. Under these conditions R15's rhythm runs down in a few cycles, but extracellular recording from other cells in the same ganglion shows sustained rhythms for weeks, each cell with its own period close to 24 hours. [Strumwasser, 1973.]

nomical cycles? Second, if there are many cells or other bodily structures with circadian rhythms, what keeps them working in synchrony? We know something in the way of an answer to the first question and can speculate on the second. The rhythm of the *Aplysia* cell, like those of other tissues or whole organisms, can be increased or decreased in frequency by certain environmental conditions. High intensities of light, for example, increase the circadian frequency of diurnal animals (active during the light) and decrease that of nocturnal ones (active during the dark). When the environmental stimulus (Zeitgeber = time giver) is itself periodic, as with solar illumination, the effect on the endogenous rhythm is to speed it up or slow it down for only part of the day. This affects the phase of the endogenous rhythm with respect to the environmental one by a certain amount every day. If the amount of advance or retardation is a suitable function of the phase of the endogenous rhythm at which the

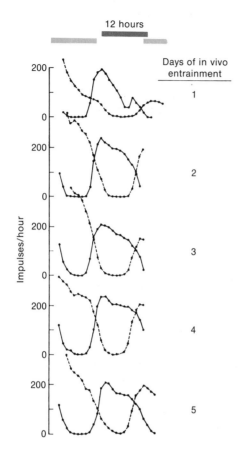

12 hours

Days of in vivo entrainment

1

2

3

4

5

Impulses/hour

Figure 8.31

"Entrainment" of a circadian rhythm, or getting into step with the environment. Activity, expressed as impulses per hour, in neurons of eyes of *Aplysia.* Solid curves are rhythms of eyes taken from control animals exposed to light/dark (L/D) cycles in phase with the pretreatment L/D cycle; broken curves are rhythms from animals exposed to cycles that were phase-advanced 13 hours relative to the pretreatment L/D cycle. The dark brown bar at the top of the figure represents the projected light portion of the control L/D cycle; the light brown bars represent the projected light portion of the phase-advanced L/D cycle. The top curves in this figure were obtained after the experimental animals had been exposed to one phase-advanced cycle, and the next curves down were obtained after 2, 3, 4, and 5 exposures to the phase-advanced L/D cycle. Even after one cycle entrainment or phase-capture by the L/D cycle is conspicuous. [Eskin, 1971.]

stimulus is presented, **entrainment** is possible. By entrainment we mean that the endogenous, but dependent, oscillation is synchronized in some steady-phase relationship with respect to the stimulus rhythm. (In the next section, notions of oscillation coupling and phase locking, or entrainment, are presented in the context of neural or simple behavioral functions.) A free-running endogenous circadian rhythm and its entrainment in the presence of periodic environmental stimuli are illustrated in Figure 8.31.

Since single pacemaker neurons can be phase-locked to environmental stimulus rhythms of slightly different frequency, it would not be difficult to imagine a whole population of internal oscillators with similar circadian rhythms coupled to each other, either by parallel action of the environmental stimulus on each of them, or by interactions among them so as to keep them working in synergy to effect a daily rhythm in the whole-animal activity. There is little evidence on this hypothesis as yet, but it is probably a reasonable answer to the second question. The several outputs or expressions of the basic rhythm can be desynchronized, as when sleep, hunger, bowel, and other circadian activities get out of step during adjustment to jet travel. This is thought to result from unequal readout delays with a single master clock.

C. Absolute and Relative Coordination

We can use the word coordination here in a special sense, referring to the working together of rhythmically active body parts. Hence the study of coordination is, in part, study of the coupling of oscillatory mechanisms. Von Holst, whose studies on the fin movements of fishes are the classics of this field, distinguished two main classes of coordination of parts, absolute and relative coordination. By absolute coordination he meant those cases in which two or more oscillating members have the same frequency and, over long periods, hold a stable phase relationship with respect to each other (e.g., the two legs in walking). Under relative coordination he placed those cases in which the two members may have somewhat different frequencies, but are not quite independent of each other (e.g., the legs of the father and small child walking hand in hand); some phase relationships, such as being "in step," tend to last longer than others. Examples of **absolute coordination** include the walking limb movements of most animals, and wing movements of birds and insects during flight. In animals that locomote terrestrially and have few limbs, stability requires absolute coordination. Animals with large numbers of limbs might be stable under many patterns of limb movement, but other considerations may dictate absolute coordination. For example, the metachronous rhythm of leg movements in a centipede or millipede assures that one limb does not collide with the next even though the strides overlap. Frequency differences or incorrect phasing may be intolerable for mechanical reasons.

Relative coordination occurs between arms and legs when an untrained swimmer uses the crawl stroke. It is most common in aquatic organisms, such as fishes, which do not use the limbs for antigravity support, but occurs also in some many-legged animals. Tarantula spiders have a metachronal rhythm of leg steps, but the two sides may have

diverse phases or even different frequencies. Since there are always four or more legs on the ground, there is no stability problem, and legs on opposite sides cannot collide. Even so, the two sides are not quite independent, but bilaterally paired legs show a statistical tendency toward working in antiphase. In a case such as this, the observed coordination may give us access to some basic neurological mechanisms of subtle integration.

Certain species of fish exhibit long episodes during which the rhythmic movements of the dorsal and pectoral fins are at different frequencies; one may be at 3/sec and the other at 4/sec. Nevertheless, they are not quite independent. Two types of one-way dependence have been specifically recognized.

Von Holst called the interaction of pulling one frequency toward the other the **magnet effect;** he showed examples of its operation in which, though transient and unstable, these influences could be significant and last for some time. Another type of interplay he called the **superposition effect.** Here there is no interaction of the frequencies of the two oscillators but the output amplitude of rhythm A shows cyclic modulation at the difference frequency, as though the rhythms were beating. The output of oscillator B shows no influence of A.

Such influences of independent oscillators, weaker or stronger, symmetrical or asymmetrical, affecting amplitude or frequency, are probably rampant in the nervous system from the level of single cells (p. 333) and populations (perhaps in EEG, p. 227) to the control of slow behavioral rhythms.

Behavior in these cases provides an unusually good handle on underlying neural mechanisms. We are dealing with

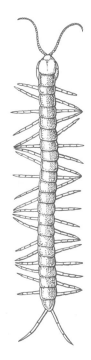

Figure 8.32
Delay function.

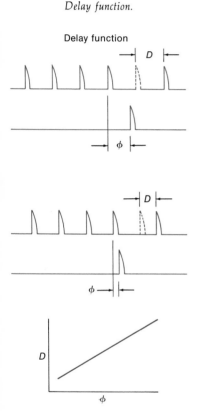

Delay function

A train of events in the output (upper line) and the delay (D) from the expected occurrence as a result of an inhibitory input event (lower line) at a time (φ) after the most recent spontaneous output event. D increases with φ in some systems. [Schulman, 1969.]

a class of relations called **coupled oscillators.** In the last chapter we stated that the neural oscillatory mechanisms for control of rhythmic limb movement are not known. We presume that they are based on spontaneous neurons with superimposed network properties like positive feedback and reciprocal inhibition, or possibly on proprioceptive feedback loops. We also know that there are separate neural oscillators for the limbs of different segments or sides. Since we do not know the cellular or network oscillator mechanisms we can study their coupling only in abstract, nonneural terms. In this abstract realm the following discussion might apply to any physical or biological set of oscillators, from pairs of neurons to networks, and to the coupling of circadian behavioral rhythms, from mechanical pendulums connected by springs to TV synchronization circuits. Although the problem is an old one, it is unfamiliar to contemporary physiology; nevertheless, it is one of the promising opportunities for systems-related cellular neurobiology. Some descriptive considerations might prime our intuition about the phenomenology to be expected, and eventually explained.

In order to abbreviate the discussion we will consider only one type of example. This is the case of two oscillators with a connection from one to the other, but no feedback—an independent oscillator driving a dependent one. Each oscillator has an inherent frequency, and each can be said to have some phase relationships with respect to the other. If the two were not coupled and their frequencies differed, then the phase would drift on successive cycles. If they are weakly coupled, then the phase may drift, but not uniformly, so that there are

semistable epochs, a type of relative coordination. This relationship can be revealed by the phase histogram (Figs. 6.9, 6.10). Another relationship that expresses the coupling between a driven and an independent oscillator is the delay function (Fig. 8.32) for the inhibitory influence. Depending upon the phase of the driven oscillator, it may be accelerated or retarded by the driving one or not affected at all. If the effect of the driver on the driven is strong enough, then there can be a range of phase over which the advance or retardation just equals the difference in the inherent oscillatory periods. The driven oscillation then locks into synchrony with the driver; the coordination is absolute.

Even with a given coupling strength as expressed in the delay function, two oscillators may exhibit different sorts of coordination depending upon the differences in their inherent frequencies. If the frequencies are similar, even weakly coupled oscillators may synchronize; if they are very different the coordination will be weak or unmeasurable. A range of relationships between weakly coupled oscillators is possible.

One example will be mentioned that is similar to the neuronal case (Fig. 8.32) and the circadian rhythms referred to just above. We choose an example that is intermediate and probably similar to the fish-fin rhythm interactions of von Holst, but more amenable to analysis, since we can control one of the interacting oscillators. The **mammalian heart** has its intrinsic pacemaker and is under the influence of inhibitory and acceleratory input from the parasympathetic and sympathetic nervous systems, respectively. Figure 8.33 shows the heart rate as we vary the frequency of stimulation of

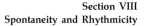

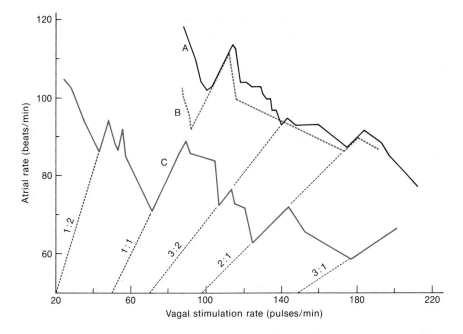

Figure 8.33

Phase locking of a heartbeat rhythm. The heart rate of a cat, expressed as atrial beats per minute, at different rates of stimulation of the whole vagus nerve, a parasympathetic inhibitor of the heart. Curve *A*: vagus nerves and stellate ganglia (a sympathetic excitor of the heart) intact; each has a certain low level of tonic influence on the heart. The atrial rate without stimulation is, in the intact condition, 144 per min. Curve *B*: vagus nerves severed above the stimulating electrodes, removing the central tonic outflow; atrial rate without stimuli, 150 per min. Curve *C*: both vagal nerves and stellate ganglia disconnected from the central nervous system, removing both tonic outflows; unstimulated atrial rate, 124 per min. Much of each curve consists of stable zones in which the atrial rate shows a paradoxical acceleration with increased vagal stimulation rate, following the latter in simple ratios. Presumably this is undesirable and avoided by adaptive irregularity and nonsynchrony in the brain-driven vagal activity. [Reid, 1969.]

the inhibitor nerve. Note that the general downward trend that gives the inhibitory input its name is not monotonic; there are anomalous zones of acceleration, each with a fixed ratio of input to output frequency. The importance of such a phenomenon, although surely not the normal way the vagus slows the heart, is the insight it gives for the general problem of interacting neuronal rhythms.

Many if not most neurons fire more or less rhythmically, and receive input from others doing likewise or, in circadian oscillators, from the Zeitgeber. If the effectiveness of the input increases to a maximum late in the cycle, entrainment is likely. If entrainment is undesirable, as in the heart, it can be prevented either by desynchronizing the input (destroying its rhythm) or by adding jitter to input or

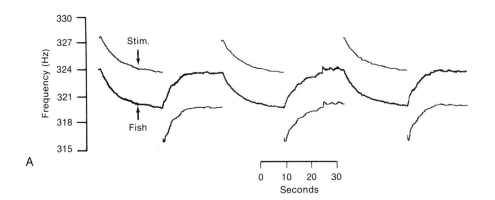

A

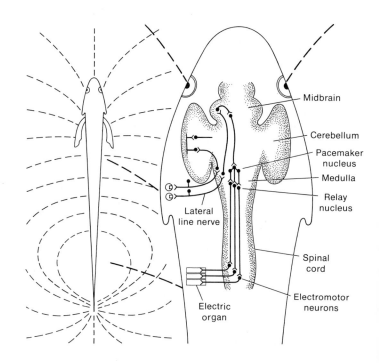

B

Figure 8.34

A normal behavioral response dependent in a complex way upon the exact frequency of the natural stimulus. The jamming avoidance response (JAR) of the high-frequency weakly electric fish *Eigenmannia virescens*. **A.** The response to an artificial stimulus, computer controlled to maintain a difference of 4 Hz whatever the fish does to try to increase the difference. The program reverses the sign of the difference every 25 sec. The fish reliably and without habituation shifts its frequency away from the stimulus as long as the difference (ΔF) is not too large or too small. **B.** Schema of the behavior system with a pacemaker nucleus acting as a single unit, two classes of receptors (*T* and *P*) coding in different ways, and two chief destinations (cerebellum and midbrain) extracting different features

output rhythm. Here is biological usefulness for randomness.

Another type of case in which neither entrainment nor desynchronization is desirable and which is more complex in instructive ways is the following. Certain South American **weakly electric fish,** like the gymnotoids *Eigenmannia* and *Apteronotus,* discharge their electric organs at a quite constant frequency of about 300 Hz and 1000 Hz, respectively, night and day, as part of an active object-detecting and social communicating system. The frequency is so constant that it seems fixed and uninteresting. But if nearby fish of the same species have frequencies within a few Hertz, both will shift in the direction that increases the difference. If a fish is discharging its electric organ at 300 Hz and we introduce a stimulus that simulates a neighbor at 299 Hz, even at extremely low voltage ($<1\ \mu V/cm$, peak to peak), the fish shifts to about 300.5 to 305, depending on the voltage. This can be shown to have normal behavioral significance as a jamming avoidance response (Fig. 8.34). Close analysis shows that the receptors and brain together are measuring the voltage, the frequency difference, ΔF (beat fre-

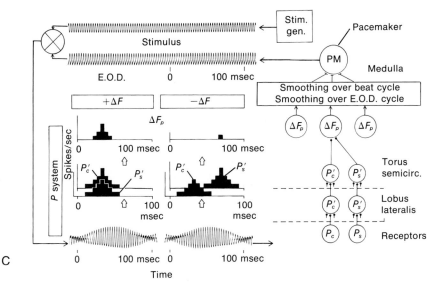

C

from the same input. **C.** The P system from receptors to fourth-order neurons that unequivocally signal the sign of ΔF and could cause normal JAR by variable inhibition of the pacemaker (*PM*). P_s cells fire impulses during a certain phase of the beat cycle, regardless of the sign of ΔF; P_c cells fire at a different phase for $+\Delta F$ and $-\Delta F$. Fourth-order "ΔF_p" detectors in the torus (midbrain) act as though firing to coincidences of input from the P_s and P_c cells. The parallel T system is omitted for simplicity. All the cells indicated are known, but the connections between them are hypothetical. [Scheich and Bullock, 1974.]

quency), and the sign of the difference (plus or minus). A sequence of neurons has been found, from first-order afferents to fourth- or higher-order midbrain units, that have properties such as to measure these parameters, give differential weight to various values, and superimpose a smoothly graded acceleration or deceleration upon the command to the electric organ. The command center is a true biological clock comprising pacemaker neurons electrotonically coupled into a single functional unit that commands every discharge of the electric organ, 1:1. The same principles apply that we have encountered above (Fig. 6.4 and p. 330), but here the desired result is quite different and can be accounted for by a plausible connectivity and dynamics of neurons actually found. Instead of phase locking, the interaction is such as to drive the frequencies apart, according to rules appropriate to the biology of the species.

SUGGESTED READINGS

Hinde, R. A. 1970. *Animal Behaviour: A Synthesis of Ethology and Comparative Psychology.* McGraw-Hill, New York. [This provides an introduction to the ethological approach to behavior.]

Ingle, D. 1968. *The Central Nervous System and Fish Behavior.* Univ. Chicago Press, Chicago. [An interdisciplinary symposium; it therefore includes selected topics, and is not comprehensive.]

Quarton, G., R. Melnechuk, and F. O. Schmitt. 1967. *The Neurosciences: A Study Program.* Rockefeller Univ. Press, New York. [Two blocks of chapters deal with behavioral correlates.]

Roeder, K. 1967. *Nerve Cells and Insect Behavior.* Harvard Univ. Press, Cambridge, Massachusetts. [A small volume with selected "stories" of established physiology underlying natural behavior.]

Schmitt, F. O. 1970. *The Neurosciences: Second Study Program.* The Rockefeller Univ. Press, New York. [Several groups of chapters deal with integration at the behavioral level.]

Schmitt, F. O., and F. G. Worden, 1974. *The Neurosciences: Third Study Program.* M.I.T. Press, Cambridge, Massachusetts. [Like the other two volumes in this series, this is a collection of chapters unusually planned, worked over, and edited, hence the work is more synthetic than normal symposia.]

Thompson, R. 1967. *Foundations of Physiological Psychology.* Harper & Row, New York. [The mechanistic side of that approach to behavior associated with the term psychology.]

DEVELOPMENT AND SPECIFICATION OF CONNECTIONS IN THE NERVOUS SYSTEM

I. INTRODUCTION

The human nervous system consists of about 10^{11} neurons, surrounded by several times that many glial cells. As we have seen in Chapters 2 and 3, these neurons are of many distinguishable types, and they are organized into clearly differentiable nuclei (cell groups) and tracts, each of which is the same in every individual of a species and has homologues in similar structures throughout the phylum. Indeed, in many animals, a good number of individual neurons can be identified. This is especially the case in well-studied invertebrates, such as *Aplysia* and leeches, but even in vertebrates this may be the case—for example, the paired giant Mauthner neurons in fish and in larval urodele amphibians, and the giant motor neurons of lampreys (Fig. 9.1, pp. 54, 99). Around the Mauthner neurons, even certain of the glial cells are recognizable from one specimen to another. It seems possible that with further research we will come to appreciate that a significant fraction of the cells of the nervous system, even in man, are in fact individually distinguishable from all others, and identifiable in each specimen.

Most of the billions of neurons synapse with hundreds or thousands of other neurons in an awesome maze of connections. The function of the nervous system depends upon the precision and appropriateness of these connections. In this chapter we will consider some of the more obvious developmental and time-dependent questions posed by this combination of complexity and precise reproducibility: What controls the proliferation and differentiation of nerve cells into given numbers of distinct types? What is the role of genetic programming, and what is the role of "experience" in governing the establishment of the correct connections? How precise are these connections and how plastic (modifiable) are they? What influence do nerve cells exert on the development and maintenance of other nerve cells or target cells? What controls the growth of axons and dendrites and the migration of cells or their processes to their appropriate locations? What factors govern the close association of neurons and neuroglia? How does regeneration differ from initial development?

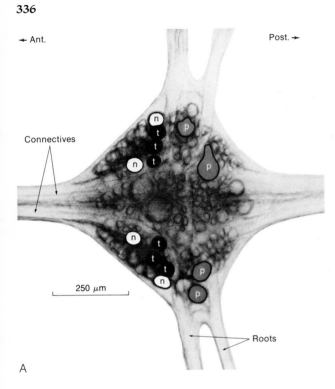

← Ant.   Post. →

Connectives

n
t
t
t
n

p

p

250 μm

n

t
t
t
n

p

p

Roots

A

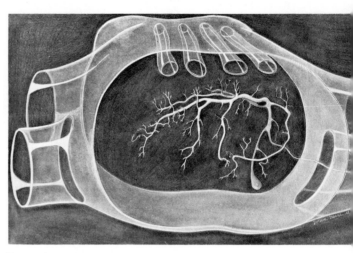

B

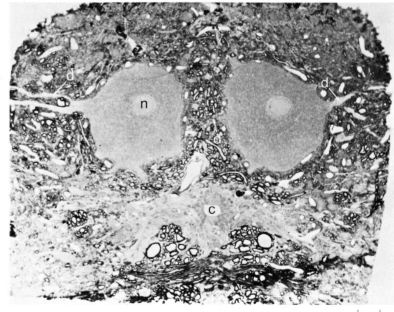

d   n   d

c

20 μm

C

Figure 9.1 (*facing page*)
A. Leech ganglion, with a few of the identifiable neurons encircled. Each ganglion contains about 350 cells, many of them large enough to be penetrated easily with microelectrodes and recognizable in each ganglion. The outlined cells are all the central cutaneous sensory cells for the body segments innervated by this ganglion: *t,* touch-sensitive cells; *p,* pressure-sensitive cells; *n,* cells responding to noxious stimuli. Each of these cell types is recognizable by its location within the ganglion, its synaptic connections, its response area and specificity, and its firing pattern. [Nicholls and Baylor, 1968.] **B.** An exact three-dimensional reconstruction of a motor neuron within the third abdominal ganglion of a lobster, made possible by the injection of procion yellow dye into the cell. This dye quickly diffuses throughout the cell and fluoresces on illumination with UV light. When two cells are injected, the areas of contact between them can be identified. The same cell can be recognized in each individual of the species studied, and is seen to have a very similar branching pattern, leading to the conclusion that there is much more consistency and order in the neuropile of invertebrate ganglia than was once believed. [Davis, 1970.] **C.** Cross section through the spinal cord of an electric catfish, *Malapterurus,* showing the two giant electromotor neurons, each of which controls the entire electric organ on one side of the body. In this species, and in a growing number of others, individual vertebrate neurons are easily recognizable and critical to function. *n,* nucleus; *d,* dendrites; *c,* central canal. (See also Figs. 9.6; 10.60.) [Bennett et al., 1967.]

Some simple examples bring these questions into relief.

Within minutes of birth, a young wildebeest is able to move with the herd, run from predators (for example, a hyena), and generally behave with most of the neurophysiological competence of an adult. It is obvious that its nervous system is nearly fully developed, with most of the necessary neurons present and correctly connected to other cells, before there is any opportunity for normal use of the pathways. Without any prior visual experience, the millions of cells that constitute the visual pathways are able to analyze visual input appropriately in order to recognize objects (predator, mother, obstacles) and to coordinate motor activity accurately.

In contrast, a human infant, much smaller than the young wildebeest but with a larger brain and having been in gestation much longer, is essentially helpless at birth. Only over a period of 2–3 years does it gain the coordination and sensory analytic ability to react to the environment with anything approaching adult capabilities. It is conspicuous to any parent that this coordination comes only with months of experience in use—in trial-and-error perfection of sensory analysis and efferent outflow. It is not safe to conclude, however, that experience is necessary in all such cases. Much of the improvement in performance may merely be the result of maturation of the nervous system. For example, it is generally said that birds "learn" to fly. Yet if hatchling pigeons are raised inside tubes so small they cannot spread their wings, they are able to fly adequately on release from the tubes at the same time unrestrained siblings have finally "learned" to fly. Clearly, behavior is a thorough blend of genetically determined factors, and behavior perfected by experience and learning. Changes in behavior with ex-

338

Chapter 9
Development and Specification of
Connections in the Nervous System

perience must represent some refinement of structure or function within the nervous system, superimposed upon basic structures and physiology determined by purely genetic, developmental processes.

The importance of genetic control on behavior is perhaps most clearly seen in the many single-gene behavioral mutants of *Drosophila*. It is also seen, however, in single-gene mammalian mutants. For example, there are at least three forms of mouse cerebellar mutants (weaver, staggerer, and reeler), which result from specific misdevelopment of granular cells and Purkinje cells at different stages of morphogenesis, even up to several weeks after birth. Study of such behavioral mutants shows that mutation of a single gene affects the development of large populations of neurons, and that many different genes influence a given class of cells, both simultaneously and in sequence. With 10^{11} neurons and several orders of magnitude more of synapses, all critical to nervous function, it is not realistic to postulate unique determination of each connection by a different gene. Instead, as will be seen below, it is necessary to conclude that single genes govern the formation of whole populations of connections and interact with other genes to establish gradients of proliferation, differentiation, and synapse formation.

II. MORPHOGENESIS OF THE NERVOUS SYSTEM

The development of the nervous system, like that of other systems or organs, happens in an orderly, patterned way, guided by mechanisms that are still mostly unknown. There are recognizable spatial and temporal sequences, with evidence of recurring waves of cell proliferation or differentiation sweeping from one part of a presumptive cell group to another, from anterior to posterior, ventral to dorsal, or the equivalent.

No attempt will be made in this book to review the details of development of all areas of the nervous system. A good embryology text should be consulted for this information. It is our intention, however, to discuss the general principles involved, and cite a few examples as illustration, with emphasis on the interactions between cells that guide formation of the correct synaptic connections. To a remarkable degree, the study of neural development and synapse formation has been dominated by a few men: S. Ramon y Cajal, R. G. Harrison, S. R. Detwiler, P. Weiss, V. Hamburger and R. W. Sperry. Much of what will be said below is either directly attributable to them or an outgrowth of their findings.

A. Induction

One of the first clear examples of embryonic induction was the demonstration by Spemann that the roof of the archenteron induces or stimulates the overlying ectoderm to form a neural tube in the gastrula stage. Subsequent investigations have established countless other examples, general and specific, of one cell population inducing or affecting the development of another. Whatever is responsible for this induction is unknown in most instances. Simple gradients of O_2, CO_2, or pH might be sufficient to affect genetic expression but usually a more specific influence is postulated, mediated probably by large molecules. Induction does not *require* direct contact between cells. Whatever is required has been shown in at least certain cases to take place across millipore filters of 0.5 μm or larger pore size with no loss of effect. A pore size of 0.1 μm causes conspicuous reduction of effect. On the other hand, since all of the cells studied in early embryos have been shown to be electrotonically coupled, perhaps so tightly that cytoplasmic connections exist, we should not quickly dismiss the possibility that important inductive influences are exerted via these connections, which are lost in the course of development. Nerve cells and skeletal muscle fibers are unusual in losing these connections; most other tissues remain coupled, even in the adult.

B. Proliferation

Most of the nervous system develops by proliferation of cells at the ependymal lining of the neural tube. The cells that form there migrate to their appropriate position and undergo differentiation. Proliferation takes place by the repeated asynchronous division of neuroepithelial germinal cells. These cells go through a curious oscillatory movement, away from and then back to the edge of the ventricle at the inside of the neural tube (Fig. 9.2). When away from the ependymal lining, they undergo DNA synthesis; when retracted to the lining, they divide. One daughter cell then migrates away into the mantle zone while the other repeats the mitotic process. Neurons and glia both arise in this way, but the glia are formed only after most of the neurons are in place.

The introduction of **pulse-labeling,**

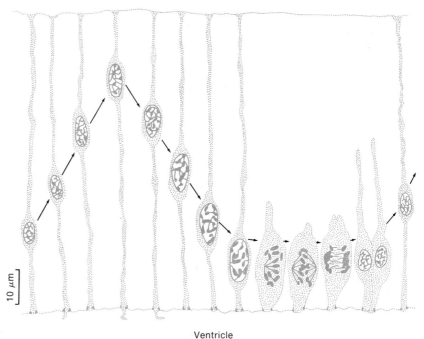

Figure 9.2

Neuronal proliferation sequence, showing the position of a single chick neuroepithelial germinal cell at intervals of approximately one-half hour throughout the mitotic cycle. [Jacobson, 1970; after Sauer, 1935.]

principally with H³-thymidine, which is incorporated into the newly synthesized DNA, has aided greatly in revealing the origin, migratory pathways, and final location of cells formed at different times. A pregnant female, or the fetus itself, is injected with H³-thymidine and fetuses are sacrificed at various times thereafter (from a few minutes to several days) to show the number and location of labeled cells. Several interesting generalizations have emerged.

Parts of the brain that are oldest phylogenetically, such as the medulla and diencephalon, are the **first to develop** and differentiate, whereas more recently evolved structures, such as the neocor-

Figure 9.3
Vertebrate brains, dorsal view.

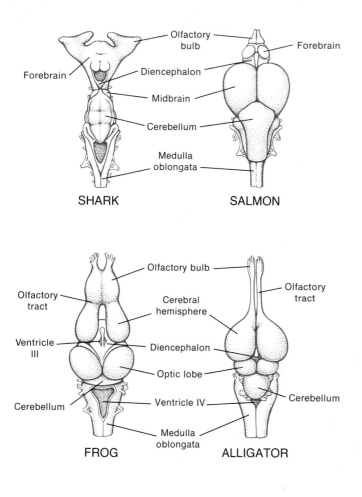

SHARK   SALMON

FROG   ALLIGATOR

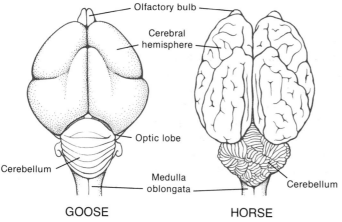

GOOSE   HORSE

tex, appear only late in development (Fig. 9.3). Within a given functional part of the central nervous system, differentiation occurs sequentially from one region to another. For example, the optic tectum differentiates first at its ventral-lateral edge at the rostral end, and progresses in a dorsal-medial-caudal direction. The mouse diencephalon differentiates in a dorsal-medial-rostral direction. Figure 9.4 shows the pattern of development of the mouse thalamus. The starting point and direction of development differ for various parts of the brain, but immediately suggest some kind of gradient of inducer, or else a chain reaction form of differentiation, triggered at one spot by a local phenomenon there. In general, adjacent parts of the nervous system that differ in cytoarchitecture and function arise from different germinal tissue or at different times. Within a given nucleus, large neurons differentiate first, intermediate-sized cells later, and small ones, especially the short-axon Golgi Type II cells that are involved in complex interactions within the nucleus, last. Within the spinal cord, motor neurons develop first, interneurons later. Retinal ganglion cells are the first cells to appear in the retina, mitral cells in the olfactory bulb, Purkinje cells in the cerebellum, and pyramidal cells in the cortex. Interestingly, these big cells are all efferent cells. Typically, their axons make synaptic contact with target cells before, or approximately simultaneously with, the first arrival of afferent input. In a chick, the first synapses on motor neurons are seen on day 5 of development, approximately simultaneous with the arrival of motor neuron axons at the muscle and of sensory axons at the skin.

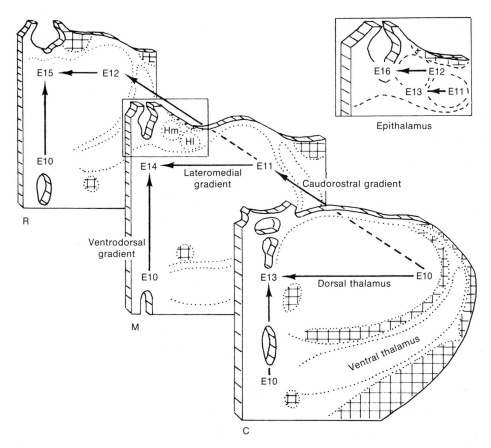

Figure 9.4
Sequence of development of different parts of the mouse thalamus. Between days 10 and 15 of gestation, as shown by the *E* numbers, massive proliferation of neuroblasts proceeds from the ventral to dorsal, lateral to medial, and caudal to rostral parts of the thalamus. Letters (*R*), (*M*), and (*C*), denote rostral, medial, and caudal transverse sections. *Hm,* medial habenular nucleus, *Hl,* lateral habenular nucleus of the epithalamus. [Angevine, 1970.]

C. Cell Migration

The development of the neocortex in rats illustrates the same principles while exhibiting another remarkable phenomenon—the migration of one population of cells right through another population (Fig. 9.5). All cells arise from the germinal epithelium lining the ventricle, and then migrate to their final location. These cells are **produced in waves,** each wave migrating past the previously produced population to form a new surface layer. Eventually there are six layers, the largest cells that arose first forming the deepest layer and the small cells produced last forming the most superficial layer. Cells of the deepest layer first

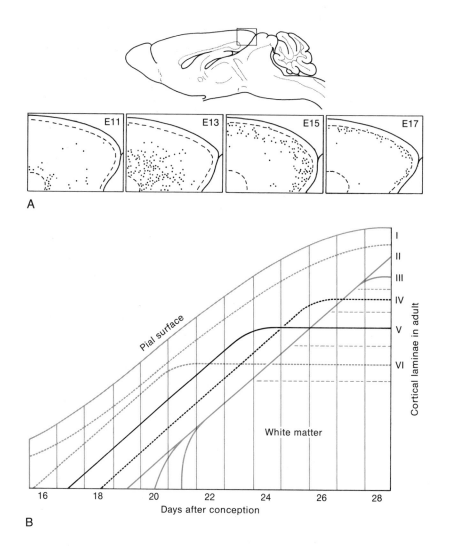

Figure 9.5

Inside-out development of the cerebral cortex of a mouse. **(A).** Autoradiograph draw-ings showing the location of labeled neurons in four different mice injected with triti-ated thymidine on days 11, 13, 15, and 17 of embryonic life. All were tested 10 days after birth when the neurons were at their final positions. [Angevine and Sidman, 1961.] **(B).** Diagram showing the time of origin and migration pattern of neurons in the rat cerebral isocortex, determined by labeling with tritiated thymidine as in part (A). [Berry et al., 1964.]

appear on day 16 after conception and reach their cortical position within 4–5 days. Those of the superficial layer appear on day 21 and reach the surface after about 8–10 days of migration past neurons that developed earlier. Neuroglia first appear on day 21.

The migration of neurons from the germinal epithelium to their final location, sometimes several centimeters away, immediately poses this problem: **What directs the migration?** When, as happens in the cortex, large numbers of cells migrate through and past other populations, it is obvious that there is no irresistible mechanical force pushing or pulling everything in the direction of migration. An even more convincing case is that described by Levi-Montalcini in the developing chick medulla. The VI and VII cranial nerve nuclei differentiate adjacent to each other near the ventricle. Then, simultaneously, some of the cells in the VI nucleus migrate laterally to form one of the V nuclei while a large population of cells in the VII nucleus migrates through the same part of the medulla in a perpendicular direction to the ventral part of the brainstem. It is not known what factors are responsible for this guided migration, just as there is really no satisfactory explanation for the growth of axons to their appropriate destination (see below). It seems most likely, however, that the migrating cells are genetically programmed to move (in unknown ways) toward an appropriate chemical environment, along a chemical gradient or series of gradients, diffusible or built into the cells passed en route. The phenomenon may not be greatly different from that of chemotaxis in leucocytes or bacteria.

D. Differentiation

The brain of a human infant at birth already has virtually its whole complement of nerve cells. Yet the brain doubles in weight during the first six months after birth and doubles again within the first four years. Thereafter, until about age 10, there is only a slight increase each year, with essentially no increase beyond age 10. A part of this growth is due to the continuing proliferation of glial cells. A large fraction of it, however, represents neuronal differentiation. During this time, cells increase in size, some becoming polyploid (4n and 8n), and there is a tremendous outgrowth of processes, especially dendrites, with resulting increase in the number and complexity of interneuronal connections. During the period of this growth, the nervous system is much more plastic than later in life, capable of correcting for such developmental defects as "squint" or congenital maldevelopment of even major brain tracts, such as the corpus callosum. Apparently in precocial mammals, such as the wildebeest mentioned earlier, these stages of differentiation are mostly complete before birth. It is obvious, on the other hand, that some forms of plasticity, probably associated with subtle changes in neuron structure or connectivity, persist throughout life, even in relatively simple animals.

Outgrowth of Axons. Very early in the course of differentiation, axons begin to grow out from the cell somas. When they appear, they usually emerge from the cell on the appropriate side and grow in the appropriate direction to make the

344

Chapter 9
Development and Specification of
Connections in the Nervous System

correct connections. For example, virtually all retinal ganglion cell axons emerge toward the center of the eye and grow medially from their earliest appearance. In the rare cases in which this does not hold, the axons quickly turn and grow in the correct direction. This phenomenon of directed outgrowth has been shown very beautifully in experiments with the large paired Mauthner neurons of fish. These normally send axons medially where they cross the midline of the medulla and traverse posteriorly. If that segment of the medulla is reversed or if another segment is implanted in reversed polarity before axon emergence, the axon first grows out in the direction appropriate to its immediate surroundings (180° off with respect to its new location) but then quickly reverses direction and grows posteriorly when it encounters part of the brainstem that has not been reversed (Fig. 9.6). These neurons, like most of the optic nerve axons (only 50% in primates with their highly developed binocular vision, but 100% in most vertebrates) and, indeed, most sensory and motor nerves, cross to end on the opposite side of the brain or body. What causes this decussation is not understood.

At least two factors have considerable experimental support as mechanisms for the guidance of axonal growth. One of these is mechanical, or **contact, guidance,** first recognized by Harrison and championed by Paul Weiss; the other is a form of chemical "neurotropism" or differential chemoaffinity, first strongly argued by Cajal. It is clear that outgrowing nerve fibers, in vivo or in tissue culture, will follow interfaces and natural mechanical guides, such as blood vessels and borders between muscles. Indeed, in tissue culture, some sort of substrate or inter-

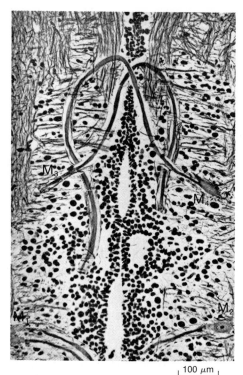

$\llcorner$ 100 μm $\lrcorner$

Figure 9.6
Growth pattern of Mauthner cell axons in the salamander *Pleurodeles waltii*. M_1 and M_2 are from the original Mauthner cells, sending their axons contralaterally and posteriorly in the normal fashion. M_3 and M_4 are axons of supernumerary Mauthner cells, resulting from a grafted segment of the medulla, implanted backwards into the host animal. Their axons begin in the normal direction (contralaterally and posterior with respect to the axes of the grafted segment) but then turn and run in a posterior direction appropriate to the axes of the host animal. [Hibbard, 1965.]

face is necessary for any outgrowth. On the other hand, mechanical guidance, by itself, can be only a small part of the answer. The initial outgrowth of axons at the correct pole occurs without apparent mechanical cues, and there are numerous other observations that are indicative of some sort of chemical guidance. Typical of these is one of Cajal's examples—the large nerve that enters the area of the tongue and which contains fibers of the hypoglossal, trigeminal, facial, and glossopharyngeal nerves. At different points during the outgrowth of the nerve, each type of fiber leaves the main trunk, the hypoglossal to innervate tongue musculature, the trigeminal to innervate ordinary papillae, the facial (chorda tympani) and glossopharyngeal to innervate different groups of gustatory papillae. Another category of evidence consists of the many examples, in invertebrates and vertebrates, of experimentally displaced tracts or nerve cells that send their axons to the correct location via totally abnormal pathways. One of the most dramatic examples is that of innervation of the optic tectum by the displaced optic nerve (see later sections). Another classical example was the demonstration that spinal nerves will grow through abnormal pathways to innervate a displaced limb bud, following a route that would suggest chemical guidance operative even at a significant distance.

Although it is generally accepted that there must be **chemical guidance** of some sort, no such factors have been isolated or described, except perhaps for nerve growth factor (see below). There is at present no convincing evidence for a diffusible guidance substance. The behavior of a growing nerve terminal—first studied in tissue culture by Harrison, but verified in vivo in tadpole tails by Speidel—suggests "recognition" of a target cell by something built into its membrane and selective adhesion to that cell (Fig. 9.7). The nerve fiber extrudes processes in all directions as it proceeds, constantly retracting some and sending out others, seemingly "palpating" the surface of objects near its path. Certain pathways are accepted, others rejected, and eventually one type of cell may be accepted for innervation and all others passed over, apparently on the basis of chemical recognition. There may even by a "sidedness" cue, helping explain decussation. In the development of the auditory pathways, for example, cochlear nucleus cell axons grow past cells of the ipsilateral medial trapezoid nucleus and olivary nuclei to synapse with their contralateral counterparts. Alternatively, the ipsilateral cells might not yet have developed the needed chemical properties when the fibers grow past.

Whatever guides axons to their correct approximate location, however, clearly need not be the presence of the cells they will eventually innervate, for frequently the axons providing input to a nucleus are present and in location before the cells they innervate migrate into that part of the brain. It is quite possible, on the other hand, that the axons are somehow guiding the migration and differentiation of postsynaptic cells. This possibility will be discussed further below.

In at least certain cases, growing neuronal processes appear to be **guided by glial cells.** In the development of the cerebellum, for example, granular cells migrate along processes of the Bergmann glial cells from the surface past the Purkinje cells, making contacts with the Purkinje cells as they pass. (This is a second migration for the granular cells, which, like other neurons, originally migrate from the ependymal surface to the pial surface.) The cerebellar mutants (e.g., "weaver"), in which the migration and progressive formation of connections does not take place, have no Bergmann glial cells, or they are drastically reduced in number.

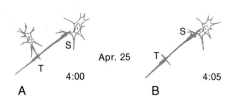
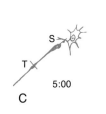
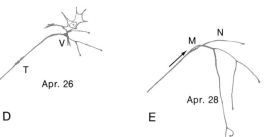

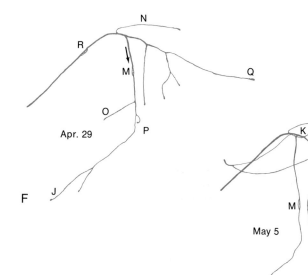

Figure 9.7

Growth pattern of a nerve terminal in the tail of a tadpole, recorded during one month of observation. **A.** While progressing through stages *a* to *d*, the growing terminal, ending in a growth cone (*S*), has passed one process of a fibroblast (*T*), at which it developed a varicosity and continued on to contact another fibroblast. Here it paused for many hours before it finally moved on (leaving another varicosity at the site of the fibroblast contact) and rapidly sent out several processes. **E.** Two days later, several processes had grown considerable distances. One small branch (*P*) was encircling another fibroblast. A sheath cell (*M*) had attached to the fiber. **F.** One day later, another small sheath cell (*R*) was present, plus several new branches. **G.** During the next six days, several more branches appeared, while others (*P* and *O*) retracted and disappeared. *Q* became the most prominent process. Distance from *K* to *Q* was approximately 750 μm, representing two week's growth. [Speidel, 1933.]

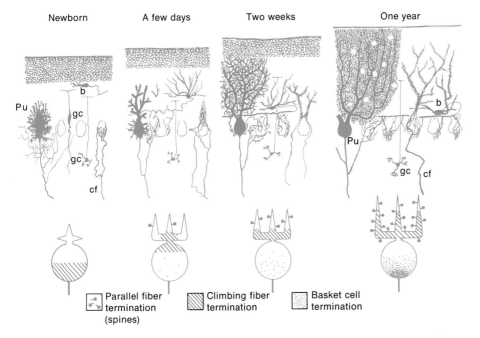

Newborn A few days Two weeks One year

Parallel fiber
termination
(spines)

Climbing fiber
termination

Basket cell
termination

Figure 9.8
Histogenesis of the cerebellar cortex. (See Chapter 3 for a full description of the adult morphology
and circuitry.) The differentiation of the cerebellar neurons can be seen to parallel the arrival and
elaboration of the afferent systems at each cell. *Pu,* Purkinje cells; *gc,* granule cells; b, basket cells; cf,
climbing fibers. The lower part of the figure illustrates the distribution of these three afferent inputs
to Purkinje cells. [Andersen and Eccles, 1965.]

As axons proceed from one spot to another in nerves or tracts, they sometimes retain a very orderly arrangement within the bundle and sometimes seem to be randomly distributed. In the ascending somatosensory tract, for example, topographic representation is maintained, whereas in the optic nerve disarray may be the rule.

Growth of Dendrites. Dendritic outgrowth appears always to take place after the outgrowth of the axon, and even after the formation of synapses by the axon. When dendrites do grow out, they show profuse branching, with **most branches later being reabsorbed.** Dendrite forma-

tion has been studied principally in motor neurons and large cells of the cortex and cerebellum. A diagrammatic summary of stages in the development of cerebellar dendrites and afferent innervation is shown in Figure 9.8. In all cases the maintenance of dendritic branches depends on the presence of innervation. There may be some sort of trophic phenomenon operative even at a distance, since the first dendritic outgrowths tend to be in the direction of the relevant afferents. If afferent axons are removed, the dendrites develop abnormally or not at all. In the visual cortex, for example, each pyramidal cell first develops the basal portion of the apical trunk of its

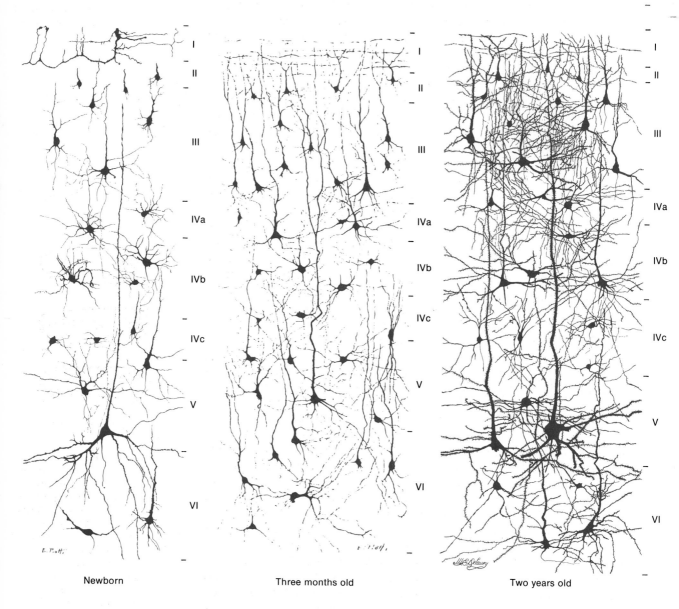

Newborn

Three months old

Two years old

Figure 9.9
Development of dendritic arborizations in the human visual cortex, as revealed by the Golgi stain technique, from newborn, three-month-old, and two-year-old infants. Depth of cortex normalized to the same height. [Conel, 1959.]

dendritic tree, where synapses form with incoming axons from the lateral geniculate. As the cortex develops and short axon interneurons appear postnatally, lateral dendritic branches sprout to accommodate intracortical connections. Figure 9.9 shows several stages in the development of the human visual cortex.

Development of Excitability. There has been little study of the ontogeny of excitability. Although some egg cells have been shown to undergo a single large change in potential (and permeability) on fertilization, only a few have been studied during cell division and differentiation. Among the few that have been examined are echinoderm and tunicate eggs. In tunicates, presumptive nerve and muscle cells can be followed from the egg stage to the formation of adult, differentiated tissue. Even immature egg cells in the animal have good resting membrane potentials and, surprisingly, show electrical excitability. In fact, they appear to have a greater variety of ionic channels or permeability properties than at later stages; for example, there are both Ca^{++} and Na^+ spikes, the Na^+ spike gradually disappearing in the course of differentiation. Rat muscle cells in the myotube stage show both Na^+- and Ca^{++}-excitable inward current channels. With subsequent differentiation, the Ca^{++} channel is lost. Perhaps it will prove a general finding that egg cells and other highly undifferentiated cells have more of their genome expressed and **gradually lose certain properties** during development. This clearly does not apply in some particulars, however, for in the tunicate and starfish eggs studied, the voltage-dependent delayed K^+ conductive increase (delayed rectification) develops only in later embryonic or larval

stages. Spikes in eggs or presumptive muscle cells are initially extremely prolonged; as delayed rectification develops, the spike becomes brief, finally assuming adult form.

Different tissues develop excitability at different times. Some even lose their excitability in the course of development. The skin cells of late embryo and early tadpole stage amphibians, for example, are electrotonically coupled and produce all-or-none cardiac-type action potentials (Fig. 9.10). When one patch of cells is stimulated, a wave of excitation propagates over the surface of the animal. In some unknown way, the skin is tied to the primitive nervous system, and excitation can initiate reflex responses. Eventually this excitability is lost as more specific peripheral sensory nerves appear.

Spontaneous Activity in the Developing Nervous System. Nerve cells in developing embryos exhibit a great deal of spontaneous activity that is not initiated or driven by sensory input. This has been seen in tissue culture, but is most apparent in the spontaneous, uncoordinated movements of most embryos. Motor connections are formed long before the sensory circuits are completed, hence reflex responses appear rather late in development. From the first stage at which neuromuscular connections are made, however, embryos show spontaneous, jerky movements, often in cycles of activity (Fig. 9.11). This spontaneous motility reaches a **peak just before the sensory connections** onto the motor neurons are made, and then gradually subsides. Probably spontaneous activity occurs before neuromuscular innervation as well. The importance of this activity to development is not known, but the

Figure 9.10
Propagation of excitation in skin.

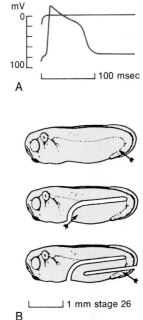

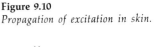

Electrically excitable skin in early tadpoles of Xenopus laevis. A. Sample intracellular action potential from the skin of a stage 26/27 embryo. Amplitude (20 mV/div) is reduced because of slow recorder response. [Roberts and Stirling, 1971.] B. Experimental skin cuts showing that the wave of excitation causing a response in rostral myotomes can apparently spread in all directions through the skin. [Roberts, 1971.]

350

Chapter 9
Development and Specification of
Connections in the Nervous System

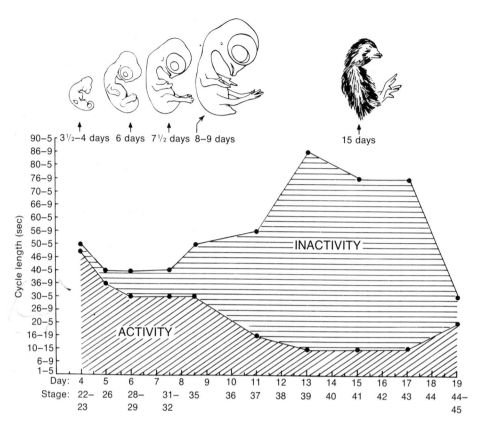

Figure 9.11

Spontaneous motility rhythms in developing chick embryos. The mean duration and cycle lengths of activity and inactivity phases are shown for different developmental stages. [Hamburger, et al., 1965.]

phenomenon may be related to the unexplained rhythms of activity seen both in late fetal stages and in postnatal humans in the form of REM (rapid eye movement) sleep cycles.

Myelination. A large proportion of the axons, both central and peripheral, become myelinated (see Figs. 2.45, 2.46, 2.48, 9.12). This is a relatively late phenomenon in development, usually occurring only after an axon has innervated its target cell and continuing until about

age 10 years in man. The association between an axon and the Schwann cells (peripheral nervous system) or oligodendrocytes (central nervous system) that surround it with myelin is a very interesting one. The sheath is found only around certain axons—not others that look identical, not dendrites, collagen bundles, or blood vessels. (There is, however, a form of oligodendroglial tumor in which the glial cells will form myelin over anything.) When the axon is cut and degenerates, the myelin sheath also degenerates. Clearly, the **axon and its sheath cells normally interact** in important ways that have not yet been recognized. An understanding of demyelinating diseases, such as multiple sclerosis, must await further understanding of these interactions.

Influence of Nutrition, Hormones, and NGF. During the time that the brain is growing rapidly and is most plastic (up to 4 years of age in humans), it is extremely dependent on adequate **nutrition,** especially on essential amino acids. Eventual brain weight is decreased and cell numbers reduced by even short periods of malnutrition in prenatal or early postnatal stages, in contrast to similar malnutrition suffered later in life. The deleterious effects on behavior are not easily assessed in most animals, but are readily discernible in humans as a reduced IQ. Even the relatively mild competition for nutrients by twins is reported to result in perceptibly lower average IQ.

Several **hormones** also play an important role in normal development. Growth hormone, the sex hormones (especially testosterone), and thyroxin have all been shown to stimulate proliferation and differentiation, whereas corticosterone, in early developmental stages, reduces body and brain weight. In most cases the effects are sufficiently broad and nonspecific that it is difficult to attribute them to some direct action on neurons. The sex hormones, however, have more specific effects. Testosterone causes selective differentiation of cells responsible for male behavior. Injected into a female or castrated male rat before age ten days, testosterone will result in typical male behavior.

Only one truly specific trophic factor is known in any detail, however. This is the **nerve growth factor** (NGF), first discovered in certain mouse sarcoma tumors when investigators noticed these tumors being invaded by a great profusion of sympathetic fibers growing out from hypertrophied sympathetic ganglia. NGF was shown to be a diffusible nucleoprotein that, in vivo or in tissue culture, causes a tremendous increase in protein syntheses and outgrowth of processes in sympathetic ganglion cells and, during embryogenesis, of the dorsal-medial cells of dorsal root ganglia (Fig. 9.13). There is no apparent effect on other parts of the nervous system. NGF is not yet a well-defined entity. Early attempts to degrade it with phosphodiesterase from snake venom led to the unexpected discovery that NGF is present in high concentrations in the venom of several different poisonous snakes. Study of analogous glands of mice revealed that the factor is present in still greater concentration (about 10,000 times as concentrated as in the original sarcomas) in the submaxillary salivary glands of adult male mice (but not rats). A thorough physical characterization has been done on this salivary gland NGF, which appears to have a molecular weight of about 140,000–150,000. At high or low *p*H this molecule dissociates into three different subunits, α, β, and γ, of which only the β unit, with a molecular weight of 30,000, has a trophic effect similar to that of the original complex. Antibodies to the β unit, like that to the whole NGF molecule, cause destruction of most of the sympathetic ganglion cells. NGF and anti-NGF are thus useful tools for study of the development and function of the sympathetic nervous system. Another tool recently introduced

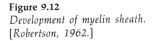

Figure 9.12
Development of myelin sheath.
[*Robertson, 1962.*]

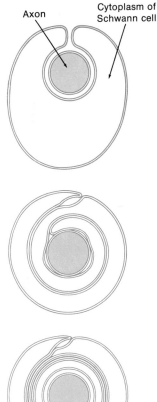

Axon

Cytoplasm of
Schwann cell

352

Chapter 9
Development and Specification of
Connections in the Nervous System

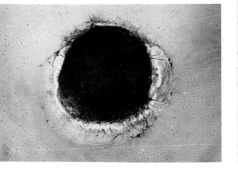

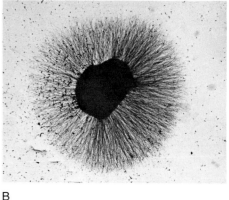

A

B

Figure 9.13
Effect of nerve growth factor (NGF) on outgrowth of processes from a chick sensory ganglion. **A.** Whole mount of control 8-day ganglion, incubated 24 hours in a semisolid control medium. **B.** Eight-day ganglion incubated 24 hours in semisolid medium containing NGF at a concentration of 0.01 μg/ml. [Levi-Montalcini and Angeletti, 1968.]

is 6-OH-dopamine, which, when injected into an adult mouse, stops sympathetic transmission by selectively eliminating synaptic vesicles, leaving other pre- and postsynaptic structures intact. Injected during the first few days after birth, 6-OH-dopamine causes destruction of the whole postganglionic sympathetic nervous system.

The role of NGF in normal development is not yet clear. Its presence in small amounts in the blood of all mammals tested, and the disastrous effect of anti-NGF injections, argue that it may have an important function in normal growth of the sympathetic and sensory systems.

E. Cell Death and the Influence of Target Cells on Neuronal Development

Most neuronal proliferation appears to be genetically programmed, continuing even in the absence of other parts of the nervous system or peripheral structures that would eventually be innervated. The new neurons migrate to their destinations, undergo considerable differentiation, and even send out their axons to the appropriate area for synapse formation. If the appropriate target cells are present, synapses are formed, and "fulfilled" neurons complete differentiation, with accelerated growth of the innervating process and retraction of other axonal branches. If the target cells are not present, however, the neurons eventually degenerate and disappear. Thus the **postsynaptic cells are vitally important** to the maintenance of neurons. The mechanism of this influence is not known, but it may be very similar to the trophic influences exerted in the other direction by nerve cells on their target cells (see below).

Surprisingly, there is a tremendous amount of cell death during normal development. During innervation of an amphibian limb, for example, motor neuron axons grow out in great numbers, innervating first the muscles of the proximal portion of the limb, then the more distal muscles. Nearly 10 times as many neurons send axons into the limb as eventually remain to innervate it. Thus during the process of innervating the limb, the vast majority of neurons degenerate. As Figure 9.14 shows, the period of peak degeneration coincides with the time of formation of neuromuscular junctions and the onset of movement. It is known that once a muscle fiber has been innervated by one nerve fiber, or in some cases two, innervation by additional fibers does not occur. Apparently, therefore, the first axons that grow into the muscle successfully innervate it, and axons that arrive later either find uninnervated muscle fibers somewhere else to innervate or else degenerate ("retrograde degeneration"). Indeed, careful counts of the number of motor neurons show that approximately the same number of motor neurons differentiate at all levels of the spinal cord. A large fraction of them survive as somatic motor neurons in the brachial and pelvic plexuses, but most of them disappear from the thoracic and sacral segments, where there is much less muscle to innervate. (Some of the latter do not die, but migrate toward the ventricle and become visceral motor neurons, preganglionic cells of the sympathetic nervous system.) Similarly, approximately 1400 neurons differentiate in the chick trochlear nucleus. Normally about 50% of these survive by successfully innervating ocular muscles. If the optic vesicle is removed, approximately 85% degenerate.

The same phenomenon of **excess production and subsequent death** governed by the amount of peripheral tissue pres-

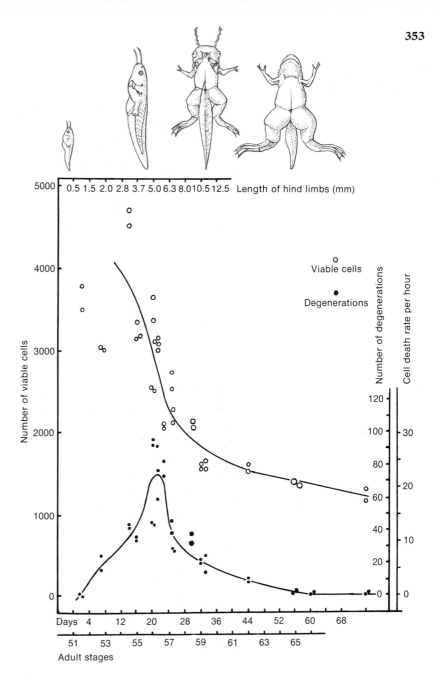

Figure 9.14
Cell death during development of the amphibian lumbar spinal cord (*Xenopus laevis*). The peak rate of degeneration, calculated on the basis of an average degeneration time of 3.2 hours, occurred at approximately the time of formation of neuromuscular junctions. If the hind limb on one side is removed during this process, the number of cell deaths is sharply increased. In the presence of a supernumerary limb receiving innervation from the same portion of the spinal cord, the number of cell deaths is much reduced. [Hughes, 1961.]

Chapter 9
Development and Specification of
Connections in the Nervous System

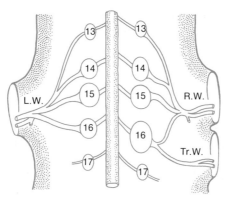

Figure 9.15
Effects of a supernumerary limb graft on the
dorsal root ganglia of a chick embryo. The six-
teenth ganglion provided most or all of the sensory innervation. The letters *L.W, R.W,* and
Tr.W. refer to left, right, and transplanted wing
buds. [Hamburger and Keefe, 1944.]

ent is seen in the dorsal root ganglion
and in the surviving proportion of sensory axons that grow out to the periphery. Two out of every three dorsal root
ganglion cells in *Xenopus* degenerate
during development.

The opposite phenomenon also occurs. In amphibians and birds, in which
such experiments can be done successfully, implantation of an extra limb in an
embryo leads to pronounced hyperplasia
of spinal sensory and motor centers, not
by the proliferation of an abnormally
large number of neurons, but by a relative reduction in cell death (Fig. 9.15).
More neurons successfully innervate
muscle cells of one limb or the other,
and hence are themselves maintained.

It is most probable that the same
dependence exists within the central
nervous system, although much less
study has been devoted to central trophic
interactions. Two cases can be cited. (a)
When the optic vesicle of a growing

chick is removed, for example, the ciliary
ganglion undergoes severe retrograde
degeneration because of removal of its
peripheral field. When the ciliary ganglion cells degenerate, the neurons of the
accessory oculomotor nucleus, which
innervate the ciliary ganglion, also degenerate. This phenomenon has been
called "retrograde transneuronal atrophy." (b) Similarly, if the optic cortex is
damaged, lateral geniculate cells atrophy
or degenerate, and retinal ganglion cells
show retrograde transneuronal atrophy.

III. TROPHIC INFLUENCE OF NEURONS ON THE CELLS THEY INNERVATE

Nerves exert a powerful influence on
their target cells. They are necessary for
the differentiation of many tissues, for
the physiological properties of others,
and for the general maintenance of postsynaptic cells. In some cases these effects
are apparently independent of impulse
activity in the neurons; in others this
may not be so.

A. Importance in Guiding Differentiation of Sensory Structures

It is well known that the afferent nerve
must be present before secondary sensory structures differentiate. Among
systems in which this has been demonstrated are the lateral line hair cells,
muscle spindle organs, various specialized skin structures, and taste buds. The
taste buds are an interesting case in
point. The normal life-span of a taste-
bud cell in a mammal is a few days,
approximately 10 days in man. If the

tongue is never innervated, the taste buds do not differentiate. If an innervated tongue is denervated, the taste buds degenerate as they would have anyway, but new ones do not differentiate from the tongue epithelia. It is now possible to test the effectiveness of different types of nerves in inducing the formation of new taste-bud cells. In such experiments it is found that only sensory nerves that normally carry gustatory information are capable of this induction. The normal tongue nerves, the glossopharyngeal and chorda tympani, are effective, as is the part of the vagus that normally innervates gustatory receptors in the pharynx and larynx. On the other hand, a nongustatory sensory nerve, the auriculotemporal, is ineffective, as is the purely motor nerve, the hypoglossal. Such tests have been extended to test the specificity of induction influence. The taste buds of the front of the tongue are normally innervated by the glossopharyngeal nerve, the back of the tongue by the chorda tympani. These taste buds differ both in chemosensitivity and in response to cooling. When the nerves are crossed, the chorda tympani, which now innervates the front of the tongue, carries information characteristic of that part of the tongue rather than characteristic of the nerve. The glossopharyngeal carries information characteristic of the back of the tongue. Apparently the taste-bud characteristics are determined by the part of the tongue in which they differentiate, rather than by the nerve. Moreover, single fibers in regenerated nerves often innervate taste buds that are widely separated from each other on the surface of the tongue, yet all taste buds affecting a given nerve appear to have the same response specificities. Either the nerve is selectively finding taste buds with the same response specificity or it is inducing the formation of identical ones.

B. Trophic Influence of Nerve on Muscle

Nerves are required for the development, maintenance, and properties of muscle fibers as well. Skeletal muscle develops by the condensation of many myoblast cells into myotubes, and of myotubes into myofibers. Contractile fibrils normally appear first in the early myotube stage. The nerve is usually present well before this, and it has been shown in tissue culture that nerves can form functional junctions with myoblasts, before any contraction is possible; but the function of such junctions, if they form in vivo, is not well known. In amphibians, muscle can fully differentiate even in the absence of the nerves, although regeneration requires the presence of nerves (Fig. 9.16). In birds and mammals, on the other hand, muscle differentiation stops at the myotube stage if the nerve is not present. Thus, in higher vertebrates, the nerves acquire a critical function in governing the differentiation of muscle.

Not all skeletal muscle fibers are alike. In most vertebrates there are at least two or three different types, with extremes commonly designated as "fast" and "slow." "Fast" fibers, in addition to having a faster contraction velocity, tend to be pale in color, have less myoglobin and poorer capillary circulation than "slow" fibers, are usually higher in anaerobic metabolic enzymes, and have a different molecular form of myosin ATPase. Fast fibers are innervated by large-diameter axons with phasic firing

Chapter 9
Development and Specification of
Connections in the Nervous System

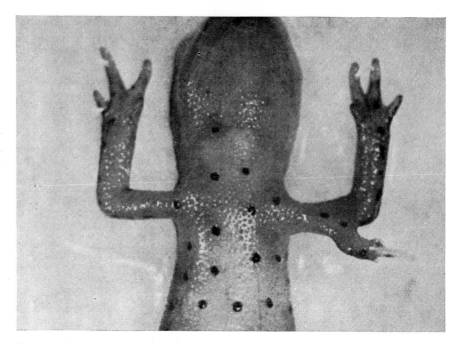

Figure 9.16
The role of nerves in regeneration. In amphibians, nerves are necessary for limb regeneration, and can even induce the formation of an extra forelimb. In this salamander the limb nerve was moved from its normal position to a new site on the surface of the upper arm. [Singer, 1958.]

patterns, slow fibers by small, tonically active axons (see p. 244). During embryogenesis, both muscle types start as slowly contracting fibers that gradually increase in contraction velocity until innervation, following which one population becomes slower again as the other continues to increase in velocity. If innervation is delayed, development of both types follows a nearly parallel course (Fig. 9.17). These observations suggest that the **nerve may be playing a dominant role** in determining the properties of the muscle fibers it innervates. This suspicion is strengthened by the finding that if the nerves that go to both a "fast" muscle and a "slow" muscle (such as the flexor hallucis longus and the soleus of mammals) are cut and crossed, so that the nerves regrow into a muscle of the opposite type, the muscles change properties. The "fast" muscle acquires most of the characteristics of the "slow" muscle, and vice versa, both in enzyme levels, molecular species of myosin ATPase, and contraction speed. The nerves do in fact appear to be governing the properties of the muscle.

Clearly, there are **two categories of explanation** for this phenomenon. It may be that each type of nerve releases particular trophic substances that, either independent of impulse activity or in proportion to it, induce muscle proper-

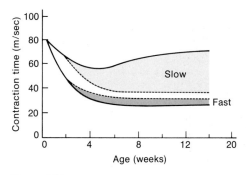

Figure 9.17
Time course of differentiation of fast and slow muscles of kittens (solid lines). If the muscles are kept inactive, the contraction speed differs less during development, tending toward that shown by the broken lines. If the nerves to the two muscle types are crossed, even in adult cats, their contraction velocities change in a way appropriate to the new nerve, the slow muscles becoming faster, the fast muscles slower. [Buller, et al., 1960.]

ties appropriate to that nerve. Alternatively, the overall amount of spike or contractile activity or pattern of activity may determine muscle properties, independent of the type of nerve, or even the presence of a nerve. This is an important but difficult question to resolve, since it is almost impossible to separate presynaptic impulse traffic and possible neurosecretion from postsynaptic activity.

It is clear that trophic substances by themselves are not an adequate explanation, since changes in amount of muscle use, with no change in innervation, can dramatically alter muscle characteristics. For example, prolonged exercise or an increase in tonic activity forced on a "fast" muscle by denervating synergistic tonic muscles will cause changes in contraction velocity, ATPase activity, and concentration of oxidative enzymes, making the muscle much more like typical "slow" muscles. Nevertheless, the

changes are not quite as great as when the muscle is actually innervated by a nerve of the opposite type. Moreover, in frogs, in which "slow" muscle fibers do not produce action potentials, denervation "induces" the property of impulse generation, which remains if the muscle is reinnervated by a "fast" nerve. Reinnervation with a "slow" nerve, however, leads to displacement of a "fast" nerve and the return of electrical inexcitability. Thus there is evidence that the properties of muscle fibers can be **governed both by their level of activity and by trophic substances** released by the nerve that innervates them.

Similar questions about the relative role of trophic substances and activity have centered around the effects of denervation. When the nerve to a muscle is cut, the muscle gradually atrophies, and there are concomitant changes in enzyme activities, myoglobin content, membrane electrical properties, and contraction velocity. When the nerve reinnervates the muscle, these changes are reversed. There are other effects of denervation that happen much more quickly. For example, in a mammalian "fast" muscle, within about 12–14 hours after denervation, miniature e.p.p.'s disappear, preceded in some fibers by an increase in miniature e.p.p. frequency. The fibers depolarize by 5–10 mV soon after the miniature e.p.p.'s disappear. After about 24 hours, there is the beginning of "denervation hypersensitivity," a phenomenon in which sensitivity to ACh develops over the whole surface of each muscle fiber. After 2–3 days, the first signs of changing membrane electrical properties appear.

Denervation hypersensitivity to ACh is one of the most studied results of denervation, and has served as a focus

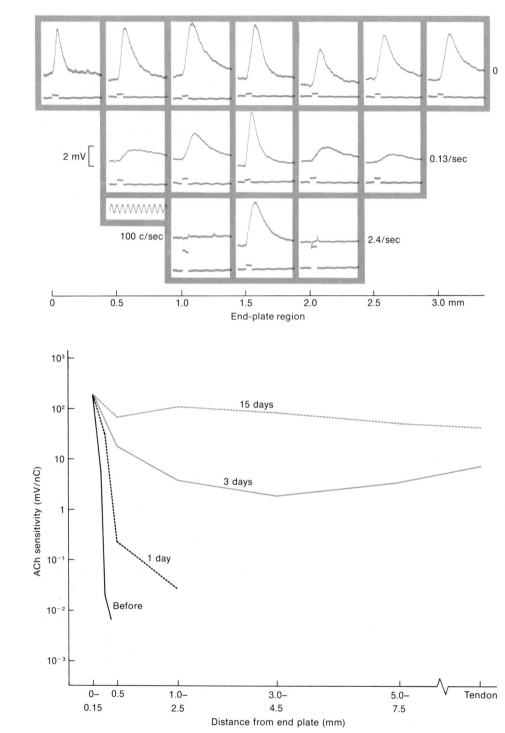

A

2 mV

100 c/sec

0

0.13/sec

2.4/sec

0   0.5   1.0   1.5   2.0   2.5   3.0 mm

End-plate region

B

10^3

10^2

10^1

1

10^{-1}

10^{-2}

10^{-3}

ACh sensitivity (mV/nC)

15 days

3 days

1 day

Before

0–   1.0–   3.0–   5.0–
0.15   2.5   4.5   7.5   Tendon
0.5

Distance from end plate (mm)

Figure 9.18 (*facing page*)
Development of acetylcholine hypersensitivity. **A.** ACh responses in three different fibers in a cat tenuissimus muscle poisoned with botulinum poison 3 weeks before. ACh is electrophoretically ejected onto the muscle membrane at different points from a microelectrode filled with a concentrated solution of the substance. Where there are receptors on the membrane, depolarization results, with an amplitude and time course that depend on the sensitivity of the membrane—that is, the density of receptors and their coupling to membrane conductance channels. The lower records are typical of innervated fibers, with no responses except at the end-plate. The middle records are from a fiber in which sensitivity was more widely spread, with slower responses to either side of the end-plate. The upper records are from a fiber that still showed complete denervation sensitivity, with sensitivity at all points essentially as great as at the end-plate. Miniature e.p.p. frequency is shown to the right, reflecting the fact that the block was partly reversed in the middle records and essentially gone in the lower fiber. [Thesleff, 1960.] **B.** Mean ACh sensitivity of surface fibers of a mammalian extensor muscle before and at different times after denervation. All distances are with respect to the recognizable end-plate. Sensitivity is plotted in mV depolarization per nanocoulomb of positive current from a micropipette filled with ACh. [Albuquerque and McIsaac, 1970.]

for arguments in the debate over mechanisms of neural influence—that is, trophic substance versus activity. In mammalian muscle the entire surface rapidly becomes nearly as sensitive as the end plate normally is (Fig. 9.18); in frog muscle the extrajunctional sensitivity is less, and spreads more slowly. (Interestingly, a comparable spread of ACh sensitivity has been shown following denervation of the cholinergic synapses of preganglionic parasympathetic fibers onto cells of the heart septum in frogs, as is illustrated in Fig. 9.19.) The receptors involved are not simply distributed from the old end-plate site, since an uninnervated fragment of muscle, severed from the rest of the fiber, develops denervation sensitivity. Apparently the receptors are either newly synthesized or else normally present but unmasked by the effects of denervation. The observation that cyclohexamide and other substances that block protein synthesis prevent denervation sensitivity without affecting ACh sensitivity at the old end-plate argues strongly that the receptors are newly synthesized. Curiously, as it becomes more sensitive to

ACh, the muscle membrane (in mammals, but not in birds or amphibians) also develops a tetrodotoxin-resistant spike. Whether there is a causal relationship is not known.

When a denervated muscle is reinnervated, the sensitivity changes are reversed, and sensitivity eventually becomes restricted again to the site of the new nerve termination. This restriction of sensitivity appears to be a recapitulation of what happens during normal development, in which (as has been inferred from studies of newly innervated fibers in rats, and directly demonstrated in embryonic tissue cultures) uninnervated fetal muscle remains uniformly sensitive over its surface until initial innervation is effected, after which sensitivity is restricted to the site of the ending. This restriction of sensitivity during development is accompanied by a concentration of cholinesterase at the end-plate. Moreover, concurrent with the restriction of sensitivity is a loss of the ability to be innervated by another nerve. The correlation is reinforced by the observation that if an innervated muscle fiber is damaged, ACh sensitivity

360

Chapter 9
Development and Specification of
Connections in the Nervous System

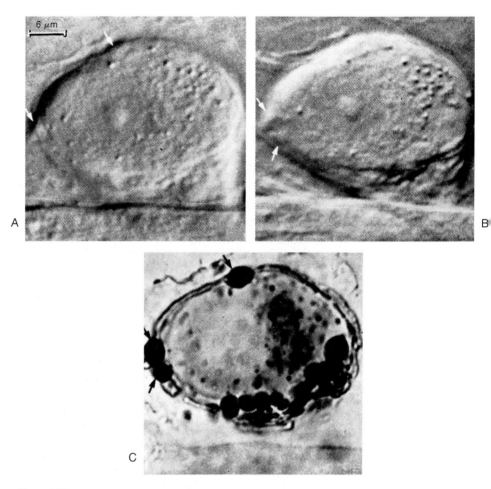

Figure 9.19
A and **B.** Photographs, using Nomarski optics, of living parasympathetic ganglion cells in the heart septum of a frog, with arrows pointing to recognizable synaptic boutons, confirmed by zinc staining (**C**). This reveals other boutons near the axon hillock. ACh sensitivity is high near the boutons, low elsewhere. Following denervation, the entire surface becomes approximately as sensitive as the original bouton areas. [McMahan and Kuffler, 1971.]

develops in the membrane at the site of damage, and a second nerve can successfully innervate the fiber at that point. Denervation sensitivity has also been shown in autonomic ganglia, and may be characteristic of most synapses, even noncholinergic ones.

The strong influence of innervation on postsynaptic cells has led to the proposal

that some **chemical trophic factor** is released by the presynaptic terminal. An obvious candidate is ACh itself. Indeed, botulinum toxin, which blocks the release of ACh, induces denervation hypersensitivity (see Fig. 9.18). (The end-plates are apparently irreversibly blocked; function is restored by formation of new junctions by axonal sprouting from the old nerve or by the ingrowth of new nerve fibers.) Tetanus toxin apparently acts in much the same way. There is other evidence, however, that ACh is not the factor, because denervated or embryonic muscle fibers, cultured in a medium rich in ACh, nevertheless develop and retain hypersensitivity. On reinnervation, the restriction of sensitivity toward the new endplate is reported to begin before there is any sign of ACh release.

On the other hand, serious doubt has arisen recently whether *any* trophic substance is necessary. A cuff of local anesthetic placed around a nerve in a rat leg—thus blocking **impulse traffic** for many days—has been shown to result in denervation sensitivity, as if the nerve had been cut (Fig. 9.20). Normal miniature e.p.p.'s are seen. Local anesthetics, however, also block axoplasmic transport. TTX, which blocks conduction, does not affect transport, and causes partial development of extrajunctional sensitivity. Perhaps the continued presence of miniature e.p.p.'s partially inhibits the development of extrajunctional receptors, but there is other evidence that miniature e.p.p.'s, by themselves, are not sufficient to deter denervation sensitivity. Such hypersensitivity is seen in the muscles of mice with congenital motor end-plate disease, in which motor nerves release miniature e.p.p.'s at normal frequency, but produce no e.p.p.'s.

There is, in addition, however, convincing evidence that direct muscle stimulation can mimic the effects of innervation. In an elegant series of experiments, Lømo and Rosenthal and their colleagues have shown that denervated rat skeletal muscle can be kept from developing denervation hypersensitivity if it is regularly stimulated directly, and that stimulation of a sensitive muscle will cause rapid restriction of sensitivity to the site of the old end-plate (Fig. 9.20,C)! This strongly suggests that inactivity, by itself, might be responsible for the buildup of extrajunctional receptor or for the failure to degrade it. Moreover, by appropriate direct stimulation patterns, slow muscle can be made fast, and fast slow.

Other arguments have been offered for the existence of nerve-released trophic substances. For example, many workers have observed that the early effects of denervation, including muscle fiber depolarization, the disappearance of miniature e.p.p.'s, and the development of denervation hypersensitivity, manifest themselves at different delays proportional to the length of severed axon left attached to the muscle. The longer the attached nerve, the more delayed are these changes, the delay being approximately 2 mm/hour (see Fig. 9.21). But the longer the nerve stump, the longer it takes to begin to degenerate; and it has recently been shown that pieces of cut nerve, or even thread, on the surface of a normally innervated mammalian muscle somehow cause the temporary development of significant ACh sensitivity at that location. Thus some degradation product, or cells associated with inflammation, may be inducing the sensitivity. Furthermore, it has been demonstrated that colchicine, which is known to block **axoplasmic transport** but which does not block spike activity or transmitter release, can induce the same postsynaptic changes as denervation. A colchicine-impregnated silastic cuff attached around a nerve causes muscle depolarization, spread of ACh sensitivity, and development of TTX-resistant action potentials, even though the muscle maintains apparently normal contractile activity. Thus in the presence

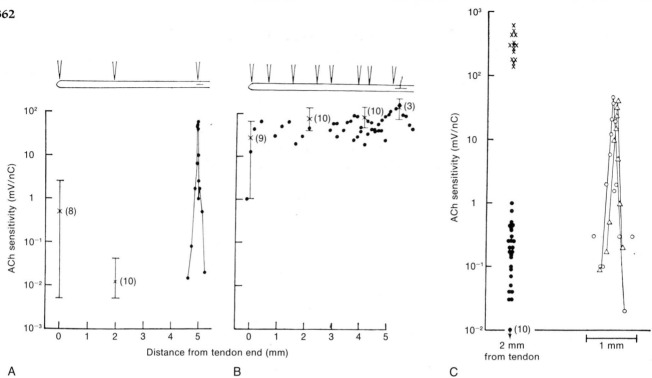

Figure 9.20

Effects of use and disuse in governing ACh sensitivity of mammalian muscle. **A.** Normal distribution of sensitivity along an innervated soleus muscle. Note the sharp localization of sensitivity within the fraction of a millimeter on either side of the nerve terminal and the slight increase in sensitivity near the tendon end. (See also Fig. 9.19.) **B.** After 7 days of blockage of transmission in the nerve by local anesthetic, the muscle surface is uniformly sensitive, as if it had been denervated. The anesthetic does not affect miniature e.p.p. release or block the production of action potentials and e.p.p.'s by stimulation of the nerve distal to the block. **C.** In fibers denervated on the seventh day, at which time the ACh sensitivity is high everywhere, sensitivity can be sharply restricted again, as shown in this figure, by seven days of direct muscle stimulation. The open circles and triangles show the sensitivity distribution along two fibers near their loci of high sensitivity in such a stimulated muscle. The closed circles show the sensitivity of a number of such fibers 2 mm from the tendon end, compared with the high sensitivity shown by fibers in other muscles (indicated by X's) that had not been stimulated. Apparently muscle excitation and/or contractile activity can restrict sensitivity to the end-plate as successfully as reinnervation. [Lømo and Rosenthal, 1972.]

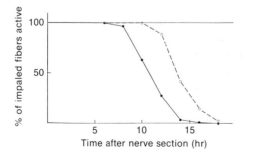

Figure 9.21

Time course of disappearance of miniature e.p.p.'s in rat diaphragm fibers when the phrenic nerve was sectioned at its entry into the diaphragm (closed circles) and in the neck, 4 cm from the diaphragm (open circles). [Miledi and Slater, 1970.]

of normal release of ACh and normal muscle spike and contractile activity, something is missing. This would appear to be strong evidence for the existence of some transported substance (other than ACh). Its persuasiveness has been reduced, however, by the demonstration that systemically applied colchicine, at concentrations that do not effectively block transport, can have a direct affect on muscle, inducing denervation hypersensitivity.

Thus muscle activity and patterns of activity are demonstrably important in governing a whole spectrum of muscle properties, whereas evidence for a motor-nerve-released trophic substance is much less compelling. Nevertheless, trophic substances almost certainly exist in other systems, and clearly there are ways of inducing extrajunctional ACh sensitivity without interrupting normal activity. The interaction of these various factors is still not understood. A powerful new tool has recently been introduced, however, that may help answer some of these questions. The venom of a Formosan cobra has been found to contain a mixture of neuromuscular blocking agents. Two major protein components have been described; one of them blocks transmitter release by motor nerves, and the other blocks the action of ACh postsynaptically. One of the components, **α-bungarotoxin,** is a basic polypeptide (molecular weight about 8000) that binds irreversibly and with great specificity to the ACh receptor. This permits the extraction and characterization of the receptor protein, study of how it spreads and how it is synthesized in denervated muscle, and localization of similar receptor molecules throughout the nervous system.

C. Axonal Transport of Substances

If trophic substances other than the normal transmitter are being released, one might hope to detect them in perfusates or in the postsynaptic cell. It is well known that at sympathetic terminals release of norepinephrine is coupled with release of ATP and certain proteins characteristic of the presynaptic vesicles. Reports have now appeared of protein release at neuromuscular junctions as well. It has not been established whether in either case the released proteins are capable of exerting any effect on the postsynaptic cell.

Although some transmitter synthesis and mitochondrial protein synthesis is apparently accomplished at the terminal, most substances necessary for axonal maintenance or interaction with other cells must be synthesized in the presynaptic cell soma and transported to the terminal. Since with the work of Cajal, many investigators have studied the transport of substances along axons. Weiss showed that cytoplasmic material accumulated proximal to a ligature. The growth rate of axons implies the centrifugal movement of axonal material at a rate of 1–2 mm/day. Recent experiments have employed labeled amino acids or glycoprotein precursors (applied close to or inside neuron cell bodies) to measure the movement of label along the axon (Fig. 9.22). These studies show that the label is incorporated into protein or glycoprotein precursors (applied close to and can then move either slowly or quickly toward the axon terminal. A large fraction of it appears to be bound to membranes. In most preparations there is a **slow component** moving at only 1–2 mm/day, and a **faster compo-**

Chapter 9
Development and Specification of
Connections in the Nervous System

ate to those muscles, or are they already "specified," and simply grow out earlier than other motor neuron axons? Since both cells are differentiating in parallel, the question is difficult to answer.

Most of our discoveries about the precision of neural connections have therefore come from experimental manipulation of the system, usually in adult or late larval systems, in which one can obtain successful regeneration of axons and the formation of new synapses in lower vertebrates. Because of their accessibility and the ease of direct behavioral testing or regeneration, the systems on which most of this research has been done are the neuromuscular and visual pathways.

B. Specificity of Innervation of Musculature

We have described above the way in which a limb is initially innervated. An excess of motor fibers grows out; the first axons to reach the limb innervate the most proximal muscles, and the next ones innervate the muscles just further out the leg, and so on, each successful innervation somehow excluding further connections with that muscle. It is not surprising that if the nerve to one of these muscles is later cut, it will regrow into that muscle and reestablish functional contacts. It is somewhat surprising, however, that, in lower vertebrates, if the whole nerve growing into the limb

Box 9.1 Tissue Culture Techniques

Many laboratories have turned, recently, to cell and tissue culture techniques as an approach to answering the questions posed in this chapter. In-vitro studies of neural growth are not new. Harrison and several other workers had remarkable success with such techniques early in this century. Within the past decade, commercially available culture media, antibiotics, and other recent improvements have made such studies much easier. Both tissue explants and dissociated nerve cells, and even certain strains of cloned neuroblastoma cells, can now be cultured routinely, and the processes of growth, migration, selective adhesion, and contact inhibition plus the development of electrical activity can all be studied under direct observation. It is even possible to get synapse formation in vitro, opening the possibility of understanding much better the mechanisms governing selectivity of synapse formation and the nature of trophic influences between cells.

Also of great interest are experiments showing the reaggregation of dissociated embryonic cells into cell masses whose organization is distinctly like that of the same cells in vivo. For example, dissociated embryonic hippocampal cells have been shown to aggregate in a curved layer with different cell types aligned with respect to each other as they would be in the brain, with axons emerging and forming nerve fiber bundles in the normal way. In comparable cerebellar cell cultures, genetically abnormal cell populations show disruptions of organization in reaggregated cultures that closely resemble the abnormalities seen in vivo. Although cell and tissue cultures often behave quite abnormally, probably because of the failure to reproduce in-vivo conditions, it is clear that these techniques provide immensely powerful tools for the study of cellular differentiation and interaction.

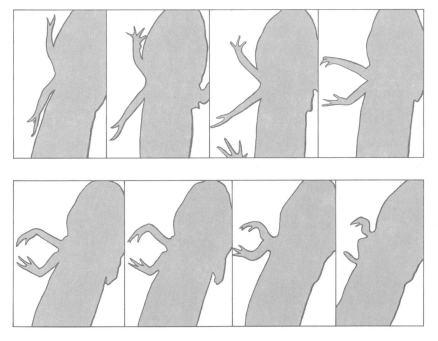

Figure 9.23
Evidence of myotypic specificity. Several frames from a motion picture showing mirror-image movement of two limbs of opposite polarity at the left shoulder of a salamander. [Weiss, 1950.]

is cut, the nerves all appear to grow to their appropriate muscles, for normal coordinated movement returns as the nerves regenerate neuromuscular junctions.

In the larval or embryonic forms of lower vertebrates, it is possible to explore the situation further by moving a limb bud from one point on the body to another or by **implanting a supernumerary limb** or limb bud. In the latter case, the extra limb, if it is near the original limb, is innervated by motor neurons from the same part of the spinal cord, and shows movements that exactly parallel the movements of the original limb. This "homologous response" was shown by Weiss not to be the result of

learning or functionally adaptive innervation, for he found that if an extra limb were grafted backwards onto a salamander, it developed normal movements that were always in opposition to the movements of the normal limb (Fig. 9.23). The flexors of both worked simultaneously, and the extensors simultaneously, but the legs were so oriented that when one was pushing the animal forward, the other pushed it backwards. If all four limbs are reversed, the adult walks backwards when stimulated in a way that would normally cause it to walk forward. Neither learning nor proprioceptive feedback has been shown to correct the maladaptive coordination. If sensory input from the extra limb is cut,

368

Chapter 9
Development and Specification of
Connections in the Nervous System

it has no effect on the movement. Apparently, in these lower vertebrates, motor coordination is relatively independent of sensory feedback and cannot be grossly modified by experience.

It is clear that not all motor neurons are capable of establishing a homologous response. After the earliest stages of development, if the extra limb is implanted too far rostrally or caudally from the spinal segments that normally innervate the limb, the axons that innervate it are not capable of working in synchrony with those of the normal limb or, indeed, in any coordinated fashion. If a hind limb is implanted in place of a forelimb, it is innervated and does show coordinated movement, but now the movement is characteristic of the forelimb. Thus it appears that the spinal cord is functionally specified at an early stage. Soon after the closure of the neural tube in newts, each part of the spinal cord is already specified. Nevertheless, this **specificity may still develop only after the formation of connections** with peripheral musculature, which happens at approximately the same time.

In the adult animal, the ability to regenerate normal movement in a denervated limb or homologous response in a grafted supernumerary limb is characteristic only of fish and urodele amphibians. Anurans and higher vertebrates are capable of this in larval or embryonic stages, but not later in life. In birds and mammals, such capability is known only in early embryonic stages. Regeneration can still occur, even in adult mammals, but it is not specific. Some nerves that originally innervated flexors innervate extensors, and vice versa, so that movement is not coordinated normally. As we have already seen, a "fast" nerve can be induced to

innervate a "slow" muscle, and vice versa. In both cases, the nerve appears to affect the properties of the muscle.

Weiss first concluded that the regrowing nerves randomly innervate whatever muscle they reach, and that each nerve transmits impulses in patterns specific to all muscles, each muscle being capable of reacting specifically to the impulse pattern appropriate to it. This hypothesis was soon eliminated by electrophysiological demonstration that action potentials in each nerve are essentially identical in form and firing pattern and that a muscle is activated whenever there are action potentials in its nerve. Myotypic specificity was then proposed by Sperry to result from a readjustment of motor neuron connections in the spinal cord in a way to make the function of that motor neuron appropriate to the muscle it had innervated. Indeed, some form of central rearrangement of connections would seem necessary if one accepts the experimental findings of Weiss that (a) muscles innervated by crossed nerves from antagonistic muscles contract as they normally would, and (b) a whole supernumerary limb, innervated by a single distal muscle nerve (the inferior brachial) from another leg, can show a complete coordinated movement homologous to that of the original limb.

Several more recent experiments, however, have suggested that in most, if not all, cases, homologous response after regeneration is a result of **specific regrowth of each nerve to the muscle** of the type it would normally innervate, rather than a rearrangement of central synapses. For example, if the nerves to the fish extraocular muscles are crossed and if care is taken to ensure that the wrong nerve does in fact innervate a given muscle and not regrow to its correct muscle, the muscle will move abnormally—in a way appropriate to the nerve's original function. However, the nerves tend to regrow via circuitous routes to their original muscles, and the original nerve is somehow able to displace the "incorrect" nerve from its muscle. A comparable displacement phenomenon has been reported following innervation of fish fin muscles. The fin adductor and retractor nerves tend to regrow to their appropriate muscles, but if induced to innervate antagonistic muscles they will do so, and backward movement persists indefinitely. If the nerves are then sectioned near the spinal cord, they return to their original muscles to re-establish normal coordinated movement.

Moreover, in a recent careful repeat of Weiss's classic experiments on nerve crossing in amphibian limbs, it has been shown that the results can best be explained by specific regrowth of axons to their appropriate muscles. The crossed nerve appears to grow as directed into the antagonistic muscle, but movements are "normal." Stimulation near the muscles shows that the implanted nerves are successfully innervating them. However, stimulation of the nerves near the spinal cord, central to the point of crossing over the antagonist nerve, shows that the effective fibers are in fact going to the muscles they originally innervated. They are crossing back at some point near or central to the point at which the nerves physically cross. This was confirmed by sectioning one or the other of the nerves central to the cross and looking at responses and degeneration patterns. The second experiment, producing homologous movement in a supernumerary limb innervated by the inferior brachial nerve, is also unconvincing as evidence

for myotypic specification, since it has been found that this nerve normally innervates both flexors and extensors, and could innervate most of the transplant without having to form synapses with nonhomologous muscles.

There is other evidence that makes it difficult to accept the conclusion that there is peripheral "modulation" of central synapse formation, even during initial development. The nerves innervating a frog's leg, for example, are already subdivided into all major branches in a limb bud stage before there is any differentiated muscle, or even myoblasts. It seems unlikely that such nerve branching, quite uniform in all individuals, could be the result of random outgrowth of nerve fibers that innervate the first muscle they come to and then have their central connections governed accordingly. The opposite might seem more reasonable, that the nerves specify the type of muscle to be differentiated. Yet this too can be shown not to be so. Under special conditions, a limb will differentiate almost normally even when the nerve is prevented from growing into it. Apparently the development both of spinal cell populations and peripheral tissues is genetically programmed in a fixed temporospatial pattern. Even in very early stages, when there is only undifferentiated mesenchyme in the limb, there are cues for governing nerve growth. Further evidence for this conclusion comes from experiments in which spinal cord segments were interchanged at different developmental stages. It was possible, in these animals (again amphibians), to cause the motor axons that innervate the limb to grow right past the uninnervated proximal musculature and synapse with the distal muscles of the limb.

370

Chapter 9
Development and Specification of
Connections in the Nervous System

An example of **specific regrowth in the visceral motor system** of mammals is seen in the connection of preganglionic sympathetic fibers to their original ganglia. The cat superior cervical ganglion contains cells controlling, among other things, the smooth muscle of the iris and of peripheral blood vessels of the head. This ganglion is innervated by preganglionic fibers from the first through seventh thoracic segments (T_1–T_7). Stimulation of fibers from T_1 causes dilation of the pupil but little if any vasoconstriction, whereas stimulation of T_4–T_7 causes strong vasoconstriction of ear vessels but no change in pupil diameter. It was known from very early work by Langley and Cannon that the cervical ganglion could be experimentally innervated by inappropriate nerves, such as the vagus or phrenic, which then caused pupillary or circulatory responses to the wrong stimuli. With the milder manipulation of simply sectioning T_1–T_3, it can be shown that the more posterior inputs (T_4–T_7) quickly take control (probably via collaterals that sprout to synapse with the denervated ganglion cells). Stimulation of T_4–T_7 then causes pupillary dilation. However, as the fibers from T_1–T_3 regenerate into the ganglion, they gradually displace the abnormal T_4–T_7 junctions, and again gain control of iris function. Similar specificity of regeneration has been reported for the preganglionic autonomic fibers innervating the chick ciliary ganglion. The ganglion is composed of two cell populations, one controlling smooth muscle in the choroid, the other the iris and ciliary body. When the axons innervating the ganglion are cut, they regenerate to restore the correct connections.

In the case of motor connections, therefore, most evidence suggests that regeneration of coordinated responses is due to specific regrowth of nerves to the muscles they originally innervated or to a homologous muscle in another limb. This implies some sort of **chemical identification or recognition**, presumably specified at or before the time of initial innervation, during embryogenesis.

C. "Repressed" Synapses

In the nerve-cross experiments described above, large numbers of inappropriate nerves appeared to be innervating muscles without causing contraction. Similarly, in the cat superior cervical ganglion, the appropriate preganglionic fibers appeared to be able to "displace" collaterals of the inappropriate preganglionic neurons. This phenomenon of inactivation or displacement has been analyzed further by Mark and his colleagues, using a fish ocular muscle preparation; they found that nerve-crossing could temporarily cause reversed eye-movement reflexes, but that in most cases collaterals of the nerves regrow to their appropriate muscles and "displace" the inappropriate nerves. It is reported that the "displaced" nerve fibers often remain in place, continue to show electrical activity (which, however, no longer drives the muscle fibers), and has normal-looking terminals and postsynaptic specializations. Everything looks normal, but they are somehow repressed.

Another experiment, on axolotl limb innervation, adds considerable support to the view that synapses can be inactivated without morphological displacement. Axolotl hind limbs are innervated predominantly by three spinal nerves (16, 17, and 18), each of which contains the fibers driving approximately one-third of the limb musculature. The muscle field driven by nerve 16 is bordered on one side by muscles driven by nerve 15; on the other, by the field of nerve 17. If nerve 16 was cut, its muscle field was temporarily denervated, but nerves 15 and 17 gradually, over a period of 2–3 weeks, expanded their fields to innervate the denervated muscles. After about 30 days, nerve 16 began to regrow into its original field, and the adjacent nerves were gradually "displaced." However, even several months later, if the regenerated nerve 16 was again sectioned, the paralysis due to denervation of its field lasted only 2 days. After that, the adjacent "displaced" nerves suddenly were able to drive the whole field, as if they had completely reinnervated it. The rapidity of reappearance of functional synapses from these adjacent nerves implies that they had remained in place when nerve 16 reasserted control, and that their **repression could be reversed** in only 2–3 days. Figure 9.24 schematically outlines these events. Interestingly, the same spread of influence of nerves 15 and 17 occurs if axoplasmic transport is stopped in nerve 16 by local block, despite the continuance of normal impulse activity in nerve 16. It seems possible that repression of synapses may be a common and important mechanism of accomplishing functional specificity of regeneration, at least in neuromuscular connections.

D. Specificity of Regrowth of Sensory Connections

Skin-Sensory Connections

The same problems of peripheral modulation versus specific regrowth have arisen in studies of sensory connections. The evidence for **peripheral modulation** is stronger in these, but still far from conclusive. One of the most simple preparations, the sensory innervation of skin, has been central to the debate. The skin is innervated by a large variety of sensory fibers, each responding to a particular type of stimulus: touch, pressure, changes in temperature, and pain. The location of a stimulus is signaled by the position of the sensory fibers that it excites, and reflex reactions to the stimulus (or central interpretations of it) are governed by the synaptic interconnections between sensory afferents, motor neurons, and their associated interneurons. In a frog, even one in which the brain has been pithed, a mild noxious stimulus to the skin at a given point causes a reflex movement of the nearest limb to brush off the stimulus. (This reflex first appears at metamorphosis.) Stimulation of the skin of the belly leads to a wiping of that area; stimulation of the skin of the back causes wiping there. What happens, then, if the skin of the belly is interchanged with skin from the back (Fig. 9.25)? If this is done in the adult or in the late tadpole stages, the skin is reinnervated by the sensory nerves that were originally present, and the frog responds as if nothing had changed. But if the reversal is done in early tadpole stages, the metamorphosed frog shows reversed reflexes; when the back skin on the belly is stimulated, the frog scratches its back, and vice versa. There is no evidence for circuitous innervation by efferents that would normally have innervated that type of skin, hence it has been concluded that, instead, the skin somehow causes the rearrangement or specification of spinal synapses so that the nerve fibers become

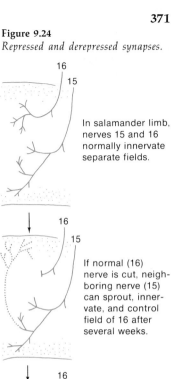

Figure 9.24
Repressed and derepressed synapses.

16
15

In salamander limb, nerves 15 and 16 normally innervate separate fields.

16
15

If normal (16) nerve is cut, neighboring nerve (15) can sprout, innervate, and control field of 16 after several weeks.

16
15

When normal nerve regenerates, junctions of 15 in field of 16 are repressed (i.e. cannot drive the muscle) but appear histologically intact.

16
15

If 16 is cut again or merely blocked, the junctions of 15 are "derepressed" and can drive the muscle within 3 days.

Schematic diagrams showing the sprouting of fibers of one nerve to innervate nearby denervated muscle, followed by the postulated "repression" of these new synapses when the original nerve regenerates and their "derepression" when that nerve is again sectioned.

Chapter 9
Development and Specification of
Connections in the Nervous System

ple concentric excitatory receptive field, the center of which is either excitatory or inhibitory, with the surround having the opposite effect (Fig. 9.26). Few ganglion cells (p. 250) depend on rate or direction of movement of a border, or on previous response pattern of that cell. (Certain mammals—e.g., rabbits and squirrels—have more of the complicated, movement-sensitive, ganglion cells, for reasons that are not yet clear.) Most optic nerve axons project directly to several cells of a diencephalic center, the lateral geniculate nucleus, in which each cell shows response patterns similar to those of the retinal ganglion cells. Each lateral geniculate cell in turn projects to many cells in the visual cortex, and it is at the cortical level that most cells first gain specificity comparable to that of the retinal ganglion cells of anuran amphibians.

Frog retinal ganglion cells, by contrast, show highly specific stimulus requirements. Five types have been described, one of which projects to the lateral geniculate and four of which go to the tectum. The geniculate-destined optic nerve fibers are simple "on"-responding units, myelinated and most sensitive to blue light. The four tectum-destined types of units have been outlined in Chapter 7 (p. 251). The vast majority are thin, unmyelinated, and probably appear late in development. These types are intermixed at all points in the retina, but they project to four different layers in the tectum, each of which maps the retina, with each map in register, so that an electrode penetrating from the surface downward records from elements (probably fiber terminals) that respond to stimulation of the same part of the retina, but to different features of the same stimulus, in the order 1 to 4 (Fig. 9.27).

In most vertebrates, optic fibers from each eye cross (decussate) entirely to the opposite side of the brain (see p. 475 for exceptions). In certain mammals, however, especially arboreal or fast-moving species in which rapid, precise, visual estimation of distance is required, and in which there is at least partial overlap of visual fields, about 12 to 50% of the optic fibers do not decussate. In primates, half of the optic nerve fibers decussate and half do not. The right temporal and left nasal half-retinas project to the left. The lateral geniculate is composed of several layers of cells, alternately receiving input from the contralateral and ipsilateral eyes, with the layers (each mapping a half-retina) again being in register. Many lateral geniculate cells appear to be excited by one eye and inhibited by the other, but only at the level of the primary visual cortex are cells excited by inputs from both eyes. More will be said about this below.

Regeneration of function. When the optic nerve of a fish or amphibian is sectioned, normal vision is regenerated within a few weeks, as judged by visual reflexes. This implies (a) that each fiber finds its way back to the same population of cells it innervated originally, (b) that regrowth is not specific but the tectum can be respecified by the regenerating optic nerve fibers, so that wherever they terminate, the cells they innervate will make new synaptic connections appropriate to their new afferents, or (c) that regrowth is random and the tectum is gradually "rewired" by experience, so that wherever incoming fibers synapse, the output becomes functionally adaptive. The last hypothesis is clearly excluded by the dramatic finding that if one rotates the eye by

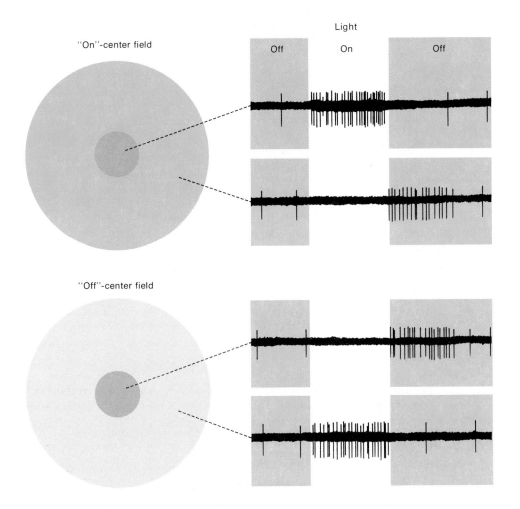

Figure 9.26

Response characteristics of retinal cells of a cat or a primate. Receptive fields are concentrically organized, with a small central region in which light stimulation causes either excitation or inhibition, and a larger surround area that has the opposite effect. The size of the receptive field may be as much as 1–2 mm, reflecting the great convergence on a single ganglion cell of influences from large numbers of other retinal neurons. The center has slightly more influence than the surround, so that diffuse light causes slight excitation of on-center cells and inhibition of off-center cells. A light covering the center-plus-one side of the receptive field causes a much larger effect than light covering the whole receptive field; thus it is apparent that cells even at this level can provide information about stimulus position and movement. Receptive fields and responses of lateral geniculate neurons are almost indistinguishable. [Hubel, 1963.]

376

Chapter 9
Development and Specification of
Connections in the Nervous System

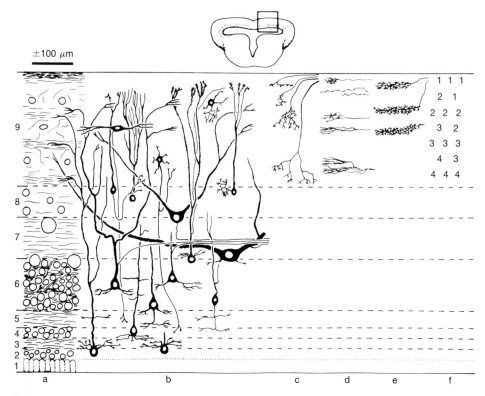

Figure 9.27
Diagrammatic cross-section of the frog tectum, showing the cell layers (*a*), Golgi-stained neurons (*b*), optic nerve terminations according to several authors (*c, d,* and *e*), and the depths at which Maturana and his colleagues recorded from their four types of units (*f*). Note that the recordings correspond best to the sites of optic nerve terminations. One of the types of inputs, probably 1, is from a population of optic ganglion cells that is not present in tadpoles but develops at metamorphosis. [Jacobson, 1970.]

180° at the same time that the nerve is cut, upside-down and reversed vision are regenerated (Fig. 9.28). There is no correction with experience. The same behavior is seen if the eye is rotated without cutting the nerve.

Histological and electrophysiological techniques have established that the optic nerve fibers do, in fact, regrow to approximately their original locus in the tectum. Recordings of responses of tectal units in such regenerated systems are essentially the same as in the intact animal. There is a time during which the pattern is abnormal, but as the regeneration nears completion, the **original response pattern is restored,** even to the detail of having the usual layers of differently responding units in register.

If an electrode is advanced slowly through the optic nerve of a fish or amphibian, successively recorded units may have receptive fields anywhere on the retina. Histological studies confirm that

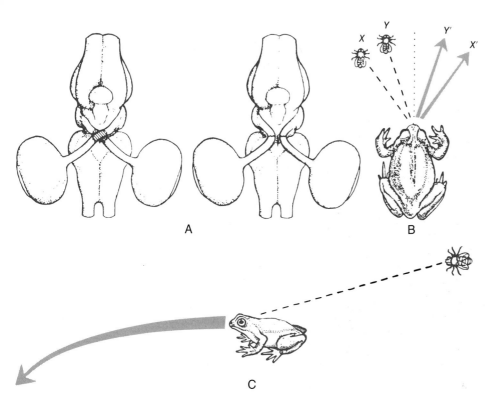

Figure 9.28
Behavioral evidence for specific reconnection of eye and tectum. **A.** Ventral view of frog brain, show-ing normal optic chiasm (left) and cross-united experimental animal (right), in which the optic nerve was cut and redirected into the ipsilateral tectum. **B.** After regeneration of the optic nerve, cross-united animals behave as if everything seen by one eye (*X,Y*) were actually being viewed by the other (*X',Y'*). **C.** Behavioral result of simple 180° rotation of the eyes, or 180° rotation coupled with crushing of the optic nerves. Following regeneration, vision is restored upside down and back-ward. There is no correction for this defect with time. [Sperry, 1951.]

the optic fibers are not ordered in any apparent way. Yet as the optic nerve approaches the contralateral tectum, it splits into two bundles, one of which carries information from the ventral part of the eye and grows over the dorsal-medial surface of the tectum, innervating cells in that part of the nucleus; the other carries information from the dorsal part of the eye and innervates the ventro-lateral part of the tectum (Fig. 9.29). The fibers somehow sort themselves out as they approach the tectum. Attardi and Sperry, experimenting with fish, showed that regenerating optic nerve fibers were able to overcome major obstacles or dis-placements to find their way to the appropriate part of the tectum. They found that if the two branches are cut and crossed, the regenerating fibers usu-ally manage to find their way back to the correct pathway near where the cut was made (i.e., where the optic nerve reaches the tectum). In some cases, however, the

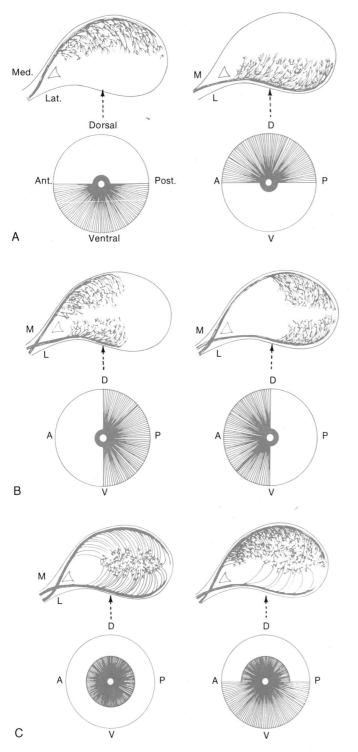

Figure 9.29
Pattern of optic nerve regeneration into the optic tectum of goldfish following nerve section and partial ablation of the retina, **(A)** leaving ventral and dorsal half-retinas, **(B)** leaving posterior and anterior half-retinas, and **(C)** removing varying amounts of peripheral retina. *M*, medial; *L*, lateral. [Attardi and Sperry, 1963.]

displaced bundles grew over the inappropriate part of the tectum and were then seen to grow entirely through that part of the tectum to reach the population of cells they originally innervated. Since the other bundle did the same, the two populations of fibers must have grown right past each other. This finding strongly suggests the **existence of some guiding factor** or factors that permit growing nerve fibers to recognize the correct pathway. In other experiments the system was tested even more severely by deflecting the cut optic nerves into the oculomotor and V nerve pathways. Regeneration took an entirely abnormal course, but the nerves finally found their way back to the tectum. In this case, however, each nerve had found its way back to the wrong tectum—the ipsilateral tectum. This experiment and others in which the nerves are cut at the optic chiasm and directed ipsilaterally instead of contralaterally show that each nerve can discriminate between different parts of the tectum, but not as successfully, if at all, between ipsilateral and contralateral tectum. Each uncrossed fiber simply grows to the homologous part of the wrong tectum, and backwards vision is "restored."

It is clear from this and many other observations that the **retinal ganglion cells are "specified"**; that is, those at one point in the retina differ somehow from others elsewhere, and are able to find their way to the particular part of the tectum that is appropriate for them. In amphibians and chicks this specification, or "polarization," has been shown to develop at approximately the time of the last cell division (last DNA synthesis) in the center of the retina and at approximately the time of initial axon outgrowth. Cells at the periphery continue to differentiate long after this. Interestingly, the specification of future connectivity is made **first in one axis** of the eye, later in the perpendicular axis. The retina normally projects to the tectum in such a way that cells at the anterior (nasal) edge of the retina terminate at the posterior end of the tectum; those at the posterior (temporal) edge, at the anterior end of the tectum. Fibers from the ventral part of the retina terminate in the dorsal region of the tectum; those from the dorsal retina, in the ventral tectum (Fig. 9.30). While the retinal ganglion cells are still dividing, rotation of the eye causes no change in these projections, even though all retinal cells are reversed in position. They are not yet "specified." Immediately after the last cell division of ganglion cells near the center of the retina (the first part to differentiate), reversal of the eye results in normal projection of dorsal retina to ventral tectum and ventral retina to dorsal tectum. The retina has not yet been specified in the dorsal-ventral axis. The anterior and posterior projections are now reversed, however, that is, the old anterior cells of the retina, now posterior cells, have already been specified, so that they grow to the posterior part of the tectum. Thus there is a short period (about 12 hours in *Xenopus*) when the retinal cells are specified in one plane and not in the other. After that time, they are entirely specified, and rotation of the eye leads to a totally reversed projection pattern. The two-phased specification suggests the

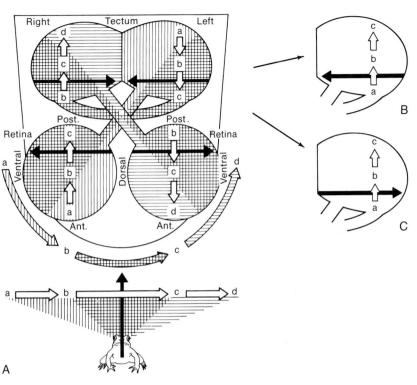

Figure 9.30
A. Normal visual projection pattern from a frog, showing the relationship between visual field and monocular and binocular retinal projections. Corresponding points in the field in which there is binocular overlap are connected by intertectal pathways. When one eye (e.g., the right) is rotated 180° before stage 29, this pattern still results. **B.** If the eye is rotated at stage 29, there is normal dorsal-ventral projection from the retina to the tectum, but reversed anterior-posterior projection, implying that the retina is already "specified" in this axis. **C.** After stage 29, eye rotation leads to a totally reversed tectal projection, indicating that the retina is polarized in both axes. [From Jacobson, 1970.]

existence of **two different gradients** or types of identifiable factors governing the growth and termination of optic nerve axons.

As in the case of specific regeneration of neuromuscular connections, the specificity of regeneration of optic nerve connections presupposes that the tectal cells have a built-in matching specificity, or polarity. This specificity may be de-

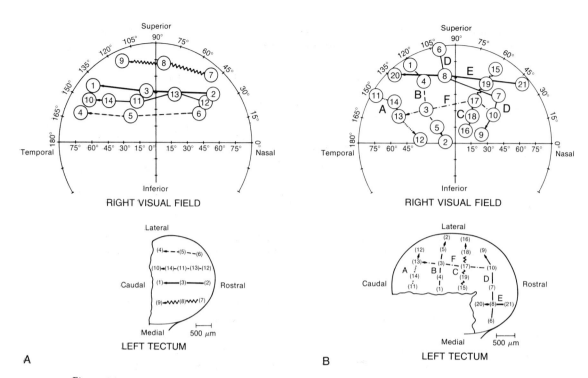

Figure 9.31

A. Retinotectal connections when the caudal half of the goldfish tectum is removed and the contralateral nerve crushed. The regenerating optic nerve fibers all end in the remaining half-tectum, with the usual projection pattern but compressed approximately twofold. **B.** When only one quadrant of the tectum is removed, the retinal projection can be compressed different amounts for different axes in adjacent parts of the retina. In this experiment, for example, the nasotemporal projection was normal for most of the retina but compressed nearly threefold in the very superior part of the visual field. The projection varies from normal to an approximately twofold compression in the inferior-superior visual axis. Both animals 160–170 days postoperative. [Yoon, 1971.]

veloped only after initial innervation by optic nerve fibers during embryogenesis. If the tectum is rotated before the optic nerve fibers reach it, the fibers simply end in their usual pattern, and normal vision results. If part of the tectum is rotated after initial innervation (a very difficult operation to do successfully), the regenerating fibers show a corresponding rotation in their projection.

Many experiments with fish, larval *Xenopus,* and tree frogs have documented the fact that once retinotectal connec-

tions have been established, they become specified in a point-to-point way. Thus several workers have shown that if half of the tectum is removed and the optic nerve cut, only the appropriate half-retina re-establishes connections with the remaining half-tectum; if half the retina is removed and the remaining optic nerve fibers are crushed, they regenerate to innervate only the appropriate half-tectum.

Nevertheless, equally convincing experiments, done with the same species

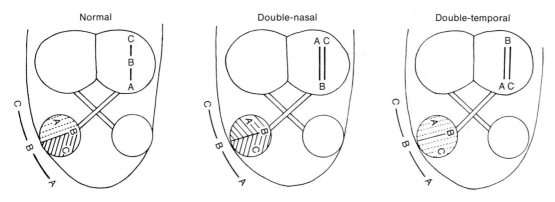

Figure 9.32
Retinotectal projection pattern in *Xenopus* with normal, double-nasal, and double-temporal retinas. In the compound retinas, the projection from each half-retina spreads over the whole contralateral tectum. [Gaze et al., 1963.]

and by the same and different investigators, give opposite results, suggesting that a great deal of plasticity survives. For example, in these cases, if the whole optic nerve is allowed to regenerate into only a half-tectum, the whole nerve innervates it, with the normal spatial projection pattern but twice the normal density of innervation. This **compression of the visual projection** occurs equally well in the anterior-posterior or dorsal-ventral axes (Fig. 9.31,A). Moreover, if any one quadrant of the tectum is removed, the projection to a given part of the tectum may be normal in one axis, compressed in the other (Fig. 9.31,B). Thus the retinal fibers that do not find their original half of the tectum innervate the inappropriate half; but they do so in a way that reproduces the normal visual projection—that is, the normal spatial sequence of fiber endings in each axis, spread over as much tectal tissue as exists in the corresponding axis. The retinal fibers behave as if they interpret the remaining tectum as being an intact, albeit small, tectum.

Moreover, if a "compound" eye is created at a developmental stage after retinal specification by removing half of the normal eye and substituting for it the opposite half of another eye (forming an eye composed of two nasal (anterior) half-retinas, or two temporal (posterior) half-retinas), the resulting eye innervates the tectum as shown in Figure 9.32. Both half-retinas would normally have innervated just half the tectum, but both the same half (on opposite sides). When the projections were mapped in these animals, it was found that both half-retinas had innervated the whole tectum in parallel, with each point on the tectum receiving fibers from homologous points in the two half-retinas (Fig. 9.32). Apparently in the absence of part of the retina, with its particular properties and capacity to preempt part of the tectum, the remaining retinal fibers can make use of all of the tectal tissue, the main restriction being that the existing retinal fibers always **maintain the normal topological relationships** with each other. It is as if each half-retina had reorganized itself to act as a whole, albeit small, retina. Perhaps in these experiments, as well as in those in which a whole retinal projection is compressed into only part

382

Chapter 9
Development and Specification of
Connections in the Nervous System

of a tectum, either the retina or the tectum (or both) still retains the embryonic character of acting as an intact population—that is, being respecified to conform to a changed morphological condition. In the other experiments, in which point-to-point regeneration is seen, this plasticity has apparently been lost. Clearly the field is still in a condition of some confusion.

In the double nasal or double temporal-eyed animals described above, another interesting question can be asked. Is the tectum that is innervated by the compound eye, with its abnormal innervation pattern, a normal tectum, or has the appropriate half-tectum vastly hypertrophied at the expense of the half for which there is no appropriate retinal input? Is the abnormal pattern established as a fixed specificity thereafter? The test is to create an animal with one normal and one compound eye. After establishment of connections with the tectum, so that one tectum is normal and the other "compound," the optic nerves were cut at the optic chiasm and deflected into the ipsilateral rather than the contralateral tectum. It was found that the compound eye innervates the "normal" tectum in the same way it had earlier innervated the "compound" tectum, with parallel projection to the whole nucleus. The normal eye projects in normal fashion to the originally "compound" tectum. Thus the tectum seems not to be specified with an invariant point-to-point map of the retina, but is instead specified only in terms of anterior-posterior and dorsal-ventral gradients (with perhaps the necessity to postulate a gradient in depth as well). The ingrowing nerve fibers assort themselves in an orderly way along each gradient, but not necessarily in a point-to-point

correspondence to a rigidly specified tectal population.

It would be unwise to de-emphasize the needed **exactness of connections.** To determine what part of the retina is projecting to a given part of the tectum, simple sampling of evoked or unit responses is sufficient. But to restore full functional vision, the regeneration must be much more exact than that. Fibers must be able not only to find their way back to the appropriate location on the tectum, but to the same layer of tectal cells, and within that layer to the correct cells to analyze again the correct rate and direction of movement, color, and prior stimulation pattern. It is perhaps unsafe to conclude that a given fiber must find its way back to the tectal cell it originally innervated. Indeed, it seems self-evident that there is not adequate genetic material to specify each neural connection individually. Nevertheless, it is improbable that two or even three guiding "gradients" are sufficient. That many might guide a growing fiber to the correct approximate location, but further refinements of connectivity may depend on the interaction of many more influences.

Indeed, almost all studies of optic nerve regeneration and tectal projections suffer from a common lack of certainty about what is being studied. Histological tracing of projections reveals only the fiber pathways and terminations of ingrowing axons, nothing about the function of the synapses formed or even whether synapses are formed. Electrophysiological mapping is interpreted by most workers to reveal points of axon termination (see Fig. 9.27, p. 376), but again, not whether functional synapses are formed. Thus there is major uncertainty about what structures are producing the responses being detected, as well as a startling paucity of behavioral testing to establish the degree of abnormality of function restored.

Plasticity of connections; effects of use and disuse. The mammalian visual system is much more complicated, and does not show significant capacity for regeneration. Study of it might therefore seem unpromising. On the other hand, responses up to the level of association cortex are in many ways simpler than those of frog retinal ganglion cells. In a series of fascinating studies, Hubel and Wiesel have used this system to show some of the subtle changes in connectivity or function that accompany use or disuse of visual pathways.

Hubel and Wiesel first analyzed in detail the response patterns of cells at different neural levels in cats and monkeys. Lateral geniculate neurons respond with concentric receptive fields almost indistinguishable from those of retinal ganglion cells. At the primary visual cortex, cells respond in more complicated ways, indicative of convergence of inputs from several lateral geniculate cells (Fig. 9.33). These cortical cells respond selectively to linear stimuli having a certain orientation, sometimes restricted to one position of the retina ("simple" cells), sometimes anywhere within a broad receptive field ("complex" cells). At higher neural levels, cells respond selectively to stimuli having two or three or more edges, each at a specified orientation and falling within a certain large area of the retina ("hypercomplex" and "higher-order hypercomplex" cells in the terminology of Hubel and Wiesel). About 80% of the cortical cells are driven approximately equally well by both eyes, with the two inputs having receptive fields that view the same point in space and have the same requirements for stimulus orientation, size, and direction of movement (Fig. 9.34,A).

When Hubel and Wiesel studied **responses in newborn kittens** with no visual experience, they found that cells at each level respond much as in the adult. Orientation specificities are often less sharp, and the responses are somewhat more "sluggish" than those of adult cells, but this sluggishness is apparently only a reflection of the cells' immaturity rather than the lack of visual experience, for responses become fully as vigorous as in adults by 3–4 weeks of age, whether the kittens have had any visual experience or not. Thus some fraction of the neural connections necessary to produce the specific responses seen in adult cats are apparently genetically determined, present at birth, and formed without visual experience.

On the other hand, if a kitten is blindfolded for several weeks after birth, it grows up functionally blind. Up to 4 weeks of age there is no apparent effect. Between 4 and 6 weeks of age, however, with peak sensitivity at approximately 28 days and decaying effect to about 3 months, there is a **critical period** during which use is of utmost importance. If one eye is blindfolded with an opaque cover during this time, or even a few days during the fourth or fifth week, the eye loses virtually all of its effect on cortical cells (Fig. 9.34,B). The retina still responds normally, and so do the lateral geniculate cells, although the layers of the lateral geniculate innervated by the blindfolded eye are much atrophied. But virtually all of the cortical cells sampled responded only to the normal eye. If the blindfold is translucent, eliminating only patterned vision, the atrophy of the

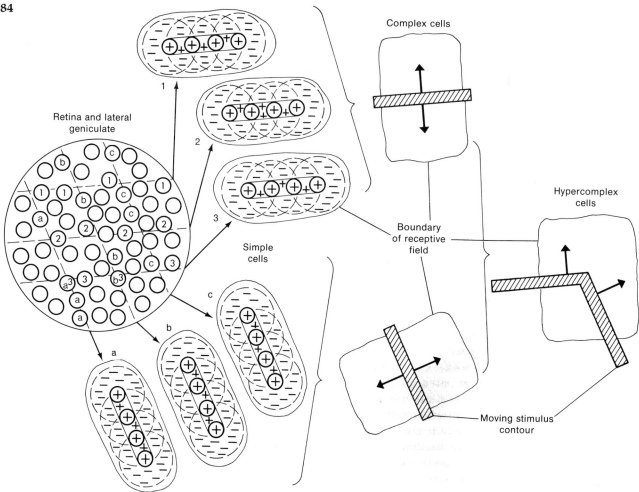

Figure 9.33

Response patterns and extraction of signal pattern in the visual CNS. Many units are shown at the level of the lateral geniculate body (at which responses are essentially like those at the retina.) Here each cell has its small concentric field, as in Figure 9.26. It will be assumed that all units designated *1, 2, 3, a, b,* or *c* are on-center units with the centers of the fields shown. If all cells marked *1* converge on a single *simple cortical cell,* for example, this cell will have a larger linear receptive field, specifically responsive to a thin bar of light or a border at the orientation shown, with light falling on the centers of the fields of presynaptic units and darkness covering only part of the surrounding field. Stimuli at other orientations are relatively ineffective or inhibitory. Other simple cells are shown receiving convergent input from groups *2, 3, a, b,* and *c.* The same lateral geniculate cells could be contributing at the same time, in different orientations. If several simple cells, each responsive to stimuli of the same orientation, converge on a *complex cell,* the resultant receptive field would be larger and specific for stimulus orientation and, often, movement or direction of movement, but the stimulus does not have to fall on any particular part of the receptive field to be excitatory. Convergence of outputs from two or more complex cells may give rise to *hypercomplex* cell response patterns, in which stimulation is most effective when there are two or more borders whose orientation is determined by the inputs. With continuing convergence of this sort, higher order cells can be programmed to respond only to highly specific and complex shapes, falling anywhere within a large retinal area, irrespective of the absolute size of the image.

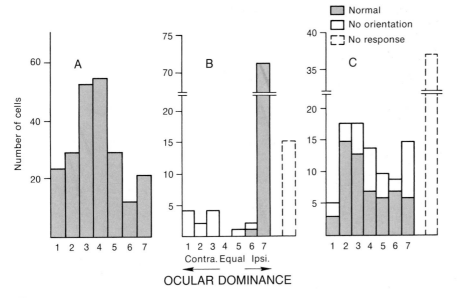

Figure 9.34
Binocular convergence on cat visual cortical neurons. **A.** Normal ocular-dominance distribution. Cells in category 1 receive excitatory input only from the contralateral eye; those in group 7, only from the ipsilateral eye. Most cells receive binocular input, with most being approximately equally well driven by both eyes. **B.** Ocular-dominance graph for cortical neurons in a kitten deprived of vision in one eye from about 4 weeks to 8–14 weeks of age. Note that the vast majority of cells are driven only by the eye that received input. **C.** If both eyes are sutured closed during the same period, however, the ocular-dominance pattern is much more normal. Such kittens nevertheless act as though blind in both eyes. [Part A, Hubel and Wiesel, 1962; B and C, Wiesel and Hubel, 1965.]

lateral geniculate is less, but there is still almost total loss of effective input to the cortex. Thus during the critical period, cortical synapses must be used or they will become ineffective. The critical period during which use is important in monkeys extends from birth to six months and even beyond. In man, the period is about four years, as judged by the serious effect of cataracts or squint before that age, and the irreversibility of their effect thereafter.

The phenomenon is actually more subtle than this, for if both eyes are simultaneously blindfolded during the critical period, approximately half of the cortical cells continue to respond normally to both eyes, even though the animal acts blind (Fig. 9.34,C). The malfunction must occur at a higher neural level not studied to date. When Hubel and Wiesel cut one of the eye muscles in a newborn kitten, causing a condition of "squint," in which the two eyes do not focus in the same direction, they found that even though both eyes continue to be used, most cortical cells cannot be driven binocularly. Unless the two eyes respond to the same stimulus simultaneously, one or the other input is lost, even though both eyes are still useful to the animal. The need for "simultaneous"

Chapter 9
Development and Specification of
Connections in the Nervous System

Figure 9.35
*Shift in vertical angular disparity of
binocular inputs to cortical visual
neurons.*

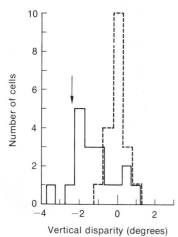

Normally (broken line), the mean disparity is zero; that is, cells receive inputs from units in the two retinas that view exactly the same point in space ($\pm$ about 1°). If goggles are applied during the critical period, so that the visual field of one eye is displaced by 2.3° with respect to the other (arrow), the disparity shifts by almost that amount (solid historgram). [Shlaer, 1971.]

use was shown very elegantly by a further experiment in which one eye was blindfolded and then the other, on alternate days. Even though both eyes were being used throughout the critical period and both were focusing on the same visual space, the cortical cells could, again, be driven only by one eye or the other, not both. It is of interest that recent experiments suggest that switching the blindfold from one eye to the other every 10 sec gives the same result as once per day. At 1/sec, there is little loss of binocular driving. There are probably countless other cases, in sensory pathways and elsewhere, in which use—perhaps use that interacts with simultaneous input on other pathways—must occur during critical periods if those synapses are to be maintained. This phenomenon is very likely involved somehow in imprinting.

The mechanism of "displacement," or inactivation, of some synapses while others are retained has not been elucidated. There is good evidence that the inputs to a cortical cell from different lateral geniculate cells somehow compete with each other. Those that succeed in driving the cortical cell apparently receive some sort of positive trophic feedback, for they are large and apparently healthy, whereas corresponding lateral geniculate cells that do not drive the cortical cells, are atrophic. It is as if the approximately simultaneous presence of activity (spikes?), both in the terminal of the geniculate axon and in the cortical neuron, somehow helps maintain that synapse and the cell providing it, whereas inputs that are not active simultaneously with the postsynaptic cell during the critical period gradually lose their effectiveness.

Why should there be such a "critical period"? It is possible that this is a mechanism for permitting correction of slight developmental abnormalities. For example, if all visual connections are present before visual experience, but one eye somehow develops with a slight displacement of axis relative to the other, this plasticity might allow a correction. If the two eyes view space too far apart, their inputs are obviously simply separated, perhaps to prevent interference. If the deviation is only a few degrees, however, the plasticity may permit a slight shift of synapses so that the two retinas are again perfectly aligned with one another.

This capacity has been tested by raising kittens with exactly fitted goggles that **displace the visual fields** of the two eyes by 2.3° (4 diopters) in the vertical axis. Normally, even at birth, if the two eyes are focused on the same object in space, cortical cells receive binocular input from cells in the two retinas that view the same point in space with an accuracy described by a normal distribution with extremes of disparity no greater than about $\pm$ 1.2°. (Horizontal disparity may be important for depth perception, since, because of it, some cells are maximally excited by objects in front of or behind the plane of focus.) The 2.3° displacement falls far outside the normal range of variability; yet it is found that after a few weeks of experience with such displaced vision, the inputs are displaced by approximately 2°, nearly enough to compensate for the shift (Fig. 9.35). These experiments are still very preliminary and have not yet been expanded to determine how great a displacement can be corrected; but it appears either that new synapses can be formed, appropriate to function, or that a peripheral part of the pre-existing synaptic network can be selectively retained while most of the remaining inputs are lost. At present, the latter hypothesis appears most likely.

Recent experiments imply that this **plasticity extends even to the orientation** specificity of cortical neurons. Hubel and Wiesel find that most cortical visual neurons in newborn kittens show orientation specificities, implying that the connections develop innately according to genetic programming. The specificity is less sharp than in visually experienced kittens, and there are some neurons that show no apparent orientation preference; but all orientations are represented. In contrast, if a kitten is raised from birth in an environment in which the only visible patterns have one orientation or if it is fitted with goggles that admit light only through slits of one orientation, virtually all of the visual cortical neurons develop a preference for that orientation (Fig. 9.36). It has even been reported that this specification can be accomplished with as little as an hour of experience, restricted to stimuli of one orientation, during the height of the critical period. Although this result needs to be confirmed, it seems probable that, in these animals, the restricted experience is somehow able permanently to activate pre-established connections appropriate to stimuli of that orientation and to inactivate connections excited by patterns of the other orientation.

Morphological changes associated with use and disuse. Presumably there are clear-cut morphological correlates for loss of effectiveness of certain inputs, but there is little solid information about such changes, partly because so little is known

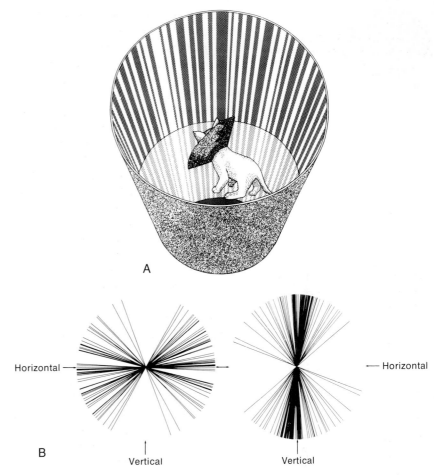

Figure 9.36
A. Visual display consisting of an upright plastic tube, about 2 m high, with an internal diameter of 46 cm. The kitten, wearing a black ruff to mask its body from its eyes, stood on a glass plate supported in the middle of the cylinder. The stripes on the walls were illuminated from above by a spotlight. The luminance of the dark bars was about 10 cd/m^2, and that of the bright stripes about 130 cd/m^2: the stripes were of several different widths. Not shown are the top cover and the spotlight, which have been removed from the tube. **B.** These polar histograms show the distributions of optimal orientations for 52 neurons from a horizontally experienced cat on the left, and 72 from a vertically experienced cat on the right. The slight torsion of the eyes, caused by the relaxant drug, was assessed by photographing the pupils before and after anaesthesia and paralysis. A correction has been applied for torsion, so that the polar plots are properly orientated for the cats' visual fields. Each line shows the optimal orientation for a single neuron. For each binocular cell the line is drawn at the mean of the estimates of optimal orientation in the two eyes. No units have been disregarded except for one with a concentric receptive field and hence no orientational selectivity. [Blakemore and Cooper, 1970.]

Chapter 9
Development and Specification of
Connections in the Nervous System

Figure 9.37
*Loss of dendritic spines in lateral
geniculate neurons (lgn) of
light-deprived dog.*

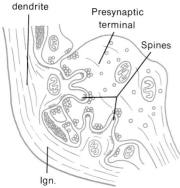

Postsynaptic
dendrite

Presynaptic
terminal

Spines

lgn.

*Normal has many spines of post-
synaptic dendrite projecting into
presynaptic terminal.*

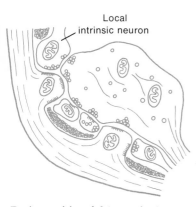

Local
intrinsic neuron

*Dark-reared has deficiency of spines.
Local intrinsic neurons are less
affected.* [Hámori, 1973.]

about the normal synaptic connections in the visual cortex (or, for that matter, in most parts of the CNS).

It is known that visual **deprivation produces gross changes** at the retinal and lateral geniculate levels. Dark-reared chimpanzees, for example, lose a large fraction of their retinal ganglion cells. Cats and rats that are raised in the dark exhibit little or no loss of retinal cells, but show pronounced deficits in ability to respond to repetitive stimuli. Light deprivation in one eye causes severe atrophy of the layers of the lateral geniculate driven by that eye. The number of neurons is not decreased, but each is smaller and there are major distortions in neuronal cytoarchitecture. In particular, there appears to be a marked reduction in the number of dendrite spines (Fig. 9.37). Deprivation of pattern, but not of light, causes similar but less severe atrophy.

At the **cortex,** however, the effects of light deprivation are much more subtle. Various workers have reported three different effects: (a) a decrease in total dendritic length of the common stellate cells, thought by most workers to be interneurons between lateral geniculate afferents and the large cortical pyramidal cells (the decrease in dendritic length is caused mainly by a loss of second and higher order branches); (b) a significant loss of dendritic spines along the central 3/5 of the apical dendritic shaft of the pyramidal cells, the area shown by lesion studies to be innervated, directly or indirectly, by the lateral geniculate efferents; and (c) a distortion of the shape of the dendritic spines along the same central 3/5 of the shaft of the pyramidal cell apical dendrite, particularly a loss of

the terminal enlargements of the spines. Whatever the morphological effects of the disuse, there is still no understanding of the necessary interaction between terminal and postsynaptic cell that must be responsible for maintenance of the endings that are successful in driving the cell.

E. Plasticity and Changes in Connectivity After Maturation of the Nervous System

Sensory-Motor Plasticity. During the critical period there is plasticity within a narrow range. After the brief critical period, synaptic connections appear to be effectively permanent. In man and most other mammals, however, there is a remarkable degree of perceptual plasticity throughout life, responsible for the learning of new patterns of motor coordination and capable of correcting for sensory distortions. This "sensory-motor plasticity" has been studied most extensively by Kohler in Austria and by Held and Hein in the United States. A classic demonstration of the phenomenon is seen when a subject wears over one or both eyes a prism that displaces his visual field. At first, and as long as the subject is asked to point to an object in the field without being able to detect his error, he points in the wrong direction, missing by approximately the displacement of the field. But if he is able to observe his own movements and, by visual feedback, correct his errors, he quickly **learns to compensate** for the displacement of field. If he points only with the right hand, and so can correlate proprioceptive and visual feedback with only that arm, the perceptual correction occurs only for that arm. When asked to point to the target with the left hand, he once again misses by the angle of visual displacement. Thus at a neural level at which visual and proprioceptive feedback converge, well beyond the primary visual cortex, sensory-motor coordination can be altered drastically to correct for erroneous input on one of the channels. For a short time after the removal of the prism(s), responses are again incorrect by the amount of the previous displacement, but in the opposite direction. With experience, certain individuals are able to compensate for prisms that totally reverse the visual field. In fact, they are said to become capable of voluntarily reversing or righting the perceived visual field even without prisms. Thus the correction is transient and almost certainly does not involve major changes in synaptic connectivity. It is instead a learning phenomenon, especially important early in life for the minor adjustment of motor patterns or the learning of new ones. The importance of this process in initial development of coordination has been well demonstrated by raising kittens from the time of eye-opening with a ruff that prevents them from seeing their feet. They move about perfectly freely, but do not learn to use vision to coordinate accurately their foot position in a complicated environment. If they are able to observe their feet with one eye, that eye quickly becomes capable of guiding limb position, but the other never does.

In lower vertebrates, such as frogs and chicks (see p. 302), there is no behavioral correction for displaced vision, implying that the capacity for such sensory-motor

plasticity does not exist, at least to the same degree.

Compensation for Injury. To a limited extent, nerve cells continue, throughout life, to be able to change their innervation pattern to compensate for injury or for altered conditions. Much of this chapter has dealt with regeneration, both peripherally and centrally, in lower vertebrates. Even in adult mammals some capacity for regeneration exists. If peripheral axons are cut or crushed, regeneration is frequently successful, although there is much less specificity of reconnection than in lower vertebrates. Within the CNS, regeneration is extremely limited or nonexistent, but perhaps mainly because of the difficulty that nerves have in growing through scar tissue. In several cases it has been shown that when central neurons lose some or all of their synaptic input, other axons in their vicinity sprout collaterals that innervate them, apparently selectively at the sites of old synapses. If part of one retina is lesioned—for example, by destroying the input to a restricted region of one layer of the lateral geniculate—collaterals from optic nerve fibers that innervate adjacent portions of the same layer, as well as axons that innervate the overlying layer from the other eye will spread into the denervated area to innervate those cells. Neuron cell death occurs at some rate throughout life. Obviously, not all of these cells are superfluous; their function must, in a large fraction of cases, be assumed by other cells. One example of such cell death is the gradual loss of somatic motor neurons. Apparently when a motor neuron dies, and its motor unit is denervated, branches

390

Chapter 9
Development and Specification of
Connections in the Nervous System

sprout from remaining axons to innervate the denervated muscle fibers, enlarging the remaining motor units. This is one of the factors responsible for decreasing motor coordination in older animals.

Memory and Learning. At this point, one can only speculate about mechanisms of memory and learning. It seems most probable that subtle changes in synaptic connections or effectiveness can be responsible, at least in some forms of learning. There are too many possible changes and too little solid evidence for any of them, however, to justify a discussion of possible models at this time.

SUGGESTED READINGS

Bennett, M. V. L., ed. 1974. *Synaptic Transmission and Neuronal Interaction.* (Soc. Gen. Physiol. Series vol. 28), Raven Press, New York. [Contains a number of valuable papers on tissue culture approaches to neuronal and neuromuscular interactions.]

Drachman, D. B., ed. 1974. *Trophic Functions of the Neuron.* Ann. N.Y. Acad. Sci., vol. 228. [Comprehensive collection of papers on development of neural connections, trophic interactions and their possible mechanisms, and clinical disorders implying trophic interactions.]

Eccles, J. C. 1973. *The Understanding of the Brain.* McGraw-Hill, New York. [Contains excellent introductory chapters on neurogenesis and neural plasticity, as well as most other aspects of central cellular neurophysiology.]

Gaze, R. 1970. *The Formation of Nerve Connections.* Academic Press, New York. [A detailed analysis of the problem of how specific connections are formed, especially in the context of amphibian neuromuscular and retinotectal pathways.]

Gaze, R. M., and M. J. Keating, eds. 1974. *Development and Regeneration in the Nervous System.* British Medical Bulletin, **30:**105–194. [A one-topic issue of the journal including excellent papers on neuronal specificity, developmental plasticity, and effects of environment on neural development.]

Hughes, A. 1968. *Aspects of Neural Ontogeny.* Logos Press, London. [A survey of several aspects of nervous system development, especially strong in the areas of neurogenesis, the role of cell depth during development, and the ontogeny of behavior.]

Jacobson, M. 1970. *Developmental Neurobiology.* Holt, Rinehart & Winston, Inc., New York. [An excellent, broad survey of the whole field of neural development, with special emphasis on problems of specificity. Contains the most extensive bibliography available.]

Kandel, E., ed. 1976. *Cellular Biology of Neurons* (Handbook of Physiology. The Nervous System, vol. 1). Williams and Wilkins, Baltimore. [Up-to-date, authoritative survey of the field, with relevant chapters on "Trophic Interaction Between Neurons and Between Neuron and Muscle" (J. Rosenthal), "Specificity of Neurons and their Interconnections" (A. D. Grinnell), "Neuronal Plasticity and the Modification of Behavior" (E. Kandel), and "Nerve Cells in Tissue Culture" (G. Fishbach and P. Nelson).]

Lømo, T., and J. Rosenthal 1972. Control of ACh Sensitivity by Muscle Activity in the Rat. *J. Physiol.* **221**:493–513. [A now-classic paper showing the effectiveness of muscle activity in retarding or reversing the effects of denervation previously attributed primarily to trophic substances.]

Purves, D. 1976. Long-term regulation in the vertebrate peripheral nervous system. *In* R. Porter (ed.), *Neurophysiology II* (International Review of Physiology Ser., Vol. 10). University Park Press, Baltimore. [A thorough review of trophic influences between cells and the specificity of regeneration of connections; especially in the periphery.]

Schmitt, F. O., ed. 1970. *The Neurosciences: Second Study Program.* Rockefeller Univ. Press, New York. [A large volume, including sophisticated papers on subjects spanning the whole of neurobiology, but with an excellent section on development (pp. 51–160).]

Wolstenholme, G. E. W., and M. O'Connor, eds. 1968. *Growth of the Nervous System* (A Ciba Foundation Symposium). Little, Brown & Co., Boston. [Important papers by leaders in the field on the development of specific connections, development of movement, biochemical factors that affect development, and the evidence for trophic interactions.]

I. INTRODUCTION

In this chapter we summarize the principal features of the nervous systems in the major groups of animals in order to place the principles of earlier chapters into context and to get some feeling for a major element missing so far—the functional anatomy of whole systems. Further details can be found in the treatises by Bullock and Horridge (1965) and Kappers et al. (1967).

II. PROTOZOA

By the definition of the nervous system given in Chapter 1, an organized constellation of specialized cells, the Protozoa cannot possess a true nervous system. But the amazingly vigorous skitterings to and fro of some ciliates strikingly resemble those of small metazoans—rotifers, turbellarians, crustaceans —which live right beside them and do possess true nervous systems. Whether protozoans have some kind of specialized system of intracellular organelles that conducts excitation is a question that has long attracted attention and inspired new work in recent years. The answer, on the balance of evidence, is apparently "no."

Excitation, localized receptors, localized effectors, and coordination are certainly present. Electrical properties of the cell membrane show many similarities to those of nerve cells, including potential changes associated with activity and a threshold current for barely perceptible changes in ciliary beat. Change in direction of beat and in stopping and starting of beat are nearly simultaneous over the whole animal, and appear to be mediated by the cell membrane via an electrotonically spread membrane potential change together with built-in gradients of membrane properties and of spontaneous frequency of beat. A mechanical stimulus at the front end of a paramecium causes a specific increase in conductance to Ca^{++} and a depolarizing potential change with backward beating of the cilia and reversal of locomotion. Stimulation of the posterior end causes a specific increase in conductance to K^+

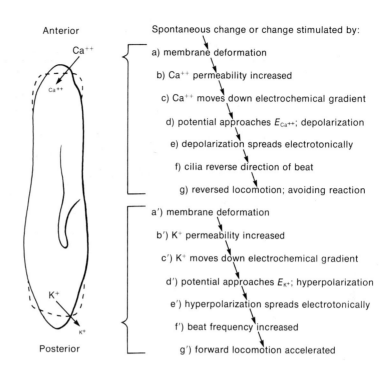

Anterior

Ca++

Ca++

K+

K+

Posterior

Spontaneous change or change stimulated by:

a) membrane deformation

b) Ca++ permeability increased

c) Ca++ moves down electrochemical gradient

d) potential approaches E_{Ca++}; depolarization

e) depolarization spreads electrotonically

f) cilia reverse direction of beat

g) reversed locomotion; avoiding reaction

a') membrane deformation

b') K+ permeability increased

c') K+ moves down electrochemical gradient

d') potential approaches E_{K+}; hyperpolarization

e') hyperpolarization spreads electrotonically

f') beat frequency increased

g') forward locomotion accelerated

Figure 10.1
Specific excitability, conduction, and response within a single cell. Summary of the steps linking mechanical stimuli with motor reactions in *Paramecium*. These steps lead either to the avoiding reaction or to accelerated forward locomotion, depending on which end of the cell is stimulated and which ion permeability is therefore specifically increased. [Naitoh and Eckert, 1969.]

and a hyperpolarizing membrane potential shift with accelerated forward beating of the cilia (Fig. 10.1).

The mechanism of coordination of the metachronal wave is more difficult to discern. Evidence now favors the theory that each cilium mechanically stimulates the next, which responds actively, like a cilium-to-cilium reflex. No system of formed elements in or under the surface appears necessary or adequate to account for the observed directions of metachronal wave propagation. The Protozoa, lacking a nervous system, stand as impressive testimony to the capacity of intracellular functional specialization to achieve prompt, coordinated, locomotor response to stimuli.

III. PORIFERA

Sponges exhibit several forms of response. Rarely is there any sign of conduction beyond the probable area of action of the stimulus. One such rare sign is reported to be a very slow general body contraction in response to pricking the surface of a colony of *Tethya*. Electrical stimulation occasionally causes a

response, but not more than 3 mm away from the site of the stimulus. No correlation is apparent between responses of pores and oscula or between either of these and the beating of choanocyte flagella. Although there are claims that sponges have nerve cells, the criteria used are inadequate, and the resemblances to nerve cells are superficial. In particular, there is insufficient evidence that these cells are functionally connected with each other, as well as with receptor and effector cells. We conclude that Porifera lack a true nervous system.

IV. COELENTERATA

The nervous system of coelenterates occupies a strategic position in our understanding of neurological evolution. This phylum contains the most simply organized animals that have a true nervous system; such a system appears here fully formed, comprising a diverse set of connected neurons. Our knowledge of coelenterate nervous systems has advanced greatly in recent years, and still further important changes can be expected as new tools are brought to bear.

The **structural and functional elements**—nerve cells, synapses, and nerve impulses—are well developed; simple ganglia are achieved in some forms, but there is no central nervous system. Many of the neurons are organized into nerve nets (Fig. 10.2); others are not so organized, and the organization in many cases is not understood. A nerve net is a system of functionally connected nerve cells and fibers anatomically dispersed throughout some considerable portion of an animal and so arranged as to permit diffuse conduction. That means there are multiple paths available, as can be seen

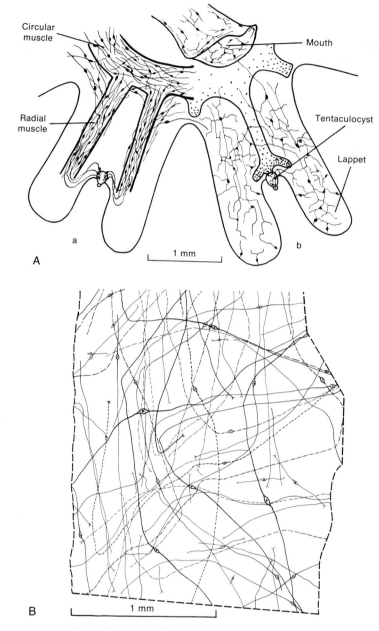

Figure 10.2

Anatomy of nerve nets. **A.** *Aurelia aurita* ephyra. Two arms of the bell, showing the main structures related to the nerve nets. *a*, the muscle strips (heavy lines) and bipolar nerve net system. *b*, cells of the multipolar nerve net and the underlying gastric cavity. [Horridge, 1956.] **B.** Plan of the nerve net on one face (retractor muscle face) of a mesentery of *Metridium senile*. A whole mount of the mesentery was stained with silver and every nerve cell and fiber traced with a camera lucida. Natural ends of nerve fibers are shown by forked tips; these all dip down into the muscle layer. Dots and dashes are only for convenience in following elements; relative thickness of fibers is indicated but not to scale. [Batham et al., 1960.]

Figure 10.3
Experimental evidence for a nerve net.

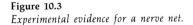

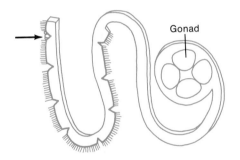

Medusa, intact except that seven of the eight marginal bodies have been extirpated.

Marginal body

Gonad

When the body is cut spirally, waves still travel from the remaining marginal body (arrow) to the other end.

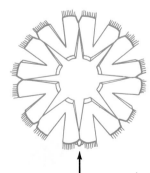

When the body is cut like this, waves still travel from the remaining marginal body (arrow) to both left and right, usually clashing and cancelling on opposite sides but occasionally going only one way and starting a long-lasting trapped circuit wave. [Romanes, 1885.]

by the tolerance of all kinds of incomplete cuts (see Fig. 10.3). The diffuse conduction—and hence its substratum, the nerve net—can be either quite unoriented (Fig. 10.2, upper right) or somewhat preferentially oriented (upper left), but generally lacks well-formed tracts. Discontinuity of neurons, and hence the presence of synapses, is the rule. In *Hydra* and some other coelenterates, however, it is possible that secondary fusion of neurites of initially independent neurons may occur.

Most synapses are unpolarized, but some are polarized; many synapses normally transmit one-to-one, but others require facilitation by successive impulses at an appropriate frequency. Anatomically they are simple crossings, intertwinings, or contacts-in-passing (Fig. 10.2, lower). Dendrites are not differentiated, and most nerve fibers appear axon-like. Neurons are chiefly multipolar and bipolar isopolars (see Fig. 2.1, p. 10). Some are probably interneurons, but these are not differentiated from motor neurons. Primary afferent neurons are distinct and exceedingly abundant. Several kinds of differentiated sense organs are developed. Neuroglial cells do not appear to be differentiated.

In different parts of different species it may be the epidermal or the gastrodermal subepithelial plexus that is better developed. This and other **forms of differentiation** between regions (Fig. 10.4) and among species show the evolution of the primitive nervous system to adapt it to the habit of life. The forms of differentiation include preferred orientation or conduction, preferred polarity, nearly 100:1 differences in overall conduction velocity (including junctional delays) (Fig. 10.5), occasional tracts in which most fibers are parallel, and the development of simple ganglia

Figure 10.5
Differentiation in the nerve net occurs in velocity of propagation. Figures are in meters per second for an anemone. [Pantin, 1935.]

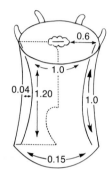

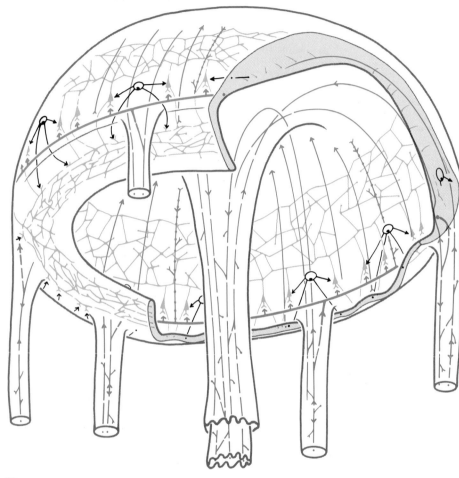

Figure 10.4
Differentiation of parts of the nerve net. Diagram of the nervous system of *Geryonia,* based on physiological experiments by Horridge (1955). The networks and density of fibers are drawn quite arbitrarily. Light brown = swimming beat system and associated local system in the velum. Dark brown = tentacle-manubrium system. Black = statocysts. The light brown triangles represent pacemakers of the rhythmic swimming beat. Their frequency can be increased by tilt of the statocysts and by tentacle stimulation or subumbrella stimulation (via beaded pathways and heavy ring). This is an irreciprocal influence of one system on the other. The pacemaker and statocysts also excite the velum to symmetrical and local contraction, respectively. Broken lines are local pathways, conducting decrementally; solid lines are through-conducting. Note that the dark brown radial lines conduct in only one direction.

Figure 10.6

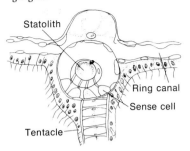

Statolith

Ring canal

Sense cell

Tentacle

Statocysts, which signal gravity, are complexes of tissues in medusae, hence true organs. [Hertwig and Hertwig, 1878.]

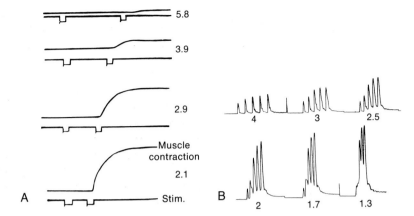

5.8

3.9

2.9

Muscle contraction

2.1

Stim.

A

B

4   3   2.5

2   1.7   1.3

Figure 10.7
Facilitation, a basic property of neuromuscular junctions in primitive nervous systems.
A. Records from the tonic muscle of an anemone. [Pantin and Vianna Dias, 1952.] **B.** Records from phasic muscle of *medusa*. The numbers are intervals between stimuli, in seconds. [Bullock, 1943.]

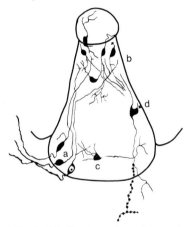

Marginal bodies in medusae innervate statocysts and eyes; their several neuron types (a, b, c, d) and irreciprocal connections make them the first ganglia. [Horridge, 1956.]

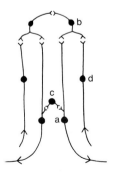

A diagrammatic interpretation of the neuron types shows input from the multipolar nerve net and output to the bipolar net. [Horridge, 1956.]

in medusae (Fig. 10.6). All-or-none nerve impulses have been recorded in many coelenterates. Graded and decrementally propagated forms of activity have been suggested as representing possible stages in the evolution of the impulse. There is no evidence of such stages, however, and the only reason for believing that they may exist is that the small size of some Hydrozoa should allow electrotonic spread of excitation to suffice. But the large size (up to 2 m) of closely related solitary hydroids and the coordination over considerable distances in colonial hydroids make it unlikely that electrotonic spread is generally adequate. Nonnervous spread of excitation through sheets of cells—the remarkable epitheliomuscular cells—is often an important additional form of communication.

Independence of the intensity of direct electrical stimuli above threshold and strong **dependence on facilitation** (Fig. 10.7)—that is, on frequency and numbers of shocks—is a general characteristic of nerve net systems. Different neuromuscular junctions have widely

differing rates of the accumulation and decay of facilitation, so that the frequency of nerve impulses in the net can control which muscles contract, and thus the character of response. These principles obtain both in phasic (fast relaxing) and in tonic (slowly relaxing) muscles—for example, in anemones and medusae, in which the presence or absence of mechanical summation makes the response appear quite different. Both have relatively fast and slow components of contraction under the control of separate conducting systems, pointing to double innervation of the muscle.

Normal mechanical, chemical, and other stimuli operate by evoking a certain number and temporal pattern of impulses in a net (Fig. 10.8). Sense cells probably act only in a graded fashion, their actions summing to determine firing in the net. Adaptation is important; mechanical stimuli that cause high frequencies of firing also cause rapid adaptation, so that the number of impulses is small. Since clear responses with low thresholds for frequency re-

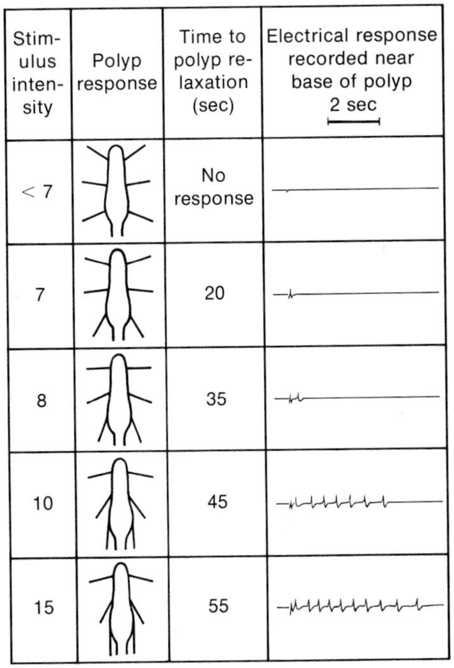

Stimulus intensity	Polyp response	Time to polyp relaxation (sec)	Electrical response recorded near base of polyp 2 sec
< 7	(diagram)	No response	
7	(diagram)	20	(deflection)
8	(diagram)	35	(deflection)
10	(diagram)	45	(repetitive spikes)
15	(diagram)	55	(repetitive spikes)

Figure 10.8
Comparison of the response of a polyp of the colonial hydroid *Cordylophora* and the electric potentials recorded near the polyp stalk. Only tentacle depression is shown in the diagrams of the responding polyp; hydranth shortening, which also occurs, is not shown. The times to relaxation are approximate. The deflection in the first electrical record is the stimulus artifact. Note repetitive firing of impulses following a single shock slightly above threshold; this provides a gradation of response with intensity and with distance. [Josephson, 1961b.]

quire higher numbers of impulses, this helps in selectivity of response. The formation of an impulse group or burst by the sense-cell-nerve-net junction is probably the first mode of nervous integration that evolved.

Chemical stimuli apparently never excite sense cells capable of firing a through-conducting net (true at least for food). Mechanical stimuli in some species can do so only if injurious; there is a whole conducting system solely developed for response to outright injury, particularly in some colonial forms (Fig. 10.9).

Physiological and anatomical evidence make it clear that in many epithelia at least **two nerve nets must coexist,** making only indirect functional contact with each other, responding to different kinds of stimuli, and eliciting different forms of response (Figs. 10.2; 10.4; 10.9). These nets may include one or more extensive or confined through-conducting pathways. They may rule many muscles, widespread throughout the anemone, jellyfish, colonial hydroid or coral; or they may be restricted to the symmetrical responses of single coral polyps or a few muscles in the solitary forms. Through-conducting nets generally conduct faster, and facilitation takes place only at neuroeffector junctions (Fig. 10.10). Non-through-conducting systems vary widely in extent and properties, as well as in their relation to the through-conducting system. One type is highly local and unable to spread excitation to any considerable distance. Another can be converted to through-conduction by repetitive stimulation; some kind of integrative process occurs in the net itself, presumably at the neuro-neural junctions. One type spreads excitation with successive stimuli in approximately uniform increments of radius, whereas others spread it with smaller and smaller increments, and still others with increasing increments. These facts have inspired formulation of hypotheses of the spread of excitation in non-through-conducting nerve nets. One is based on the assumption that only a certain proportion of the available neurons in a given area are excited by a stimulus and that the number originally excited is a prime factor in determining

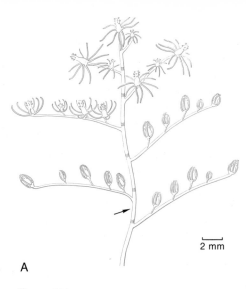

A

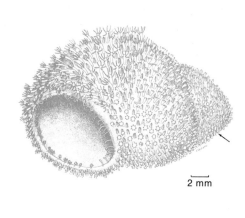

B

2 mm

2 mm

Figure 10.9

A. Colony of the hydroid *Pennaria,* following a shock of not much above threshold intensity applied to the skin at the point marked by the arrow. The closing response has spread a few millimeters and is graded in degree with distance. Repetitive firing of impulses to a single shock is probable, and conduction barriers permit fewer and fewer impulses to reach the polyps farther away. Excitation spreads more readily distally than toward the base. **B.** Colony of *Hydractinia* living on a snail shell, as the colony would appear shortly after stimulation at the arrow. A wave of polyp contraction that will affect all the polyps is shown sweeping across the colony. The dactylozoids (around aperture of shell) are lashing out from their coiled resting position. These two kinds of response travel in separate nerve nets. [Josephson, 1961a.]

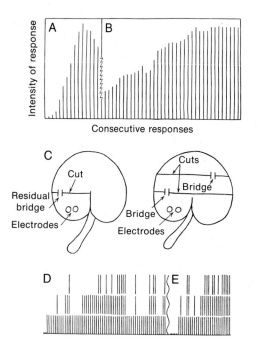

Figure 10.10

Facilitation in the luminescent response of a soft coral, *Renilla köllikeri* (Pennatulacea), to electrical stimulation. **A.** Response to a burst of shocks at a frequency of 1 per second, recording from the entire animal; the first response occurs on the second stimulus. **B.** Responses to a burst of 42 shocks at a frequency of 42 per minute; slit recording. Measurable responses above background were evident only on the sixth shock. **C.** Diagram of patterns of incisions made in the rachis of *Renilla* to investigate facilitation. **D** and **E.** Responses of an animal prepared as in the right-hand diagram in C; the bottom row represents responses in the proximal region (under the electrodes) to repetitive stimulation; the middle row, responses in the central piece; the upper row, responses in the distal region. The small pips at the bottom (E) are initial stimuli before the onset of responses. In C, frequency of stimulation is 42 per minute; in D, 1 per second. [Nicol, 1955.]

spread. Another is based on interneural facilitation, which in certain instances may involve repetitive discharge to single shocks.

Interaction between conducting systems may occur (a) only at the level of the muscle, (b) within the nervous system, or (c) not at all. An example of the second case occurs in medusae in the marginal ganglia or nerve rings, which therefore represent the first integrating concentrations of nervous tissue in the animal kingdom. No corresponding concentration is known in polyps. It is especially interesting that the interaction is only one-way: the slow system influences the occurrence or the expression of activity in the fast. The fast systems are regarded as evolutionarily derivative, the slow as primitive. Thus the most primitive system appears to be one with integrative processes at its junctions with receptors, and at its own junctions, and at neuroeffector junctions. The differentiation of nets into long pathways (e.g., tentacle to mouth stalk) represents the beginning of local sign or labeled lines in nerve fibers.

True **spontaneity**—intermittent discharge in the absence of any stimuli except the steady-state milieu—occurs in many coelenterates, and can be assigned to nerve cells. In pulsating jellyfish, multiple pacemakers are in the marginal ganglia or nerve ring; control is handed from one to another of the marginal pacemakers, which are mutually accessible—that is, able to fire and reset each other, thus assuring that the fastest rhythm will dominate. The pacemakers are irreciprocally accessible to influence from other conducting systems and from sense organs; they are internal clocks subject to modulation by adequate phasic and tonic stimuli.

In their highest form of behavior, medusae and anemones show a succession of **"moods,"** or phases of activity, that last for minutes or hours and occur both spontaneously (Fig. 10.11) and after stimuli. In contrast to the long-held, purely stimulus-response view of coelenterate behavior, our present knowledge upholds Jenning's (1906) view that fluctuations of internal state are of profound importance in these simple nervous systems. The release of a complex predetermined pattern of movements by any of several stimuli is suggested—and the same pattern can be initiated spontaneously. A number of examples of natural behavioral sequences have been analyzed in terms of nerve nets, facilitation, and nervous organization, including escape swimming of an anemone, and feeding in siphonophores.

V. PLATYHELMINTHES

The flatworms—at the bottom of the Bilateria—are strategic as the first animals of organ-grade construction and the first to have a real central nervous system (Fig. 10.12), including a distinct brain. Unipolar nerve cells are present (Fig. 10.13), and the ganglia separate neuropile from the cell body rind. Even small chromatin-rich globuli cells, characteristic of the highest invertebrate centers, are found in the brain in some species (Fig. 10.14).

The phylum is quite heterogenous, but the gross plan is a brain and a set of longitudinal medullary cords connected by commissures, all differentiated out of a superficial plexus. At least six types of superficial sense organs are differentiated. Some groups, including the parasitic forms, are neurologically very

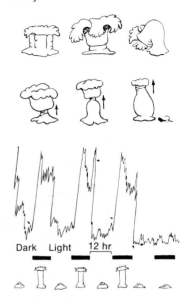

Figure 10.11
One form of complex behavior in anemones is the succession of moods or postural phases. [*Batham and Pantin, 1950.*]

Figure 10.12
Some flatworms have a gross plan called an orthogon, from which others may have been derived. [*Reisinger, 1926.*]

Bothrioplana
(Alloeocoela)

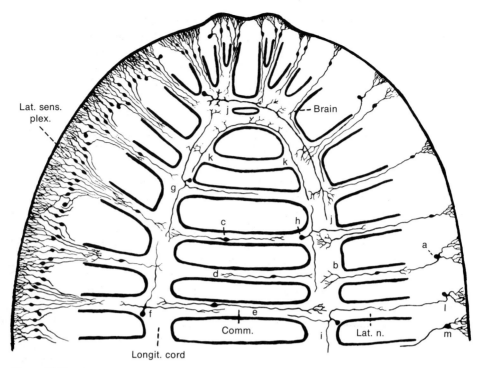

Figure 10.13
The central and peripheral neurons in a flatworm (*Bdelloura candida*). Drawn from Golgi preparations of the anterior end. *Comm*, commissure; *Lat. n.*, lateral nerve; *Lat. sens. plex.*, lateral sensory plexus; *Longit. cord.*, longitudinal cord; *a–m* and *1-3*, individual neurons mentioned in the original description. [Hanström, 1926.]

Figure 10.14
Transverse section through the brain of the polyclad *Notoplana*. Note the grouping of fibers in the core, forming distinct tracts, and the mass of globuli cells contrasting with medium-sized and large ganglion cells in the rind. [Hadenfeldt, 1929.]

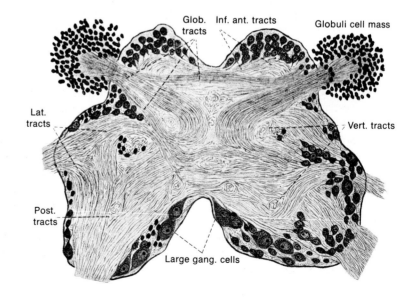

simple; polyclads are the most elaborate; triclads, including planarians, are intermediate.

VI. ANNELIDA

The basic plan of the nervous system among all the animals in the annelid-arthropod line (sometimes called the Articulata) is impressively consistent, making it one of the most widely ranging common morphological schemes among organ systems in the animal kingdom. Actually, the similarity may be thought of as a special case of the underlying similarity, perhaps even homology, of all nervous systems. The overall plans in all groups are hypothetically derivable from a common "orthogon" pattern of longitudinal cords and commissures (Fig. 10.12).

The plan that we deduce is basic to the articulates includes a ventral rope-ladder-like system (Fig. 10.15) of paired longitudinal cords, and metameric pairs of ganglia, each with a commissure. This system communicates with a rostral, dorsal brain via paired circumesophageal connectives. Among the more profound shared similarities are special sensory portions of the brain, three pairs of nerves per segment (of which the second is chiefly sensory), and a typical histological composition of central neuropile and peripheral cell body layer. The cells are mainly unipolar. Nerves from the brain and from the first ventral ganglia supply the anterior portion of the gut and its intrinsic ganglia; this is called the stomodeal system.

The **ventral nerve cord** is primitively paired and superficial—that is, situated in the outer body epithelium. In most forms, however, it has moved internally and fused into a single midventral cord (Fig. 10.17). The leeches (Hirudinea) are distinctive among the three main classes of annelids in having long, cell-free connectives that separate bead-like, sharply demarcated ganglia (Figs. 10.16, 10.17). Oligochaetes and polychaetes, in contrast, have ill-defined segmental ganglia that virtually run into each other (Fig. 10.17,B). Such cords are called "medullary" because nerve cells are scattered all along their length. Their histological structure is basically similar to that of most invertebrate ganglia, and unlike anything in the vertebrates. A central neuropile, free of nerve cell bodies, occupies a large part of the ganglion (Figs. 10.16; 10.17). Tracts differentiate in the better-developed species. The rind of cell bodies is generally loosely packed, and often surrounded by layers of longitudinal and circular muscle. The somas lack synapses.

Neuronal composition, cell groups, and pathways of the ventral cord are known in some detail, from Golgi impregnated preparations, reduced silver, and methylene blue intra-vitam stains (Fig. 10.18). Afferent fibers are usually small and numerous, and enter the cord through all the nerves; they usually make T-shaped branches from which processes go up and down the cord for short distances ipsilaterally, in three longitudinal neuropile strips. Afferent cell bodies are, with a few exceptions, located in the periphery, in or under the skin. Internuncial and motor neurons of the CNS vary in several ways, including combinations of the following: ipsilateral and contralateral axons, short intrasegmental and long intersegmental axons, large and small somas, unipolar (the great majority) and multipolar

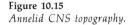
Figure 10.15
Annelid CNS topography.

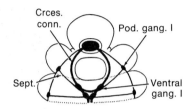

Polychaetes, anticipating arthropods, sometimes show the first ventral ganglion part way up the circumesophageal connectives (crces. conn.), where they join the brain. Some also show presumably primitive lateral cords with podial ganglia. Note the brain (dark mass) and mouth (just behind). [Gustafson, 1930.]

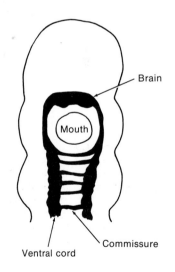

Oligochaetes have a poorly differentiated brain and ventral cord and irregular commissures.

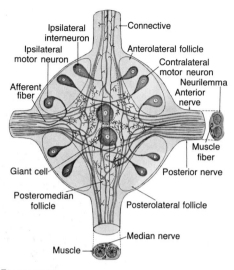

Figure 10.16
Slightly schematized typical ventral ganglion of a
leech, showing the main nerve cells and fibers.
At right and below are cross sections of the segmental nerve and the intersegmental connective,
respectively. [Scriban and Autrum, 1932.]

Figure 10.17
The central and peripheral nervous systems of annelids.
A. Anterior end of the nervous system of the leech *Hirudo,*
dorsal view; *1–6*, presumed segments of the ganglia.
[Liwanow, 1904.] **B.** Ventral cord and nerves of the
earthworm *Lumbricus.* [Hess, 1925a.] **C.** Ventral cord of
Lumbricus in transverse section (Bodian protargol stain),
showing three giant fibers dorsally, cell somas ventrally and
laterally, neuropile centrally, muscular sheath and neurofibrils
in the cells and large fibers. [Bullock and Horridge, 1965.]

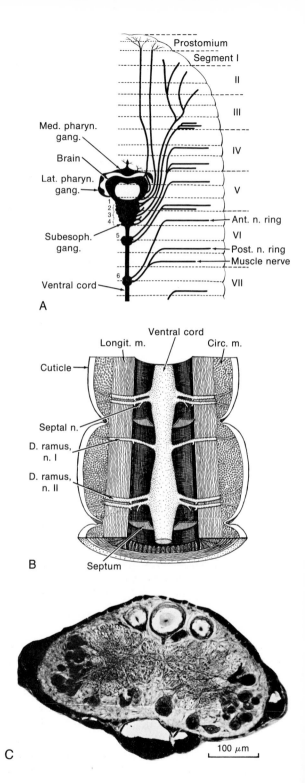

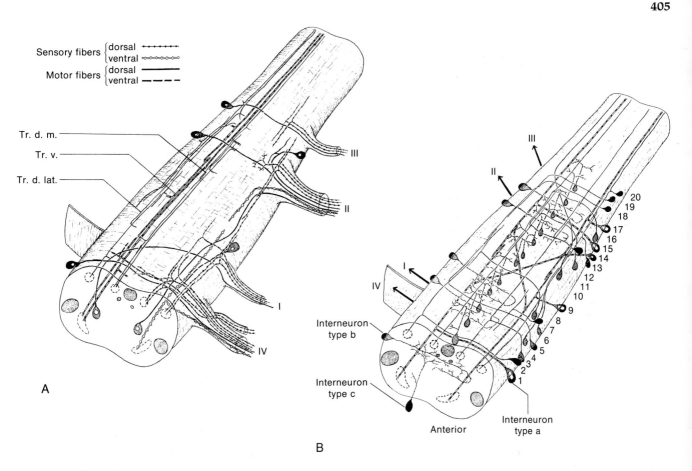

Figure 10.18
Schematic stereograms of typical ventral ganglia of a polychaete, *Nereis*. **A.** The sensory and motor fibers; the anterior end is toward the viewer, the cross section showing the giant fibers (stippled) and tracts of small fibers (broken line). I–IV, segmental nerves; *Tr.d.lat.,* dorsolateral tract of fine-fiber interneuron axons; *Tr.d.m.,* dorsomedial tract; *Tr.v.,* ventral tract of the same. **B.** The principal interneurons and fibers. Type a interneurons, which supply the longitudinal fine-fiber tracts, are shown on the right, numbered 1–20. Type b interneurons, with horizontally branching axons, are on the left; type c interneurons, with vertical axons from ventral somas, are in a row down the center. All three systems are actually bilateral. [Smith, 1957.]

somas. Probably many of the cells are consistent and identifiable from specimen to specimen. Motor neurons are few. A typical midbody segmental ganglion of the earthworm *Pheretima* contains about 1000 nerve cells; of the leech *Hirudo*, 350.

A constituent of the nerve cord in many polychaetes and oligochaetes, but not leeches, is the system of **giant fibers.** A few giant fibers are final motor axons, but most are purely central and internuncial; a few have segmental septa (Fig. 10.19). When there are several giant fibers, they differ in afferent and motor connections. The limited evidence supports a generalization that giant systems mediate an abrupt symmetrical overall

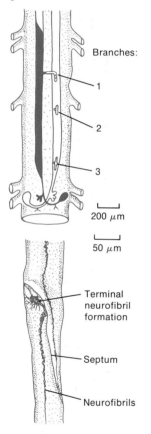

Branches:

1

2

3

200 μm

50 μm

Terminal
neurofibril
formation

Septum

Neurofibrils

withdrawal response to startle stimuli. It is not certain that high conduction velocity is the only significance of the axon diameter; possibly extracellular spike current is also important. Velocity appears to vary as some function of diameter closer to the square root than to simple proportionality. The special synapses are treated in Chapter 2.

Neuroglial elements in annelids are not clearly different from ordinary connective tissue cells. A nerve cord sheath is well marked in some earthworms, and a framework of glial cells and fibers forms more-or-less distinct partitions between rind and core and between small masses of cells and fibers. Larger cells and fibers have individual sheaths. The glial cells form spectacularly thick sheaths around earthworm giant fibers, perhaps even nodes of a kind; the lamella are quite uniform and, except in patches, free of cytoplasm. Among invertebrate sheaths, these are the closest to true myelin, but the lamellae are thicker, the nodes are not circular interruptions in the sheath, and the birefringence properties are distinct from those of true myelin, which is found only in vertebrates. The large and ramifying glial cells of leeches have been analyzed with microelectrodes, as reported in Chapter 4.

The **brain** is quite diverse, and in some polychaetes it attains a remarkably high level of complexity, exceeding that in the simpler arthropods and molluscs. Within this phylum, regions of the brain show differentiation, raising the difficult question of whether regions in different families are homologous or not. A forebrain, a midbrain, and a hindbrain are easily recognized in more advanced polychaetes (Fig. 10.20, B, C), and there is some reason to consider that they may

be homologous in different families, perhaps even with subdivisions of the arthropod brain. The forebrain includes palpal and stomodeal centers and anterior roots of the circumesophageal connectives. The midbrain includes antennal and optic centers and the posterior roots of the connectives. The hindbrain contains centers for nuchal sense organs, which are possibly chemoreceptive and greatly developed in some families. Superimposed on these three divisions are some advanced structures. Certain polychaetes possess the first optic ganglia—distinct masses between the eye and brain, foreshadowing a typically arthropod feature. Advanced polychaetes even more commonly have corpora pendunculata (Fig. 10.20, B), the highest structures in the system, perhaps homologous to structures of the same name in arthropods. Somewhat less specialized are the median masses of the brain and the glomerular neuropiles found in better-developed palpal and nuchal centers, but not in the antennal or optic centers. Various commissures are identified; they appear to be synaptic neuropile bands rather than mere tracts. The giant axon system is sometimes represented in the brain by cells of origin, sensory connections, and a decussation where the axons make functional contact with a two-way synapse. Motor neurons are present but few in number; the brain is mainly a receiving area for sensory nerve fibers and a mass of interneurons. Brain nerves are variable in number and not readily homologized. The first ventral ganglia can be seen in various stages of migration up the circumesophageal connectives to approach the brain (with which they fuse in arthropods).

Histologically the brain shows more advanced features than the cord in both

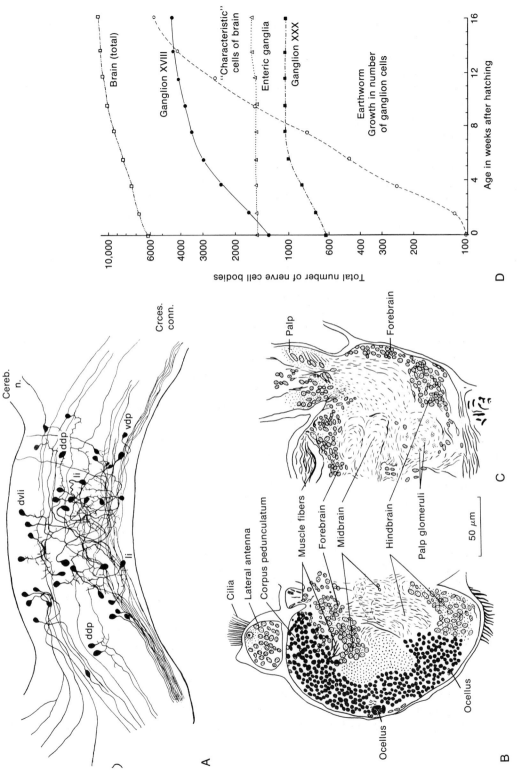

407

Figure 10.20

Nerve cells in the brain of annelids. **A.** Some of the neuron types in the brain of an earthworm, *Pheretima*, representing a relatively low level of development for the phylum. *Cereb. n.*, cerebral nerve; *Crces. conn.*, circumesophageal connective; *ddp*, dorsal descending projection neuron; *dvli*, dorsoventral local interneuron; *li*, local interneuron; *vdp*, small ventral descending projection neuron. [Ogawa, 1939.] **B.** The brain of an advanced polychaete (*Harmothoë*) in frontal section, at a dorsal level, and **C**, at a more ventral level. [Korn, 1958.] **D.** Growth in numbers of nerve cells in an earthworm, *Pheretima*, in the first several weeks after hatching. The small "characteristic" cells of the brain continue to increase in number long after ventral ganglion cells have reached their plateau. [Ogawa, 1939.]

Figure 10.21
Though most annelid receptors are single elements, a few proper sense organs do occur, such as the sense buds in the skin of Lumbricus *and the eyes of* Alciope. *[Above: Langdon, 1895. Below: Milne and Milne, 1959.]*

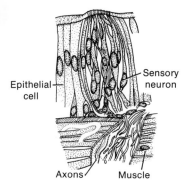

Epithelial cell — Sensory neuron — Axons — Muscle

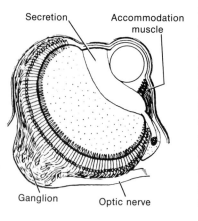

Secretion — Accommodation muscle — Ganglion — Optic nerve

rind and core (Figs. 2.56, C; 10.56). In particular, the brain includes clusters of small chromatin-rich, plasma-poor globuli cells (Fig. 10.20, B) and the corresponding dense, fine-textured neuropile masses. Such "characteristic cells" of the brain are relatively late developing in ontogeny (Fig. 10.20, D).

Because of the great diversity among the polychaetes, comparison of families is instructive. The brain, by contrast with its highest development, can be exceedingly simple; in some species it is merely a thickened basiepithelial plexus in correlation with the virtual absence of special sense organs.

Receptors include, prominently, the generalized epidermal sense cells, which occur singly and in smaller and larger sense organs (Fig. 10.21). Deep sense cells with long, branched, free nerve endings are common, and are thought to be mechanoreceptors. Counts of integumental sense organs are available for some earthworms. Photoreceptors are widely distributed in isolated cells and in organized clusters and, in a few families, in moderately good eyes (Fig. 10.21). Experiments have suggested that there are separate systems of receptors for shadow and light detection. Shadow reception is prominent in some tube dwellers. Statocysts are known in a number of polychaetes; their axons go to the subesophageal ganglion, not to the brain, and they signal slow events, not vibrations. Nuchal organs innervated from the brain are dorsal pits, grooves, or holes at the posterior edge of the prostomium; they may be elongated and extend back over many segments; they are reduced or absent in some families of polychaetes and elaborated and even erectile in others. The function of chemo-

reception is merely inferred from the histology.

In leeches, identifiable afferent neurons belonging to three modalities have been studied; consistent receptive fields can be mapped on the skin, and their constant synaptic connections with central neurons have been determined in the ganglia. Knowing precisely the normal anatomical and functional patterns has permitted studies of development and of regulation after imposed perturbation by lesions (Fig. 10.22).

The **stomodeal system** is well developed in annelids; it is composed of a set of nerve cells and fibers that form a plexus on the wall of the anterior parts of the alimentary canal, and connect primarily to the brain and the first ventral ganglion. Limited knowledge of function in earthworms shows that two sets of fibers exert antagonistic effects on gut muscle tone and that one set causes secretion of digestive enzymes.

Turning to **physiological studies** on central organization, the ganglia and cords are continually active electrically, chiefly in the form of spikes, and are largely independent of the removal of segments and receptors (see p. 228). Movements in complex behavior appear to be composed of the following simpler coordinated activities. (a) Peristaltic and antiperistaltic creeping is normally coordinated through the nerve cord. The sequence of reciprocal excitation and inhibition seems to be inherent in the cord and requires only tonic input from receptors, perhaps contact receptors or stretch receptors. Reflexes enhance or reduce these movements. (b) Parapodial creeping is even more dependent on central autorhythmicity and less dependent on reflex enhancement. (c) Walking

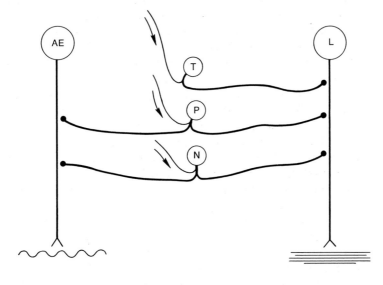

| Synapse | Normal | | Properties after removal of connectives* |
	Transmitter and influence	Effect of repetitive stimulation	
T–L	Electrical, excitatory	No facilitation	Chemical and electrical
N–L	Chemical, excitatory	Small facilitation, then depression	
N–AE	Chemical, excitatory	Large facilitation	Inhibitory
P–AE P–L	Chemical, excitatory		Inhibitory

*Approximately 5 weeks after transection of
connectives just above and just below ganglion.

Figure 10.22
Some identified cells and synapses in the leech (*Hirudo*), their normal function and mode of transmission, their differences in response to repetitive
stimulation, and their alterations after recovery from the loss of connections
with the rest of the nerve cord. *T, P,* and *N* are constant, identifiable, primary
sensory neurons sensitive to tactile, pressure, and nociceptive stimuli, respectively; they are unusual in having their cell bodies inside the CNS. They synapse with motor neurons that mediate shortening (*L,* longitudinal muscles)
and wrinkling (*AE,* annulus erector muscles). [Based on results of J. Nicholls.]

by leeches involves the alternate use of the suckers and is dependent more upon a chain of reflexes than (a) and (b). Leech muscle shows a "catch action," an apparently frozen state of muscle contraction, but this is under constant reflex control. (d) Swimming is probably initiated by an intrinsic central rhythm. (e) Writhing may actually include several distinct types of movement; in the present context they are interesting because they show the presence of fast intersegmental paths other than giant fibers. (f) The twitch reflex is the response to giant fiber activation; as a result of afferent input and of highly labile, integrative central activity, a specialized intersegmental response briefly overrides segmental individuality. (g) Reflex arrest of peristalsis and antiperistalsis have fast pathways.

The important general conclusion from analysis of these activities, and nonlocomotor activities as well, is that the simple ganglia of the annelid ventral cord are able to manifest phenomena of reciprocal excitation and inhibition, intrinsic spatial and temporal patterns, local and chain reflexes, occlusion, facilitation, central and peripheral inhibition, fast and slow pathways, afterdischarge, and other features familiar in higher forms.

The **role of the brain** has been studied mainly by total ablation. Many activities survive its extirpation, including locomotion, coitus, righting, maze learning, feeding, and burrowing. The main effects of ablation are heightened excitability, restlessness, and sensory deficits attributable to cephalic receptors. Some restricted brain lesions have been reported; one conclusion for *Hirudo* is that crossed and uncrossed pathways descending from the brain must be of nearly equal

value. The brain plays some role, for a critical initial period, that permits regeneration of lost tail segments. This and other signs suggest neurosecretion. Removal of the subesophageal ganglion reduces spontaneity, muscle tone, search movements, and the ability to creep backwards.

Some **phylogenetic remarks** should be made to place the annelid nervous system in perspective. As in platyhelminths and other groups, the main cords can be considered as derived from an ancestral orthogonal array, and therefore as homologous with each other as well as with those of virtually all higher phyla. The original orthogonal array is supposed to have been developed before the platyhelminths, at least as early as the ctenophores. In contrast to this sweeping permissiveness, we should be very cautious about making homologies between parts of the brain or stomodeal nervous system of annelids and lower phyla, even though there are many striking analogies. Between annelids and arthropods, however, there is more expectation of relationship. Two main views on the morphological composition of the brain in annelids and its relationship to arthropods have developed. One is the theory of an essentially unitary, prostomial origin of the brain (Fig. 10.23); the other is that of step-wise increments from segmental sources.

VII. ARTHROPODA

We now turn to the largest and best-known invertebrate phylum. We will take up in order the gross and microscopic layout, the nervous control of effector organs, and some aspects of receptors.

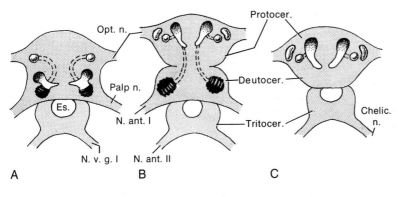

Opt. n.
Protocer.
Palp n.
Deutocer.
Es.
N. ant. I
Tritocer.
Chelic. n.
N. v. g. I
N. ant. II

A   B   C

Figure 10.23
Scheme of the relations of parts of the brain in (**A**) polychaete, (**B**) mandibulate
arthropod, and (**C**) chelicerate arthropod. Visual masses shown in black, corpora
pedunculata by stippling, palpal and antennal glomeruli in gray. *Chelic. n.*, nerve to
chelicerae; *Deutocer.*, deutocerebrum; *Es.*, esophagus; *N. ant. I* or *II*, nerve for first
or second antennae; *N.v.g.l.*, nerve of appendage of the first ventral ganglion;
Opt.n., optic nerve; *Palp n.*, nerve for palp; *Protocer.*, protocerebrum; *Tritocer.*, tritocerebrum. [Hanström, 1928a.]

Figure 10.24
Centralization in Crustacea.

*The brain and ventral cord of a prim-
itive crustacean, a fairy shrimp
(Anostraca).* [*Hilton, 1934.*]

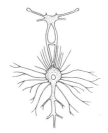

*The brain and cord of a higher crus-
tacean, a crab (Decapoda).* [*Balss,
1944.*]

A. Arrangement of Nervous Elements

The gross plan is a ventral cord with a
pair of ganglia in each segment connected by commissures and giving off
nerves to a peripheral nervous system, a
dorsal anterior brain with circumesophageal connectives putting brain and cord
into contact, and a distinct stomatogastric system supplying the anterior
alimentary canal. It seems best to conclude from a long historical dispute that
there is no nerve net in the skin in
arthropods, in probable contrast to some
annelids.

Microscopic Layout of the Ventral Cord. The
segmental pattern is usually clear, with a
pair of ganglia in each embryonic segment joined longitudinally by connectives to the next ganglia and transversely
by commissures. Fusion of ganglia is

considerable in higher orders and families, especially toward the anterior end
(Fig. 10.24). The subesophageal is the
first ganglion of the ventral cord and is a
composite derived from two, three, or
more segments that supply the mouth
parts. Each segmental ganglion has
nerves to its appendage, to the dorsal
musculature, the sense organs of its
segment, and to the heart. Nerve fibers
enter and leave in bundles or nerves
which are mostly mixed, containing both
sensory and motor axons. There are only
a few dozen motor axons per segment,
but the number of sensory fibers runs
into several thousand. Between the segmental ganglia the longitudinal connectives carry several thousands of fibers
and, in lower groups, scattered cell
bodies, but in higher groups the connective is essentially cell free. A median
nerve in each segment is found in many
groups; it typically branches into a

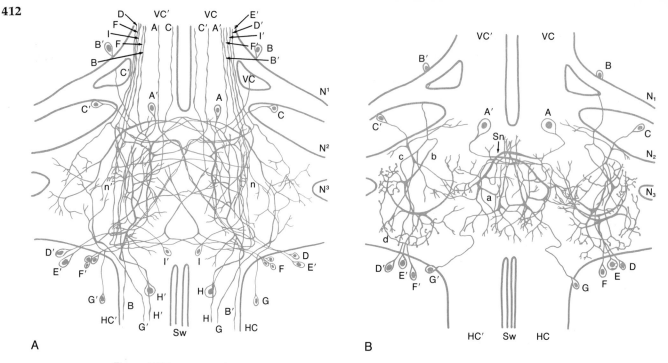

Figure 10.25
Cell types in a ventral ganglion, the second thoracic ganglion of the larval dragon fly, *Aeschna*, as revealed by methylene blue. **A.** Dorsal area of the central neuropile. **B.** Inner neuropile: *hc, vc,* posterior and anterior connectives; *n₁* to *n₃*, segmental nerves; *sn*, sensory neuropile; *sw*, median nerve. Other letters identify recognized neurons. [Zawarzin, 1924.]

transverse nerve that runs on either side to the paired spiracles and the paired heart nerves.

The **types of neurons** in the ventral ganglia (Fig. 10.25) are essentially those outlined above for Annelida. In an increasing number of species a growing list of "identifiable neurons" is now being catalogued, the criteria being position and cytological characters of the soma, the position of the main process in the central connectives, and the connections made with other neurons, both afferent and efferent, including the dynamic properties of the junctions. Since the functional connectivity is more consist-

ent than would be expected from the degree of variability of the branching of short, receptive processes (Fig. 10.26), it must be the weighted product of synaptic contacts and their coupling functions that is well specified for the species. Little suspected a few years ago, this has been a major trend of discoveries. To judge from the representativeness of the species already studied, we are driven to believe that very many of the neurons in most species of animals are in fact constant and specified.

In arthropods, intersegmental neurons have been particularly well studied; they usually run through several ganglia in characteristic positions and send arborizations into each of several segmental ganglia. Some are known to have centers of impulse initiation in several ganglia. These and intrinsic or intraganglionic neurons doubtless account for the considerable local control of movement and local association of sensory impulses at the ganglionic level of the cord. Many interneurons can be excited by touching a given skin locus, and single interneurons have receptive fields that vary from a fraction of one joint of one leg to all the legs on one side and even larger areas; the receptive field is a consistent feature of each of many central interneurons.

Giant fiber systems occur in most forms with an elongated abdomen: among lower crustacea, in shrimps and lobsters, in scorpions, and in some insects. They generally mediate a rapid tail flick or a leaping movement. Higher macrurous crustaceans have one or two pairs (Fig. 10.27), of which the lateral are formed by end-to-end contact of segmental units, between which are septal synapses of the nexus (electrical) type. The median pair consists of two single large axons, each running from one cell in the brain. In insects, giant fibers are specialized for more diverse functions, and commonly occur in a range of sizes, grading into ordinary fibers (Fig. 10.28). They may be enlarged premotor neurons ascending in the abdominal cord or descending in the circumesophageal connectives to coordinate rapid escape movements. In cockroaches and dragon fly larvae there are many; in dipterans one pair of giant fibers runs from the brain at least as far as the thoracic ganglia, and possibly as far as the thoracic muscles, where they can presumably initiate sudden take-to-flight responses.

Figure 10.26
The degree of consistency and variability in branching patterns of the same cell in four individuals.

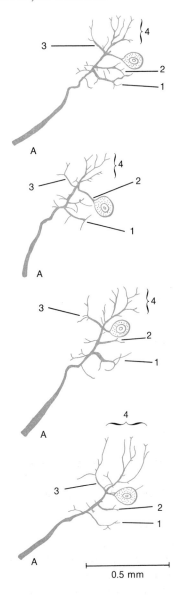

Right motor neuron 28 in metathoracic ganglion of cockroach. (Cobalt-filled; drawn from whole mounts.) [*Tweedle et al., 1973.*]

Figure 10.27
Abdominal ganglion, showing the medial (Med. g. f.) and lateral (Lat. g. f.) giant fibers of segments 5 and 6, and the flexor motor neuron of segment 5 (Mot. N. 5), exiting in the third segmental nerve root (N. 3) [*Johnson, 1924.*]

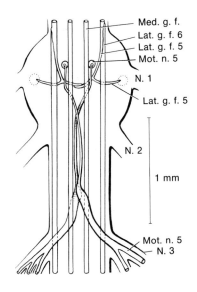

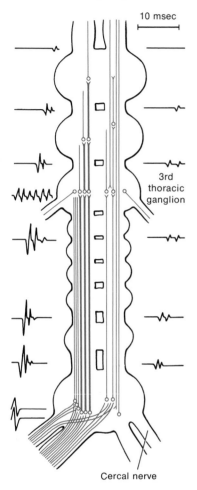

10 msec

3rd
thoracic
ganglion

Cercal nerve

Ascending large fiber pathways that prepare for or permit the evasion response in the cockroach. Silent until activated by mechanoreceptors in the cercal nerve, they conduct forward to the brain, the compound action potential at each level appearing as shown to the side. The multiple firing of motor neurons in the thorax requires the mediation of other, smaller fibers. [Roeder, 1948.]

The giant fiber usually makes synapses with motor fibers of the ganglion to which the effector muscles belong. The synapses between giant interneurons and certain motor fibers in crustaceans are peculiar in that the actual current of the presynaptic fiber is adequate for electrical excitation of the postsynaptic fiber; transmission is electrical and polarized (p. 178).

The main feature of the organization of neurons in the ventral ganglia is the presence of formed pathways, circuits, or connections so specified as to bring about appropriate patterned discharge to the muscles, either in response to adequate input or, under the right conditions, spontaneously.

The Brain. Three main regions are characteristic of the arthropod brain (Fig. 10.23). (a) The protocerebrum consists of several neuropile masses: the paired optic lobes, the median mass, the central body, and the median protocerebral bridge are the chief of these; in addition, there may be association neuropile, usually called corpora pedunculata (Figs. 10.29, 10.30), with its groups of globuli cells, calyx, stalks, and lobes. The corpora pendunculata are fine-textured integrating centers, comprising large numbers of specialized synapses, including glomerular complexes, that bring together input from the anterior sense organs, especially the eyes; from the antennae and the ventral cord with its intrinsic neurons; and from efferent neurons to many parts of the brain. These are regarded as the highest centers in the brain. They may be regions in which the most complex forms of behavior are initiated, as during "spontaneous" changes in mood. They are much

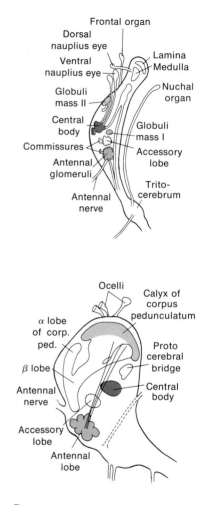

Figure 10.29
Brains in arthropods. **Above.** The brain of a primitive crustacean (*Branchiopoda*) in side view, with its main parts. [Holmgren, 1916.] **Below.** The brain of a typical insect (*Orthoptera*). [Huber, 1960.]

better developed in some insects (e.g., the social insects) than in others; yet they are particularly large in *Limulus*, and have no known behavioral correlate. (b) The deutocerebrum contains neuropile association centers for the first antennae.

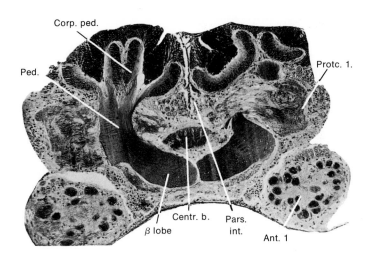

Corp. ped.

Ped.

Protc. 1.

Centr. b.   Pars.
β lobe   int.
                    Ant. 1

Figure 10.30
The brain of an insect. Cross section through the level of the central
bodies of the roach *Periplaneta,* stained with a common histological stain.
Ant. l., antennal lobe (= deutocerebrum, with antennal glomeruli);
β lobe, beta lobe of peduncle; *Centr. b.,* central body; *Corp. ped.,* corpus
pedunculatum ("mushroom body"); *Pars int.,* large cells of the pars in-
tercerebralis; *Ped.,* pedunculus of corpus pedunculatum; *Proc. l.,* protocerebral lobes. [Hanström, 1928b.]

(c) The tritocerebrum, the remainder of
the brain, contains nerves to the anterior
alimentary canal and the upper lip, plus
association neuropile for the second
antennae, where those are present.

The segmentation of the brain has
long been a topic of discussion. The most
compelling view now is that the tritocerebrum is the ganglion of the first body
segment and the mandibular ganglion
(part of the subesophageal) that of the
second; the protocerebrum and deutocerebrum are divisions of a presegmental
anterior head ganglion.

Despite the underlying constancy of
the plan, diversity of brain development
is great in this large and diversified
phylum. The corpora pedunculata are
not necessarily homologous in all the

groups, but probably are at least in
Crustacea and Insecta. The tritoce-
brum is meagerly developed in insects
and the deutocerebrum even less developed in arachnids.

The brain comprises some dozens of
different, characteristic nerve cell types,
distinguished on the basis of size and
position of soma, direction and destination of axons, and branching of receptive
processes (Figs. 10.31; 10.32). The numbers of neurons in the brain range from a
few thousand in minute insects up to
about 100,000 in larger crustaceans (see
Chapter 2).

Neurosecretion is highly developed in
arthropods; details can be found in Bern
and Hagadorn's chapter in Bullock and
Horridge (1965).

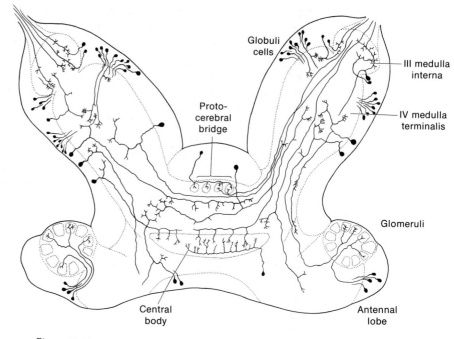

Figure 10.31
The brain of a crustacean. Cross section through the brain of a sand crab, *Emerita* (Anomura), stained by the Golgi impregnation, showing selected neurons. [Hanström, 1924.]

Figure 10.32
Neuron of a crab with the cell body in the brain and the axon going to the ventral ganglion. Stained with methylene blue in life. [Bethe, 1897.]

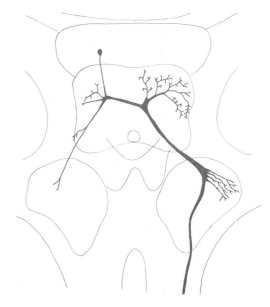

B. Control of Effector Organs

In the earlier chapters we have seen many details drawn from favorable arthropod nerve-muscle preparations, but it is well here to give an overview of the general features of nervous control of effectors. It seems likely that arthropod neuromuscular control is more representative of the general situation in animals than the more familiar kind of vertebrate neuromuscular control, which now appears to be a specialization.

Only a **few motor axons** run to each muscle block. For example, the entire flexor of one segment of a limb usually receives from two to five axons. Of these, one may cause a fast contraction, the other a slow contraction, largely in the same muscle fibers—from which properties the axons get their names "fast" and "slow." These two are typically the complete complement in insect, myriapod, and chelicerate. In crustaceans, each muscle has one or two inhibitor axons as well, and there may be three motor axons (Fig. 10.33). A given axon may run to more than one muscle; in fact, one of the inhibitor axons in the limbs of decapods is common to all the muscles of the limb. Each axon may innervate from 2% to 100% of the muscle fibers of a muscle. Each axon has many endings on each muscle fiber it supplies. Thus there is both multiple (polyneuronal) innervation (several axons per muscle fiber) and multiterminal innervation (many junctions per axon on each muscle fiber).

Only in rare muscles does a twitch follow a single impulse in a fast motor axon. **Facilitation** is usually required, and the response, both electrical and mechanical, is completely graded; thus threshold frequency and duration of impulse train are characteristic of each neuromuscular junction. Slow fiber junctions need several hundred impulses in a second to achieve normal large tension; fast axon junctions may require, for example, six impulses in 0.1 sec. Slow fibers are typically tonic and postural; fast fibers are utilized for escape, prey-grabbing, and the like.

These properties mean that there is extensive **peripheral integration.** Although this is best known in arthropods, peripheral integration must be regarded as more typical of neuromuscular control in the animal kingdom than the control of muscle in the vertebrate system with one axon, one junction, all-or-none muscle fiber response and gradation chiefly by central recruitment. In arthropods and their kin it is also more conspicuous that contraction is often not directly linked to the polarization of the surface membrane. The remarkable phenomenon of peripheral inhibition at the neuromuscular junction is especially well developed in Crustacea. One form tends to clamp the membrane near the resting potential, and thus resist excitatory depolarization, as in the classical i.p.s.p. There is also an inhibition of the release process at the nerve terminal via presynaptic inhibition. So favorable are these muscles and junctions for experimental work that much of our knowledge of i.p.s.p.'s, reversal potentials, presynaptic inhibition, facilitation, miniature end-plate potentials, and inhibitory transmitters derive from crustacean preparations (Fig. 10.34).

Time of arrival of inhibitory and excitatory impulses is often critical because of the requirements of presynaptic interaction. The time limits are different in different junctions. This is true not only for inhibitory and excitatory input, but in some of those cases in which there are

Figure 10.33
The distribution of axons to the muscles in the leg of a crayfish.

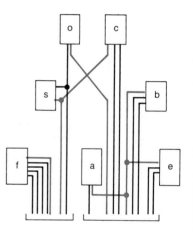

a, *accessory flexor*; b, *bender*; c, *closer*; e, *extensor*; f, *flexor*; o, *opener*; s, *stretcher. Black lines represent motor excitors; colored lines, inhibitory axons.* [Wiersma and Ripley, 1952.]

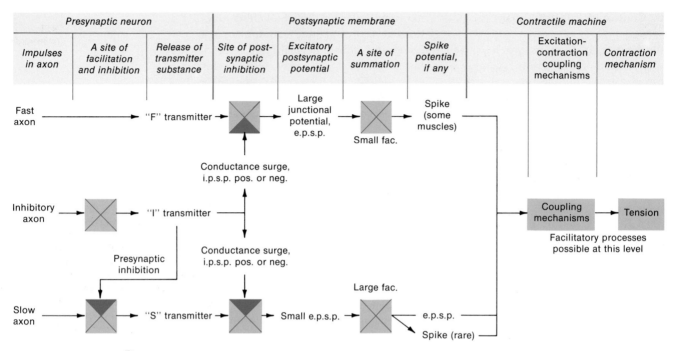

Presynaptic neuron			Postsynaptic membrane					Contractile machine	
Impulses in axon	A site of facilitation and inhibition	Release of transmitter substance	Site of post-synaptic inhibition	Excitatory postsynaptic potential	A site of summation	Spike potential, if any		Excitation-contraction coupling mechanisms	Contraction mechanism

Figure 10.34
The chain of events at the neuromuscular junction in Crustacea. Three classes of axons interact to produce tension. Solid triangles indicate inhibition. (The "S" transmitter, at lower left, has not been shown to be different from the "F" transmitter; they are probably the same in some muscles, but not in others.) [Bullock and Horridge, 1965; modified with assistance of H. L. Atwood.]

two different motor axons to the same muscle (Fig. 10.33), heterofacilitation is said to occur, with a critical dependence on timing.

Insects are notable for certain **specialized muscles.** The indirect flight muscles of higher orders (Diptera, Hymenoptera, Coleoptera, and Hemiptera) do not contract in synchrony with the impulses that reach them in their motor axons. Instead, the impulses set the muscle into an excitable condition such that a sudden stretch causes contraction and a sudden release of tension causes relaxation. The wing frequency is then controlled by a combination of peripheral, mechanical factors, such as skeletal rigidity, wing inertia and wind resistance. It is typically several hundred per second in the presence of five or ten muscle action potentials per second. This is a kind of stretch "reflex" inside the muscle fiber, without Ca^{++} cycling for each contraction.

Arthropod muscle has also been useful in revealing a general property of hysteresis of tension in isotonic or moderately loaded contraction. The tension achieved at a certain frequency of stimulation is quite different if the frequency is increasing than if it is decreasing, even at very slow rates of change. A single impulse injected into a train at a steady frequency can change tension by a large amount for a long time.

Other effectors, such as the luminescent organs of crustaceans and fireflies, the muscles of the alimentary canal and those of spiracles, present many special features of interest, but are beyond the present scope.

The crustacean heart, which has been well studied, possesses an intrinsic nervous system, the **cardiac ganglion** (Fig. 10.35), that initiates bursts of impulses causing a short, graded tetanus in the heart muscle; this is the heart beat. Extrinsic regulatory nerves come from the ventral cord, and in decapod crustaceans these contain a single inhibitor and two accelerator axons on each side. The effects of inhibitory and excitatory fibers are not equal and opposite, especially in their time course. In lobsters the cardiac ganglion consists of nine neurons with different functions; probably no more than two are normally potential pacemakers. The cluster is well organized and its activity is self-sustained but externally modifiable. It shows many features of more complicated ganglia, including pattern generation, rhythmicity, integration, facilitation, antifacilitation, inhibition and acceleration, pattern sensitivity, electrotonic interaction, low-pass filtering, independent spikes in the same neuron, stretch sensitivity, and superimposed humoral effects.

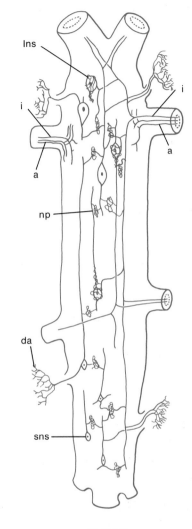

Figure 10.35
Cardiac ganglion of spiny lobster. The 12-mm, 9-celled cardiac ganglion receives one inhibitory (*i*) and two accelerating (*a*) fibers from the CNS. The five large neurons (*lns*) are followers; one of the four small cells (*sns*) is the pacemaker; *np* is a small neuropile; *da*, a dendritic arborization, possibly sensory. [Maynard, 1955.]

C. Receptors

Arthropods have developed a greater variety of morphologically distinct receptors than any other group, including vertebrates. Sensory neurons have their cell bodies in or close to the sensory surface and not generally gathered into major sensory ganglia. Type 1 are bipolars with but one peripheral sensory process, usually short (Fig. 10.36). Type 2 are usually multipolar and have long, branched sensory processes with free endings innervating an area of cuticule

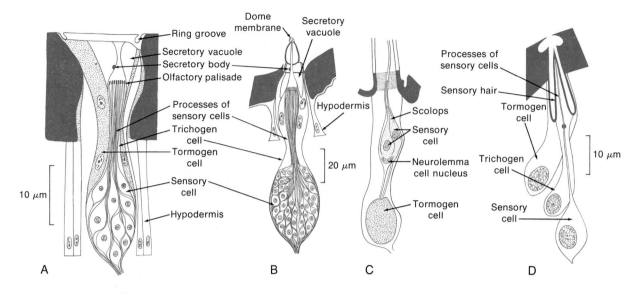

Figure 10.36

Four of the types of cutaneous sense organs (sensilla) of insects. **A.** Sensillum placodeum (Olfactory plate) of *Vespa*. **B.** Sensillum basiconicum (olfactory cone) of *Vespa*. **C.** Antenna hair of *Amorpha* (Lepidoptera). **D.** Sensillum ampullaceum of *Apis*. [Parts A, B, and D, Vogel, 1923; C, Hsu, 1938.]

or of specialized muscle fibers or ligaments (Fig. 10.37).

Table 10.1 gives a condensed and simplified idea of the kinds of receptors that are anatomically known, together with their presumed or identified functions. The above remarks about sensory neurons apply to all of them. Some have a single sensory neuron, others many; the sensory dendrites may penetrate far out into the tips of long hairs or end abruptly at the base of a hair or other structure. Although there are often accessory cells that contribute to forming specialized sense organs, it is believed that none is a sense cell unless it is also a nerve cell. That is, in arthropods (and in invertebrates generally) there seem not to be the secondary sense cells that we find in the hair cells of the acoustico-lateralis system or the taste buds of vertebrates.

Statocysts are common in crustacea as equilibrium sense organs, though they are absent in most other groups of arthropods, which apparently do the same thing with cuticular, muscle, or ligament receptors. A statocyst is formed in a sunken vesicular cavity of the exoskeleton, in which several special types of cuticular sense hairs are deflected by the movement of the water in the cavity and by the weight of grains or secretions cemented into a mass, the statolith. As with other mechanoreceptors, some of the hairs are specialized for tonic signalling of position, some for phasic signalling of movement or acceleration (often directional), and some do both. Some are sensitive to such rapid movement, so

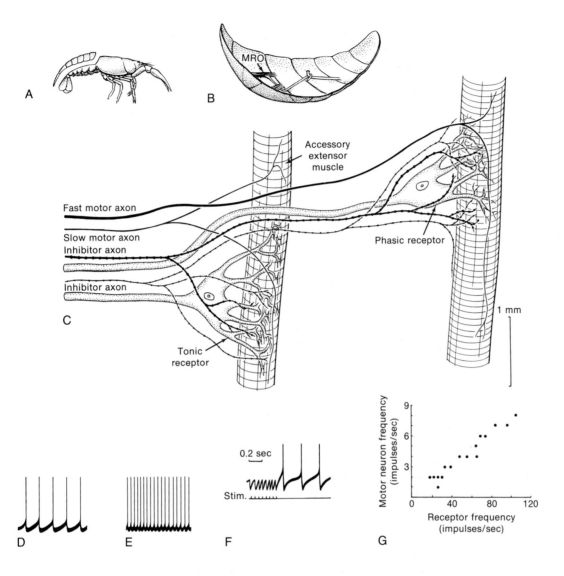

Figure 10.37
One of the pairs of muscle receptor organs (*MRO*) of the crayfish or lobster, with some typical responses. **A.** Removal of the dorsal side of the abdomen. **B.** The boat-shaped preparation with one remaining accessory extensor muscle bearing a sensory neuron and other nerves. **C.** The two modified slips of accessory extensor muscle from one side of one intersegmental joint, as in B, showing the separate neurons. **D.** Typical intracellular response of the slow (tonic) receptor neuron to a small degree of stretch. **E.** The same as in D, to a greater stretch. **F.** Suppression of activity by stimulation of the thick inhibitor axon at 20 per sec. **G.** The reflex relationship between the frequency of impulses in the tonic receptor and the impulses in a tonic extensor motor neuron of the same segment; that is, stretch of the sense organ causes a reflex contraction that relieves the stretch. [Parts A and B, Florey, 1957; D, E, and F, Kuffler and Eyzaguirre, 1955; G, Fields and Kennedy, 1965.]

Table 10.1
Types of Arthropod Receptors in the Skin

Structure	Morphological character	Function, known or presumed
Sensilla trichoidea	Sensory hairs and setae	Mechanoreceptors, proprioceptors, sound, contact chemoreception, humidity, olfactory in different places
Sensilla chaetica	Sensory spines and bristles	Mechanoreceptors, proprioceptors
Sensilla squamiformia	Sensory scales	Mechanoreceptors
Sensilla basiconica	Short, thick hairs, few to many neurons	Mechanoreceptors, contact chemoreceptors, olfactory, humidity, osmotic, temperature receptors
Sensilla coeloconica	Sunken cuticular cones	Olfactory or humidity receptors
Sensilla ampullacea	Sensory tubes	Olfactory receptors
Sensilla campaniforma	Cuticular domes	Directional strain gauges
Sensilla placodea	Cuticular plates and pore plates	Unknown; abundant on bee antennae

Figure 10.38

A type of chordotonal organ, a complex sense organ for mechanoreception in insects. [Gray, 1960.]

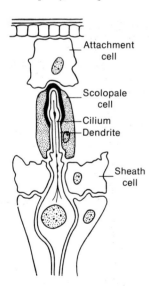

Attachment cell

Scolopale cell

Cilium

Dendrite

Sheath cell

that they are effectively substrate vibration detectors. Statocysts are apparently not the detectors of fluid-borne sound.

Receptors on ligaments are numerous and complex. At the joints of the legs in decapod crustaceans, there may be as many as four different receptor organs, some with special muscles, each with several neurons; some are tonic, others phasic with various rates of adaptation, some respond to movement of a joint one way, others the other way, some to any movement. Some respond to rate of movement rather than acceleration, some respond to movement within one part of the range, others over a different part, which may be wide or narrow.

For example, specialized muscle bundles on the dorsal side of the main extensors of the abdomen of decapods have a single large multipolar sensory neuron acting as a stretch receptor, analogous to the muscle spindle in vertebrates. In the crustacean **muscle receptor organ** (MRO), there is only one sensory neuron, but more than one motor neuron supplies the muscle fibers (Fig. 10.37). An important added feature of control is the presence of one or more inhibitor axons, which reduce the discharge of the sensory neuron to a given stretch. This has been a favorite object of study, not only for sensory physiology, but for the properties of an isolated nerve cell. Simple inhibitory synapses, pacemaker mechanisms, sensory transduction, adaptation, the interaction of an input frequency (in the inhibitor) with a background frequency (in the receptor); these are among the topics that have been discussed in other chapters, in connection with MRO's.

Chordotonal organs are a distinct type of internal mechanoreceptor, usually acting as proprioceptors. They have a complex structure and accessory nonnervous cells (Fig. 10.38). The receptor cell has a modified cilium in the receptor process. In some insects certain chordotonal organs form hearing organs,

(Fig. 2.79); others are receptors for wind direction, flight speed, substrate vibration, gyroscopic haltere vibration, direction of gravity, and velocity of water flow in swimming insects. Responding to displacement rather than to air pressure, some hearing organs with a slack drumhead provide a more directional sensitivity than vertebrate ears but a lower frequency selectivity and perhaps more discrimination of the frequency of periodic amplitude modulation.

The special advantages offered in this array of receptors has made possible a vast range of **physiological studies.** For example, contact chemoreceptors, best known in insects, include the sensory terminals of separate neurons sensitive selectively to water or to certain sugars, ions, amino acids, or plant products. Adequate excitation of a single chemosensory neuron in the housefly can initiate a normal behavioral response of the whole animal. Odor receptors of insect antennae (Fig. 2.70) can be extraordinarily sensitive to particular compounds. Some male moths detect the female by odor; a few molecules make an effective stimulus, far less than one per receptor, suggesting that a single molecule may be sufficient to excite a receptor neuron if there is not clumping of molecules. The dramatic story of this discovery should be read in one of the reviews of Schneider—for example, that of 1970.

Many structurally distinct receptors have not yet been assigned a function on the basis of adequate physiological tests, and some known sensibilities in arthropods are not yet assigned to specific structures—for example, sensibility to temperature, radiant heat, humidity, and uniform external hydrostatic pressure.

Photoreceptors are of two main types in arthropods. (a) Simple eyes with a lens and cup-shaped retina occur in chelicerates and in dorsal ocelli in insects. (b) Compound eyes composed of many neighboring ommatidia (Fig. 10.39) occur particularly in insects, crustaceans, and *Limulus.* Photoreceptors of these and other types that do not fit into either class have primary sensory neurons, called retinula cells, with an axon and a rhabdomere. The rhabdomere, occupying a region of the retinula cell membrane, is a highly elaborated system of tubes or microvilli packed tightly together at right angles to the surface of the cell and frequently at right angles to the direction of the incident light (Fig. 2.62). The rhabdomere is thought to be the primary photoreceptor structure—it contains the photopigment—and in special cases is capable of discriminating polarized light.

The reader is referred to the Suggested Readings at the end of the chapter for works that deal further with the analysis of the structure and mode of action of the compound eye and its ommatidia.

From the point of view of organization, the compound eye presents an interesting question—that is, whether the brain has available to it a representation of the visual world with the grain of the ommatidia or with a finer grain. The answer seems to be the latter. In some forms (e.g., flies), the individual retinula cells, about six in each ommatidium, see a somewhat different but largely overlapping field of view and send axons to different places in the array of second-order neurons, suggesting some preservation of the difference. Furthermore, different retinulae of the same ommatidium have been found to have distinct spectral sensitivities and to prefer certain planes of polarized light. In other forms (e.g., *Limulus*), only one axon appears to

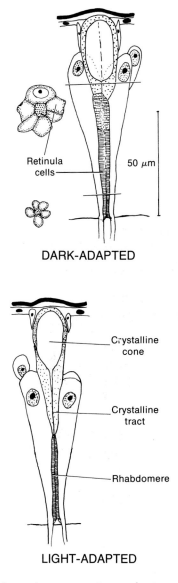

Figure 10.39
Ommatidia of a lower crustacean, Artemia, *in two states.*

Retinula cells

50 μm

DARK-ADAPTED

Crystalline cone

Crystalline tract

Rhabdomere

LIGHT-ADAPTED

Insets show cross sections made at positions indicated by broken lines. [*Debaisieux, 1944.*]

Figure 10.40
Responses of receptor cell and second-order cell in Limulus eye.

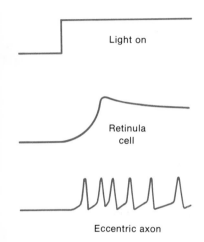

Light on

Retinula
cell

Eccentric axon

carry impulses from each ommatidium, and this represents the pooled excitation of all the retinula cells in that unit. The single axon is that of an eccentric cell; it is regarded as a second-order neuron postsynaptic to the retinulae, which themselves respond in a graded manner, without spiking (Fig. 10.40). Collaterals of the eccentric axon send and receive impulses to and from neighboring eccentric cells to exert the lateral inhibitory influence, sharpening boundaries and enhancing movement detection (Chapter 7).

The ommatidia have receptive fields largely overlapping those of their neighbors, and a question of interest is whether the brain can avoid the fuzzy view that such overlap could produce. The answer is that in principle it can, with central processing based on lateral inhibition. It is not yet clear whether the brain of *Limulus* actually does this. Even if it does, the compound eye provides much less resolution than the eyes of cephalopods and vertebrates.

It is sometimes said that this eye is a specialization for motion detection; perhaps it is, but it does not follow that the cephalopod or vertebrate eye is poorer for motion detection. The compound eye is, however, much more favorable for our analysis (Chapter 5) of motion processing, largely because of the smaller number of receptors and the more discrete layers of neuropile that follow.

Behind the ommatidia are several layers of neurons (Fig. 10.41). In typical invertebrate fashion the cell bodies are segregated at the outside in synapse-free masses like the rind, and the business goes on in concentrated cell-free neuropiles that are typically three in number in higher arthropods. Beginning just behind the retina is the *lamina*, followed by the *medulla* and the *lobula* (in crustacea, commonly called the "internal medulla"). These are connected by well-formed chiasmata or crossings of axons, oriented in one plane across an axis of the eye. Each neuropile mass is composed of (a) vertical fibers at right angles to the surface of the eye that run through the mass and make synapses with sharply limited layers of (b) structurally diverse fibers tangential to the eye surface. The vertical fibers have specialized side processes of various kinds, and doubtless some are receptive and some efferent at definite positions in the tangential layers. The tangential fibers presumably carry excitation or readout of a certain degree of abstraction sideways from the neuropile and convey it to deeper centers. Small internuncials without assignable axons are numerous. The number of distinguishable neuron types characterized by the position, form, and connections of their endings is well in excess of the thirty described long ago by Cajal (Fig. 2.37). Recent analyses recognize many more (Fig. 10.42), and the degree of specification of connections is only minimally indicated by such a number. In the only case carefully reconstructed from serial electron micrographs (Fig. 10.43), the second-order neurons in the lamina of the eye of a fly make precisely specified contacts with each of six different ommatidia and with a particular geometrically defined retinula cell from each of the six (all viewing the same point in space), at the same time receiving two sets of six each of centrifugal fibers coming out from an origin deep in the brain. Such centrifugals are a prominent feature of the eyes of arthropods (and of cephalopods and at least some vertebrates). We must conclude that the connections are highly specified and give a substratum for very particular abstracting of defined kinds of visual stimuli.

In agreement with this, microelectrodes have picked up units in various parts of the postretinal neuropile masses that have fairly complex firing criteria. Most have large receptive fields, not readily subdivided into excitatory and inhibitory fields. Many require or respond best to movement and may be specialized for slow, intermediate, or fast movement, unidirectional, or bidirectional. Some fire particularly in response to novel movements that have not recently occurred, and others are what might be called movement-gated mechanoceptors—that is, they respond to certain mechanical stimuli if there is at the same time visual movement in the right place and direction.

Many insect and crustacean eyes are "slow," as defined by flicker fusion frequency, which is in the same range as our own, about 30 to 50 per second in bright light. Others are extraordinarily rapid, such as the "fast" eyes of certain day-flying insects (flies, bees, dragonflies, some beetles, but not butterflies or orthopterans) and fast running crustaceans, like the high-tide isopod *Ligia.* Fast eyes show flicker fusion frequencies up to three hundred per second or higher, both electrophysiologically and behaviorally, as well as certain differences in the electroretinogram. There is apparently a match of the time constants in the peripheral receptor events and the central processing.

Insect **ocelli** function remarkably, in that the very short axons of the retinula cells end just below the retina on large receptive processes or second-order brain neurons and exert an inhibitory effect without developing impulses. The

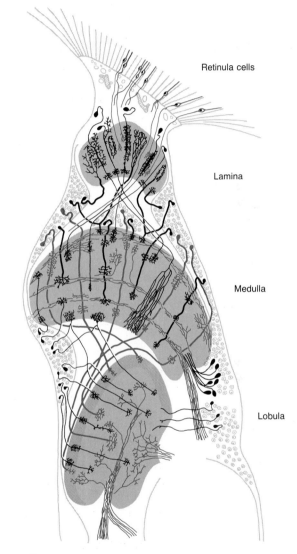

Retinula cells

Lamina

Medulla

Lobula

Figure 10.41
Representative neurons of the optic lobes of the bee, *Apis,* Golgi stained. [Cajal and Sanchez, 1915; modified by Bullock and Horridge, 1965.]

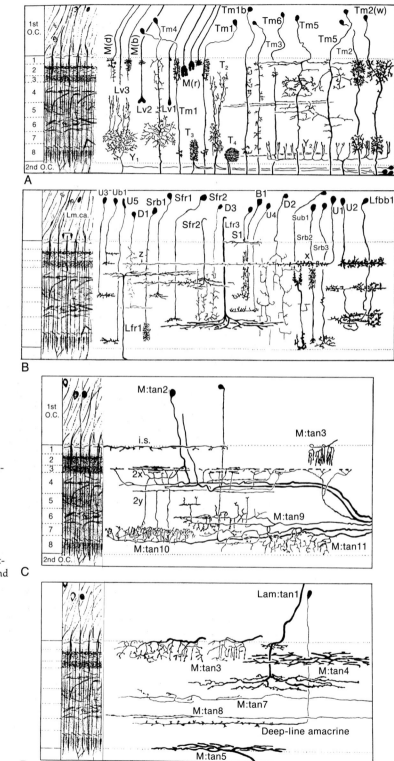

Figure 10.42
Neuron types, in the medulla of the fly *Calliphora*. The figures relate the strata of the neuropile seen in a reduced silver stain (left) to the Golgi-stained cells and fibers. Each of the four parts of this figure shows a certain set of elements in the same neuropile. These sets, and others, constitute the medulla. The symbols stand for some of the neuron types. This is a modern analysis that reveals even more consistency and differentiation of types of neurons and of contacts than were known to Cajal and Sanchez. [Strausfeld, 1970.]

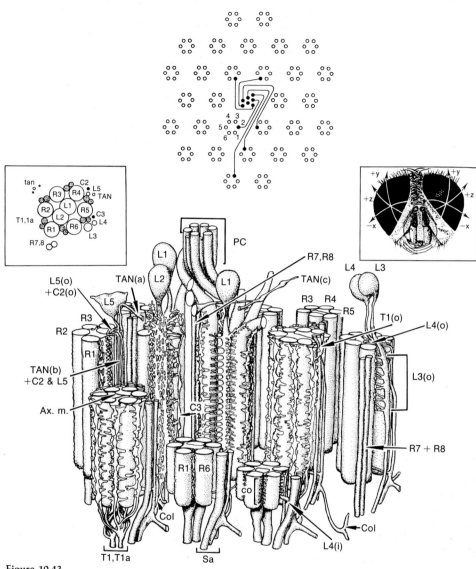

Figure 10.43

Constant neurons and connections. The projection of axons from an ommatidium in the eye of the fly. Six of the receptor (retinula) axons are shown in the upper diagram; they go to six different synaptic complexes, called cartridges, in the next layer, the lamina. Each fiber reaches a cartridge that is exactly defined and a position within its cartridge that is exactly defined. The inset of the fly's head shows the axes that define the ommatidia. The inset on the left is a cross section of a cartridge, showing the six axons from the retinula cells of the ommatidium, labeled R1–R6; the two second-order neurons upon which they all synapse, labeled L1 and L2; and other identified and consistent local and centrifugal axons, indicated by the other symbols. Strausfeld requests that newer findings be noted, showing that what is labeled T1a is actually derived from an amacrine cell in the lamina, and terminates in the first optic chiasm, unlike T1, which does invade the lamina and invests the cartridge (Campos-Ortega and Strausfeld, 1973; Strausfeld, 1976). [Strausfeld, 1971; insert from Braitenberg, 1970.]

onset of illumination terminates spontaneous activity in second-order neurons, which have axons running down the ventral cord.

Many other features of the physiology of the central nervous system of arthropods have already been dealt with in earlier chapters or can be referred to in Bullock and Horridge (1965) and elsewhere. These include the organization of simple and compound reflexes; the character, origin, and influences upon spontaneous activity in the ganglia; the organization of giant fiber systems and features of the connectivity between interneurons and motor neurons therein; the nongiant secondary sensory interneurons and command interneurons that trigger extensive stereotyped movement patterns; the synaptic bases of habituation; and similar topics.

VIII. MOLLUSCA

The phylum of the chitons, clams, snails, and squids presents an enormous range of complexity in nervous organization, as it does in behavior and habit of life. The simplest amphineurans have nervous systems not obviously more elaborated than those of some flatworms. The cephalopods are at the level of fish in structural and functional complexity. A characteristic of the phylum is to superimpose on the usual conservative morphological set of ganglia and connectives a plasticity of topographical relations. This is manifested in an exceptional freedom of ganglia to shift, fuse, and even cross the midline.

The **basic plan** of all classes is a series of about six pairs of well-defined ganglia—a pair each of supraesophageal, pedal, pleural, buccal, branchial, and visceral ganglia. Others, in addition to the basic list, are developed in certain classes. All ganglia communicate by a system of connectives and commissures, which complete a nervous ring around the esophagus and a loop into the visceral mass. We will speak further only of gastropods and cephalopods, the former being one of the most diverse of all classes of animals and the latter one of the most homogeneous.

Major subdivisions of the **Gastropoda** (Fig. 10.44) are defined by their nervous anatomy, which suggests a hypothetical ancestor having a symmetrical set of ganglia. The lowest living gastropods have departed from this form by a remarkable torsion of the whole visceral mass, resulting in crossing of the connectives (called chiastoneury), making a figure eight of the visceral loop. This is the condition of most marine snails, limpets, abalones, and their allies. In groups supposed to be more derivative or advanced (higher prosobranchs, most

Figure 10.44 *(facing page)*
Survey of gastropod types with respect to the general plan of the central nervous system. Shown in dark brown are the ganglia and connectives of the visceral loop (pleurals, subintestinal, supraintestinal, visceral, and parietals, when present as distinct ganglia). Shown in light brown are the pedal ganglia and commissures. Cerebral ganglia, uncolored. Buccal ganglia not shown in most examples, and accessory ganglia shown in only a few: *comm.*, commissure; *conn.*, connective; *g.*, ganglion; *n.*, nerve. [*Haliotis, Pomatia, Pterotrachea, Aplysia, Cavolinia,* and *Lymnaea* modified from Spengel, 1881; *Patella,* Bouvier, 1887; *Buccinum,* Simroth, in Bronn, 1896; *Gastropteron,* Vayssiere, 1879–1880; *Coryphella,* Dreyer, 1910; *Latia,* Pelseneer, 1892; *Helix,* Bang, 1917.]

PROSOBRANCHIA

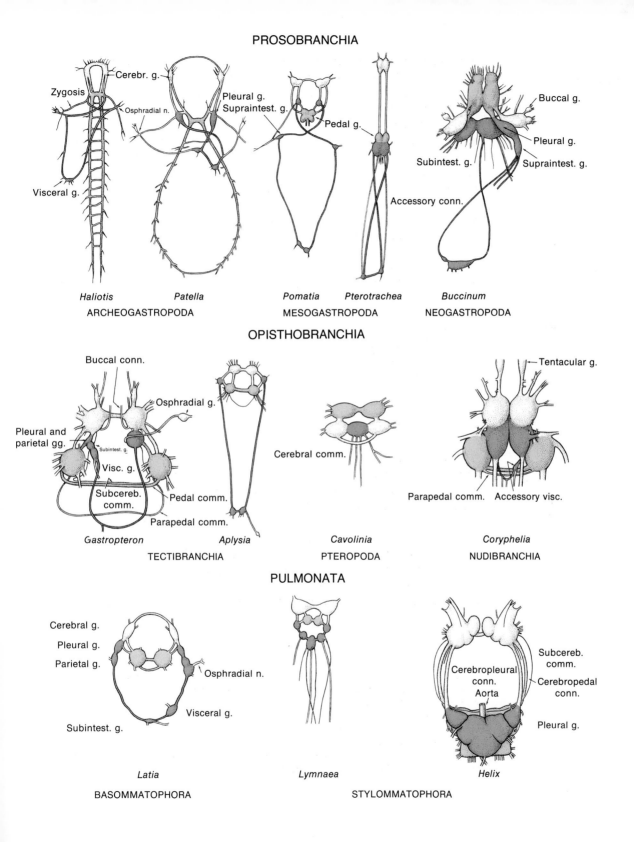

Cerebr. g.

Zygosis

Osphradial n.

Pleural g.
Supraintest. g.

Pedal g.

Buccal g.

Pleural g.

Subintest. g.   Supraintest. g.

Accessory conn.

Visceral g.

Haliotis   *Patella*   *Pomatia*   *Pterotrachea*   *Buccinum*

ARCHEOGASTROPODA   MESOGASTROPODA   NEOGASTROPODA

OPISTHOBRANCHIA

Buccal conn.

Osphradial g.

Tentacular g.

Pleural and
parietal gg.

Subintest. g.

Visc. g.

Subcereb.
comm.

Pedal comm.

Parapedal comm.

Cerebral comm.

Parapedal comm.   Accessory visc.

Gastropteron   *Aplysia*   *Cavolinia*   *Coryphelia*

TECTIBRANCHIA   PTEROPODA   NUDIBRANCHIA

PULMONATA

Cerebral g.

Pleural g.

Parietal g.

Osphradial n.

Visceral g.

Subintest. g.

Cerebropleural
conn.
Aorta

Subcereb.
comm.

Cerebropedal
conn.

Pleural g.

Latia   *Lymnaea*   *Helix*

BASOMMATOPHORA   STYLOMMATOPHORA

Figure 10.46
Innervation of the snail foot.

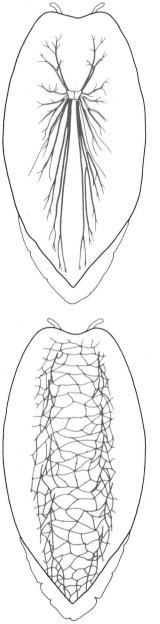

The main pedal nerves (*above*) leading to the plexus deeper in the sole (*below*). [*Schmalz, 1914.*]

Figure 10.45
Brain lobes and cell types in a snail (Helix).

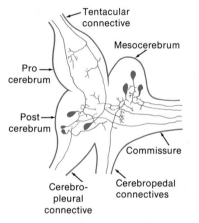

Tentacular connective

Mesocerebrum

Pro cerebrum

Post cerebrum

Commissure

Cerebro-pleural connective

Cerebropedal connectives

The left half of the brain, with tentacle connectives extending upward and cerebropleural and cerebropedal connectives extending downward. [Hanström, 1925.]

opisthobranchs, and pulmonates), this is replaced by euthyneury, which is an uncrossed condition believed to have evolved from the crossed condition independently at least twice and in different ways. Separately from these changes, amalgamation of ganglia into composite masses has occurred in all degrees and several times independently; extreme concentrations are reached in nudibranchs and some pulmonates.

Histological structure is typical for higher invertebrates, as described elsewhere. Differentiation attains a moderate level but not as high a level as in polychaetes and a far lower one than in arthropods or cephalopods. Some extraordinarily large nerve cell bodies are common in opisthobranchs and pulmonates, reaching a diameter of 0.8 mm; the axons do not exceed 35 to 50 micrometers in diameter and 3 m/sec in conduction velocity. The highest level of microscopic differentiation is attained in the garden snail, *Helix* (Fig. 10.45), which displays fine-textured neuropile associated with dense masses of globuli cells in the procerebrum.

Something special happens in the **foot** of gastropods, in which a remarkable variety of locomotor waves is observed among the many species. Snails, for example, can crawl along the edge of a razor blade or on a window pane while locomotor microwaves flow either in the direction of travel or against it. The special physiology of the foot is only incompletely understood. There is an extensive plexus of nerve cell bodies and axons (Fig. 10.46) apparently capable of autonomous reception, intermediation, and motor command, though normally under the control of the pedal and, indirectly, the cerebral ganglia. Peripheral reflexes may play a part. Special mechanical properties of the muscle fibers are also important.

Receptors of all grades abound, from unicellular to complex sense organs. Subepithelial plexuses include not only sense cells and axons but intrinsic ganglion cells. The evidence, however, is against peripheral nerve nets in the strict sense of a diffusely conducting system.

A stomatogastric nervous system is well developed.

The combination of features in gastropods has made several species of this group important in general neurobiology, including the genera *Aplysia, Tritonia, Pleurobranchea, Helix, Helisoma,* and others. Elsewhere in this book, studies of these and related forms are mentioned concerning identified cells and circuits, neurogenic rhythms, simple behavior, and habituation (Figs. 10.47, 10.48, 7.31).

Cephalopods are the most active molluscs. They live by hunting prey, chiefly by sight, and possess very highly developed nervous systems. Quite characteristic are a wide repertoire of behavior, recognition of complex objects, rapid muscle, color, and luminescent responses, and rapid learning.

The central nervous ganglia are all concentrated in a mass around the esophagus (Fig. 10.49), and include the ganglia of the common molluscan plan with several additions: brachial, optic, olfactory, or peduncle, plus several in the periphery notably the stellate. The histological differentiation, especially in higher lobes of the brain, is far beyond that of other molluscs, and on a par with that of the highest invertebrates and of fish. Fortunately, special stains work well, and a great deal of detail is known (Fig. 10.50). Besides a rind of nerve cells, as is general for invertebrates, islands of cells in the neuropile form a complex

synaptic field suggestive of vertebrate gray matter. Small globuli nerve cell clusters occur in the higher regions with a dense and ordered neuropile. Giant cells are present in certain specific places. Though most cells are unipolar, (Fig. 10.51), cephalopods have multipolars in certain sites, and these are often vertebrate-like in being heteropolar. Glial cells are well developed and quite specialized. Tightly wrapped myelin and nodes of Ranvier are unknown. The blood supply to the brain is rich, and vasomotor control is active.

Regional differentiation of brain lobes is extensive, and much is known of the inputs and outputs connected to these lobes. There are regions connected as centers for visceral control, others as lower motor centers for head, eyes, funnel, and arms. There are lobes chiefly related to special sense organs, lobes of intermediate sensorimotor level, and large, subdivided, intricate lobes of higher level, related especially to learning.

The **optic lobe** has been especially well studied, and in some forms is larger than all the rest of the brain. It has a cortex differentiated into at least five sublayers that receive first-order visual cell axons from the retina, and second- and higher-order processes. The output of the cortex projects into the medullary part of the optic lobe, and here a wealth of cell islands, neuropile masses, and intrinsic and projection fibers achieve a complexity close to that of the highest centers. A prime feature is the large number of centrifugal axons from the brain to the optic lobe cortex, and even to the retina.

The eye muscles are richly innervated, and accommodation involves two sets of antagonistic nerves and muscles. The

Figure 10.47
Defensive withdrawal reflex of gill and siphon in Aplysia.

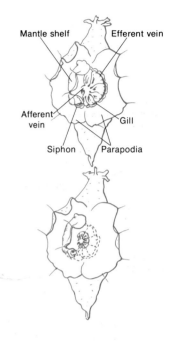

Before and after tactile stimulation of the siphon or mantle shelf.

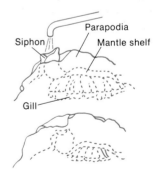

Side view shows siphon contraction and gill contraction. The neuronal circuit for this behavior is almost completely known. [Kandel, 1974.]

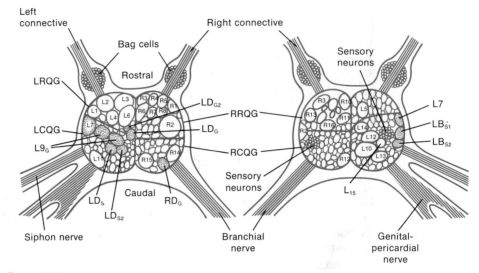

DORSAL SURFACE **VENTRAL SURFACE**

Figure 10.48
The parietovisceral (abdominal) ganglion of *Aplysia* and some of its identified cells. Dorsal and ventral views. The motor neurons for siphon and gill and the two sensory clusters involved in their responses are stippled. [Frazier et al., 1967.]

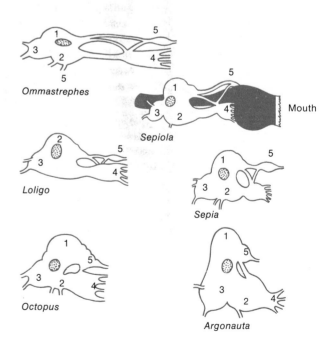

Figure 10.49
Central nervous system of several cephalopods (optic lobes removed), showing degrees of condensation. 1, cerebral; 2, pedal; 3, visceral; 4, brachial; 5, superior buccal ganglia. Stippling marks optic tract. For *Sepiola*, the buccal cavity and esophagus are shown in black. [Pelseneer, 1888.]

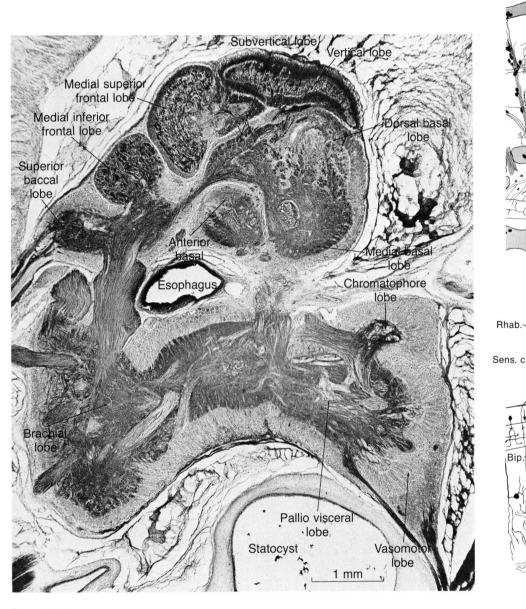

Figure 10.50
The brain of a cephalopod. Sagittal section of the brain of *Octopus* (by Cajal reduced silver method). [Young, 1971.]

Labels on Figure 10.50:
Subvertical lobe
Vertical lobe
Medial superior frontal lobe
Medial inferior frontal lobe
Dorsal basal lobe
Superior baccal lobe
Anterior basal
Medial basal lobe
Esophagus
Chromatophore lobe
Brachial lobe
Pallio visceral lobe
Statocyst
Vasomotor lobe
1 mm

Labels on Figure 10.51 (above):
Cell rind
Vertical lobe
Basal lobe

Labels on Figure 10.51 (below):
Lim. mem.
Rhab.
Sup. cell
Sens. c.
Centrif. f.
Retina
Bip.
C.f.
M.d.
Tan.
C. aff.
Cortex
Optic lobe
Subcortical medulla

Figure 10.51
Above. Neurons of the brain of Sepia. [*Thore, 1942.*]
Below. Neurons of the retina and optic lobe of Octopus. [*Young, 1971.*]

largest eyes known occur in giant squids, in which they reach at least 24 cm in diameter. The retina is fine grained and often shows a rectilinear ordering of the rhabdomeres (the part of the receptor cell membrane that contains the visual pigment). Statocysts are elaborate, and their reflexes are important; body proprioceptors and eyes cannot substitute. The slit pupils are held in a constant orientation with respect to gravity as the head changes position, but this requires the presence of at least one statocyst.

A **giant fiber system** is highly developed in decapods, more so in some than others (Fig. 10.52). In *Loligo,* a pair of first-order giant multipolar neurons in the brain send their axons, which anastomose across the midline, to the posterior part of the brain to synapse with about seven pairs of second-order giant unipolar neurons. Most of these send axons to muscles of the head and funnel, but one runs to the stellate ganglion and makes excitatory distal giant synapses with the third-order giant fibers, one such in each stellar nerve. Each arises as a confluence of processes from many small nerve cells in a giant cell lobe (Fig. 10.53), ends by profuse branching, and innervates a large area of mantle musculature. An additional set of junctions, called proximal synapses, is made by a second preganglionic giant fiber from the brain onto the confluence of the third-order giant axons and is also excitatory.

Considerable detail is available on the visceral nerve supply; on the nerve elements in muscle, especially in the arms, whose brachial nerve cords (Fig. 10.54) in some species aggregate more nervous tissue than the brain; on control of chromatophores; on the functions of ganglia; and on the divisions of the brain (for

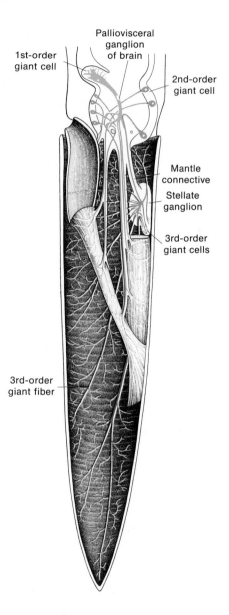

Figure 10.52
The giant fiber system of a squid, *Loligo.* Light brown, first-order giant neuron. Medium brown, second-order neurons. Dark brown, third-order giant neurons. The cells and fibers are not to scale. [Young, 1939.]

details reference can be made to Bullock and Horridge, 1965).

The **highest functions** of the brain have been studied by ablation and learning tests. Removal of the vertical lobe leaves behavior superficially normal, neither hyperexcitable nor profoundly depressed. But whereas normal *Octopus* learn a tactile discrimination between objects of differing roughness (but not between objects with vertical grooves and objects with horizontal grooves of the same total roughness), the discrimination is impaired by vertical lobe lesions. This, however, is compensated for by more training per day. This lobe seems to assist retention of learned associations by enhancing the establishment of a memory. It is not necessary for the formation of associations or for reversing previously learned responses. Visual discriminations behave similarly, and much is known of the qualities that matter in recognizing shapes. Short-term and long-term memories can be distinguished. Measurable time in hours is needed for an intensive training on one side to transfer to the other side of the brain.

IX. VERTEBRATA

A. Differences Between Invertebrates and Vertebrates

The nervous system of the lowest living forms of vertebrates, the cyclostomes, and even the elasmobranchs and teleosts, is not clearly more complex than that of the highest invertebrates, especially the cephalopods and perhaps also the most advanced arthropods. In ancestral vertebrates it was no doubt still simpler. Nev-

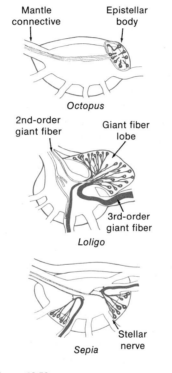

Figure 10.53
The stellate ganglion in three members of the class Cephalopoda. [Young, 1936b.]

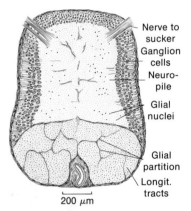

Figure 10.54
The brachial cord.

Drawn from a cross section of an arm of an octopod. [Colosanti, 1876.]

436

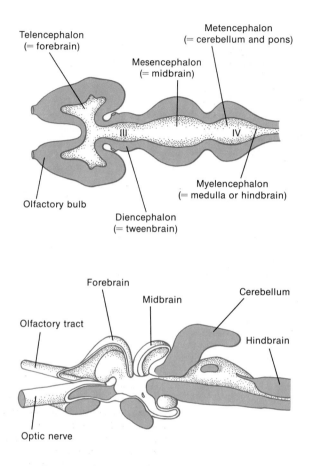

Figure 10.55
Neuromeres and ventricles of the vertebrate brain.

Telencephalon
(= forebrain)

Metencephalon
(= cerebellum and pons)

Mesencephalon
(= midbrain)

III

IV

Olfactory bulb

Myelencephalon
(= medulla or hindbrain)

Diencephalon
(= tweenbrain)

Forebrain

Midbrain

Cerebellum

Olfactory tract

Hindbrain

Optic nerve

ated with relaying sensory input, mainly rostralwards; the cells of the ventral half, the basal plate, are primarily motor neurons or interneurons connected with them. The consequent separation of sensory from motor roots of the segmental nerves is for the physiologist a most convenient difference. The sensory nerve cell bodies are gathered together into dorsal root ganglia (except the olfactory and optic). A similar concentration into segmental ganglia occurs for motor cell bodies that supply visceral structures, basically smooth muscle and glands; these are the autonomic ganglia of the prevertebral chain. In many invertebrates there are scattered neurons or ganglionic collections on or in viscera; these are apparently similar to the plexuses in the wall of the vertebrate gut. The cells of the neural tube of vertebrates come from its inner lining, and the tracts of fibers for long-distance projection occupy a position superficial to the cellular core. This is not simply the reverse of the situation typical of invertebrates, as we see in the next section.

Histologically, there is a profound difference between the possession of gray matter (a synaptic region including nerve cell bodies) in vertebrates and the possession of pure neuropile (lacking cell bodies) in the invertebrates (Figs. 2.53; 2.54; 2.55; 2.56; 2.57; 10.56). Somas are relegated to the surface of the ganglion in the invertebrates—to a layer called the "rind," which is typically free of nerve endings and synapses. In the vertebrates, true neuropile is usually of microscopic dimensions, and lies between the neuron somas of the gray matter. Only in special places does it become extensive—that is, with somas few and far apart. Both groups have regions of pure fiber matter containing bundles of fibers without

ertheless, a number of features are shared by lower and higher vertebrates that distinguish them from virtually all invertebrates.

Morphologically the whole central nervous system is developed from a dorsal tubular invagination, and the brain from a linear series of expanded vesicles of this tube with variously thickened walls, floor, or roof (Fig. 10.55). The cells of the dorsal half of the tube, the alar plate, are primarily associ-

instances in the invertebrates in which mechanisms are found without counterpart in the vertebrates, especially in transmitters and ionic permeabilities.

B. Spinal Cord

The evolution of the spinal cord—from a highly autonomous, segmental reflex organ in lower classes to one that is much subordinated to superimposed levels of control in higher groups—presents us with a kind of capsuled version of vertebrate central development (Fig. 10.59; Table 10.2). Meninges are present in the most primitive groups as a single-layered meninx primitiva. In addition to this, elasmobranchs have a plexus of perimeningeal blood vessels. In the teleosts and higher forms, there is added a distinct dura mater and, beneath that, differentiated leptomeninges comprising an arachnoid layer in the center and pia mater innermost, immediately covering the parenchyma of the central nervous tissue. In amphioxus and lampreys there are no intramedullary blood vessels in the spinal cord. Differentiated neuroglia cells are rare in amphioxus, few and primitive in cyclostomes, but the basic vertebrate types—astrocytes, oligodendroglia, and probably also microglia—are present from elasmobranchs up. They are said to be more abundant in higher forms, but this is not clearly established. A blood-brain barrier of sorts is already present in fish. There are a number of special features in the spinal cords of cyclostomes, elasmobranchs, and teleosts concerned with alternation of dorsal and ventral roots, the presence of intra- and supramedullary ganglion cells, a marginal plexus, and giant fibers

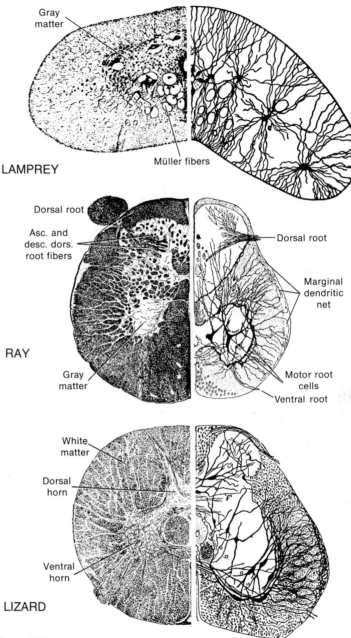

Figure 10.59

The spinal cord. Three classes of vertebrates, each shown by a general stain (*left*) that emphasizes myelin, where present (absent in lamprey), and a Golgi impregnation (*right*). [Upper, Kappers et al., 1936; middle, Cajal, 1909; lower, Lenhossék, 1895.]

Table 10.2
Summary of the Spinal Cord of Vertebrates

Structure	Amphioxus	Cyclostomes	Elasmobranchs	Teleosts	Amphibians	Reptiles	Birds	Mammals
Meninges	Meninx primitiva		Meninx primitiva, perimeningeal vessels	Dura mater, leptomeninx (arachnoid + pia)	Same; more trabeculae	General increase	Same	Good cerebrospinal fluid
Intramedullary blood vessels	Absent	Lamprey: absent Myxine: present	Present					
External dorsal root ganglia	Absent	Present						
Dorsal roots	Alternate left and right	Same level left and right; alternating with ventral roots	Same level; may alternate with ventral roots	Same level; alternate less with ventral roots	Both emerge at same level			
Spinal nerves by dorsal and ventral root fusion	Not formed	Lamprey: no Myxine: yes	Formed					
Intra- or supramedullary ganglia	?	Present (intramedullary; segmental)	Absent	Present (supramedullary; nonsegmental)	Absent			
Glia	Not differentiated; no myelin	Few, primitive; no myelin	True astrocytes, oligodendrocytes; myelin	More abundant	Increasing in differentiation			
Giant fibers	Rhode's	Müller's and Mauthner's	None	Mauthner's	Mauthner's (in embryo)	None		
Marginal plexus ("dendritic net")	Probably	Present				Reduced	Embryo only	Absent
Segmental enlargement, cell pattern	None			Motor cell pattern at fin level	Cervical and lumbar enlargement	Reflex limb development	Increased nuclei;	Well-developed lateral horn

(Figs. 10.60; 10.61). The lamprey cord is remarkable in lacking all trace of myelin, even on the largest axons. Elasmobranchs and teleosts have full fledged myelin; giant fibers 50 μm in diameter conduct at about 50 m/sec instead of 5 m/sec as in cyclostomes.

Chief among these myelinated giant fibers are the **Mauthner's fibers** of teleosts and aquatic amphibians, which arise from a single pair of cells in the medulla, decussate and descend in the contralateral cord (Fig. 10.62) as penultimate motor pathways, and activate motor neurons for muscles that flip the tail. They mediate the first phase of the startle response to vibration, and are intimately connected with VIIIth nerve input, presumably from vibratory receptors. They inhibit each other reciprocally,

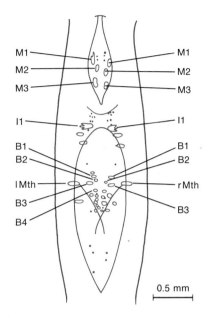

Figure 10.61
Stereophotomicrographs of giant cells of Müller in the lamprey. Dorsal view of whole mount of Müller cells M2, M3, and I (filled and stained with cobalt). [Courtesy of M. J. Cohen.]

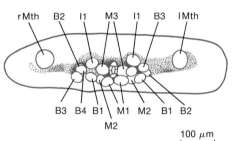

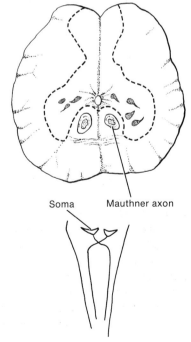

Figure 10.60
Müller and Mauthner cells in the lamprey. **Above.** Plan view of the brain to show the distribution of identified Müller-cell somas (*B* symbols) and Mauthner cells and axons (Mth), reconstructed from serial sections of a larva. **Below.** Cross section of the spinal cord. [Rovainen, 1967.]

Soma   Mauthner axon

Figure 10.62
Mauthner cells in the carp. Transverse section of the spinal cord. [Cajal, 1909.] Plan view of the medulla and cord.

and hence fire on one side only or alternately on the left and right. These cells are among the best known anatomically; their synapses include chemical excitatory, electrical excitatory, chemical inhibitory, and electrical inhibitory junctions (see also Chapters 2 and 5).

But the principal developments in the evolution of the spinal cord are **two major changes** associated with emergence from aquatic to terrestrial life and the vast increase in the number of muscles and joints in the limbs, followed, probably independently, by the increase in cephalization. (a) The intrinsic mechanisms in the lower swimming vertebrates mainly involve coordination of locomotor rhythms in fins requiring various patterns of separate and coordinated movement (as diverse as the undulations of eels and the rigidity of seahorses, which have independent rhythms in the several fins). In tetrapods there are still rhythms, and coordination is intersegmental, but in addition (b) there is coordination locally, between the joints of each limb and between the digits, in a wide range of movements. These, together with higher control of locomotion, have brought on a great increase in long ascending and descending pathways.

The **intrinsic mechanisms** that determine temporal and spatial pattern of commands to muscles are one of the chief components of the cord. We deal elsewhere (Chapter 8) with the general phenomenon of nervous involvement in rhythmic movements and the major alternatives of central versus peripheral (reflex) determination of the timing of cycles. It is probable that at least many of the rhythms (e.g., locomotion and respiration) arise from intrinsic pacemakers in the segmental levels of the central nervous system, with more or less modulation of the form by proprioceptive reflexes. It is not clear what difference there may be in this respect between aquatic and terrestrial, lower and higher, finned and tetrapod vertebrates (Fig. 10.63).

There is probably an evolution from chiefly multisynaptic, phasic, and propulsive reflexes in lower forms to the monosynaptic, tonic postural reflexes in vertebrates that bear their own weight on land, but this is not yet clearly documented.

The **gamma loop** control of the muscles (p. 267) is probably an invention of higher vertebrates. With the evolution of muscle spindles and of separate (gamma) motor neurons to control their sensitive range, it is easy to see how higher neural levels, such as cortex, reticular formation, and cerebellum, would add control of the main muscle contraction by commanding the gamma discharge.

In some fish there is evidence that a small but important fraction of the **dorsal root fibers are efferent** to peripheral effectors. In mammals these are apparently much reduced, represented in the ordinary spinal nerve only by a small number of vasodilator fibers. Perhaps more of the visceral efferent fibers of cranial nerves derived from dorsal roots (e.g.) the trigeminal, glossopharyngeal, and vagus) represent this component.

Beginning in teleosts there is a well developed substantia gelatinosa; the functional significance of this region of the dorsal horn is not understood, but presumably it represents some aspect of improved processing of sensory input.

Although the spinal cord is quite autonomous in lower forms and capable

Spinal transections

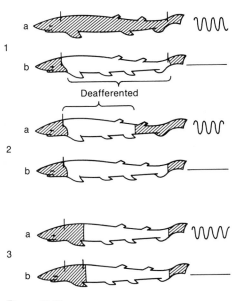

Figure 10.63
Dependence of the swimming rhythm of a shark upon sensory input. In shark 1, two spinal transections (vertical bars) do not stop the shark from swimming (a), but deafferenting, by cutting dorsal roots of the segments between the transections, does (b). In shark 2, deafferenting fewer segments and leaving a number of posterior segments with sensory input intact allows the rhythm to persist (a), but extending the dorsal root section abolishes it (b). In shark 3, anterior segments with dorsal roots intact can support the rhythm (a), but a new cord transection that isolates the deafferented region stops it (b). These experiments do not prove that input must be rhythmic, and others show that nonrhythmic input can sustain a rhythm, although rhythmic input can entrain and dominate. [Lissmann, 1946.]

of considerable autonomy in higher groups in visceral control and crude somatic reflexes, **cephalic dominance** is a major trend in evolution. Phylogenetic trends like this may be less clear because of the separate influence of body size

(Fig. 10.64). Higher forms show less normal autonomy of the cord, more traffic to and from the brain, and more dependence on the brain in the operations of the cord. A sudden transsection of the spinal cord in higher mammals is followed by a state of "spinal shock" below this level, during which both inactivity and unresponsiveness are profound. This passes off in days or weeks in man, hours in the cat, and is hardly demonstrable in amphibians and fish.

The earliest mechanisms for brain control over the cord are **pathways** made up of short relay neurons. Collectively the ascending components may be called spinobulbar pathways, whereas the descending components include both reticulospinal pathways from the patchy gray matter of the medulla and vestibulospinal pathways from the eighth nerve nuclei processing information from the equilibrium receptors of the labyrinth. These elements are already present in cyclostomes and only become more elaborate and faster, with long intersegmental fibers removing the dependence upon short relays. Ascending spinocerebellar tracts first become evident in elasmobranchs, providing the cerebellum with sensory input from the proprioceptors of the body wall. The dorsal funiculus of white matter in these lower fishes is a mixture of ascending and descending fibers. Beginning with reptiles, but particularly in mammals, ascending first-order axons from somesthetic receptors dominate the dorsal column, relaying in the gracile and cuneate nuclei in the lower medulla with second-order neurons en route to the diencephalon (Fig. 10.65). Another principal ascending pathway, probably more primitive, is the spinothalamic. This is

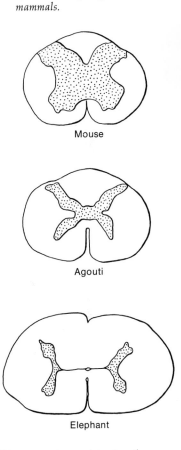

Figure 10.64
Spinal cords of large and small mammals.

Mouse

Agouti

Elephant

Drawn nearly equal in size, the greater proportion of white matter in larger species is clear. [Kappers et al., 1936.]

Box 10.1 The Visceral Nervous System

Although some nerve supply to the viscera is found in the lowest vertebrates, as well as in invertebrates, it increases greatly in extent, complexity, and control among the higher classes, especially birds and mammals. The efferent supply is called the **autonomic system** and comprises (a) **preganglionic** neurons with somas in the CNS and axons coursing out to autonomic ganglia to synapse with (b) **postganglionic** neurons whose somas in these ganglia send axons to the viscera. (c) In addition to these two well-known types of neurons it now seems clear that there are **intrinsic neurons,** confined to the autonomic ganglia, that receive input from preganglionic neurons and from sensory neurons and give output inhibitory to postganglionic neurons, sometimes by direct axons and sometimes by liberating a transmitter-like substance from terminals not closely apposed to any cell. Such an instrinsic neuron is intermediate between ordinary neurons and neurosecretory neurons (see Box 6.2, p. 216).

The relayed motor arrangement is found rarely elsewhere. Preganglionic outflow through cranial nerves III, VII, IX, and X, together with that through sacral spinal nerves (see figure in the adjacent column) and the corresponding postganglionic neurons is collectively called the craniosacral or **parasympathetic system** because of commonalities in physiologic effects and in transmitter released (the postganglionic neurons are generally cholinergic). Preganglionic outflow through thoracic and lumbar segments, and their corresponding postganglionic neurons, is collectively called the thoracolumbar or **sympathetic system** and shares certain physiologic effects and, usually, the postganglionic transmitter epinephrine or norepinephrine(= adrenalin or noradrenalin); hence the neurons are called adrenergic. The sympathetic system is typically more active in behavioral states of strong emotional content, such as fighting and fleeing, and adjusts the actions of the heart, blood vessels, organs of digestion, and temperature control adaptively to the emergency. The parasympathetic system innervates nearly all the same viscera and exerts an opposing effect. Details may be found in standard texts of mammalian physiology or neurology. Autonomic outflow controls smooth muscle contraction, blood flow, and some glandular secretion. It can also enhance the sensitivity of receptors—for example,

to touch and taste. Possibly pain and other modalities are also subject to influence.

The lowest vertebrates show little or no double innervation, antagonism, or differentiation of either the sympathetic or the parasympathetic system (see

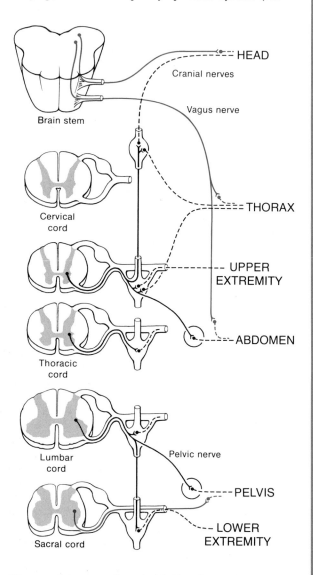

Diagram of the sympathetic (in black) and parasympathetic systems (in color). Preganglionic axons in solid lines; postganglionic axons in broken lines.

Box 10.1 (*continued*)

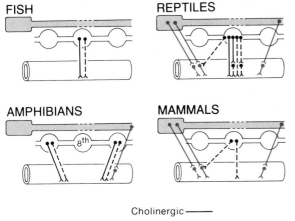

FISH

REPTILES

AMPHIBIANS

MAMMALS

Cholinergic ———
Adrenergic - - - -

Diagrams of the autonomic innervation of the intestine in several vertebrate classes. The brain, giving origin to the cranial (vagal) outflow, is indicated by the swollen (left) end of the CNS (shaded), the paravertebral chain ganglia by the three connected circles; sympathetic fibers by black lines, parasympathetic in color. [Burnstock, 1969.]

figure above). In cyclostomes, the sympathetic system is rudimentary, and the vagus alone supplies the gut and heart. Blood-borne adrenergic influence from diffuse "chromaffin" tissue may represent the sympathetic. Elasmobranchs are rather primitive; some teleosts show a high degree of differentiation. From amphibians on there is much less change. Little is known of the ongoing background level of activity or the increase above this in different states among the vertebrate classes. It has been discovered in frogs that arousal of the sort that would induce the animal to jump is typically accompanied or preceded by a massive sympathetic barrage. We do not know whether this is true in other vertebrates.

In addition to the pre- and postganglionic and intrinsic neurons there are in some viscera, notably the gut, peripheral neurons forming enteric plexuses, both between layers of smooth muscle (the myenteric plexus) and between the mucosa and the innermost muscle (the submucous plexus). These comprise excitatory and inhibitory motor neurons, interneurons, and sensory neurons or their axons and endings. The plexuses are largely autonomous, governing the smooth muscle by local reflexes and spontaneous rhythms (see

also Box 7.1, p. 248). They are also under central control, both parasympathetic excitatory and sympathetic inhibitory (see figure below).

The visceral afferent neurons supply a rich input to the central nervous system from widespread receptors, only rarely reaching the level of conscious sensation, but important for regulatory reflexes. Some of the afferent somas are in dorsal root ganglia or their cranial nerve equivalents. Others are apparently in the periphery, including the enteric plexuses. Their connections are both central and peripheral, so that smooth muscle response to visceral stimuli can occur not only via central reflexes, but also directly, via sensorimotor synapses in the autonomic ganglia, and even within the enteric plexuses.

The central representation of visceral control extends from the intermediate gray column of the spinal cord upward through the reticular formation by short relays to the main coordinating level for autonomic integration, the hypothalamus. Even higher limbic and certain frontal cortical regions exert effects upon the viscera, as in emotional expression. Little is known of the evolution of central control of viscera but this should be a fruitful subject in view of the increasing knowledge of reciprocal influences of viscera upon affect, somatic performance, disease, and conditioning—both of and via the viscera.

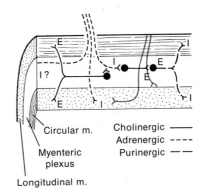

Circular m.

Myenteric plexus

Longitudinal m.

Cholinergic ———
Adrenergic - - - -
Purinergic — — —

Diagram of nerve components in the wall of the mammalian rectum. Adrenergic nerves are concerned largely with extrinsic modulation of gut motility, and not with sustained propulsive function. Afferent components are not shown. *I*, inhibitory; *E*, excitatory; *Long. m*, longitudinal muscle layer; *Circular m.*, circular muscle layer. [Burnstock and Costa, 1973.]

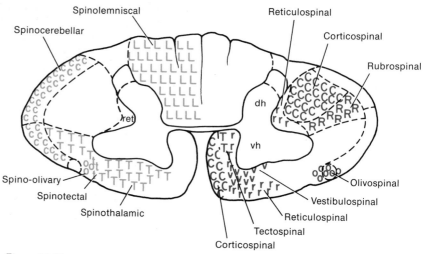

Figure 10.65
Schematic distribution of long tracts in the spinal cord of mammals. Descending tracts are shown on the right and ascending tracts on the left. *dh,* dorsal horn; *ret.,* reticular substance; *vh,* ventral horn.

characterized by relaying and crossing in several segments adjacent to the level of entrance of the dorsal root afferent. The spinothalamic pathway carries largely nociceptive and temperature input, but in man considerable tactile input as well.

In lower classes there are probably no descending direct connections from the forebrain to the spinal cord. Even the midbrain probably communicates by relayed pathways until (evolutionarily) the tectospinal tract becomes distinct in amphibia and a primitive rubrospinal tract begins in reptiles. A direct tract from forebrain to cord has been demonstrated only in mammals and birds. Within the mammals the corticospinal pathway shows great diversity, from a scattered group of fibers with feeble effect (and quite dependent upon antecedent influence of reticulospinal or vestibulospinal impulses to achieve control of the motor neuron) to a large

bundle of fibers forming the medullary "pyramids," hence called the "pyramidal tract" of primates, which can by themselves command some spinal motor neurons. Even here it is still a relatively slower and more heterogenous population of fibers than either the reticulospinal or the vestibulospinal tracts. It is very small in cetaceans; small and reaching only to cervical levels in ungulates; and progressively larger in rodents, carnivores, and primates (which, of course, are not a series phylogenetically).

A well-defined **sympathetic nervous system** is present even in elasmobranchs, but it does not supply the skin or the head. There is also little or no double, antagonistic innervation of the viscera, with opposing actions by sympathetic and parasympathetic systems. That antagonism and the contrast between sympathetic and parasympathetic systems is first well developed in the amphibia, which therefore basically resemble mammals in respect to the autonomic system (see Box 10.1). Teleost fish are intermediate.

C. Myelencephalon and Metencephalon

The medulla and pons are in part similar to the spinal cord, as expected from their role as the segmental supply to the head somites. This part includes somatic and visceral efferent nuclei for those effectors in the head or derived from it, internuncial cells and related circuits, and afferent nuclei receiving from the receptors of the cranial nerves V–XII. In addition, as in the cord, ascending and descending tracts traverse these levels as well as arise and terminate in them. But there are in the medulla and pons certain other novelties that markedly enhance their complexity, importance, and evolutionary modifications.

A scheme to systematize the **functional components** of the cranial nerves and their central connections is helpful, particularly in the light of comparative neurology and ontogeny (Figs. 10.66, 10.67). The medulla, with its thin nonnervous roof, consists of columns of neurons that, in the order from most ventral and medial toward most dorsal and lateral, are for somatic efferent, visceral efferent, visceral afferent, and somatic afferent functions (Table 10.3). Somatic efferent neurons consist of motor neurons of the nuclei of the eye muscle nerves, VI, IV, and III (here we reach forward into the midbrain to include all the segmental innervation), and in the higher terrestrial classes the hypoglossal, XII, for the tongue (Fig. 10.68). The visceral efferent column has to be subdivided because of the peculiar development of some branchial arch musculature, regarded as visceral, into striated voluntary muscles of the jaws and face, pharynx and larynx. These components of nerves V, VII, IX, X, and XI are hence treated as special visceral efferent motor neurons (Fig. 10.68). The general visceral efferent components are for the iris and ciliary muscles of the eye, the salivary glands and the viscera, which have migrated back from the throat region (heart, stomach, etc.), carrying their wandering or vagal nerve supply. Taste is somewhat arbitrarily considered visceral afferent because its nucleus of second-order neurons is in the same column as those of typical visceral afferents from the abdominal organs. The ordinary or general somatic afferent component of the cranial nerves comprises the proprioceptors in the striated muscles and the cutaneous supply of the face. This leaves a special somatic afferent category for the organs of special sense, which make up the acousticolateralis system.

The **intrinsic mechanisms** of the medulla and the pons, like those of the cord, include many interneurons, more or less closely associated with the columns of motor neurons and secondary sensory neurons. These include the cells of the reticular formation, a region that is inexactly delimited. As a descriptive term, it refers to the zone of transition from gray to white matter, in which trabeculae of the one cut up the other like the meshes of a net. This term is textural, and does not correspond to a natural morphological or functional entity. It has acquired another definition, based especially on functional properties, setting these neurons apart from the early sensory and final motor cells, as explained in Box 10.5 (p. 492).

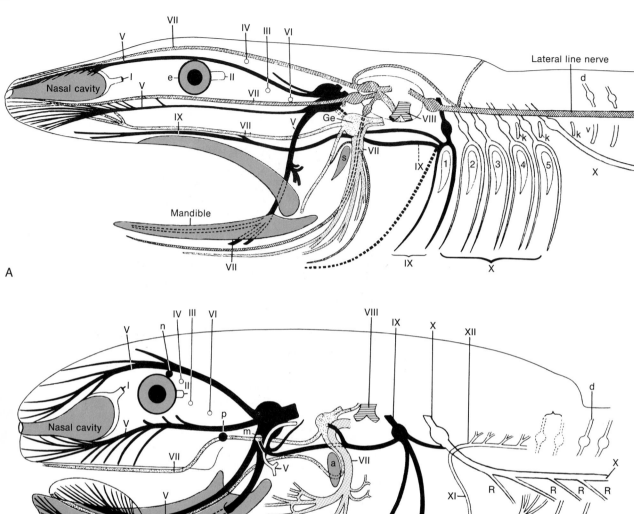

Figure 10.66
The cranial nerves in lower vertebrates (**A**), which respire with gills, and in higher vertebrates (**B**), which breathe air. 1–5, gill clefts; I–X, cranial nerves; *a*, auditory canal; *d*, dorsal root of first spinal nerve; *e*, eye; *k*, spinal nerves that pass out through the skull and possibly correspond to the hypoglossal nerve (XII) of higher vertebrates; *m*, otic ganglion; *n*, ciliary; *p*, spenopalatine; *r*, submaxillary ganglia of the parasympathetic system; *s*, spiracle; *v*, ventral root of the first spinal nerve. The eye-muscle nerves, III, IV, and VI, are not drawn, but their positions are indicated by the leaders ending in circles. [Wiedersheim, 1897.]

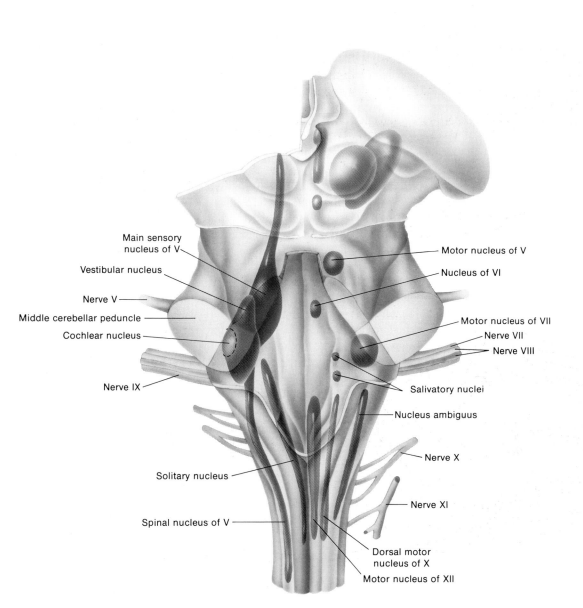

Main sensory nucleus of V

Vestibular nucleus

Nerve V

Middle cerebellar peduncle

Cochlear nucleus

Nerve IX

Solitary nucleus

Spinal nucleus of V

Motor nucleus of V

Nucleus of VI

Motor nucleus of VII

Nerve VII

Nerve VIII

Salivatory nuclei

Nucleus ambiguus

Nerve X

Nerve XI

Dorsal motor nucleus of X

Motor nucleus of XII

Figure 10.67
The medulla, based on the mammal. Motor (right) and sensory (left) nuclei are schematically indicated; the tracts and reticular substance are not.

450

Figure 10.68
Motor nuclei of the cranial nerves.
[*Kappers, 1929.*]

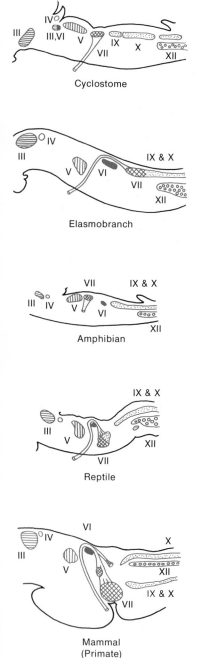

Cyclostome

Elasmobranch

Amphibian

Reptile

Mammal
(Primate)

The main **ascending and descending tracts** are the same as those in the cord, often shifted in position in the cross section. Most notable are (a) the large number of corticobulbar fibers (especially in higher mammals) that give voluntary supply to the muscles of the face, tongue, and larynx, and (b) the still larger number of cortocopontine fibers (peculiar to mammals and birds and especially important in primates) that relay cortical input to the cerebellum. The bulbar contributions to ascending paths are those for taste, cutaneous sensation from the face, the lateral line in fish, and (especially in tetrapod classes) hearing. The first central relay nucleus for the lateral line input is widely different among fish, becoming a large structure, the lateral line lobe, formidably complex histologically and physiologically. Its equivalent, in part, in the higher vertebrates is the set of cochlear nuclei. These also vary widely in size and detailed make up; only a few of the specializations are presently understandable in terms of the habits of the species. The vestibular inputs are chiefly distributed locally within the bulb and metencephalon (which includes the cerebellum) and downward to the cord.

The bulb is a **"vital center"** because its interneurons include the centers that drive the ventilatory movements of respiration, whether they be opercular movements, as in fish; mouth movements, as in frogs; or intercostal or diaphragmatic movements, as in higher classes. (There is some opinion that control of the blowhole in cetaceans is more rostral.) In addition, other visceral interneurons mediate the control of blood pressure and heart rate in response to a variety of chemical and mechanical inputs. At least in higher vertebrates, there are distinguishable centers for reciprocal effects: inspiratory and expiratory, increasing and decreasing of blood pressure, enhancing and diminishing of reflexes of the limbs.

Beginning in fishes, several masses of cells differentiated out of the reticular formation in the lower medulla, becoming distinguished as a heterogeneous cluster, the **olivary complex,** which in higher classes develops into a number of separate and often quite large and elaborate clumps of gray matter (Fig. 10.69). In primates, the largest member of the inferior olivary complex is a structure called the principal olivary nucleus; in cetaceans, it is another member of the complex. The afferent connections of the inferior olive are not well understood, but appear to be from internuncial systems both above and below; the efferent connections are chiefly axons that terminate as climbing fibers in the cortex of the cerebellum; each fiber powerfully excites a set of Purkinje cells distributed one in each of many folia.

The medulla is also the level of the great **decussations.** In mammals these are (a) the corticospinal or pyramidal tract and (b) the medial lemniscus, made up principally of ascending second-order neurons of proprioceptive and tactile modalities, which arise in the nuclei gracilis and cuneatus on top of the dorsal horns at the caudal end of the medulla. Ascending and descending decussations are ancient (they are already common in invertebrates), but their concentration into one level is a feature of the highest vertebrates, especially of some mammals. It may be associated with the enlargement of these two tracts as cephalic dominance and fine distal movements

Table 10.3

Summary of the Brain Stem Cranial Nerve Nuclei (*An abbreviated tabulation showing the nuclei and the main destinations of motor components and the main functions or sources of sensory components. Blanks mean the cranial nerve has no components of that type.***)**

MEDIAL   LATERAL

Cranial nerve	Somatic efferent	Special visceral efferent	General visceral efferent	Visceral afferent	General somatic afferent	Special somatic afferent
III	III nucleus (extrinsic eye muscle)		Edinger-Westphal nucleus (iris and ciliary muscle)		Unknown (proprioceptive to eye muscle)	
IV	IV nucleus (extrinsic eye muscle)				Unknown (proprioceptive to eye muscle)	
VI	VI nucleus (extrinsic eye muscle)				Unknown (proprioceptive to eye muscle)	
V		Motor nucleus of V (masticatory muscle)			Main sensory and spinal nucleus (skin) Mesencephalic nucleus (proprioceptive to masticatory muscle)	Lateral line nucleus* (exteroceptive)
VII		Motor nucleus of VII (facial muscle)	Salivatory nucleus (salivary glands)	Solitary nucleus (taste and face)		Lateral line nucleus*
VIII						Vestibular nucleus (labyrinth) Cochlear nucleus (cochlea)
IX		Nucleus ambiguus (branchial muscle and stylopharyngeal muscle)	Salivatory nucleus (salivary glands)	Solitary nucleus (taste and pharynx)		Lateral line nucleus*
X		Nucleus ambiguus (branchial muscle of pharynx and larynx)	Dorsal motor nucleus of X (viscera)	Solitary nucleus ("vagal lobe") (taste, adbominal viscera)	Spinal nucleus of V (skin)	Lateral line nucleus*
XI		Nucleus ambiguus (branchial muscle of pharynx and larynx) Ventral horn, cervical 1–6 (trapezius and sterno-cleido-mastoid muscle)	Dorsal motor nucleus of X (viscera)			
XII	Hypoglossal nucleus (tongue muscle)				Unknown (proprioceptive to tongue)	
Spinal Nerves	Ventral horn		Lateral column (ventral roots, thoracic 1 to lumbar 2 = sympathetic; ventral roots = parasympathetic; sacral 2–4)	Dorsolateral nucleus of lateral horn (visceral)	Nucleus dorsalis (proprioceptive) Nucleus gracilis and nucleus cuneatus (proprioceptive) Dorsal horn (exteroceptive)	

* The lateral line nerves and their sensory ganglion cells are so distinct from the classical components of VII, IX, and X that we must either broaden the definition of those nerves or regard this major cranial inflow as supernumerary to the series of numbered cranial nerves.

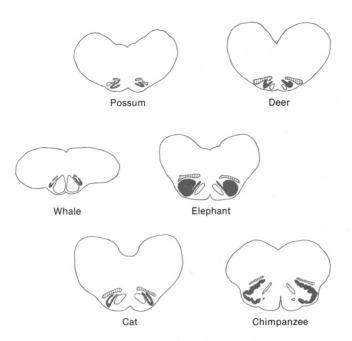

Possum   Deer

Whale   Elephant

Cat   Chimpanzee

Figure 10.69
The inferior olivary complex in various mammals. Sections are representative of the relative sizes of three subdivisions of the complex seen in its rostral portion. Black areas, the principal olivary nucleus; hatched areas, dorsal accessory olivary nucleus; open areas, medial accessory olivary nucleus. [Courtesy C. Watson.]

of the extremities become elaborated.

The **centers for special senses** in the medulla account for a major part of its bulk and diversity among species. In many fishes the taste centers are enormous, producing conspicuous bulges on the external surface and supplying receptors widely distributed over the skin and its appendages. The great system of lateral line sense organs characteristic of fish and aquatic amphibia (but lost in all higher forms, except for the homologous structures of the inner ear) are supplied by the lateral line nerves, closely associated with the Vth, VIIth, IXth, and Xth cranial nerves, but best regarded as distinct from them (see note, Table 10.3). Most lateral line fibers end by synapsing

in the nuclei of the lateral lobe, belonging to the special somatic afferent column, but some project directly to the archicerebellum.

We defer to the consideration of the cerebellum some of the indications of extreme specialization in the **lateral line system.** In most fishes it is principally mechanoreceptive for low-frequency, nearby movements in the water. Other forms of effective stimuli, including certain ions and temperature changes, have not been shown normally to be adequate stimuli on the basis of behavioral criteria. Only for electrical sensitivity in elasmobranchs and certain families of teleosts has behavioral evidence been adduced to permit the conclusion that the "ampullary" and "tuberous" derivatives of lateral line receptors found in these groups are really specialized for a different modality, electroreception (p. 84).

Accompanying the evolutionary loss of the lateral line in terrestrial vertebrates, there may perhaps have been some increase in the relative importance of vestibular input and its nuclei in the descending control of the spinal cord, and possibly in cerebellar input. Even more striking is the development of a new sense organ, the **cochlea,** and of the associated nuclei of second-order neurons. These become large and specialized in structure, particularly in certain groups, such as bats and toothed whales. Unlike the vestibular component, which connects mainly locally and only modestly upstream beyond the midbrain, the cochlear component cascades upstream by projecting to progressively larger masses. Especially in amniotes the cochlear input goes with the development of the inferior colliculus and auditory thalamus (medial geniculate of mammals, ovoidalis of birds, reuniens of reptiles).

The cochlear division of VIII begins feebly in amphibia, like a gnostic addition to the older lateral line modality for detecting and localizing sound sources. It blossoms out in higher classes, differentiating in reptiles, birds, and mammals into distinct nuclei of dorsal and ventral groups, with quite different kinds of synapses, cells, and responses. As with the other first-order to second-order relay nuclei, many fibers end here from the reticular and perhaps other higher systems that exert a modulating effect upon the sensory input. This modulation is already present in teleosts. There is direct evidence in the lateral line system of efferent fibers that modulate the sensitivity of the receptors. These may be the forerunners of the important olivocochlear bundle in mammals, which profusely innervates the hair cells of the organ of Corti, exerting selective inhibitory effects.

The sensory component of the **trigeminal nerve** is believed primitively to serve essentially a protopathic cutaneous function for the skin of the face—that is, a relatively crude, largely nociceptive reception—and to project into the descending sensory nucleus of V. Already in lower classes, but with increasing importance in the higher classes, there is in addition a component that may be more involved in fine discrimination ("epicritic sensation") and, at least for the human, in appreciation or recognition of stimuli ("gnosis"). This fraction projects to the ascending limb of the sensory nucleus of V (analogous to the nuclei of gracilis and cuneatus) and then upward, through the trigeminal lemniscus, to the new structures of the thalamus and cortex.

A considerable body of comparative neurological literature concerns differences in position and in relative sizes of the cranial nerve nuclei in different groups of vertebrates. Some authors believe these correlate with habit of life and with the special sensory capacities of different species. This correlation led Dutch neuroanatomist Kappers to propose that, during evolution, nerve cells have tended to migrate toward their principal source of input, thus shortening their dendrites. This inferred process he called **neurobiotaxis.** The correlation has not impressed most recent workers, and the concept is out of fashion. It is usual merely to note that equivalent cell groups lie in different positions.

D. Cerebellum

The roof of the metencephalon becomes the cerebellum. The complex layered cortex of the cerebellum is, relative to other brain structures, notably uniform among vertebrate classes in types of cells, architecture, and intrinsic connections, although developing greatly in surface area and in associated deep nuclei. Its input and output connections show drastic evolutionary modification. It varies in size (Fig. 10.70) from a mere transverse ridge in cyclostomes and amphibians to 20% of the brain in large whales (twice its relative size in man) and even more in mormyrid fish. It is remarkable that this great suprasegmental organ can be so differently construed as to be called by Fulton a motor component of the extrapyramidal system and by Sherrington the head ganglion of the proprioceptive (sensory) system. We shall see that neither is a satisfactory characterization, though the second is closer. This organ, with its enormous input and much smaller output, is largely

454

Figure 10.70
The brain in several classes, for comparison of cerebellum. [*Nieuwenhuys, 1969.*]

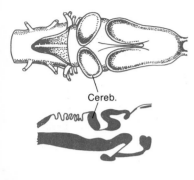

Cereb.

Amphibian (*Rana*)

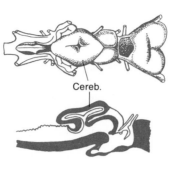

Cereb.

Elasmobranch (*Squalus*)

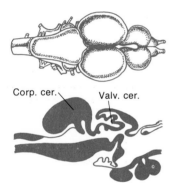

Corp. cer.   Valv. cer.

Fish (*Salmo*)

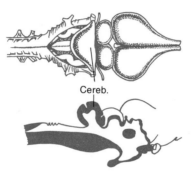

Cereb.

Reptile (*Alligator*)

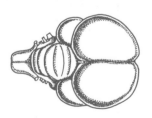

Cereb.

Bird (*Columba*)

input from vestibular portions of the VIIIth nerve. (b) Much of the anterior lobe and posterior vermis, often called the paleocerebellum, receive input from the somatic proprioceptors of the segmental levels via the spinocerebellar tracts. (c) The lateral lobes, the great bulk in mammals, are called the neocerebellum and receive their afferent fibers from the cells of the pontine nuclei, and through them from the cerebral cortex. The neocerebellum is a mammalian feature that is almost unknown in lower groups. Even the paleocerebellar spinal proprioceptive connections are possibly secondary to an original dependence upon exteroceptive input from the lateral line receptors, which are serial homologs of the vestibular hair cell receptors of the VIIIth nerve. What was then a second- and higher-order sensory processing mechanism for exteroceptive and equilibratory inputs, which had the common denominator of describing the position of the body in its near-field, extrapersonal space, became in the course of time more closely associated with the control of fine movements, especially of the appendages. It is now generally thought of as a motor coordination organ, on the basis of the principal visible deficits that follow severe lesions.

Additional afferent connections from three sources are important. (a) One is from the inferior olive. (b) A second is from sensory systems, including tactile, auditory, and visual. (c) The third is from a small nucleus in the dorsal pons, the locus coeruleus.

The functional significance of the **olivary input** is not understood, though its immediate effect is known to be powerfully excitatory to localized regions

devoted to extracting important subconscious aspects of proprioceptive, exteroceptive, and command inputs having to do with the position of the body in space and integrating these to formulate some kind of modulation of motor commands initiated elsewhere.

Let us consider first the **input connections** (Table 10.4). The cerebellum in its most evolved form—in mammals—exhibits three distinct sets of inputs going to as many distinct parts of the cerebellar cortex, only partly overlapping (Fig. 10.71). (a) The flocculonodular lobe, often called the archicerebellum, receives

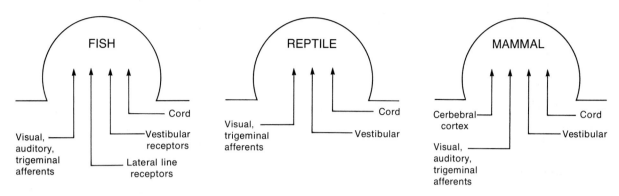

Figure 10.71
Sources of principal inputs to cerebellum. These are only some of the more conspicuous inputs, shown here to emphasize certain common denominators as well as contrasts between classes.

Table 10.4
Summary of the Cerebellum

Phylogenetic interpretation	Parts (in mammals)	Afferents to cortex	Efferents from cortex	Efferents from nuclei	Possible functional sphere
Archicerebellum	Flocculonodular	FROM VIII nerve Vestibular nucleus of both sides (to cortex and fastigial nucleus) Inferior olive	TO Vestibular nuclei Fastigial nuclei	TO Reticular formation Nuclei of extraocular muscles Vestibular nuclei (contralateral)	Equilibrium, including eye movement
Paleocerebellum	Anterior lobe lingula, centralis, culmen Posterior lobe, inferior part pyramis, uvula, paraflocculus	Spinocerebellar tracts (ipsilateral and contralateral) Cranial nerve nuclei (including lateral line in lower vertebrates) Reticular formation Inferior olive Vestibular nuclei (few) Locus coeruleus	Fastigial nuclei (first in reptiles) Globose nuclei Emboliform nuclei (dentate, special part)	Paleorubrum (contralateral)	Posture, including stance and orientation in a space represented by exteroceptive and proprioceptive input
Neocerebellum	Posterior lobe, anterior part ansiform, paramedian, tuber, declive, simplex	Pontine nuclei (contralateral from cerebral cortex) Inferior olive (contralateral) Few overlapping spinal Locus coeruleus	Dentate nuclei (first in reptiles) Emboliform nuclei	Neorubrum (contralateral) Lateroventral thalamus (contralateral)	Synergy of ipsilateral, voluntary, fine movements (lesions can produce tremor, adiadokokinesis, hypermetria, hypotonia, asthenia, nystagmus)

of the cerebellar cortex via the one-to-one synapses of climbing fiber on Purkinje cells. The sensory inputs are weaker and more complex in influence upon the cortex; they show topographic representation—for example, areas for hind leg, foreleg, and head and for auditory and visual evoked potentials upon stimulation of the relevant receptors. These areas agree with those defined by stimulation of the correspondingly labeled regions of the cerebral cortex, that cause evoked electrical activity in the cerebellar cortex or movements in these parts of the body. The movements are modified by simultaneous stimulation of the appropriate region of the cerebellar cortex.

The input to the cerebellar cortex **from the locus coeruleus** is remarkable in several ways. The axons arise in a very small group of cells, branch extensively, project to the whole cerebellar cortex, and end diffusely and sparsely in the vicinity of the Purkinje cell dendrites. The cells and axons are noradrenergic, hence brilliantly visible by fluorescence histochemistry in thick sections. In spite of their sparseness, they exert a prolonged suppressing effect on Purkinje cell firing. This is the only inhibitory input to the cerebellum; moreover, it is not mediated by conventional i.p.s.p.'s but by a peculiar form of hyperpolarization without conductance increase, not yet understood. Axons from the locus coeruleus also project widely in the brain stem and cerebral cortex.

As one might expect from the connections, **lesions** in the archicerebellum cause difficulties with equilibrium. Those in the paleocerebellum cause symptoms in the postural sphere, and those in the neocerebellum cause them most conspicuously in voluntary movements of the distal members (e.g., terminal tremor, over- and underreaching, and difficulty in rapidly alternating movements of the hand). Tectocerebellar inputs were formerly thought to be important, especially in fish, but new work indicates the need to re-examine these animals with improved methods.

The **output connections** are not quite so distinct. The archicerebellum projects back to the vestibular nuclei and via deep cerebellar nuclei to other lower brain stem centers. The paleocerebellum likewise projects to brain stem centers chiefly of the nature of upper motor neurons and to nuclei differentiated from the reticular substance (e.g., the red nuclei). A few fibers go directly to motor neurons of the eye muscle nuclei. In higher forms a major outflow goes directly to the thalamus and from there relays onto the cerebral cortex, completing the loop that began with cortico-pontocerebellar projections. The correlation is not nearly so clear between major regions of the cerebellum and these three principal efferent destinations (vestibular, reticulorubral, and thalamic).

In connection with inferring the normal **role of the cerebellum** from deficits caused by lesions, it is important to keep in mind that large lesions can sometimes be sustained without noticeable symptoms, especially if the lesion had a gradual onset or occurred early in life. Careful study of such individuals would probably reveal telling signs or novel strategies for coordination. Experimental, reversible block of cerebellar function in monkeys by chilling the deep nuclei with a controllable thermode causes symptoms typical of large lesions; but after several cycles of chilling and rewarming, the symptoms abate. Careful study shows that the motor performance after such compensatory learning is accomplished by an altered strategy.

In the same vein, it is difficult to accept the usual argument that the cerebellum is well developed in birds in association with their well-coordinated locomotion. Von Holst showed that flying models of birds, with no control system, can even compensate for turbulent air. The cerebellum is large in sharks, especially in large species, whose locomotion is less complex than that of many teleosts; it is small in trout, huge in mormyrid fishes (in which it overhangs all the rest of the brain and adds so much to its volume that the brain-weight/body-weight ratio can reach 4.4%, higher than in adult man). Although it is small in ganoids, *Lophius,* frogs, and snakes, it is as large or larger in gymnotid fish, catfish, alligators, turtles, ostriches, sloths, and whales as it is in more active relatives.

The suggestion that seems better than mere "coordination of movement" and which correlates with connections as well as phylogeny and physiology is that the cerebellum serves the unconscious appreciation of the spatial relations of the body and its members, elaborately computing and representing in analog form on the cortex some resultants of present and projected positions of members in relation to each other. This input-processing function mainly involves sensory afferents in the lower vertebrates, who invented the cerebellum, and afferents from the cerebral cortex in mammals; such processing is likely to exert an influence on motor performance. Electric fish, however, such as *Apteronotus* or *Gymnarchus,* may lie motionless on the bottom yet we know that the cerebellum is actively working to report the electro-receptor input. The same must be true in other vertebrates at rest. It is the clinical deficit in motor coordination, sparing conscious sensation, that makes us emphasize the motor role of the cerebellum.

The enormous number of afferent fibers entering the cerebellum in man is many times the number of fibers in our largest nerve, the optic. The still vaster number of intrinsic elements in the cerebellum (in man the granule cells of the cortex alone, exceeding 10^{10} in number, are a large percentage of the total number of neurons in the brain) contrast strongly with the limited number of efferent fibers.

The **histological architecture** of the cerebellum (see p. 115 et seq.; Figs. 2.18; 2.34; 3.15; 3.16; 3.17; 3.19; 10.72; 10.73; 10.74; 10.75), is discernible in its basic features from the lowest vertebrates upward, and is virtually the same in fish, amphibians, birds, reptiles, and mammals. It has nevertheless evolved in details. Consider these features of the mammalian cerebellum: (1) recognizable granule and Purkinje cells, (2) three layers, molecular, Purkinje, and granular, (3) one class of afferent fibers obeying criteria for mossy fibers, (4) a second class of afferent axons obeying criteria of climbing fibers, (5) deep cerebellar nuclei that receive most of the Purkinje axons, representing the output of the cortex and in turn projecting out of the cerebellum, and (6) a certain class of intrinsic cells of the molecular layer obeying criteria of basket cells. Cyclostomes show (1) in primitive form and (2) barely or doubtfully, whereas the elasmobranchs and all higher groups show these features clearly. Teleosts exhibit (3) and (4) in

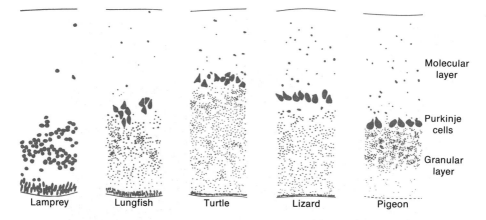

Molecular layer

Purkinje cells

Granular layer

Lamprey   Lungfish   Turtle   Lizard   Pigeon

Figure 10.72
Cerebellar cortex stained by Nissl method. [Nieuwenhuys, 1969.]

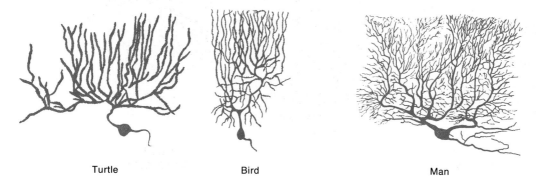

Lamprey   Shark   Teleost

Turtle   Bird   Man

Figure 10.73
Purkinje cells in several classes (Golgi stain). [Nieuwenhuys, 1969.]

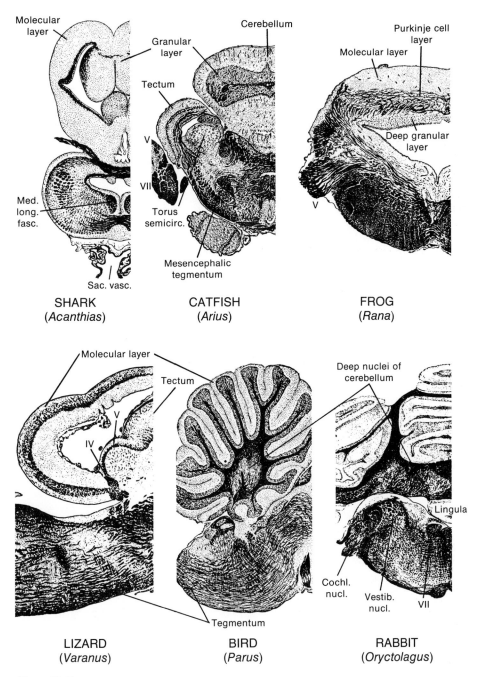

Figure 10.74

The cerebellar cortex in several classes. Sections stained by the Weigert method for myelin. *Med. long. fasc.*, medial longitudinal fasciculus; *R* VII, root of the VIIth cranial nerve; *Sac. vasc.*, saccus vasculosus; *Torus semicir.*, torus semicircularis; *Vestib. nucl.*, nucleus of vestibular part of VIII. [Kappers et al., 1936.]

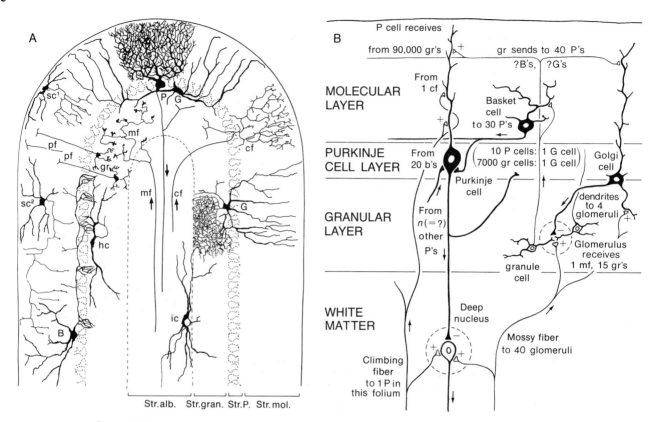

Figure 10.75
Neurons and intrinsic connections in the cerebellar cortex. **A.** A selection of cells of the main types. [Jansen and Brodal, 1958.] **B.** Schematic of the connectivity. *b,* basket cell; *cf,* climbing fiber; *G,* Golgi cell; *gr,* granule cells; *hc.* horizontal cell; *mf,* mossy fiber; *P,* Purkinje cell; *pf,* parallel fiber; *sc¹, sc²,* stellate cells; *Str. alb.,* white matter; *Str. gran.,* granular layer; *Str. mol.,* molecular layer; *Str. P.,* Purkinje cell layer. [Jansen and Brodal, 1958.]

primitive form, whereas amphibians and higher groups show them clearly. Amphibia show (5) in primitive form, whereas reptiles, birds and mammals have well developed nuclei. Reptiles are the first to show (6) in distinct but primitive form, whereas birds and mammals have good basket cells. If we had the information, it is likely that other details, such as the structure of glomeruli in the granule layer, would show similar progression.

E. Mesencephalon

Basically, the mesencephalon is the most anterior part of the neural tube that is still segmental in nature, gives off true nerves, and is divided by the sulcus limitans on the neurocoel surface into a dorsal sensory and ventral motor region. However, extensive changes have been superimposed, first by the enlargements due to optic input and second by the converging termination of local and ascending sensory input—from the surface of the head (V); from the whole lateral line system; from the vestibular system in lower vertebrates, and later the cochlea as well (VIII); and from the spinal somesthetic input, chiefly the kind called protopathic, or crude skin sense.

In early vertebrates the mesencephalon was as big as the metencephalon, and the tectum (Fig. 10.76) as big as the cerebellum. The latter perhaps performed a more detailed analysis of the lateral line and vestibular input, and the mesencephalon, with more converging modalities, performed higher sensory recognitions leading to decisions and formulation of commands of the nature of prey capture, mate clasping, threat vocalization, and the like. In the higher classes, with bourgeoning of the forebrain and its associated neocerebellum, the autonomy of the midbrain is less; the responses it mediates are no longer among the highest performed by the animal—not necessarily because it is reduced in capability but because higher centers have increased in importance. Even in the lower classes (e.g., teleosts), although the midbrain can be regarded as an important correlation center for afferent modalities (especially the optic, acoustic, lateral, and somesthetic) and as a command center for initiating complex responses, these are really at the higher reflex level. The midbrain is not competent for mood or instinct switching, motivating, reinforcing, and advanced forms of learning.

The **main cell masses** resident in the mesencephalon are shown in Table 10.5 (see also Fig. 10.77). In addition, it includes many important tracts that arise in it, terminate in it, and pass through it.

The **tectum** is already large and laminated in fish (Fig. 10.79). In lower classes

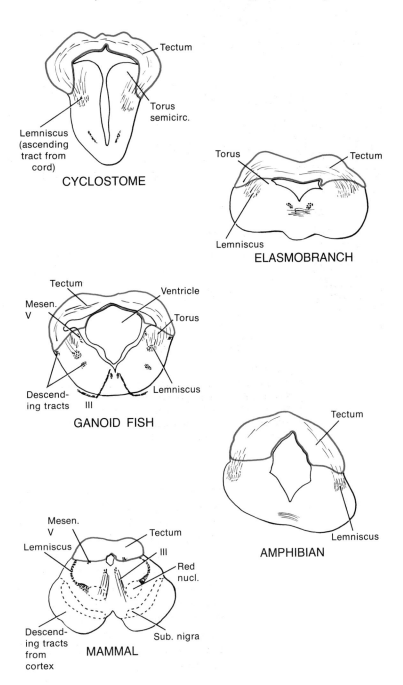

Figure 10.76
Tectum of the midbrain in several classes. [*Johnston, 1906.*]

Table 10.5
Main Cell Masses of the Mesencephalon

	Cell masses	Connections		Possible functional associations
		Afferent	Efferent	
Tectum	Tectum opticum (called superior colliculus in mammals)	II, cord, bulb, sensory nucleus of V, isthmus, torus semicircularis or inferior colliculus, pretectum, thalamus, telencephalon	Cord, bulb, periaqueductal gray, reticular formation, nucleus isthmi, thalamus (especially birds and mammals), retina (teleosts, amphibians)	Correlation of visual, auditory, and somesthetic; feature-extraction; localizing stimuli; formulation of higher reflex commands; eye and head movements, especially in orientation
Tegmentum	Torus semicircularis (called inferior colliculus in mammals)	Lateral line nuclei (fish), cochlear nuclei (tetrapods), vestibular nuclei (less in higher groups), cord, V sensory nucleus	Tectum, thalamus, reticular formation	Correlation of information on equilibrium and near-field aquatic displacements (and electric fields); sound sources; localization
	Nuclei III, IV (including general somatic and general visceral efferent)	Vestibular nuclei, cerebellum, tectum (indirectly), reticular formation	Extraocular muscles, iris, and ciliary muscle (parasympathetic)	Movements of eyes; accommodation; pupillary constriction
	Periaqueductal gray matter, tegmental nuclei, interpeduncular nuclei	Complex, including tectum, hypothalamus, habenula, cord, telencephalon	Complex, including nuclei of III, IV, VI, pons, thalamus, hypothalamus	Limbic system; affect, visceral control
	Isthmo-optic nucleus	Tectum	Retina (in birds only)	Horizontal cell response
	Nucleus isthmi (in nonmammalian forms)	Tectum, probably torus semicircularis	Tectum, torus semicircularis tegmentum, thalamus	Correlation of optic, equilibrium, acoustic influences
	Reticular formation, including tegmental reticular nuclei	Cortex, pallidum, reticular formation of other levels, cerebellum, vestibular nuclei, cochlear nuclei, tectum, cord	Reticular formation of other levels, thalamus, cord	Motor control; pupil; many functions; reticular activating system
	Red nucleus	Dentate, interposed nuclei, precentral cortex (somatotopically organized)	Cord, bulbar reticular formation, inferior olive, cerebellum, thalamus (especially from small-celled newer part of red nucleus)	Motor coordination, especially righting; flexor activity; well developed in carnivores, poor in primates
Intermediate zone	Substantia nigra (large in man, small in other mammals; only a forerunner in reptiles)	Caudate, putamen, subthalamus, pretectum,	Striate, pallidum, thalamus	Extrapyramidal motor; inhibition of forced movements; pathologic in Parkinsonism

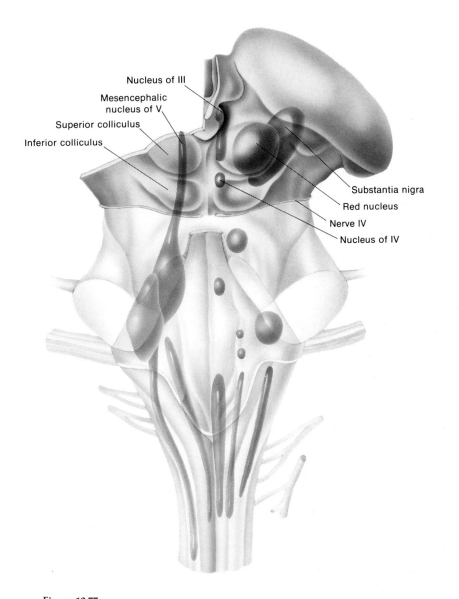

Figure 10.77
The mesencephalon. Some principal cell masses are indicated as developed in man.

Figure 10.78
Neurons of the 14-layered optic tectum of lizard (Golgi stain at right; Weigert at left). [Ramón, 1891.]

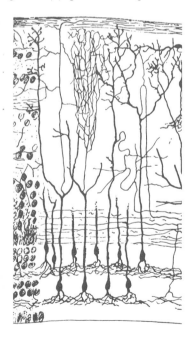

it is the main destination of optic input; only a small percentage of the optic nerve peels off to the five other destinations of retinal ganglion cell axons: the basal optic nucleus and pretectum in the transition zone between mesencephalon and diencephalon, the dorsal and the ventral lateral geniculate nuclei of the dorsal thalamus, and the suprachiasmatic nucleus of the hypothalamus. In the higher classes the trend is for more to go to the geniculate, and in mammals a relatively minor percentage goes to the tectum, which is now fused with the underlying tegmentum (by retreat of the ventricle to the midline) to form the superior colliculus. (For the interesting evolution of the percentage of decussation of optic fibers and other comparative aspects, see Walls, 1942; Ebbesson, 1970; Rodieck, 1973.) Optic input is not only topographically organized into a precise point-for-point representation of the retina, and hence of the visual field, chiefly on the contralateral tectum but also to some extent ipsilaterally in many groups. In addition different classes of retinal ganglion cells, whose connections make them responsive to different features of photic stimuli (e.g., small moving-object detectors and dimming detectors in the frog), send their axons to terminate in subjacent layers with congruent maps of the visual field.

Two or three possibly general classes of optic nerve fibers are distinguished on the basis of simple parameters of the stimuli (e.g. on, off, on-off). Further subdivision of input fibers based on more complex feature extraction involving size, motion, and the orientation and direction of motion is probably common, although the exact differentiation doubtless varies among species. It remains a guess that the variations are significantly related to the habits of life. The discovery of these types and relations is a part of the general problem of the neuronal basis of recognition, and hence of the change of meaning of signals in successive orders of neurons in pathways, which we considered in Chapter 6. Four classes of optic inputs have been discovered, confirmed, and, in considerable measure, quantitatively characterized for the tectum in the frog; five or six are qualitatively characterized in the tectum of the pigeon; and indications exist for similar classes in the tectum of lizards, rabbits, squirrels, cats and other forms. These are optic nerve fiber or retinal ganglion cell types. In addition, there are more complex types of units, fragmentarily known and intrinsic to the deeper layers of the tectum, like those described on page 254. In contrast to the other species just named, cats, as well as goldfish, newts, and monkeys, mainly employ retinal ganglion cell types with simple or concentric excitatory receptive fields. The cat, and possibly others, has the types of units that extract more complex features but in proportionately smaller numbers.

Our knowledge of **tectal function** is certainly very meager compared to the elaboration of the histology (Fig. 10.78). Up to 15 layers of differently textured neuropiles and cells occur in birds, and great variation is known, particularly among fish and reptiles. The tectum is relatively modest in amphibia and most mammals (better in squirrels and tree shrews).

Besides visual motor reflex movements of the head and eyes, evidence from the hamster indicates that the tectum is essential to orienting a learned response to the "where" of the situation. The visual part of the cerebral cortex is necessary to discriminating between certain learned patterns (the "what"). This distinction is certainly an over-simplification; it depends on the specific task and the species. The tectum receives important input from the cerebral cortex and projects to it via the pulvinar.

The **torus semicircularis** lies below the ventricle (Fig. 10.79), hence in the tegmentum in lower classes, but is believed to include a component that becomes the inferior colliculus in mammals, moving dorsally as the ventricle shrinks, till it lies just behind the superior colliculus. This is the great end-station for input from the acousticolateralis system. Hence in fish its input is chiefly from the lateral line, and in tetrapods from the cochlea. Its functional contribution is little known beyond anatomical inference. In catfish part of this structure receives input via the lateral line lobes from microampullary electroreceptors and extracts information about locus, orientation, and frequency of extrinsic sources of weak, very slowly changing current, such as small prey. In electric fish, units have been found in the torus semicircularis that analyze the wave form of electric field signals of social significance with high resolution and influence the electric organ-command center in the medulla (see p. 333).

In higher classes one sees the trend to reduce, relatively, the vestibular input and to increase the number of auditory fibers that relay in the colliculus and project to the medial geniculate nucleus of the thalamus. The **inferior colliculus** is thought to mediate reflex responses of orientation to sound sources. A great deal is known about the properties of single auditory units in cats and bats. For example, there are neurons specialized for sensitivity to the direction of a sound source, to the simultaneous presence of two frequencies, to frequency modulation, to white noise and to sounds below some inhibiting intensity. The inferior colliculus is large in Carnivora and greatly enlarged in Chiroptera (bats) and Cetacea (porpoises and whales). In porpoises, evoked potentials reveal a high specialization for sensitivity to and discrimination of the click-like, ultrasonic echo-locating signals. The lower-frequency, sustained whistle sounds, believed to be important in social communication, are conspicuously ineffective in the midbrain though effective in the cortex. The evidence suggests a distinction between two auditory analyzing systems, perhaps an object-assessing system and a social-signal-assessing system. Sea lions (Pinnipedia), living in much the same environment, are quite different—virtually without specialization in the inferior colliculus for echo-locating sounds.

The eye-muscle nuclei and their general visceral efferents to the iris and ciliary muscle, as well as the **nuclei of the gray matter** immediately below the ventricle (or those surrounding the aqueduct in higher mammals) plus a series of small nuclei in the tegmentum, including the interpeduncular nuclei, are all treated in the Table. There is much less obvious evolution among these cell masses, apart from new connections with telencephalic structures as those emerge in mammals. The range of functions is important from localized eye movements to general visceral influence; there are even powerful effects upon affect, motivation, and generalized responsiveness.

The **nucleus isthmi** is well developed in many lower vertebrates, although highly variable, perhaps not even every-

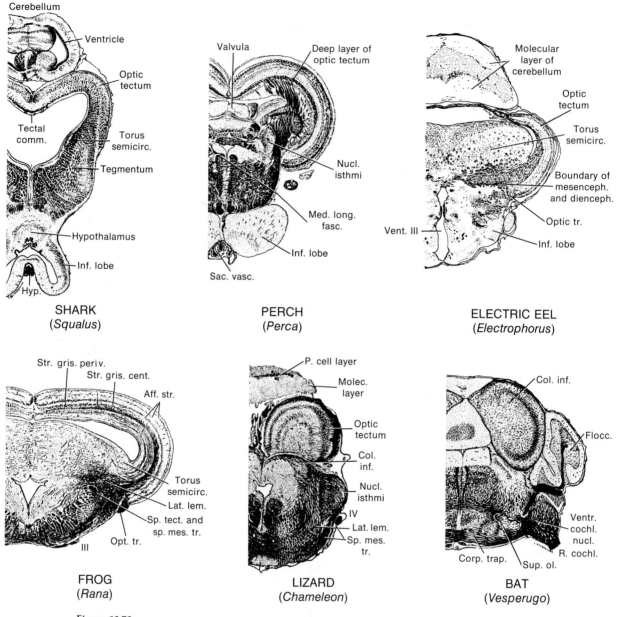

Figure 10.79

The mesencephalon in several classes. Sections stained by the Weigert method for myelin. *Aff. str.,* afferent layers of the tectum; *Col. inf.,* inferior colliculus (equivalent to torus semicircularis); *Corp. trap.,* trapezoid body (an auditory nucleus); *Flocc.,* flocculus; *Hyp.,* hypophysis; *Inf. lobe,* inferior lobe of the diencephalon; *Lat. lem.,* lateral lemniscus; *Med. long. fasc.,* medial longitudinal fasciculus; *Molec. layer,* molecular layer of cerebellum; *Opt. tect.,* optic tectum; *Opt. tr.,* optic tract; *P. cell layer,* Purkinje cell layer; *R. cochl.,* root of the cochlear division of VIII; *Sac. vasc.,* saccus vasculosus; *Str. gris. cent.,* stratum griseum centrale; *Str. gris. periv.,* stratum griseum periventriculare (actually a series of gray layers); *Sup. ol.,* superior olive (an auditory nucleus); *Tectal comm.,* tectal commissure; *Torus semicirc.,* torus semicircularis; *Sp. tect. and sp. mes. tr.,* spinotectal and spinomesencephalic tract; *Ventr. cochl. nucl.,* ventral cochlear nucleus; *vent. III,* third ventricle. [Top right, Chagas and Carvalho, 1961; others, Kappers et al., 1936.]

where homologous. Apparently it is lost as a differentiated structure in mammals or is represented by the nuclei parabigemini. Connected reciprocally with both tectum and tegmentum, it receives afferents via the optic tectum. These connections and the properties of units that have been observed in the frog point to a role in correlating optic and equilibrium responses, but its essential significance is not yet appreciated. The isthmic nucleus is large in chameleons but small in some other lizards, large in birds but lost in the reticular substance in mammals. One important output has been demonstrated clearly in birds, and is likely to be found elsewhere; these are efferents in the optic nerve that arise from a subdivision of the nucleus called the isthmo-optic nucleus (Fig. 10.80) and exert a centrifugal influence upon the responsiveness of elements in the retina.

Differentiated out of the reticular formation, which still includes in all vertebrates many cells of diverse characteristics—both in respect to their responses and to their influences upon sensory and motor functions—are two great masses or nuclei. These are the red nucleus and the substantia nigra. Antecedents of these two structures may be found in some lower groups, but they are really well differentiated only in mammals. The forerunners of the **red nucleus** are large cells with descending influence upon the bulb and cord, already visible in fishes. In mammals, there is a distinct population of such large cells, collectively called the paleorubrum. A population of small cells is believed to be recently acquired, and is hence called the neorubrum; it evolved with the neocerebellum, neothalamus, and neocortex. Both cell groups receive afferents from the deep nuclei of the cerebellum of the

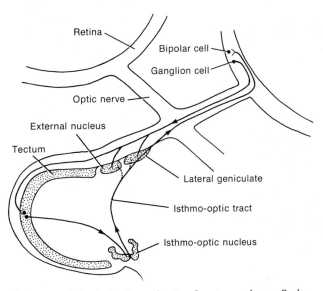

Figure 10.80
Afferent and efferent fibers in the optic nerve of pigeon.

Only some of the destinations of optic afferents are shown. Broken lines show possible but unconfirmed collateral connections. [Cowan and Powell, 1963.]

opposite side; the neorubrum also receives from the ipsilateral precentral gyrus, or motor cortex. All these connections are somatotopically organized—that is, one region of the dentate nucleus or precentral gyrus projects to a certain region of the red nucleus. The efferent connections to the lateral reticular nucleus of the bulb, inferior olive, and spinal cord, as well as to points upstream from the neorubrum, point to a function in the motor sphere. The nucleus is more prominent in carnivores than in primates. Its functions are hard to study in view of the abundance of fibers of passage, which transit the red nucleus and hence suffer from lesions and may respond to stimulation. Stimulation in the red nucleus can produce e.p.s.p.'s in flexor motor neurons, especially on a decerebrate background, and particularly if the stimulus is applied instead to the interposed nuclei of the cerebellum. Lesions in man are believed to account for some types of compulsive movement, such as involuntary tremor, writhings, and jerkings (choreiform movements).

The **substantia nigra** is the largest nucleus in the primate mesencephalon and is largest in man, small in other mammals. Its connections also point to motor functions; the evidence, chiefly clinical in man, suggests it exerts an inhibitory influence upon forced involuntary movements. It is pathological in Parkinsonism, which involves rigidity, akinesia (relatively few slow movements), and a tremor of rest.

F. Diencephalon

Since the diencephalon is entirely suprasegmental and does not send or receive any peripheral nerves (apart from some limited afferents from the parietal or the pineal eye in certain species), it poses the same question that is posed by the intrinsic structures of the lower levels of the brain stem—namely, what else is it doing other than relaying from lower to higher levels; what is it besides its common name of " 'tween brain?"

The sulcus limitans runs into the floor of the neural tube at the posterior end of the diencephalon, so that the whole forebrain is a development of the alar plate, primitively a sensory receiving and processing region, connecting midbrain and endbrain. Table 10.6 and Figure 10.81 indicate the main components of the diencephalon. The four principal divisions can be discerned in all classes of vertebrates, but conspicuous trends of development are evident, from the cyclostomes (in which the epithalamus is the dominant component), through the elasmobranchs and teleosts (in which the hypothalamus becomes most prominent), to the amphibia and higher classes (in which the dorsal thalamus increases progressively).

The functional significance of the **epithalamus,** with its associated epiphysis, is difficult to state. In some vertebrates there is a parietal eye that responds to radiant energy and mediates behavioral thermoregulation. In some forms, the pineal gland has been implicated in hormonal control of gonad function.

The **ventral thalamus,** which never becomes massive, connects forebrain and midbrain to motor control centers of the extrapyramidal system. Besides these connections, clinical evidence in man is our best clue to function. Relatively discrete lesions in the ventral thalamus cause involuntary forced and violent movements (called hemiballismus) of both upper and lower limbs and of the trunk on the side opposite the lesion. Presumably the ventral thalamus normally holds in check some circuit with a runaway or chain-reaction tendency.

The **hypothalamus** is thought to have been primitively a center for "crude exploration." In lower vertebrates, it has an appendage—the saccus vasculosus, of quite unknown function, with an infundibular sense organ, likewise not understood. These disappear in terrestrial groups. Another hypothalamic appendage, the hypophysis, early came into relation with the buccal derivative that becomes the anterior pituitary. A major function of parts of the hypothalamus is to initiate the "release hormone" commands for each of the pituitary hor-

Table 10.6
Components of the Diencephalon in its Full Development

Divisions	Nuclei		Some principal connections		Functions
			Afferent	*Efferent*	
Epithalamus	Habenula		Hippocampus Preoptic area Hypothalamus Pallidum	Interpeduncular nuclei (thence to reticular formation and cord)	Primitively olfactory re- flexes; now?: visceral and pineal control
Hypothalamus	Anterior group supraoptic paraventricular Lateral group Middle group Posterior group mammillary posterior hypothalamic area		Rhinencephalon (via medial forebrain bundle) Thalamus, medial and midline Hippocampus Amygdala Tegmen of medulla	Midbrain tegmentum Cord—indirectly via peri- ventricular gray Posterior hypophysis Thalamus, anterior medial Cortex, diffuse	Autonomic control Emotional expression Affective tone Vagal—cortical arousal
Ventral thalamus (= Subthalamus)	Subthalamic nuclei, zona incerta, tegmental fields		Pallidum	Pallidum Reticular formation	Extrapyramidal descending motor Arousal
Dorsal thalamus (= Thalamus)	Paleo-thalamus	Midline group Intralaminar Centromedian Anterior ventral	Hypothalamus, thalamus Midbrain reticular formation Pallidum, striatum Cortex, diffuse	Approximately the same as afferent connections	Subcortical and Diffuse cortical
	Neothalamus	Lateral ventral Posterior ventral Medial geniculate	Dentate nuclei Cord and bulb (via spinal and medial lemniscus) Inferior colliculi	Cortex, precentral, 4 and 6* Cortex, postcentral, 1, 2 and 3 Cortex, superior temporal, 41	Cerebellar relay Somatic relay Auditory relay
		Lateral geniculate Anterior	Retina Mammillary bodies	Cortex, calcarine, 17 Cortex, cingulate, 23, 24, and 32	Visual relay Motivational relay
		Dorsal medial	Intralaminar and lateral thal- amic nuclei; hypothala- mus; olfactory tubercle; amygdala	Cortex, prefrontal and or- bital	Visceral affective associa- tion
		Lateral	Cortex; tectum	Cortex, parietal association	Somatic association
		Pulvinar	Lateral and medial genicu- late; superior colliculus	Cortex, parastriate, posterior parietal, poster- ior temporal, frontal	Visual—Auditory association

* Most cortical and some other connections are reciprocal.

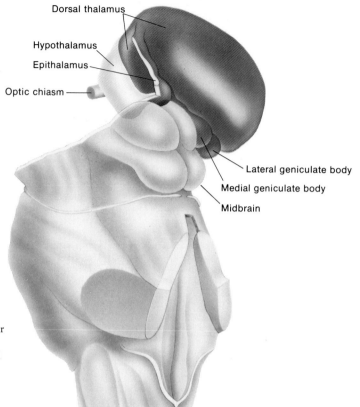

Dorsal thalamus
Hypothalamus
Epithalamus
Optic chiasm
Lateral geniculate body
Medial geniculate body
Midbrain

Figure 10.81
The diencephalon on the right side. Three of the four
principal divisions in man. The diencephalon on the
left and the whole telencephalon have been dissected
away. In this view, the fourth division, the subthalamus, is out of sight.

mones. Though we should be somewhat
suspicious of too facile a chain of reasoning, it has long been supposed that
these functions—together with the hypothalamic connections—suggest a major
role in some drives and in vegetative
functions, including the adjustment of
certain set-points. This role would make
reasonable its eventual expansion into
the head ganglion of the visceral control
or autonomic system. With a high degree
of localization of function, we find mammalian hypothalamic centers especially
concerned with compensating for low
body temperature and high body temperature, with water balance (including

thirst), with appetite and satiety, with the
secretion of hormones by the local neurosecretory cells, and, indirectly, with the
activity of many endocrine glands in the
body. In rats, one small part of the
hypothalamus, the suprachiasmatic nucleus, appears to be more directly concerned with the circadian rhythm than
other parts of the brain. It also receives
direct optic input. Thus whole complexes
of responses involved in emotional expression are controlled in the hypothalamus (Fig. 10.82), such as cardiovascular
changes; piloerection; postural, pupillary, vocalizing, and other responses in
rage, fear, eating, coition, and other

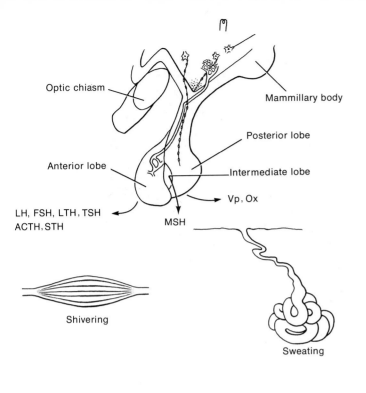

Optic chiasm

Mammillary body

Posterior lobe

Anterior lobe

Intermediate lobe

Vp, Ox

LH, FSH, LTH, TSH
ACTH, STH

MSH

THE PITUITARY HORMONES

ACTH	adrenocorticotrophic hormone
FSH	follicle-stimulating hormone
LH	luteinizing hormone
LTH	luteotrophic hormone (prolactin)
STH	somatotrophic hormone
TSH	thyroid-stimulating hormone
MSH	melanocyte-stimulating hormone (intermedin)
Vp	Vasopressin
Ox	Oxytocin

Shivering

Sweating

Sleep

Rage: piloerection, attack posture, claws extended, ears retracted,
pupil change, growling, hissing, changes in blood pressure, blood sugar.

Wakefulness

Satiety may be lost; medial stimulation
acts as though it causes hunger

Hunger may be lost; lateral stimulation acts
as though it causes satiety

Figure 10.82
Some results of stimulation of the hypothalamus.

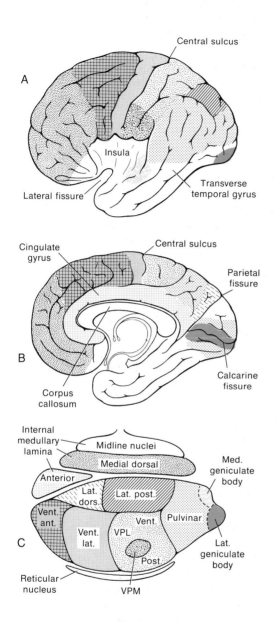

Figure 10.83
Thalamic projections. Diagram of the lateral (**A**) and medial (**B**) aspects of the cerebrum and of the thalamic nuclei (**C**), showing projections by similar symbols. [Curtis et al., 1972.]

manifestations of largely instinctive, species-characteristic spheres of action. We expect, therefore, that there are connections from visceral afferents that report the state of the functions regulated, and connections to the sympathetic and parasympathetic outflows to effectors. Upstream, as we see elsewhere, there are intimate connections with the rest of the limbic system, which determines affect, value judgements, and emotional and instinctive switching.

The **dorsal thalamus,** more commonly called simply the thalamus, is by far the largest of the four parts in amniotes, especially in mammals. Of two main components, the neothalamus, which began to emerge in amphibians, is said to predominate over the paleothalamus in mammals. The distinction between these components and the problems of homologizing subdivisions are under active study today; until new interpretations and terminology are established, the classical view reflected in these terms should be taken with caution. The great development of neothalamus begins with the flowering of ascending tracts from the dorsal thalamus to the forebrain in amphibians and the increased projections forward from trigeminal and spinal somesthetic pathways, visual, and auditory pathways beyond the tectum and into the dorsal thalamus. The early neothalamus in amphibians and reptiles sends efferents to the dorsal striatum and to the pallium of the forebrain, though there is not yet a real cortex.

In mammals, and to a somewhat lesser degree in birds, there is a full flowering of the dorsal thalamus, with its many distinct **nuclear groups** (Fig. 10.83). In mammals these are subdivided into (a) those with purely subcortical or subcortical and diffuse cortical connections, (b) those that relay from ascending afferent paths to specific sensory regions of the cortex, and (c) those called association nuclei, for want of a better term, that project to certain areas of cortex called association areas (see Table 10.6). Study of the much simplified table reveals or hints at much of the strategy of brain organization in higher vertebrates and will be more intelligible as we learn something of the cerebral cortex, of the limbic system, and of reticular activation. Even if the names of the nuclei are not remembered, and hence their specific afferent-efferent connections, the table should be read as though it represents several pages of text.

What we must add here is by way of further complication, which must be built on an initial familiarity with the basic nuclear groups and their types of connections, as shown in the table. First, we must add that many connections are reciprocal, especially those listed in the efferent column; the return connection will no doubt have some special modulating influence upon the forward projections, but this is not yet well understood. Second, several of the nuclei, especially in the relay group, form synaptic complexes called glomeruli (p. 48), which serve as junctions between sensory afferents and thalamocortical projection cells. In these complexes, one incoming afferent terminal makes contact with the dendrites of the thalamic neurons that project to the cortex, but in a knot that introduces the influence of simultaneous input from other sources: cortical, thalamic, and lower levels. In fact, a majority of the presynaptic fibers come from sources not yet identified. The two notable features are that relaying is subject to modulation and that the synapses are complexes with several intertwined inputs.

To oversimplify, in some sense we may say that the neothalamus and telencephalon took over from the midbrain in respect to higher sensory projection and analysis. The forebrain projections are there in lower classes, and the midbrain projections are retained in mammals, but there is a vast shift in their proportions and functional dependence. In reptiles and birds, the principal upward connection of the neothalamus is with certain nonlaminated gray masses derived from the pallium, which can be regarded as possible homologs of neocortical neurons in mammals. Certainly, part of the mammalian thalamus is developed in concert with the tremendous innovation of neocortex and the associated newer parts of striatum, the neocerebellum, neorubrum, and neo-olive. One cannot help but wonder whether this development goes with increased discriminative capacities in mammals, perhaps like those referred to clinically as epicritic or gnostic.

The basic distinction between two types of sensory capacity is exemplified by the difference between **two classes of units in the somesthetic** relay nucleus, ventralis posterior (Fig. 10.84). (a) Those units from the spinothalamic pathways are perhaps more primitive; the cutaneous receptive fields are large and show inhibition from adjacent skin or even from distant large areas. They do not accurately map the body surface onto the cortex. They are driven by mechanical or nociceptive stimuli or both, and even sometimes by auditory stimuli. In short, they are relatively nonspecific, both as to mode and locus. (b) The other class of

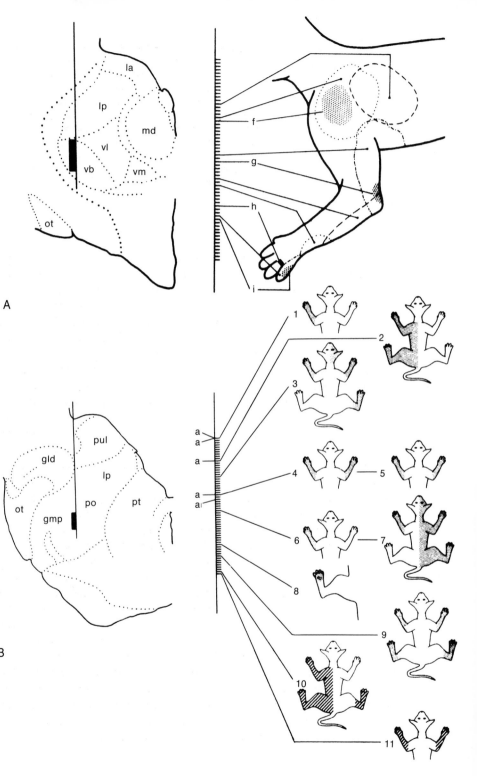

A

B

Figure 10.84 *(facing page)*
Two systems of cells in the thalamic sensory relay nuclei. **A.** Small-field, fast units in a microelectrode track (thick bar) through the ventral posterior nucleus of a cat (left). The expanded scale (50 μm intervals, at right) shows where units were encountered; the body diagram shows the location and size of their receptive fields. Units *f, g, i* were sensitive to light mechanical stimuli to the skin; *h* was sensitive to gentle rotation of the metacarpal-phalangeal joint of the third toe of the contralateral forepaw. **B.** Large-field, slow units in the "posterior nuclear group." The expanded scale (20 μm intervals) shows where units were found that responded only to sound (*a*) and to light mechanical stimuli to the skin (sketches 1–9) or to strong stimuli that were damaging to tissue (*10, 11*). Note the absence of a topographical pattern compared to part A. *gld,* dorsal lateral geniculate; *gmp,* medial geniculate; *la,* lateral anterior; *lp,* lateral posterior; *md,* mediodorsal; *ot,* optic tract; *po,* posterior nuclear group; *pul,* pulvinar; *vb,* ventrobasal; *vl,* ventrolateral; *vm,* ventromedial nuclei. [Poggio and Mountcastle, 1960.]

units, largely from the lemniscal pathways, respond to transient light touch and project focally to the cortex, making an accurate map. The response is very intensity dependent, without refractoriness, but with facilitation, and therefore has excellent temporal resolution. There is inhibition from the immediately surrounding skin; the units have a high security of discharge, are poised for action in rapid cadence, have a wide range of firing frequency, precise anatomical connections, faithful relays, and high amplification in the sense of discriminating small differences in stimulus. Even a single impulse in a single afferent fiber can fire the cortex and cause sensation. In short, such units are highly discriminative, both as to mode and locus.

In the **visual system** a major change between lower and higher vertebrates is a change from the nearly complete optic nerve decussation of all inframammalian forms to a partial decussation (ca. 50–90%) at the optic chiasm in most mammals. Uncrossed optic nerve fibers, at least to the thalamus, are also found, though in smaller numbers, in cyclostomes, chondrostean and holostean fishes, amphibians, and some reptiles (snakes and lizards), as well as in larger numbers in mammals. Among mammals the porpoises have none. Partial decussation, more directly than commisures, permits binocular projection, in that each dorsal nucleus of the lateral geniculate and each side of the cortex receives input from both eyes. In carnivores, primates, and possibly others, there is a basis for improved stereoscopic vision, in that many units receive input directly from the same part of the visual field of both eyes. Where binocular input occurs in animals with complete decussation, it depends on commissural fibers from the contralateral to the ipsilateral side.

The best way to epitomize our present understanding of the function of the **diffuse or nonspecific cortical projection nuclei** is to consider the effects of electrical stimulation. This causes remarkable changes in the electrical activity of the whole cortex—spoken of as an activation of the recruiting type. The effect is related to, although not identical to, the activation produced by stimulation of the brain stem reticular formation that resembles arousal from sleep. The diffusely projecting thalamic nuclei are considered to be a continuation of the midbrain reticular substance, and act in

Figure 10.85
Representation of body surface in thalamic nucleus of three mammals.

Rabbit

Cat

Monkey

Electrical responses in ventral posterior nuclei reflect innervation density in the skin. [*Rose and Mountcastle, 1959.*]

close concert with the hypothalamus and ventral medial thalamus. The identity and even the presence of this category of thalamic nuclei remains uncertain in nonmammalian groups.

In striking contrast, electrical stimulation of the **relay nuclei** elicits short latency and localized activity in the cortex. The relay nuclei in particular are highly organized topographically, both in respect to the sensory surface of the body from which they are responsive (Fig. 10.85) and the cortical locus to which they project.

The **association nuclei** are still less understood. It is notable that structures like the pulvinar are small in lower mammals and increase enormously in size in the primates, together with ascending input from the superior colliculus. Here we are tantalized by what must be a development of importance—one that strains our capacity to think of good candidate functions. But we must think of them, so that we can test for them. This is a familiar dilemma for the neurophysiologist in trying to understand the most complex systems nature presents to us (see p. 292).

This constellation of developments in the diencephalon may contribute to the evolution of sensation, in the sense of conscious awareness. There is much clinical evidence that consciousness may involve the thalamus at least as much as the cortex. Very small lesions in the posterolateral part of the lateral nuclei of the thalamus in man cause the "thalamic syndrome," which includes attacks of "spontaneous" or very gently elicited "central" pain, which is severe, agonizing, diffuse, irradiating, and intolerable. Pleasantness of certain sensations is also magnified. Pain is only slightly or transiently influenced by lesions of the cortex, and stimulation of the cortex does not give rise to clear cut pain experiences. Thus the thalamus is tentatively thought to mediate a crude sort of awareness, in respect to both aspects of sensation—the quality and the affect. The former is further discriminated (see Box 10.2) in the specific and association areas of the cortex and the latter in the limbic system (see Box 10.4).

G. Telencephalon

The telencephalon evolves from a thin-walled rostral bulge of the neural tube, which has a large undivided cavity, into an endbrain that is dominated by cerebral hemispheres representing the greatly thickened walls of ventricles—reduced, paired, and widely separated (Fig. 10.86). The ventricles are primitively bounded by gray matter next to the ependymal lining, and the outer surface of the brain is mainly nerve fibers. This is basically true throughout the fishes and amphibians; only in some parts of the telencephalon of reptiles does the tendency of cells to migrate outward produce a clearly superficial layer of gray matter—a primitive cortex. The consequent increase in ascending and descending projection fibers produces an enlarged white matter that here and there displaces the gray lining of the ventricles. This tendency, not much different in birds, is greatly enhanced in mammals, in which virtually the whole surface becomes a specialized, laminated cortex of gray matter, covering a voluminous white matter; embedded deep in the white matter are distinct masses of subcortical gray matter.

The differentiation among vertebrates is so great that the telencephalon pre-

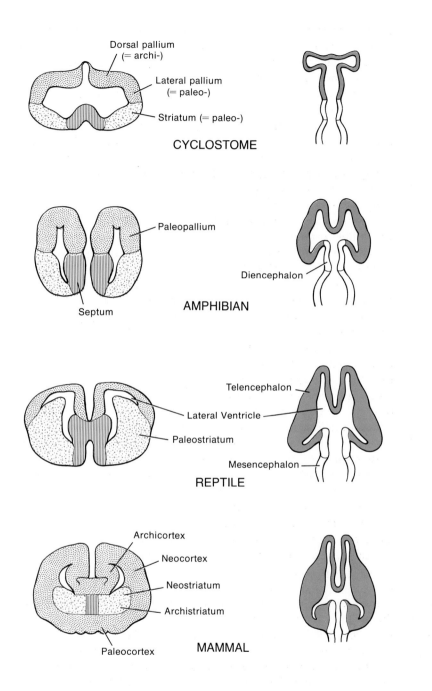

Figure 10.86
Evolution of the forebrain. Transverse and longitudinal sections, highly schematized, for four classes of vertebrates. [Kappers, 1929.]

sents a difficult problem of rationalizing its subdivisions into a natural scheme and homologizing the structures of different classes. The following is one scheme, certainly oversimplified and imposed with some violence upon the lower groups as a result of our knowledge of the fate of structures in higher groups.

Three main divisions can be discerned already in the cyclostomes, and they become quite clear in elasmobranchs, teleosts, and amphibians—groups in which they are roughly at the same level of development. The first set of structures includes the **olfactory lobe and septum,** and receives the only aboriginal peripheral input to the telencephalon: the first nerve and the nervus terminalis (a sensory nerve of unknown function, most apparent in elasmobranchs and some teleosts and discovered long after the cranial nerve numbers had become well established).

The second main division of the telencephalon is the **striatum,** typically ventrolateral or also dorsolateral; the third, typically dorsolateral, is called the **pallium,** for it evolves into a mantle that covers all the rest (Tables 10.7 and 10.8, Fig. 10.87).

Though the striatum and the pallium were long thought to have their direct, modality-specific input almost exclusively from the olfactory lobe in all the lower vertebrate groups, recent work shows otherwise. Olfactory input projects to a small part of the cerebral hem-

---

Box 10.2   Sense Modalities and Affect

Sensation covers not only touch, sight, sound, taste, and smell, but more complex phenomena just as important to the individual, like hunger, nausea, fatigue, headache, vertigo, sexual feelings, and others from deeper receptors. Input from many receptors does not normally reach consciousness. Sensation has two aspects: the quality and the affect, or pleasure-pain aspect.

Affect is not a simple dimension. Pain, for example, is distinct from suffering; they can vary independently. This is an important concept for humanitarian dealings with people and with animals (see Box 6.3, p. 221). Humans, at any rate, may suffer without frank pain or may experience painful procedures without suffering.

Modalities (i.e., any differences in "quality" of sensation experienced) probably mean either a difference in receptors or arrays of them—hence in the whole pathway—or in pathways diverging centrally on the basis of a receptor-coded difference in impulse discharge. In any event, we believe there must be a difference in the constellation of some "final" level of neurons in a sensorium or higher-level analyzing center.

The same is true for loci of stimulation, therefore including the neural representation of place, size, shape, and pitch—stimulus features that involve different constellations of receptors. This is stated as probable, not certain. That is, we believe scalar differences like intensity, brightness, and loudness may be merely differences in activity of the same neurons, whereas molar differences are differences in neurons, like visual and auditory differences. But the borderline is not easy. We believe that tickle, wetness, and the like are not primary modalities, but temporal or spatial combinations.

ispheres in elasmobranchs and others. Optic and somesthetic pathways also project to nonolfactory regions of the forebrain. At least in sharks, these ascending projections from the diencephalon are largely crossed, so that the forebrain represents chiefly the ipsilateral eye. The output of the telencephalon in lower classes, chiefly from the striatum, is heaviest to the epithalamus and hypothalamus; much less projects to lower levels directly. Venturing a guess, which is hazardous in our present state of knowledge, it may be that from the beginning the telencephalon served essen-

tially as a coordinating center for initiating changes in mood, or sphere of action; for playing a role in switching between the major instincts, hence in social behavior, migration, nest-building, hunting, and the like.

The **striatum** of fishes and amphibians is simply the ventrolateral wall of the cerebrum, and possibly corresponds only to the paleostriatum of higher vertebrates. When we come to the reptiles, there is an obvious subdivision into (a) the archistriatum, (b) the paleostriatum, and (c) an "external" or "dorsal ridge" striatum, containing neurons said to be

Table 10.7
The Striatum: Summary of Components and Connections in Mammals

Phylogenetic interpretation		Part	Connections	
			Afferent	Efferent
Archistriatum (in part)		Amygdala	Olfactory lobe and olfactory cortex Insular cortex Temporal cortex Midbrain reticular formation	Hypothalamus Septum Olfactory gray Thalamus, dorsal medial Cingulate cortex Temporal cortex Midbrain reticular formation
Paleostriatum (in birds and reptiles the homologue is called paleostriatum primitivum)	"Pallidum"	Globus pallidus	Caudate Putamen Midbrain reticular formation Subthalamus Cortex (area 6) Thalamus Substantia nigra	Subthalamus Thalamus Midbrain reticular formation
Neostriatum (in birds and reptiles the homologue is called paleostriatum augmentatum)	"Striatum"	Putamen	Cortex (areas 4, 6) Caudate Substantia nigra Thalamus	Pallidum Substantia nigra
		Caudate	Cortex Thalamus Substantia nigra	Putamen Pallidum Substantia nigra

homologous to those in the neocortex of mammals.

In mammals the archi- and paleostriatal regions of reptiles develop, as shown in Table 10.7, into amygdala, globus pallidus, and putamen and caudate. The last two are now called neostriatum. The term "basal ganglia" is often applied to the last three or to all four; occasionally even some diencephalic nuclei are included (subthalamus; hypothalamus), but this should be avoided. The trend has been a continual increase in the size and differentiation of these parts, with addition of new connections from below via the neothalamus, even directly from the midbrain. These increases are coincident with the development of increased descending influence, especially on the subthalamus, thalamus, and motor centers of the midbrain (substantia nigra). Connections are also achieved with the newly developed cortex, especially from the amygdala to the temporal and cingulate cortex and from the motor and premotor cortex to the putamen and caudate.

Table 10.7 does not give a functional column because this is extraordinarily difficult in the striatum, even in mam-

Table 10.8
The Pallium: Summary of Components and Connections in Mammals

Phylogenetic interpretation	Part	Some principal subdivisions	Some Principal Connections	
			Afferent	Efferent
Medial pallium = ARCHIPALLIUM (becomes allocortex)	Hippocampal formation	Hippocampal gyrus Dentate gyrus Subiculum	Olfactory cortex, entorhinal, septal Midbrain (catecholamine areas)	Hypothalamus-mammillary body (via fornix) Thalamus, anterior, intralaminar Midbrain reticular formation Septal cortex
Lateral pallium = PALEOPALLIUM (becomes allocortex or transitional cortex)	Olfactory cortex Septum Pyriform cortex	Olfactory tubercle Septal area Entorhinal area Prepyriform area Periamygdaloid area	Olfactory bulb Hippocampus Amygdala Midbrain reticular formation	Temporal cortex Frontal cortex Amygdala Thalamus Habenula Hypothalamus Midbrain reticular formation Hippocampus
NEOPALLIUM (becomes isocortex; some transitional)	Occipital cortex Temporal cortex Parietal cortex Frontal cortex	Primary sensory areas, 1, 2, 3, 17, 41, 42, 43 Secondary sensory areas, 18, 19; others not corresponding with cytoarchitectonic areas Primary motor, area 4 Premotor, area 6 Supplementary motor area Frontal eye fields Occipital, temporal, parietal, prefrontal association areas	Thalamus, relay nuclei Thalamus, association nuclei Thalamus, diffuse projection nuclei Amygdala Hypothalamus	Thalamus Striatum Tectum Midbrain reticular formation Red nucleus Substantia nigra Pontine nuclei Bulb, motor nuclei Cord, motor centers

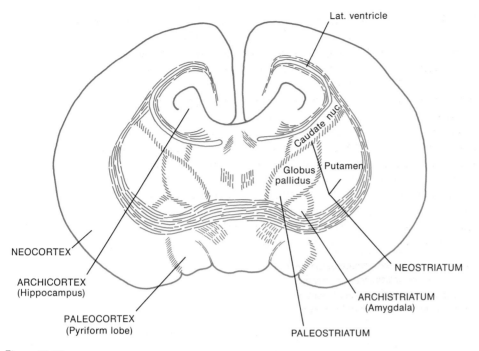

Lat. ventricle

Caudate nuc.

Globus pallidus

Putamen

NEOCORTEX

ARCHICORTEX
(Hippocampus)

PALEOCORTEX
(Pyriform lobe)

PALEOSTRIATUM

ARCHISTRIATUM
(Amygdala)

NEOSTRIATUM

Figure 10.87
The forebrain. Principal gray masses in a lower mammal (*Ornithorhynchus*, the duck-billed platypus).
[Hines, 1929.]

mals for which we have a large literature. Therefore, we cannot give the evolution of its functional significance. In man, pathology of the pallidum is associated (as are lesions in the subthalamus and substantia nigra) with forced-movement syndromes, like chorea, athetosis, and Parkinsonism. This suggests that these structures may normally exert some inhibitory influence upon commands arising elsewhere. In agreement are experiments on animals in which stimulation of caudate or globus pallidus causes little or no movement, when properly controlled against current spread, but can interrupt a cortically induced movement from simultaneous stimulation of the motor area.

The **basal ganglia** are considered to be parts of the **extrapyramidal** system—a system of structures, other than the pyramidal tract, concerned with the generation and regulation of motor commands. In our search for principles of neural strategy, this system is of extraordinary importance, but is still inadequately known. We see once more the hierarchical evolution of level upon level, from midbrain to subthalamus, striatum, and cortex. We see also the development of loops, as in the circuit from motor cortex to putamen to pallidum to thalmus to cortex. We see a large and complex motor system that is only indirectly connected with the cerebellum. It is too much to expect a valid theory or model as yet, but these are clues or constituents of an eventual picture.

The **amygdala** (archistriatum) is quite different in functional role, as might be expected from its connections. The evidence from stimulation in the unanesthetized cat and monkey is dramatic and consistent. Implanted stimulating electrodes elicit responses of sniffing, searching, licking, biting, chewing, swallowing, salivation, chop licking, gagging, and retching—a category clearly related to eating, plus retraction of ears, cessation of purring, snarling, dilation of pupils, face-twitching, pawing, panting, piloerection, grunting, hissing, growling, and attack—a category related to aggression or defense. In some animals, as in a Spanish fighting bull, when electrodes are implanted in these structures, attack behavior can be interrupted and the animal turned relatively tame by electrical stimulation. Ablation can likewise cause tameness in some animals, but savageness in others. It is not surprising to find that stimulation and ablation produce the same result or that either one produces contrary results, for in structures concerned with motivation and visceral function, oppositely acting neuronal groups, sometimes called half-centers, often lie side by side or intermingled. The point is clear, however, that the amygdala, whose principal connections are with olfactory lobe, temporal and cingulate cortex, hypothalamus and septal nuclei (Figs. 10.88; 10.89), is concerned particularly with switching the emotions and the level of motivation. The rage of a rat with amygdalar lesions is not sham rage, as it is with some hypothalamic stimulation, but has all the hallmarks of a true emotion. We shall encounter these functions again in parts of the pallium, the whole forming together a coordinated system called the limbic system (see Box 10.4). A better understanding of the comparative as-

---

Box 10.3 Bird Brains

The brain of birds is highly specialized and deserving of more investigation; it is not close to the main line of mammalian evolution, hence homologies are often difficult.

The same striatal parts seen in reptiles are present in birds, but are still more developed and form most of the bulk of the hemispheres. The archistriatum is of fair size, but the paleostriatum and external striatum are particularly large, receiving increased inputs from optic and acoustic systems by additional thalamic connections that were not yet important in reptiles. The great striatum of birds indeed has some structures that cannot be found in the striatum of mammals but have become neocortex.

It is convenient to complete the remarks on the class Aves here by referring to the pallium. Both an archipallium and a paleopallium are present, although the poor development of olfactory structures in birds is reflected in the meager rhinencephalon. An important mass called the Wulst (from German for bulge) is homologous to mammalian neopallium, but the cellular arrangement in this mass and elsewhere does not meet the defining criteria for neocortex.

What is remarkable and special for the class is a large structure usually called the hyperstriatum but really a part of the pallium. Disturbances of courting, feeding, and fighting have been observed following lesions here.

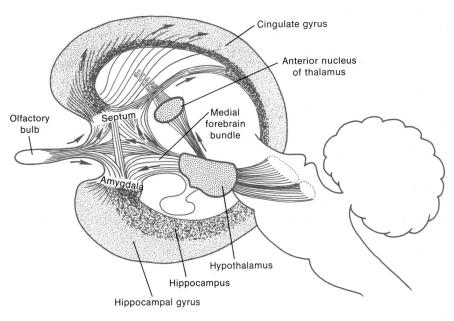

Figure 10.88
The limbic system. Diagram of the principal components. [MacLean, 1970.]

Figure 10.89
The limbic system in several mammals.

Rabbit   Cat   Monkey

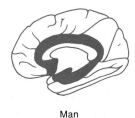

Man

Principal masses forming the limbic system in medial view. [Russell, 1961.]

pects of the amygdala and its functions is badly needed.

The concept of the **pallium** is a helpful and plausible simplifying idea that is difficult to prove or disprove, since it rests on the application of the concept of homology to intricate and changing structures. In the amphibians and even in lower forms, a medial part of the pallium can be distinguished from a lateral. The lateral is said by older authors to receive only olfactory tract input, and the medial chiefly higher-order olfactory fibers by way of the septum and lateral pallium. In the lateral or paleopallium (Fig. 10.90) the cells lie close to the ventricle, and their dendrites extend superficially into axon bundles from the olfactory tracts that run close to the surface and are unmyelinated. In the medial or archipallium the cell bodies have migrated toward the superficial fiber layer, which is here thicker and moreover myelinated.

Beginning in reptiles the **archipallium,** which is thus the first cortex, is subdivided into a large pyramidal-celled region called the hippocampal gyrus and a region of smaller cells representing the future dentate gyrus. The archipallium in mammals characteristically rolls in like a jellyroll to form a curious cortex partly buried beneath the surface and partly lining the lateral ventricle. The small granule cells of the dentate gyrus, or fascia dentata, send their axons only to their near neighborhood, whereas the pyramidal cells or Ammon's pyramids of the hippocampal gyrus send axons a long distance into the diencephalon and mes-

Box 10.4 Emotion and Motivation: The Limbic System

A simplifying concept of great power, the idea of the limbic system (limbus = border) refers to the fanciful border formed by a ring of structures around the anterior end of the brain stem and interhemispheric commissures. The apparently quite diverse structures involved (Fig. 10.88) include the subcallosal, cingulate and parahippocampal gyri, the hippocampal formation and dentate gyrus, the posterior orbital cortex of the frontal lobe, anterior insular cortex and that of the tip of the temporal lobe, the amygdala, septal nuclei, hypothalamus, anterior thalamus, parts of the striatum and of the midbrain reticular formation. Thus mesencephalic, diencephalic and telencephalic regions, archicortex, paleocortex and neocortex of the transitional type are involved. Much of this is olfactory in origin and connections, much is autonomic, and there is overlap with the reticular activating system. Most of these structures are identifiable with equivalents in birds, reptiles and amphibia, and at least some in fish, although information is too meager for many homologies there. The limbic system must be counted an ancient heritage.

The common denominators that tie together such an array of structures are certain intricate anatomical connections plus an involvement with emotion and affect. There is much localization of role, but the system works together intimately. Focal lesions can produce excessive eating or nearly fatal loss of appetite. Stimulation can do the same and can compel an animal to drink like an automaton, to exhibit signs of rage, hissing, growling, flattening of the ears, piloerection, pupillary dilation, increased heart rate, and the like, or of flight or fear, hypersexuality, or compulsive exploratory behavior. A monkey with focal limbic lesions may, quite unlike a normal monkey, approach without wariness such unfamiliar objects as a snake, or a cat or dog, and will try to mouth all kinds of animate and inanimate objects as though value judgements and distinctions between harmful and harmless, together with emotional threat, have been distorted. Elsewhere we have mentioned the influence of the amygdala upon tameness and wildness. Properly coordinated with the somatic response, gastrointestinal motility and secretion may be inhibited or activated, defecation and micturition may be induced, body temperature may change in either direction, and gonadotrophic hormone and ACTH may be released from the pituitary.

To study objectively the subjective phenomena of emotion and affect, a number of ingenious techniques have been developed. An elementary question is whether a limbic structure at issue is simply a motor center for coordinating the external expression or is far enough upstream to be assigned the real affective role. A set of criteria are used: as applied to "sham rage" verses genuine emotion one asks (a) is the aggression directed, (b) does it inhibit and replace other activity, (c) does it outlast the stimulus, (d) does the animal become conditioned against the situation in which the stimulus is repeated, (e) can the brain stimulation serve as an unconditioned response in Pavlovian conditioning? The answers for limbic stimulation in general are positive. Confirming the experiments in laboratory animals, electrical stimulation of the amygdala in man has caused subjective feelings of fear and rage; in patients with histories of episodic, compulsive fits of violence, lesions of the amygdala tend to moderate such behavior.

A useful technique has been to give the animal control of the brain-stimulating circuit by a switch it discovers by chance. Many electrode locations in the limbic system yield either avoidance of the switch or marked self-stimulation at rates up to several thousand per hour, as though stimulation had a high pleasure content. This or other stimuli of presumed motivating content can be objectively measured by titrating switching rates against, for example, electric shocks to the feet or work expended to get food or water. Electrodes can be placed in one or another of the centers of natural drives and hormones can be used to bias the "set." Thus electrically induced hunger can influence self-stimulation rates from one set of loci while androgen injection influences them from another. It is not too much to expect that we can measure neurophysiologically quantities causally related to the ethologist's older concepts of action-specific potential, releasing value, mood or set. It is not too much to say that we can measure—and manipulate—the deepest value systems and motivations, the neural mechanisms of selfish and selfless attitudes that coexist as part of our heritage. It is not too much to say, as Nauta has done, that the limbic cortex of the basal part of the frontal lobe is where we live. Ablation of the limbic cortex would destroy one's sense of self; any other part of the brain or body can be removed, or substituted for, with but limited effects on abilities or leanings but no fundamental effect on the sense of self.

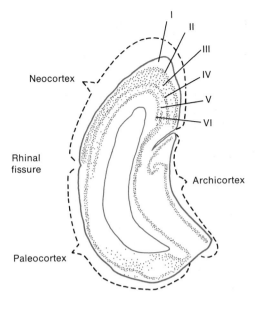

Figure 10.90
Three kinds of cortex. The forebrain of the embryo of an armadillo (*Dasypus*), showing the newly developing six-layered neocortex (isocortex) between the simpler paleocortex (allocortex) and archicortex (hippocampal cortex). The deep layer of the last named consists of large pyramidal cells continuous with layers V and VI of the neocortex. I–VI are the layers of the neocortex. [Kappers, et. al., 1936.]

encephalon. The combination is called the hippocampal formation; phylogenetically it is the archicortex, and histologically the arrangement is called an **allocortex.**

The **paleopallium** also develops allocortex in places: elsewhere it is transitional. Its complex connections have been simplified in Table 10.8, where we note that its input is chiefly from olfactory lobe, archicortex, and archistriatum, though it projects, in its fully evolved mammalian form at least, to many parts of the neocortex and neothalamus, as well as to older parts of the diencephalon, midbrain reticular formation, and, reciprocally, to the amygdala.

The **neopallium or neocortex** is virtually a mammalian monopoly. According to classical criteria, which deserve to be re-examined, no true neocortex is found in reptiles or other vertebrates, although a primordium of the neopallium can be discerned in some reptiles, in which it receives projections from the already

well developed neothalamus and lies between the paleocortex and archicortex. From this limited antecedent in reptiles, there blossoms—even in the lowest mammals—a full-blown and extensive neocortex (histologically an **isocortex**) that far overshadows the older forms of cortex. Birds (see Box 10.3, p. 482) have a homolog of neocortex, the Wulst, of considerable elaboration.

Except for limited areas of transitional cortex, the neocortex is everywhere constructed on a **basic plan of six layers** (Fig. 10.91). Reception of input is chiefly in the fourth, or granular, layer. Outflow is mainly from pyramidal cells in the fifth and sixth layers. The first, or molecular, layer is outermost, latest in phylogeny and slowest in ontogeny; it is a specialized neuropile nearly free of nerve cells. Significant regional variation into four or five types permits mapping, which, in the most commonly used form and nomenclature, is shown in Figure 10.92. The superficial three layers are the nov-

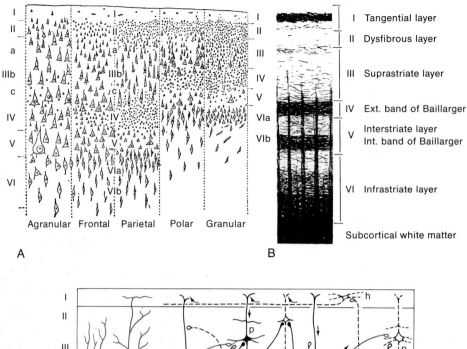

I Tangential layer
II Dysfibrous layer
III Suprastriate layer
IV Ext. band of Baillarger
V Interstriate layer
  Int. band of Baillarger
VI Infrastriate layer

Subcortical white matter

A

B

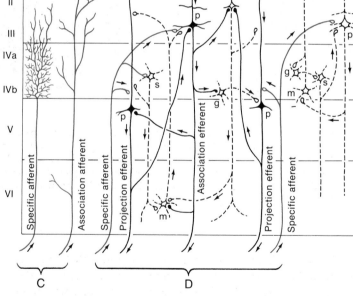

Figure 10.91

The neocortex. **A.** Five main subtypes of isocortex such as define the areas of Brodmann (Fig. 10.92) seen in the Nissl stain. [Economo, 1929.] **B.** The same in the Wiegert stain for myelin. [Brodmann, 1909.] **C.** Two main modes of termination of afferent cortical fibers. **D.** Simplified scheme of the neuron types and connectivity. *g*, granule cell; *h*, horizontal cell; *m*, Martinotti cell; *p*, pyramidal cell; *s*, stellate cell. [Truex and Carpenter, 1969, based on data of Lorente de Nó, 1949.]

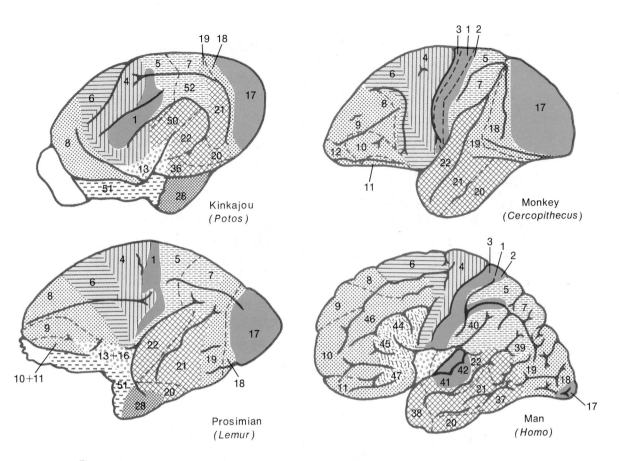

Figure 10.92

Cortical areas in several mammals. The lateral aspects of the hemispheres, with cytoarchitectonic fields (areas of the cortex, such as those in Fig. 10.91, defined by the subtype) and the numbers generally used to designate them. The primary sensory areas are indicated by solid brown; the secondary sensory areas by the brown patterns; the motor and premotor areas by the gray patterns. The white area shown in the kinkajou and prosimian examples is often called "association cortex." [Brodmann, 1909.]

elties added to the basic allocortex. They are late developing, thicker in the larger of two related species (Fig. 10.93), and receptive of association fibers from nearby and remote areas of the cortex. Underlying white matter is proportionately more voluminous (Fig. 10.94) and cortex less voluminous (Fig. 10.95) in larger species.

There is some opinion that the primary projection areas of neocortex, defined by either afferent projections from thalamic relay nuclei or efferents to lower motor neurons, are the first to develop, but even the lowest mammalian brains probably have a considerable area of association cortex defined by thalamic projections from the association nuclei.

Figure 10.93
The cortex is thicker in larger species of the same order, especially layers II + III and V + VI. [Kappers et al., 1936.]

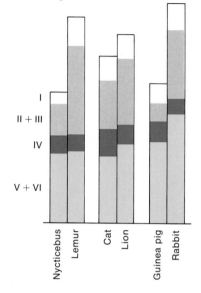

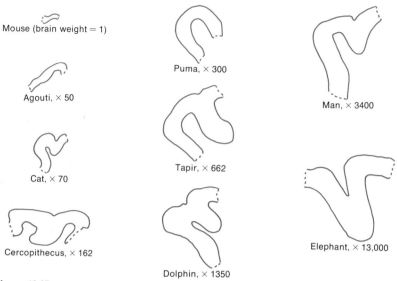

Mouse (brain weight = 1)

Agouti, × 50

Cat, × 70

Cercopithecus, × 162

Puma, × 300

Tapir, × 662

Dolphin, × 1350

Man, × 3400

Elephant, × 13,000

Figure 10.95
The thickness of cerebral cortex differs only by a factor of three, whereas brain weight differs by a factor of 13,000. (Parietal cortex, uniform magnification.) [Kappers et al., 1936.]

Figure 10.94
The proportion of white matter (stained by Weigert method) to gray matter (unstained) is greater in the brain of the larger of two related species. (Brains of cat and puma drawn to same size.) [Kappers et al., 1936.]

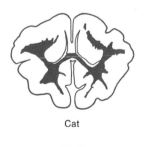

Cat

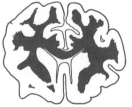

Puma

The size of the sensory and motor cortex devoted to a part of the body is correlated with the number of peripheral nerve fibers (Fig. 10.96). Thus the cortical areas for the pig's snout, the human hand, and the bat's ear are disproportionately large. The folding of the cortex begins in primary projection areas; sulci and gyri in the association areas come later and are less constant. Histologically, primary sensory areas have a well-developed granular layer (IV); motor areas are agranular, emphasizing layers V and VI; and association areas have the thickest supragranular layers (I, II, III). All the association areas increase in size disproportionately in primates and especially in man, but in particular the frontal association areas are vastly larger.

We have already considered many aspects of physiology pertinent to the **functioning of the cortex,** including principles of neural coding, labeled lines, change of label with successive stages in a pathway, filtering and feature extraction, decision-making and central command, wide and narrow receptive fields, overlap and lateral inhibition, modality and submodality, convergence and parallel processing, evoked potentials and electroencephalograms. In the following paragraphs we deal with a few more principles of neural organization in the cortex. We cannot in the present book deal with a number of phenomena, such as interhemispheric relations, dominance of one side, aphasias, prefrontal functions in personality, and the neurological bases of learning. The interested reader

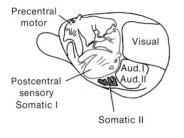

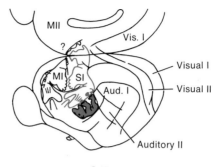

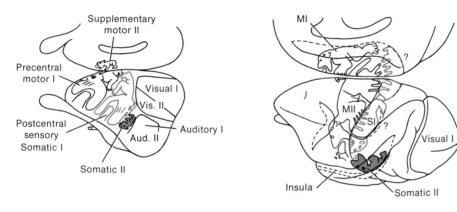

Figure 10.96

Somatic cortex in four species. The locations and general plans of the precentral motor (*MI*), supplementary motor (*MII*), postcentral sensory (*SI*), and second somatic sensory areas. The relations to auditory and visual cortex are shown, except in the monkey, in which the auditory area is hidden on the lower bank of the Sylvian fissure. The medial surface of the hemisphere is shown, swung upside down, for the rabbit, cat, and monkey. [Woolsey, 1958.]

Box 10.5 Arousal, Alerting, Attention, Sleep and Unconsciousness: The Reticular Activating System

An organizational principle of profound significance is the development of a whole system, in parallel with classical specific pathways for ascending sensory information, that receives converging collateral fibers from all pathways and blurs their specificity. This system is found in the reticular formation that hugs the core of the neuraxis, throughout its extent from spinal cord to diencephalon (see figure on p. 493). When it is inactive, the individual loses consciousness, and although the specific sensory pathways are intact, he cannot be aroused. If the reticular activating system is stimulated, the individual is aroused and the directing of attention is aided by descending impulses that inhibit competing modalities of input at each level (at least to the first relay) and, in hearing, even at the cochlea.

The EEG is strikingly different in the waking and sleeping state, permitting a convenient assay of the influence of various stimuli. The effect of alerting stimuli is generally interpreted as a desynchronization of the cortical cellular activity. The specific sensory pathways are, by themselves, incapable of awaking or alerting, whereas activity in the reticular system will do so even without sensory input.

The reticular formation is neither monolithic nor completely unspecific; there are subdivisions with different roles—for example, separable regions in the medulla that enhance and that diminish reflexes. Stimulation of the diffuse projecting nuclei of the thalamus can at certain frequencies cause, instead of EEG desynchronization, a growing ("recruiting") series of slow waves.

The **locus coeruleus,** sometimes considered to be a special derivative of the reticular formation, is a tiny cluster of catecholaminergic cells in the pons; axons with very large numbers of collaterals project to many parts of the brain stem and basal ganglia and to the entire cerebellar and cerebral cortices, probably exerting a generalized inhibiting influence. There is much to learn about the normal control and fractional behavior of the reticular system.

Impulses ascend in short relays through the reticular formation of the brain stem to the dorsal hypothalamus and the reticular and ventromedial part of the thalamus, from which they project diffusely to influence widespread regions of the cortex. General anesthetics act specifically on this system, and **sleep** is primarily an activation and deactivation of the reticular system. Study of this system has improved our understanding of sleep, which turns out to have at least four phases, based on EEG criteria (see figure on p. 494). Among them is a paradoxical deep-sleep phase, during which the EEG is as unsynchronized as it is in the waking state. This tells us that although the conscious state is integrated and controlled by the reticular formation and the nonspecific thalamocortical systems, consciousness cannot be unequivocally read from the pattern of the EEG.

The **functions of the reticular formation** are not adequately summarized in this way, however, and indeed are quite imperfectly understood. Scheibel and Scheibel (1968) have said, "These functions include: determination of operational modes; gating of all sensory influx; participation at all levels of cortical function, including read out for cortical differentiative and comparative processes; gain manipulation of motor output; multilevel control over most visceral functions; and the active manipulation of a spectrum of states of consciousness from deep coma to maximal vigilance." After reviewing the structural and physiological features of the reticular core of the brain stem, they conclude in this way: "The picture that we have tried to develop is that of a core of neurons shielded from direct contact with the environment on the one hand and from the rostral differentiative (cortical) centers on the other—yet continually washed by reflections of excitation patterns flowing from both. These samples of ongoing activity are then integrated within the elements of the mosaic. The resultant output, expressed as an intensity continuum, is discharged widely upon centers upstream and downstream and throughout the rest of the core."

Box 10.5 (*continued*)

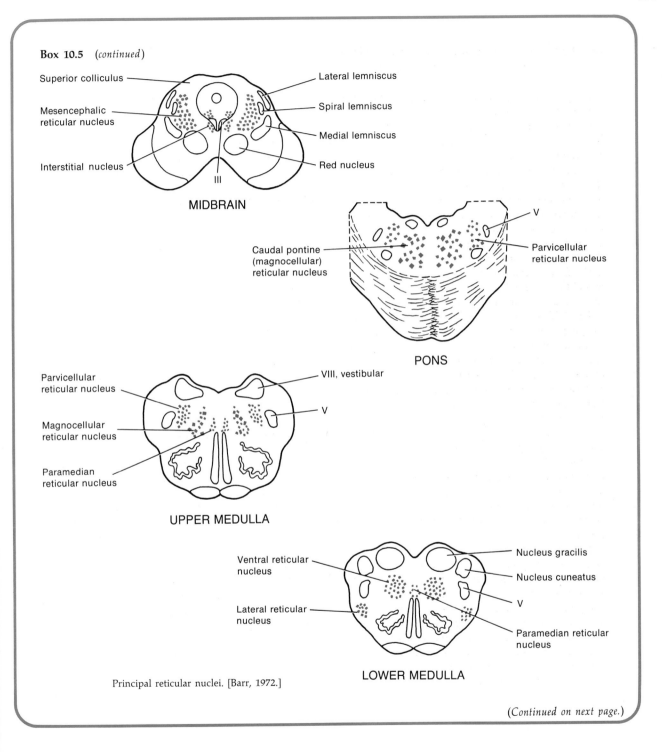

MIDBRAIN

Superior colliculus

Mesencephalic reticular nucleus

Interstitial nucleus

Lateral lemniscus

Spiral lemniscus

Medial lemniscus

Red nucleus

III

PONS

Caudal pontine (magnocellular) reticular nucleus

Parvicellular reticular nucleus

V

UPPER MEDULLA

Parvicellular reticular nucleus

Magnocellular reticular nucleus

Paramedian reticular nucleus

VIII, vestibular

V

LOWER MEDULLA

Ventral reticular nucleus

Lateral reticular nucleus

Nucleus gracilis

Nucleus cuneatus

V

Paramedian reticular nucleus

Principal reticular nuclei. [Barr, 1972.]

(*Continued on next page.*)

Box 10.5 (*continued*)

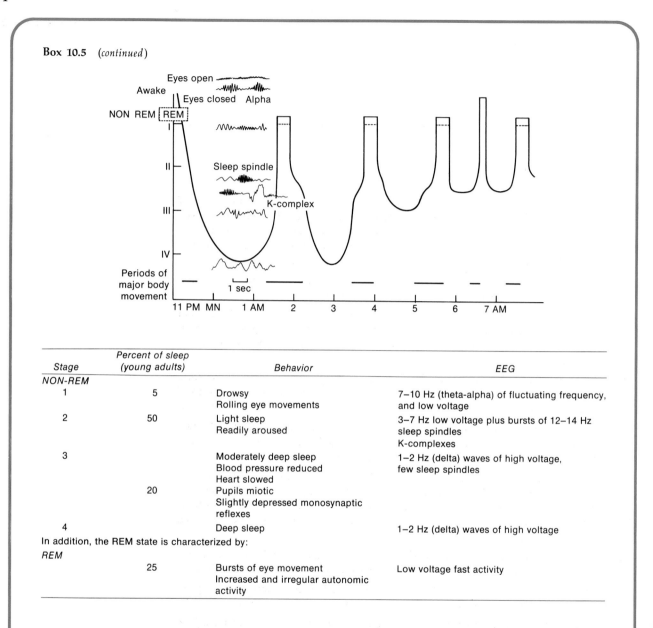

Stage	Percent of sleep (young adults)	Behavior	EEG
NON-REM			
1	5	Drowsy Rolling eye movements	7–10 Hz (theta-alpha) of fluctuating frequency, and low voltage
2	50	Light sleep Readily aroused	3–7 Hz low voltage plus bursts of 12–14 Hz sleep spindles K-complexes
3		Moderately deep sleep Blood pressure reduced Heart slowed	1–2 Hz (delta) waves of high voltage, few sleep spindles
	20	Pupils miotic Slightly depressed monosynaptic reflexes	
4		Deep sleep	1–2 Hz (delta) waves of high voltage
In addition, the REM state is characterized by:			
REM			
	25	Bursts of eye movement Increased and irregular autonomic activity	Low voltage fast activity

Stages of sleep. **Above.** The progression of stages of non-REM and REM sleep in a young adult human. The EEG characteristic of stages 1 through 4 of non-REM sleep is shown. **Below.** The states of sleep with their incidence and characteristics. [Willis and Grossman, 1973.]

SUGGESTED READINGS

Bullock, T. H., and G. A. Horridge. 1965. *Structure and Function in the Nervous Systems of Invertebrates.* W. H. Freeman and Company, San Francisco. [Reviews the literature to 1964 with chapters on principles and chapters on the groups of animals, each with a summary. More recent papers and books on particular groups or topics are numerous and readily found with standard bibliographical aids. Only a few are listed here.]

Corning, W. C., J. A. Dyal, and A. O. D. Willows. 1973. *Invertebrate Learning* (2 vols.). Plenum Press, New York. [A review of the state of knowledge, group by group.]

De Reuck, A. V. S., and J. Knight. 1968. *Hearing Mechanisms in Vertebrates.* Churchill, London (Little Brown & Company, Boston). [This is one example of many symposia on special topics published in book form; some chapters are comprehensive reviews, most are research papers.]

Hoar, W. S., and D. S. Randall. 1969–1971. *Fish Physiology* (6 vols.). Academic Press, New York. [Especially vols. 4 and 5, but also others in this series, have chapters reviewing aspects of sensory, nervous, effector, endocrine, and behavioral physiology of this diverse class.]

Iggo, A. 1973. *Somatosensory System* (vol. II of *Handbook of Sensory Physiology,* 8 vols.). H. Autrum et al., eds. Springer-Verlag, New York. [We list this example from the series of volumes (eight, some in parts), each with more or less monographic chapters.]

Kappers, C. U. A., G. C. Huber, and E. C. Crosby. 1967. *The Comparative Anatomy of the Nervous System of Vertebrates, Including Man.* Hafner Publishing Company, New York. [Reviews the morphological literature on vertebrates to the early 1930's; difficult to read, but a basis for much modern work and a guide to a substantial volume of early work. Reprinted without revision from a 1936 printing.]

Mountcastle, V. B. 1968. *Medical Physiology.* Mosby Company, St. Louis, Missouri. [This may serve as a reference for the large body of knowledge on mammalian neurophysiology.]

Pearson, R. 1972. *The Avian Brain.* Academic Press, New York. [Anatomical, physiological, developmental, and biochemical aspects of the brains of birds are treated.]

Prosser, C. L. 1973. *Comparative Animal Physiology* (3rd ed.). W. B. Saunders, Philadelphia. [The chapters on sensory, effector, and neuroendocrine physiology are uniquely valuable as surveys of scattered knowledge of vertebrates and invertebrates with respect to function from cellular to organ level.]

Sarnat, H. B., and M. G. Netsky. 1974. *Evolution of the Nervous System.* Oxford Univ. Press, New York. [Though lightly illustrated and sketchy in spots, this is uniquely useful today as a survey of the vertebrates.]

Strausfeld, N. J. 1976. *Atlas of an Insect Brain.* Springer-Verlag, New York. [A comprehensive account of the cerebral structures of a single species of advanced insect, the house fly, by an array of classical and modern methods; lavishly illustrated.]

Willis, W. D., Jr., and R. G. Grossman. 1973. *Medical Neurobiology.* C. V. Mosby, St. Louis. [This is one of a number of standard texts on mammalian, especially human, neuroanatomy, from which most of our knowledge of vertebrates comes.]

Young, J. Z. 1969. *The Anatomy of the Nervous System of Octopus vulgaris.* Oxford Univ. Press, Oxford. [An example of a detailed monograph on a relatively well-studied, higher invertebrate based on classical descriptive methods.]

GLOSSARY

Included here is a selection of terms particularly pertinent to general neurobiology, especially those widely used terms that have a special meaning not clearly given in ordinary dictionaries. Words not found here should be sought in the index. General zoological terms (e.g., "medial" versus "median"; "epidermis," "bristle," "deuterostome") and names of parts peculiar to certain groups of animals are not in general given here and should be sought in dictionaries or zoological treatises. Terms used in the study of animal behavior may be found in a useful glossary by Verplanck (1957). Words defined here should also be sought in the text, via the index, for possible additional explanation.

AC. *See* Alternating current.

Accommodation. True accommodation in neuronal physiology is a rise in the threshold transmembrane depolarization required to initiate a spike when depolarization is slow or a subthreshold depolarization is maintained. By extension, the term is also used to refer to the consequence of maintained hyperpolarization, which sometimes causes a gradual decrease in the threshold depolarization, measured as a higher than normal membrane potential at which spikes arise. Apparent accommodation (or simply "accommodation" in the older literature) is the rise in threshold measured as strength of current applied to the outside of a nerve with slowly rising current. It may be largely an artifact of the sheath, and bears little relation to true intracellularly measured accomodation. (See p. 151.)

Action potential. An older synonym for the spike potential of the nerve impulse (see Impulse and Spike potential); it should not be confused with several other potentials of activity. (See Fig. 6.1)

Active process. With reference to potential changes across the membrane of excitable cells, a process involving an energy-requiring response, usually marked by a change in the ionic conductance of the membrane as opposed to an electrotonic change (*which see*).

Adaptation. *Of excitable tissues,* a process inferred from the decline in response during a maintained stimulus; the response is typically measured as frequency of spike discharge. The term is usually applied to the response of sensory axons, because the stimulus must be known to be maintained, and is distinguished from "accommodation," in which a threshold rise occurs without any spike discharge having taken place; accommodation is one possible basis of adaptation. Also distinguished from "fatigue" (*which see*) in not requiring repeated stimulation or overstimulation, in recovering promptly after the end of stimulation, and in preserving good responsiveness to a fresh stimulus at the depth of the decline in discharge. *Of whole organisms,* a regulatory or advantageous change in response to an environmental stress by an individual or by a species in the course of evolution.

Adequate stimulus. The kind of stimulus that (there is reason to suppose) a given receptor normally detects and signals; distinguished from other stimuli that can excite under artificial conditions or cannot excite. The reasons for deciding what is normally detected must be based on behavior, not physiological sensitivity.

Adrenergic. Applied to a neuron or synapse that releases on stimulation a substance with the properties of adrenaline or noradrenaline; transmission mediated by such substances.

Afference. The influx of messages from sense organs responding to events in the environment (*compare* Reafference).

Afferent. Incoming; said of impulses or pathways carrying impulses *toward* a center of reference, usually the central nervous system.

Afterdischarge. Continuation of an impulse train following the end of a period of stimulation.

Afterpotential. Membrane potential following a spike and not yet recovered to resting values; it may be depolarized or hyperpolarized relative to normal, or both in sequence.

All-or-none. A characteristic of certain forms of response, like spikes, which behave as though they are regenerative or explosive chain reactions. Amplitude is not necessarily always the same but is always maximal. "None" does not preclude small, graded preresponses but only intermediates between a threshold and a full discharge.

Alternating current (AC), AC amplifier, AC recording. As used in electrophysiology, AC recording refers loosely to the recording of any fluctuating potentials more rapid than a sine wave of about 0.2–0.5 Hz. The AC fluctuating potentials in bioelectricity are generally transient or nonsinusoidal; rates of change up to an equivalent of about 10,000 Hz. occur in animals. An amplifier is spoken of as AC when it is sensitive only to fluctuating potentials. Whether the driving energy for the instrument is a DC or AC source of power is irrelevant. *See* Direct current.

AC amplifiers are sometimes called capacity coupled. The time constant of the coupling determines what frequencies the amplifier passes on. Generally such amplifiers are coupled so as to work in the physiological band width and yet reduce drift, hum, and extraneous potentials.

Analog. An adjective applied to a class of communication devices contrasting with digital devices (*which see*). Analog devices carry intelligence by representing some variable in the input with a corresponding quantity in the output, which varies as some continuous function of the input quantity (mercury thermometer, pointer-type voltmeter). A train of nerve impulses usually carries information by the continuously graded intervals between impulses and is therefore an analog representation, although a pulse-coded analog.

Analogous. Having a similar function, as wings of birds and wings of insects, or transducers in technology and sense organs in animals.

Analogue. Structure or process having a function similar to some other function or process. In theoretical work, a model or simulation, conceptual or physical. *See* Analogous, Homologous.

Anastomosis. Union by complete confluence; referring to neurons, fusion with cytoplasmic continuity between two nerve fibers.

Anelectrotonus. *See* Electrotonus.

Antidromic. Said of an impulse, or of the stimulation causing an impulse, that conducts in the direction counter to the normal (dromic or orthodromic) one for that fiber; usually an impulse conducted toward the dendrites.

Antifacilitation. A process inferred from the reduction in height of a second graded response elicited shortly after a first at physiological intervals, the reduction being less severe at longer intervals. Possibly due to relative refractoriness, but said of a response at intervals normal to the functioning of a particular cell; in some cases, the same cell shows facilitation (*which see*) at the same time via other presynaptic pathways.

Arousal. In neurophysiology, a process whereby large groups of neurons (as in the cortex) are brought into a state of increased responsiveness, measured either via behavior of the animal or electrical activity of the brain; one important cause is stimulation of a nonspecific activating system.

Association neuron. Older term nearly equivalent to interneuron (*which see*).

Augmenting response. A special term used in vertebrate brain physiology for certain evoked potentials that grow larger with repetition. This type, in contrast to the recruiting response (*which see*), is confined to the primary sensory cortex, is elicited by stimulation of specific afferent pathways, and appears with a short latency in response to the first shock.

Autogenic. Arising spontaneously within the system; said of pacemaker potentials in neurons.

Autonomic nervous system. A part of the nervous system, central and peripheral, arbitrarily defined as being only efferent, which supplies viscera, including especially the vascular, alimentary, and reproductive systems but also skin, skeletal muscle, excretory system, and some sense organs, such as the eye. *See* Parasympathetic, Stomodeal, Sympathetic nervous system.

Axo-axonal. Said of synapses formed by contact of two axons.

Axo-dendritic. Said of synapses formed by contact between a presynaptic axon and postsynaptic dendrites.

Axon. A process of a neuron distinguished by slow tapering; a smooth outline; branches set relatively far apart; cytoplasm (axoplasm) unlike that in the soma, without ribosomally studded endoplasmic reticulum (Nissl bodies, *which see*); and by other features. Functionally a principal output structure, but not the only one (*see* Reciprocal synapses, p. 48); usually carrying impulses; often individually sheathed; usually only one per neuron, but there may be two or none. (See p. 11.)

Axon reflex. A class of responses in which one branch of an axon acts as an afferent limb and is excited by an adequate stimulus and another branch carries the excitation to an effector—typically arteriolar smooth muscle in the adjacent skin.

Axoplasm. Cytoplasm typical of the axon (*which see*); often there is a sharp boundary between this type and the cytoplasm of the soma. (See Fig. 2.14, p. 28.)

Axosomatic. Said of synapses formed by an axon ending on a nerve cell body.

Bipolar. Referring to nerve cells having two main processes, generally from opposite ends of a spindle-shaped soma.

Boutons terminaux. A type of axonal ending with a slight terminal enlargement occupied by mitochondria and a ring of neural filaments; especially numerous on the soma and large dendrites of vertebrate spinal motor neurons.

Brain. The enlarged and specialized rostralmost or highest ganglion of the central nervous system. Not necessarily the largest nervous mass, but if the ganglion is not distinctive enough in size or differentiation or if it is too far from a distinct head, the term "brain" is not used; "cerebral ganglion" is more general and includes brains as well as some minor ganglia. *See also* Supraesophageal ganglion.

Brain waves. Fluctuating electrical potentials recorded from the brain, directly or through overlying tissues; loosely, those potentials equivalent to sinusoidal frequencies from about 0.3 Hz to a few hundred Hz, seen with electrodes that sum the activity of many cells; a record of brain waves is called an electroencephalogram (EEG).

Burst. A short sequence of spikes in the same neuron.

Catelectrotonus. *See* Electrotonus.

Cell body. The perikaryon or soma; the portion of a neuron containing the nucleus and perinuclear cytoplasm, as opposed to the portions making up processes.

Cellulifugal. Conducted away from the cell body; said of impulses or materials in the axon.

Cellulipetal. Conducted toward the cell body.

Center. As used in neurophysiology, a more or less circumscribed region of the central nervous system, consisting of a few or many cells, for which there is evidence of a special importance in some function. The operational word is "special"; mere involvement or mediation is not sufficient. Roles such as controlling, initiating, integrating, or receiving justify the term. The highest level of these roles is not necessarily implied.

Central nervous system. That part of the nervous system (*which see*) that forms a distinct principal concentration of cords or ganglia and their interconnections; contrasted with the peripheral nervous system. *See* Peripheral.

Cephalization. The presumed evolutionary process resulting in progressively greater concentration of nervous structures and functions in the head.

Cerebral ganglion. The highest ganglion (*which see*); the ganglion associated with the anterior end and the principal organs of special sense, if any, generally dorsal to the anterior end of the gut; in most animals, synonymous with the brain (except when the cerebral ganglion is poorly developed, as in some clams) and with the supraesophageal ganglion (except when the mouth is elsewhere, as in planarians).

Chemoreceptor. *See* Receptor.

Chiasm. A structure in which a large group of nerve fibers crosses a midline symmetrically while coursing mainly transversely (*compare* Decussation). The midline may be that of the body or of an organ, such as part of the optic lobe.

Chiastoneury. The condition in prosobranchs and a few other gastropods in which the visceral loop is twisted in a certain way (see p. 428 *and* Torsion); also streptoneury; contrasted with orthoneury and euthyneury (*which see*).

Cholinergic. Said of neurons and synapses that liberate acetylcholine with activity; transmission mediated by this substance.

Chromatolysis. A certain kind of change in the histological appearance of nerve cell bodies caused by injury to the axon; the Nissl bodies in particular are involved; for details and figures, see pp. 24 and 73.

Chronaxie. The duration of an electrical stimulus just sufficient to give a response when the strength of the stimulus is twice the rheobase (*which see*).

Circadian. Approximately with the period of a day; said of rhythms of biological activity with a 24-hour period or close to it.

Circumesophageal connectives. The anterior continuation of the ventral cord on either side of the pharynx in higher invertebrates, running between ventral ganglia and brain.

Clonic. Refers to movements that are repeated spasmodically, as distinct from tonic or phasic.

Coding. Applies to the translation into, or the representation of, a message by a set of signals; in the nervous system we may speak of the coding in a single element (axon, neuron, even a single junction) or in an array of units; of temporal or spatial coding; of spike or graded potential coding.

Collateral. In neurohistology, an important side branch of an axon that is of a caliber equal or nearly equal to that of the axon.

Command cell. An interneuron that, upon stimulation, reproducibly releases motor action resembling a major piece of normal behavior. Other than a neuron early in an afferent pathway or late in an efferent pathway.

Commissure. A transverse tract of nerve fibers connecting symmetrical parts of the central nervous system on the two sides.

Compound action potential. A term used for the externally recorded potential from a bundle of nerve fibers, (e.g., a nerve or tract). Typically it is used for the potential from a population of fibers heterogeneous with respect to velocity, threshold, and spike duration; the spikes of the individual fibers sum according to their amplitudes, numbers, and velocities.

Conductile membrane. Cell membrane of neurons or muscle cells capable of conducting impulses, as opposed to membrane capable only of electrotonic spread.

Conduction. In neurophysiology, reserved for the process of active propagation of nerve impulses within the same neuron, mainly in axons.

Connective. The whole of a major longitudinal fascicle of nerve fibers between adjacent ganglia (separated anteroposteriorly) of the central nervous system of invertebrates.

Convergence. Referring to central nervous pathways, the coming together of two or more different inputs onto a common neuron or center.

Coordination. The process of combining parts of a whole action into harmonious relation, whether reciprocal or successive or simultaneous. In a special usage, coordination is said to be absolute or relative or absent, referring to separate rhythmic movements that may (a) be synchronized, (b) show loose, labile coupling affecting amplitude, frequency, or both, or (c) be entirely independent.

Cord. In comparative neurology, a longitudinal concentration of nerve cells, fibers, and endings, often of sufficient importance to be part of the central nervous system, but sometimes not sufficiently well developed. Cords may be only slightly thickened and straightened strands of a general plexus or may be second only to the brain as principal parts of the central nervous system.

Corollary discharge. A term used by Teuber and others for a postulated neural message thought to be sent out by command centers that initiate voluntary movements (e.g., eye, hand, and head movements). The message is a corollary of the command, and conditions other centers with respect to the altered sensory input caused by execution of that movement. The term as used is more vague than efference copy (*which see*), and criteria have not been rigorously used to decide when a corollary discharge may be invoked; in some usages it is assumed that a corollary discharge exists for every neurally commanded movement, including reflex eye movements.

Corpora pedunculata. Certain densely staining neuropiles associated with globuli cells and representing higher centers in the brains of some polychaetes and many arthropods. Also called "mushroom bodies" for their resemblance to mushrooms in section.

Cortex. In neurology, a well-differentiated sheet of cells and synapses, usually gray matter (*which see*) disposed as a morphologically outer layer of a part of the brain; the term is usually applied only to such sheets on the cerebellum and cerebrum, and therefore only to vertebrates. However, it can also be applied to rare tissues in invertebrates, such as the covering of the optic lobe of cephalopods, because of the extensive sheet-like and laminated neuropile. Use of the term does not imply homology, even within the vertebrates.

Coupling function. A transfer function; more loosely, a quantitative relation between the signals arriving at a junction and those leaving it.

DC. *See* Direct current.

Decremental. A progressive lessening in strength with time or distance.

Decussation. A crossing to the opposite side of an ascending or descending fiber tract. Compare Chiasm, which is a crossing of tracts soon terminating and not ascending or descending any great distance.

Defacilitation. Equivalent to antifacilitation (*which see*); the latter is preferred, for the process is not undoing facilitation but is contrary to it.

Degeneration. Used in neurology for the deteriorating sequelae of injury, usually for changes that are irreversible and lead to death and loss of the given axon or cell; but sometimes changes are spoken of as degenerative that are only severe reactions, not necessarily irreversible. There are several types. (See pp. 24, 73.)

Delay, synaptic. Elapsed time between arrival of a presynaptic impulse at a junction and the first sign of postsynaptic response. The measurement is only properly available in situations that allow estimation of the moment the presynaptic impulse reaches the synapse and recording of the synaptic potential of the postunit. The concept is only properly applicable when the influence of a single presynaptic impulse is strong enough to make a distinct postsynaptic potential.

Demarcation potential. *See* Injury potential.

Dendrite. A process of a neuron distinguished by relatively rapid tapering, an irregular outline with branches at short intervals, cytoplasm like that of the soma with rough surfaced endoplasmic reticulum (i.e., studded with ribosomes) and other features. Functionally a principal receptive and integrative region of most neurons but also responsible for some output at presynaptic sites (see Reciprocal synapses, p. 48); often incapable of carrying impulses; typically without an individual sheath but often packed in satellite glial cells; usually several or many per neuron but not developed or recognizable in some primitive neurons. Since the defining criteria are cytological, not functional, the receptive processes of neurons in which cytological criteria are not adequately known to distinguish them from axonal terminals (e.g., most invertebrate neurons) should be called receptive processes, not dendrites (See p. 15.)

Depolarization. Decrease in polarization (*which see*), from any cause, to any degree, relative to the normal resting potential (*which see*).

Digital device. The meaning that has been considered relevant to neural function is that which distinguishes a class of communication devices contrasting with analog devices (*which see*). Digital devices may carry intelligence encoded in a series of discrete units of time known to the receiver, during each of which one or another of a few distinct alternative states prevails. Commonly there are two such states (binary codes). Between successive events there is a whole number of the discrete time units.

Diplotomic. That type of branching shown by two parallel nerve fibers which branch together; that is, every time one branches, the other does too.

Direct current (DC), DC amplifier, DC recording. As used in electrophysiology, DC recording refers to standing or slowly changing potentials, slower than a sine wave of about 0.2-0.5 per second. An amplifier is spoken of as DC when it is sensitive to standing or maintained potentials (zero frequency); in a simple form it is called "direct coupled." DC amplifiers are always sensitive as well to some range of AC (*see*

Alternating current) from infinitely slow up to some maximum frequency, and may completely overlap the range of sensitivity of AC amplifiers. The term has no reference to the source of power for the instrument.

Discharge. As used in neurophysiology, a firing of one or more relatively rapid events, usually all-or-none.

Dromic. Equivalent to orthodromic (*which see*).

EEG. *See* Electroencephalogram and Brain waves.

Effector. A specialized part of an organism that carries out some action relative to the environment and to behavior, such as muscular movement, ciliary activity, exocrine gland secretion, luminescence, pigment migration, or electric organ discharge.

Efference copy. Von Holst's term for a neural message postulated to be required in certain examples of behavior; a message that is quantitatively determined by the same causes as the efference (efferent command to effectors) but is retained until the altered sensory input, resulting from the execution of the command, has arrived or is due, and which is then so connected as to permit some process of matching with this reafference (*which see*). The efference copy is equivalent to a central state of quantitative expectation of the alteration of input due to an organism's own movements; in many such movements it is not indicated. See p. 310, and Corollary discharge.

Efferent. Outgoing; said of impulses or pathways conducting away from a center of reference, usually the central nervous system.

Electroencephalogram. A record of fluctuating electrical potentials arising in the brain. *See* Brain waves.

Electrogenesis. The production of electric potentials (more exactly, of potential differences); used in neurophysiology to refer to the production of either standing or changing potentials by living cells or model systems.

Electroretinogram (ERG). The potential change produced in the retina when light stimulates it, as seen by macroelectrodes influenced by the whole thickness and a considerable extent of the retina; measured with an active electrode in any of various positions: on the cornea, on the sclera, on or in the retina. The ERG is a composite of several potentials of distinct origin, time course, and properties, and the components are different in different species of animals.

Electrotonic change, electrotonic spread, electrotonic potential. Said of a change in electric potential of a membrane or tissue due to current from a source extrinsic to that portion of membrane or tissue, therefore imposed current, and involving only passive change in the standing potential. Opposed to active change, as in action potentials, in which there is an altered membrane resistance drawing current from local regions. Electrotonic spread is necessarily decremental and a function of the passive electrical properties—that is, the distributed resistances and capacities of the membrane or tissue.

Electrotonus. An old term for the changes in excitability caused by passage of a moderate current; during passage of the current, excitability is increased in the neighborhood of the cathode; and decreased in the neighborhood of the anode; the opposite is observed just after an abrupt cessation of current. Anelectrotonus is the change imposed by the anode; catelectrotonus is that under a cathode.

Encephalization. *See* Cephalization.

End buttons. *See* Boutons terminaux.

Endogenic potential. A type of membrane potential change that is due to antecedent activity in the same cell. (*See* p. 205.)

Endoneurium. Fine strands of connective tissue, collagenous fibers, fibroblasts, and fixed macrophages coursing longitudinally between individual nerve fibers in peripheral nerves, especially of vertebrates. *Compare* Epineurium, Perineurium.

End-plate. A flattened, disc-shaped terminal; characteristic of axon endings on certain types of muscle and rarely on other effectors.

Enteric nervous system. *See* Stomodeal nervous system.

Epineurium. The sheath of connective tissue around the outside of a nerve. *Compare* Endoneurium, Perineurium.

E.p.s.p. An excitatory postsynaptic potential; an active electrical response of the synaptic membrane of the postsynaptic cell, arising as a consequence of the arrival of an excitatory presynaptic impulse; a local, graded depolarization whose amplitude is smaller at lower membrane potentials, reversing in sign at about zero membrane potential. *See* Postsynaptic potential.

Equilibrium potential. The transmembrane potential expected across a semipermeable membrane on the basis of the concentration difference for a given ion. Usually calculated for the ion species separately.

ERG. *See* Electroretinogram.

Euthyneury. The condition of certain gastropods, notably most opisthobranchs and pulmonates, in which there is no crossing of the visceral loop; contrasted with chiastoneury (*which see*). The animals concerned are believed to have had streptoneurous ancestors and then to have lost the crossed condition by a secondary process, not the same process in opisthobranchs and pulmonates. *Compare* Orthoneury.

Evoked response or evoked potential. A special term in brain physiology for a change in electrical potential recorded, with macro- or semimicroelectrodes, from a nervous mass consequent to stimulation elsewhere; typically a potential produced by many neurons, lasting for tens or hundreds of milliseconds and exhibiting several phases.

Excitation. Arousal by a stimulus; a process of releasing potential energy built into a system, causing it to depart from its previous state; living cells are characteristically excitable by impinging events or conditions relevant to their survival and respond in ways that can be considered adaptive in a normal setting.

Exogenic. Originating outside the area under consideration; applied to a class of potentials of neurons due to transducing of impinging events from the environment or from presynaptic neurons (see Chapter 4).

Exteroceptor. Sensory receptor stimulated normally by external events (e.g., eye, ear, olfactory organ, cutaneous tactile receptors).

Extracellular. Referring to space or fluids outside and between the cells, hence to electrodes presumably recording from such fluids.

Eye. A photoreceptor organ composed of many photoreceptor cells, together with some accessory tissue, such as pigment, tapetum, or lens.

Facilitation. "The first stimulus sets up a nervous impulse which travels down the nerve and fails to pass from nerve to muscle, but facilitates the passage of the junction by a second nervous impulse arriving a little later" (Lucas, 1917). More recently it has become necessary to distinguish facilitation from summation. The former is shown by an additional effect of a second stimulus over and above the summed effects of the first and second stimuli if these were separate. Thus linear temporal summation of contractions or potentials is excluded from facilitation. (See entries in Index for additional details in text.)

Faradic. An older term for rapidly repetitive pulses of current, used for stimulating nerves or muscles.

Fascicle. A bundle of fibers; used mainly for bundles in the central nervous system.

Fast fiber. A motor axon that causes relatively quick contraction of a muscle. Used only when there is also at least one other motor axon causing a slower contraction of the same muscle or muscle fiber. More generally, any rapidly conducting fiber.

Fatigue. A state of depressed responsiveness due to protracted activity and requiring appreciable recovery time. Contrasted with adaptation (*which see*), which is not due to the maintenance of the activity as much as to the maintenance of the stimulus and does not require appreciable recovery time; also contrasted with accommodation, which does not require any activity.

Feedback. A fraction of the output brought back and added to the input. In the case of negative feedback, the sign of this part is reversed.

Feed forward. A type of connection in a communication system in which a line divides into parallel lines each carrying a signal forward to a point of reunion of the lines.

Filtering. Selective passage to one set of attributes rather than another set.

Final common path. The motor neuron is referred to in these terms when the aspect of convergence upon it from various sources is being emphasized.

Firing. A nerve cell or part of a nerve cell is said to fire when it initiates an impulse.

Flicker fusion frequency. The lowest frequency of stimulation at which the measurable responses to successive stimuli can on some criterion be said to have fused.

Free nerve endings. Terminations of nerve fibers in the periphery, not recognized as being associated with special cells, but ramifying and ending in a relatively unrestricted way; applied usually to sensory fibers only and therefore to their afferent ends.

Funiculus. A structure like a small cord, especially one composed of a bundle of fibers enclosed in a tubular sheath, as the funiculi of a nerve.

Ganglion. A discrete collection of nerve cells. The term is usually applied to a nodular mass defined by connective tissue in the periphery or in a chain separated by connectives. It is arbitrarily applied to some discrete nuclear masses within the brain, as in the basal ganglia of higher vertebrates.

Ganglion cell. Any nerve cell in a ganglion; by extension any internuncial or motor neuron, even isolated ones in the periphery, as in nerve nets and plexuses. The only neurons excluded are sensory nerve cells that are dispersed or solitary, not in ganglia.

Generator potential. A term that has changed meaning more than once and is not used in the same way by all authors today. Usually a graded potential change occurring in a receptor cell or sense organ associated with the transducer process and subsequent initiation of impulses by electrotonic spread to the spike-initiating locus. Often synonymous with receptor potential, but sometimes used more broadly to include summed or less-well-localized responses. The term "generator potential" has also been applied to pacemaker potentials, but since these are quite distinct it is best not to continue this usage. See p. 168 and Pacemaker.

Giant cell, giant fiber. Cells or fibers that are notably larger than most others in any given animal are called giant. Usually the term is reserved for elements considerably larger than the next in size, and few in number, but in some cases, especially in such insects as roaches, the largest of a continuous size distribution (at first called pseudogiants) have come to be called giants. Absolute size may be as small as 5 μm in unequivocal giant fibers of enteropneusts; the largest known nerve fibers are about 2 mm in diameter (large Chilean squid). Giant cell somas range from about 25–800 μm in diameter.

Glia. *See* Neuroglia.

Globuli cells. Small, unipolar, deeply staining neurons, with a dense chromatic nucleus and little cytoplasm, found in groups in the central nervous system, especially the brain, of many invertebrates. They are thought to be associated with the higher functions of these animals.

Glomerulus. In neuroanatomical usage, a small tight knot of neuropile. (See pp. 48, 68.)

Graded response. A change in membrane potential that varies continuously in amplitude over a range of stimulus strengths.

Granule cell. A type of neuron common in higher centers in vertebrates, having a densely chromatic nucleus and little cytoplasm; similar to globuli cells of invertebrates except in being multipolar, often occurring in great numbers in a well-packed layer, as in the cerebellar cortex.

Gray matter. Regions of the central nervous system that include nerve cell bodies, dendrites, origins of axons, and endings of axons (but relatively little myelin), hence the synaptic regions as well as the nutritive; regions other than white matter (*which see*); characteristic of vertebrates and almost unknown in invertebrates (approximations to gray matter occur in cephalopods).

Heterofacilitation. Facilitation of the postjunctional response to one pathway by antecedent input arriving in another pathway.

Heteropolar. Of neurons having two or more differing processes; the commonest are neurons with distinguishable axon and dendrite (or dendrites).

Higher, highest. As in higher level, structure, or center: that which is farther from receptors or effectors in terms of pathways and embraces the more complex and wider sphere in terms of function; the highest levels cannot be said to be either primarily motor or sensory. A higher level in the nervous system is one further toward the brain or more anterior in the brain. As in higher animal or group: that which requires more information to describe it fully or to distinguish it from a random arrangement or that which comprises more different parts or processes. A higher animal can reasonably be supposed to have evolved eventually from a lower animal (but this does not preclude instances of evolving toward simplification).

Hyperpolarization. Increase in polarization (*which see*) above a level regarded as normal or resting (*which see*). An increase from a depolarized state toward but not above this level is called repolarization.

Homologous. Two anatomical parts are said to be homologous when they can reasonably be supposed to have evolved from one common ancestral part. "Serially homologous" refers to structures that are in series in the same animal and which can be supposed to be morphologically equivalent in origin. Functions or actions should similarly be called homologous only when there is reason to believe they are related genetically. The term "homologous response" has been used in a special sense, referring to the finding that muscles transplanted to a new site move when the normal homologous muscles move.

Impulse. Nerve impulse is the special term applied to that form of activity of neurons that is all-or-none, rapidly rising and rapidly falling, typically propagated without decrement along axons. The term embraces all aspects of the brief, initial event, not just the electrical signs; it does not include the recovery or late phases. Arbitrarily, the impulse may be said to last until a line tangent to the rapid descent phase of the action potential crosses the base line. The only available useful sign of the impulse is the electrical change in the cell membrane, and there is reason to suppose that this is the essential functional aspect; it is often referred to as the spike potential (*which see*) or action potential. Impulses are characteristic of axons but occur in some cell somas and in some large dendrites.

Inactivation. Sodium inactivation is a process postulated in the sodium theory of electrical activity of nerve and muscle; the gradual paralysis of the mechanism responsible for transport of sodium down its concentration gradient, hence resulting in reduced sodium conductance. Depolarization of the membrane causes a transient increase in sodium permeability; inactivation is the recovery of the resting low permeability. Potassium inactivation is the slower and later paralysis of the transport mechanism for potassium. Sodium inactivation begins during the depolarized or reversely polarized phase and continues during repolarization; potassium inactivation proceeds while the membrane potential is recovering and while it is high.

Inhibition. Prevention of one activity by another activity. Usually, suppression of excitation in a cell by input from inhibitory axons. There are several different mechanisms (*see* Chapters 5 and 6), one of which is an increase in membrane permeability to certain ions such that the membrane potential shifts (inhibitory postsynaptic potential; *see* Postsynaptic potential) toward a reversal potential (*which see*) too high for firing, most commonly hyperpolarizing. For presynaptic inhibition, see pp. 39, 43, 46, 199; others, see pp. 208, 235.

Injury potential. The potential difference between a normal region of the surface of the body or of a muscle or nerve and a region that has been injured;

also called a demarcation potential. This was much studied before intracellular recording was introduced. The injury potential approximates the potential across the membrane because the injured surface is almost at the potential of the inside of the cells.

Integration. Process of putting parts together to make a whole. Process resulting in the output being some function of the input other than 1.0. Two or more separate inputs may converge to cause one or more outputs, or a single input line may cause a greater or lesser number of impulses in an output line.

Interneuron. An internuncial neuron; one that is neither sensory nor (purely) effector-innervating, but connects neurons with neurons. Some motor neurons are at the same time interneurons.

Interoceptor. Receptors for internal events and states in the body, such as blood pressure or CO_2 or O_2 or distension of viscera.

Intracellular. Referring to conditions or the recording of conditions inside the cell.

Intrafusal. Within the muscle spindle; said of structures or processes inside the sheath of this sense organ in vertebrate skeletal muscles.

I.p.s.p. Inhibitory postsynaptic potential: *see* Postsynaptic potential.

Isopolar. Describing a nerve cell having two or more equal or similar processes; therefore a neuron in which axon and dendrite cannot be distinguished; primitive isopolar neurons are regarded as having only neurites and no dendrites; amacrine cells of higher systems are regarded as having no axons.

Iteration. Repetition; usually used with reference to discrete events rather than a smooth oscillatory change.

Junction. The word is used in its ordinary sense, as in "junctional tissue" or "junctional potential" (j.p.), where for one reason or another the author has hesitated to use the term "synapse" because it would assume the point at issue about the presence of synapses or because he simply wishes a literary alternative.

K (Hill's constant of excitation). A time constant in a mathematical model of nerve designed to simulate properties of excitation, such as the refractory period, recovery, and strength-duration relation; K represents the duration of local excitatory disturbance.

Kindling. A process manifested by the slow development of afterdischarge and finally convulsions as a result of intermittent electrical stimulation of the brain at long intervals. In monkeys, one second per day, consisting of a weak train of shocks, can gradually kindle a seizure after 10–15 days. Similar trains once per hour may not do so.

Kinesis. A form of behavior characterized by the intensity of locomotor activity being under the influence of intensity of a stimulus. The direction of turning is random. The relation between stimulus intensity and activity is such that animals aggregate or spend more time at one end of a gradient of intensity. (See p. 295.)

Labeled lines. Communication channels; in the nervous system, nerve fibers. Although carrying similar signals, each has a long-term identity, built in or acquired. The identity, or label, is equivalent to information, such as: "Any signal in this line means a moving light-dark boundary in such and such a field."

Lambda, λ (Hill's constant of accommodation). A constant in a mathematical model of excitation in nerve, representing the time constant of the rise in threshold (*see* Accommodation); a measured time constant based on this model and made with extracellular electrodes, now known to be chiefly due to properties of the sheaths. When λ is high, the apparent accommodation is low. λ is also used for the space constant (*which see*).

Lamina. A layer; a neuropile sheet; in a special sense, the most peripheral neuropile mass of the optic lobe of Arthropoda.

Light compass reaction. (See p. 299.)

Lobe. A subdivision of a larger neural mass, marked by externally visible boundaries.

Local neuron. One without processes that spread to other neuropile masses, lobes, or other ganglia; an intrinsic neuron.

Local potential. A special term for a type of endogenic potential (*which see*) of activity in neurons and some muscle cells, characterized by decremental spread (largely passive), graded amplitude, and electrical excitability. Usually a response to antecedent electrical activity in the same cell—for example, synaptic or pacemaker potentials or impulses in the preterminal axon.

Lower, lowest. As in lower level, structure, or center: that which is closer to the receptor or effector in terms of afferent or efferent pathways and embraces the less complex and narrower sphere in terms of function; the lowest levels of the central nervous system are the proximal endings of the first-order sensory neurons and the somas and dendrites of motor neurons. As in lower animal or group: that which is less complex, in turn defined as that which requires less information to describe it fully or to distinguish it from a random arrangement or that which comprises fewer different parts or processes. A lower animal can reasonably be supposed to be ancestral to each higher animal (but some simplified animals may have evolved from more complex ancestors).

Magnet effect. A term used by von Holst for a phenomenon observed in some instances of relative coordination (*see* Coordination), in which the frequency of one rhythm is attracted toward that of another but not synchronized with it. (See p. 329.)

Mechanoreceptor. Any receptor for which there is reason to believe the normally adequate stimulus is mechanical. See Receptor.

Mediation. In neurophysiology, used in the sense of effecting by acting as the intervening agency; thus "chemical mediation at a synapse" and "coordination of this movement is mediated by that center."

Medulla. Inner portion of an organ; marrow or pith. As a special name in arthropods, the second optic neuropile mass; in vertebrates, the myelencephalon, the part of the brain stem between the anterior end of the spinal cord and the posterior end of the mesencephalon.

Medullary cord. A nerve cord with nerve cell bodies distributed all along it, as opposed to being concentrated into ganglia with intervening cell-free connectives.

Membrane potential. Voltage difference between the inside and outside of a cell membrane. Since the inside of a cell is typically nearly isopotential and the extracellular fluids are also, the membrane potential is equivalent in practice to the voltage difference between the cell contents and the cell milieu. In its presence, the cell membrane is said to be polarized.

Microelectrode. Any of a variety of types of electrode with an uninsulated tip smaller than about 10 μm in diameter. Somewhat larger electrodes are called semimicroelectrodes; the smallest electrodes in use (tenths of a micrometer) are sometimes called ultramicroelectrodes or hyperfine microelectrodes.

Miniature potential. *See* Quantal.

Modality. A term from subjective psychology meaning "class of sensation"; proposed by Helmholtz to replace the word "quality." By usage, a major class, such as visual, auditory, gustatory, olfactory, tactile; pain is often considered a modality. This method of classifying sense organs contrasts with that defining the adequate stimuli, such as photoreceptor, phonoreceptor, chemoreceptor, mechanoreceptor. *See* Receptor.

Monopolar. *See* Unipolar.

Motor neuron. A nerve cell that sends an axon to a muscle or, by extension, any effector. Strictly speaking, a neuron whose effect is excitatory to motor mechanisms. The larger category, including inhibitory efferent neurons, secretion-promoting or influencing neurons, and so on, may be called "effector neurons."

Motor unit. A single motor neuron and all the muscle fibers its axon supplies.

Movement receptors. Receptors especially sensitive to and normally signaling slow, steady movement rather than position or vibration or acceleration.

MRO. *See* Muscle receptor organs.

Multipolar. Describing nerve cells having more than two processes from the cell body.

Multiterminal. A term used to describe nerve fibers having many terminations; usually refers to motor axons. When used for a type of muscle innervation, it means that a given motor axon branches to supply many endings along a given muscle fiber.

Muscle receptor organs. Special receptors on muscle fibers. In arthropods and cephalopods, having a multipolar peripheral sensory cell with short dendrites closely applied to the special muscle fiber; in vertebrates, a muscle spindle or similar organ: A form of stretch or tension receptor.

Mushroom bodies. *See* Corpora pedunculata.

Myelinated fiber. A nerve fiber of which the sheath comprises many spirally wound double lamellae of glial cell membrane between which the glial cell cytoplasm has disappeared; all lamellae belong to one glial cell for a certain length of the fiber, defined by distinct nodes of Ranvier (*which see*). The term "myelinated" should not be used for loosely laminated fibers with glial cytoplasm still apparent or for overlapping glial cells or shingled layers.

Myogenic. Arising in muscle tissue; said of rhythms, such as the heartbeat in vertebrates, which arise in muscle cells or modified muscle cells. Contrasted with neurogenic (*which see*).

Myotatic reflex. The reflex, especially of extensor (antigravity) muscles, to sudden small tension stimuli applied so as to stretch the same muscle.

Nerve. An anatomically delimited bundle of nerve fibers running between a central ganglion and the periphery. When several bundles or funiculi run together in a common outer sheath, the whole group is the nerve. Contrasted with connectives, which run longitudinally between ganglia; with commissures, which run transversely; and with tracts, which run between populations of cells in the central nervous system.

Nerve net. A system of neurons dispersed, generally in a plane, and so connected either by synaptic contact or fusion as to permit diffuse conduction. To be distinguished from the more general category of plexus (*which see*), for which connectivity is not specified. Also used in a more general sense, especially in consideration of models, to mean any assemblage of connected elements more or less like neurons.

Nervous system. An organized constellation of cells (neurons) specialized for the repeated conduction of an excited state from receptor sites or from any neuron to other neurons or to effectors. Higher nervous systems also integrate the signals of excitation from converging neurons and generate new signals.

Neurilemma. A term that has been used inconsistently, sometimes referring to the Schwann sheath of a nerve fiber, sometimes to a connective tissue sheath around a bundle of fibers (*see* Perineurium), sometimes to the outer noncellular sheath of ganglia. It is not needed today and is usually misunderstood.

Neurite. The main or longest process of a nerve cell; usually equivalent to axon (*which see*) and axis cylinder.

Neurobiotaxis. A hypothetical tendency of nerve cell bodies to move in evolution and development toward their principal sources of excitation, thus shortening their dendrites.

Neurogenic. Said of rhythms that arise in nerve cells (as in the heart of *Limulus* and lobster and in mammalian breathing rhythms); contrasted with myogenic (*which see*).

Neuroglia. A class of differentiated accessory or non-nervous cells associated with nerve cells. They may occur as satellites or sheath cells, may define fascicles, or have special attachments to blood vessels, in both vertebrates and invertebrates. In higher animals there are distinct types of glia with different relations. Best known are the three types in vertebrates: astrocytes, oligodendroglia, and microglia (see pp. 61, 72). Usage varies with respect to such cells in peripheral nerves (see Schwann cells); they are often called neuroglia too.

Neurohemal organ. An organ composed of specialized endings of axons that carry neurosecretory products and at this organ release them into the blood (see p. 216).

Neurohormone. A hormone released from nerve cells at a distance from its target organ (*compare* Neurotransmitter).

Neurohumor. A term that has been variously used; in the narrower sense, a neurotransmitter (*which see*); in a wider sense, including this and neurohormones (*which see*).

Neuroid. Referring to types of conduction of excitation not involving nerve cells or fibers but somewhat nerve-like in effect.

Neuron. A nerve cell, together with all its processes; a form of cell specialized by connections and properties to receive certain forms of signals, respond with special signals, conduct excitation, and make specific functional contacts with other neurons or with effectors or receptors. Borderline, difficult, or doubtful cases include (a) cells in sponges that look like neurons but may not make enough specific functional contacts to be considered neurons, (b) neuroglial cells of which the same can be said, (c) neurosecretory cells with specializations for secreting that override some nerve cell features (here treated as neurons), (d) sense cells, such as cochlear hair cells, that conform to the criteria given except for the simple epithelial form and no specialization for conduction (here treated as nonnervous cells), (e) retinal rods and cones with forms like those of nerve cells, very short conducting processes, and (like many good neurons) no impulses (here treated equivocally as quasi-neurons).

Neuron doctrine. The conclusion (a) that the nervous system is composed of nerve cells, or neurons, which are in the main discrete units and interact principally at points of contact (synapses), (b) that all nerve fibers are processes of nerve cells, and (c) that they do not form a reticulum (see p. 103). Rare exceptions may occur.

Neuropile. A tangle of fine fibers (of dendrites and axon arborizations) and their endings, quite or nearly devoid of nerve cell bodies; found in parts of vertebrate brains and forming much of the bulk of invertebrate ganglia, where it is usually the sole synaptic field. (See p. 68.)

Neurosecretion. Product of, or the process of release of visible droplets of neurohormone from the cytoplasm of special secretory nerve cells.

Neurotransmitter. Substance released from nerve cells at synapses and presumed to mediate normal chemical transmission. *Compare* Neurohormone.

Nissl bodies. Neuronal organelles appearing in the light microscope as clumps of basophilic material in the cytoplasm of vertebrate and some invertebrate nerve cell bodies and in the larger dendrites of vertebrates. In the electron microscope, identified with clumps of endoplasmic reticulum studded with ribosomes. (See p. 21.)

Nociceptor. A receptor specialized to fire in response to injurious or noxious stimuli.

Node of Ranvier. A gap in the myelin sheath (*see* Myelinated fiber), exposing the axon at periodic intervals of approximately 1–2 mm, formed by the space between segments of the sheath. The segments are called internodes. Interruptions in the sheaths of crustacean giant fibers should not be termed Ranvier nodes because they are quite different in structure and properties, and the sheaths are not true myelin. (See p. 61.)

Nucleus. Any of numerous, small, anatomically demarcated masses of gray matter in the vertebrate central nervous system. Also the organelle of a cell in which a membrane encloses the chromosomes.

Occlusion. A process of spatial overlap of central excitation in a reflex activity, inferred where the sum of two reflex effects going on together is less than the sum of their separate effects.

Ocellus. A relatively simple small eye.

Ommatidium. A unit of the compound eye of arthropods, composed of cuticle, lens structures, pigment cells, and receptor cells.

Optic nerve. The nerve or tract between the eye and the brain; in some invertebrates, composed of primary sensory axons; in others and in the vertebrates, of higher-order axons.

Optokinetic. *See* Optomotor.

Optomotor response. The "reflex" or forced movement found in many animals of partially following with the body, head, or eyes a slow movement of the visual field, involving a large enough, sufficiently differentiated area. Some authors distinguish optokinetic from optomotor, but this seems inappropriate.

Order. As in first-order neuron, second-order neuron, higher-order neuron. A means of referring to the sequence of neurons in a pathway; therefore, the second-order neuron is one synapse beyond the first. In the common reference to sensory systems, the first-order neuron is the peripheral sensory unit.

Orientation reaction. A taxis or tropism (*see* p. 296). Also used with an entirely different meaning in behavior studies with higher animals, in which clicks, tones, lights, and the like are employed: the initial response to a novel stimulus not yet associated with reward or punishment.

Orthodromic. Synonymous with "dromic." The direction of impulse propagation normal for the fiber in question; usually the direction from dendritic region of the neuron toward axonal termination, contrasted with "antidromic."

Orthogon theory. The proposition that the basic morphological plan of the cords, connectives, and commissures in all animals possessing a central nervous system was an orthogonal one—that is, a set of longitudinal cords and transverse commissures; by fusion and loss these are considered to account adequately for any of the existing plans.

Orthoneury. The condition of the visceral loop ganglia and connectives in those gastropods having bilateral zygoneury (*which see*); the crossed connections indicating the earlier streptoneury are still evident, though sometimes reduced. The nervous system looks superficially to be straight, without crossing due to torsion.

Oscillatory potential. A regularly recurring (not necessarily sine-wave) change in potential. Two general categories are relevant to neurophysiology. Relaxation oscillations are those involving two processes so related that A builds up in a certain time to trigger B, which, whatever else it does by the way of output, restores the precondition for A to begin again. A is in the commonest neural case a pacemaker potential, B, a local potential with or without a spike. Quasi-sinusoidal oscillations continuously and smoothly alternate between states and do not require a non-linear interaction.

Osphradium. A sense organ in the mantle cavity near the ctenidia, found in many molluscs.

Otolith. Relatively solid and heavy bodies in an otocyst (*which see*) or statocyst (*which see*), which act to deflect sensory structures as a function of gravity and inertia. Ordinarily not distinguished today from statolith.

Otocyst. A statocyst-like structure associated with the ear. Nowadays this does not imply sound reception. The term is not appropriate for any of the invertebrate structures and can be replaced by statocyst (*which see*) for the vertebrate structures as well.

Overshoot. As used in neurophysiology, the reversal of the normal membrane potential during the peak of the spike.

Pacemaker. A group of neurons, a single neuron, or region of a neuron to which one can point as the ultimate source that drives a rhythm.

Pacemaker potential. The relatively slow-changing potential of a pacemaker region of a neuron that occupies the interlude between rhythmic discharges and by its slope determines the frequency. The pacemaker potential is to be distinguished from the spike and its rapidly repolarizing phase and from local potentials, which may sometimes be activated by the depolarizing action of the pacemaker potential; but the depolarizing recovery from the undershoot of the spike cannot be distinguished from a pacemaker potential and can be regarded as part of it. There are two types (*see* Oscillatory potential): relaxation oscillations and sinusoidal oscillations. Firing frequency is of course determined not only by the rate of change of the pacemaker potential but also by the threshold of the spike, which affects the length of each period in the relaxation oscillator type and the number of periods between spikes in the sinusoidal oscillator type.

Pain. A sensation characterized by rather intense discomfort and typically caused by noxious stimuli (*see* Nociceptor), injury, disease, or organic disorder. Since the essential features are sensation and its quality, we cannot directly know about the experience of pain in infrahuman species. It is generally inferred, however, that something equivalent is operating if an animal shows behavioral reactions resembling those of man under pain; the resemblance becomes less with lower animals, and the inference accepted is that something like pain evolves gradually. Pain should be distinguished from suffering or distress, since the two can vary independently in man under various conditions. Withdrawal responses can occur without suffering or even conscious pain and are therefore poor indices of pain. (See p. 221).

Papilla. A relatively small raised spot or pimple on the integument, usually sensory in function.

Parasympathetic nervous system. A part of the autonomic nervous system (*which see*) in vertebrates, having its outflow from the central nervous system in two separate regions, with certain cranial nerves and sacral nerves, respectively; having its ganglia in or near the viscera; having among its functions (effects of stimulation) slowing of the heart, dilating of visceral blood vessels, constricting the pupils, and increasing the activity of many glands and digestive and reproductive organs; its effector junctions are usually cholinergic.

Passive process. Used in contrast to active process for certain changes in potential across the cell membrane; a process adequately accounted for by the imposed current and the resting electrical properties (mainly resistances and capacities) of the cell.

Pathway. In neurological use, a route or course consisting of one or many nerve fibers between given termini; the fibers of a pathway may be gathered into a tract or dispersed; there may be several synaptic relays along a pathway or none.

Perikaryon. The cell body or soma of a nerve cell; the part of the cell immediately around the nucleus.

Perilemma. The layer of specialized glial cells that underlie the fibrous "neurilemma" in the sheath of arthropod ganglia.

Perineurium. The connective tissue sheath surrounding a bundle of nerve fibers in a nerve. *Compare* Epineurium, Endoneurium.

Peripheral. In neurology, referring to that part of the nervous system that is not central nervous system; the nerves, ganglia, plexuses, receptors, and motor endings in and between muscles, visceral organs, and skin.

Peristalsis. Coordinated movement of a muscular tube such that alternate waves of contraction and relaxation of circular and longitudinal muscle layers pass along the tube, as is common in alimentary canals and vermiform animals.

Phasic. Referring to a transitory state. A phasic contraction of a muscle is one that is not long maintained, in contrast to a tonic contraction. A phasic receptor response or component of a response is one that shows adaptation.

Plasticity. Ability to be modified or changed. Usually plasticity is an inferred property (mechanisms unspecified) of a nervous system that can make an adaptive change in activity as a result of experience or following removal of some of its parts. Quite different types have been described. In one, better called apparent plasticity, there is immediate, adaptive change in pattern of movement, suggesting that existing connections can meet new demands. In another, there is slow "relearning" or recovery believed to involve changes in central connections or the properties of junctions.

Plexus. In neurohistology, a tangle of interweaving nerve fibers or strands of them. Usually applied to such strata in the periphery (skin, gut, muscles, pericardium, and the like). Nothing is specified about the connections of the fibers or the neuronal composition; the term is therefore general and includes nerve nets (*which see*), tangles of sensory endings, as in human skin, and collections of ganglion cells. There may be synapses, or there may be only fibers without functional connection. A plexus may be a sheet of interlacing fibers or a coarse mesh of anastomosing nerves, with or without cell bodies.

Polarization. In electrophysiology, the presence of an electric potential difference between two points in a structure or across a membrane.

Polyneuronal. Describes muscle fibers that are innervated by more than one axon, whether motor or inhibitory or both.

Polytomic. A type of branching in which several parallel nerve fibers bifurcate together at the same point. *Compare* Diplotomic.

Postsynaptic. Refers to sites or events in the neuron on the receiving side of the transmitting junction.

Postsynaptic potential. The special form of subthreshold potential change occurring across the synaptic membrane of the postsynaptic neuron when an impulse arrives in the presynaptic fiber; typically delayed after, and not directly attributable to, the current flowing from that impulse; therefore, the first sign of response to synaptic transmission. There are two classes, those tending to drive the neuron toward a large depolarization (excitatory postsynaptic potential, e.p.s.p.), and those tending to drive the neuron toward a high membrane potential, more polarized than the spike threshold (inhibitory postsynaptic potential, i.p.s.p.). The conductance changes and dependence on membrane level are different for the two classes. (See pp. 181–184.)

Posttetanic potentiation (PTP). Enhancement of excitability following a long period of high-frequency stimulation; this phenomenon is known mainly in the mammalian spinal cord, where it lasts for minutes or even hours.

Potential. The charge or change in charge, measurable in volts, that exists between two points; many biologically produced potentials arise primarily from the difference between two sides of a cell membrane. *See* Action potential, After potential, Brain wave, Electrotonic potential, Equilibrium potential, E.p.s.p., Evoked potential, Generator potential, I.p.s.p., Local potential, Membrane potential, Oscillatory potential, Pacemaker potential, Presynaptic potential, Post-synaptic potential, Quantal, miniature potential, Resting potential, Reversal potential, Slow potential, Spike potential.

Predepolarization. A prepotential in the depolarizing direction.

Prepotential. Any of the several types of graded processes that can, if they reach threshold, lead to a spike; thus, synaptic, generator, local, and pacemaker potentials.

Prespike. The spike in a presynaptic neuron.

Presynaptic. Referring to the neuron, or the events, before a synapse. For presynaptic inhibition, see pp. 199, 208.

Process. A systematic series of actions or a course of events, as "the local, graded process." Also a protrusion or an extension, as "the nerve cell sends out long thin processes that become nerve fibers."

Projection neuron. An interneuron that sends an axon up or down to a different level of the central nervous system; contrasted with commissural and intrinsic or local neurons (*which see*).

Propagation. Conveyance of excitation with the active participation of the regions traversed; usually used for conduction within one axon, not for transmission across junctions; contrasted with electrotonic spread (*which see*).

Proprioceptor. A mechanoreceptor that normally signals movements or positions of the parts of the body; examples are muscle, tendon, and joint receptors; statocyst or labyrinth receptors; and, in the arthropods, chordotonal organs, campaniform sensilla, and hair plates.

Pseudogiant fiber. An axon of the largest size group in an animal with axons of all intermediate sizes. The largest fibers of the cockroach are examples. The term has largely been dropped, and these fibers are generally called giant fibers.

PTP. *See* posttetanic potentiation.

Pulse-coded. Descriptive of the special type of analog communication (*which see*) represented by axons, in which information is coded in the numbers of and intervals between all-or-none pulses.

Quantal miniature potential. A small synaptic potential of a fixed amplitude, or amplitudes attributable to temporal summation of identical events, occurring apparently spontaneously, without presynaptic impulses, and in a quasi-random distribution in time. Known at a number of neuromuscular and central junctions (see p. 184) and apparently characteristic of chemical synapses.

Radial conduction. Conduction along a radius, with reference to animals or parts that are radially symmetrical.

Reafference. A special term used by von Holst and coauthors for the sensory input signals caused by the animal's own movements; contrasted with afference (*which see*).

Receptor. A cell or part of a cell specialized and normally functioning to convert (transduce) environmental stimuli into nerve impulses or some active response in nerve cells. Most receptor cells are nerve cells, but some nonnervous cells, such as hair cells of the vertebrate acousticolateralis system and taste bud cells, are receptors. One method of classifying receptors is by the form of the adequate stimulus. Photoreceptors, chemoreceptors, osmoreceptors, thermoreceptors, electroreceptors, mechanoreceptors, phonoreceptors are normally stimulated, respectively, by light, chemicals, osmotic pressure differences, temperature, electric current, mechanical events, sound (a subclass of the preceding). *See* Nociceptor, Proprioceptor, Interoceptor.

Receptor potential. A graded change in potential in sensory receptor cells or nerve terminals in response to a natural stimuli. *See* Generator potential.

Recruiting response. An evoked potential in the cerebral cortex produced by stimulation of intralaminar thalamic nuclei and related structures, widely distributed but especially prominent in cortical areas with fewest projections and showing increased amplitude with repetition at certain frequencies. *Compare* Augmenting response.

Rectification. Used for a property of some cell membranes providing a greater resistance to current flow in one direction than in the other. Most chemical synapses and certain electrically transmitting synaptic membranes exhibit this property, which explains the polarized transmission of the latter. Delayed rectification is a term applied to the delayed increase in conductance to K^+ following depolariza-

tion (e.g., after the onset of a spike, in most electrically excitable membranes).

Recurrent collateral. An axon collateral (*which see*) that turns back and acts on the cell body or other like cell bodies.

Redundancy. Of a communication, the most that can be eliminated without loss of information; the number, expressed as a percentage, obtained by subtracting from one the ratio of the actual information content to the maximum information content. More often used in neurophysiology with respect to neurons, channels, or parts, often quite ambiguously or vaguely. It is sometimes used to mean superfluity; usually it implies existence of quite equivalent neurons. It may refer to that part of the function of one neuron (e.g., its receptive field) that overlaps with other neurons. (See p. 233.)

Reflex. In neurophysiology, a relatively simple action produced by an afferent influx to a nerve center and its reflection as an efferent discharge back to the periphery to appropriate effectors, independently of volition (*see* p. 266). Reflexes are simpler than instincts in number of muscles, of successive movements, and of specifications of stimuli; they are more readily evoked repeatedly. There is considerable overlap between the ethological element called a fixed action pattern (*see* p. 293) and the reflex.

Refractory period. The absolute refractory period is the period following excitation of an impulse during which no stimulus, however strong, evokes a further response. The relative refractory period is the period following an impulse during which a stimulus must be abnormally large to evoke a second response. The functional refractory period is the period following an impulse during which a second impulse cannot yet excite the given region.

Regeneration. Restoration of a lost part by regrowth. Also the result of a chain reaction, as in explosive firing.

Relaxation oscillation. *See* Oscillatory potential.

Relay. As used in relay center, a collection of junctions with relatively little cross connection, recombination, and higher integration; therefore passing on signals with nearly the same meaning as those received. A relay synapse is one with little or no convergence, or integration, transmitting one-to-one.

Response. From the Latin for "an answer"; therefore

properly used for an activity in reply to a stimulus.

Resting discharge. Background discharge; ongoing firing of impulses in the absence of known stimulation.

Resting potential. The potential across the membrane of a cell when it is not excited by input or spontaneous activity. The cell is not at rest metabolically, but the membrane is at a steady state, and the relevant environment is also not changing.

Reticulum. A network of anastomosing fibers (or tubes). Contrast nerve net, in which anastomosis is not specified. Reticularism is the belief that axons and dendrites fuse extensively in the gray matter, an older view, opposed to the neuron doctrine.

Retinula cell. Sensory neuron in the eye, especially applied to arthropods; the term is perforce used on historical grounds, hence nothing is specified as to the nature of the response to light.

Reversal potential. The value of the postsynaptic membrane potential at which the action of transmitter causes no potential change (p.s.p.) but a normal conductance change. At membrane potentials above or below this value, p.s.p.'s occur with a polarity that transiently deflects the membrane potential toward this value. The reversal potential for i.p.s.p.'s is usually higher than the resting potential; for e.p.s.p.'s, much lower. (See pp. 193, 196.)

Rhabdom. A composite of several rhabdomeres closely apposed and fitting together.

Rhabdomere. A specialized part of certain types of photoreceptor cells, usually a tightly packed stack of fine tubes or villi of cell membranes. The rhabdomere type of photoreceptor is found in eyes of annelids, arthropods, and molluscs.

Rheobase. The threshold for electrical current of infinitely long duration; in practice, the current strength when a cell fires at the longest latency to a step of current, or the current when longer durations of pulses lower the threshold by a negligible amount.

Rheoreceptor. A receptor normally signaling the presence or strength of water currents.

Rhopalium. A marginal sensory body in scyphomedusae; a tentaculocyst in the form of a hollow club, set in a sensory niche with a hood. Included in the nervous apparatus is a statocyst, general sensory epithelium, a thickened plexus and marginal ganglion, and nearby the outer and inner "olfactory" pits; in some forms (especially Cubomedusae) there are also ocelli or complex eyes.

Rind. The rind or cell rind is the layer of nerve cell bodies on the outer surface of the ganglia of invertebrates; as a rule the rind contains only somas, glia, and stem processes, but no dendrites, endings, or synapses.

Root. The part of a nerve close to its origin from the ganglion, cord, or brain; sometimes root refers to the proximal part outside the ganglion, sometimes to the fiber bundles of the nerve just inside the ganglion.

Run. A nerve fiber is said to run if it fires at a more or less regular frequency continuously; a run of impulses is a long continued regular series of spikes. *Compare* Train, Burst.

Safety factor. The spike threshold divided into the spike amplitude; a property of each region of an axon or other spike-supporting membrane determined by the local threshold measured in critical depolarization across the membrane and the amplitude achieved by the approaching spike, which normally will provide this depolarization by local circuits. Also, the ratio of e.p.s.p. amplitude and spike threshold in a postsynaptic cell.

Saltatory conduction. A form of axonal conduction found in myelinated nerve fibers in which the current across the axonal membrane is much more dense at the nodes than between them (see p. 163).

Schwann cell. A cell specialized to provide the intimate sheath of one or more axons in the peripheral nervous system; an elongated cell typically with longitudinal grooves occupied by axons, the groove being spirally wound for myelinated fibers. The Schwann cell extends from one node of Ranvier to the next for myelinated fibers. Originally applied only to vertebrate myelinated fibers, it has been extended to cells (some of different embryological origin, shape, and inclusions) sheathing unmyelinated fibers and inappropriately to invertebrates. The equivalent but quite different cells in the central nervous system are oligodendroglia.

Sense cell. A primary sense cell is a sensory nerve cell or nervous receptor cell; this is by far the commonest sort of sense cell. The term "secondary sense cell," formerly more in use than it is now, means a nonnervous receptor cell that secondarily excites nerve endings of the afferent neuron. *See* Neuron.

Sense organ. An ordered arrangement of several kinds of cells or tissues built around a group of receptor cells of a specialized type. To be distinguished from unicellular receptors and multicellular clusters of receptors without the ordered array of accessory tissues required for an organ.

Sensillum. A word used for simple types of sense organs involving only a few neurons. (Plural, sensilla.)

Sheath. An investment or tunic of material derived from other cells; some axons and some cell bodies have private or individual sheaths, and some axons share a single sheath cell with several or many others; central fasciculi, peripheral funiculi, nerves and connectives, ganglia, and cords have simple to complex sheaths.

Signal. An event carrying information about an antecedent event or state.

Slow fiber. A motor axon that causes a relatively slow contraction of a muscle. Used only when there is also at least one other motor axon causing a faster contraction of the same muscle fiber. More generally, any slowly conducting fiber.

Slow potential. In most usage, any changing potential whose rate of change is equivalent to a sine wave of a few cycles per second or slower; the frequency range may be quite different in different cases or contexts.

Soma. The cell body of a nerve cell; the perikaryon (*which see*).

Somatic. Said of muscles, sense organs, nerves, and other structures associated with the surface of the body or the main muscles of body form and locomotion; opposed to visceral or vegetative.

Somatoplasm. Cytoplasm of the soma of a neuron; used to contrast with axoplasm (*which see*).

Somatosensory. Somatic sensory.

Somatotopic. Organized topographically according to position on the body surface.

Space constant (λ). A measure of decrement of a passive potential change with distance along an axon or other cell membrane; the distance at which the potential has declined to $1/e$ or about 37% of its initial height.

Specific system. A set of pathways and centers identified with a relatively narrowly defined function, as the visual, auditory, or somesthetic systems; contrasted with nonspecific system, in which the structures are not identified with a function so restricted in modality or part of the body.

Spike potential. In cellular neurophysiology, a short-lived (usually in the range of 1–3 msec), all-or-none change in membrane potential that arises when a graded response passes a threshold; the electrical record of a nerve impulse or similar event in muscle or elsewhere. In gross EEG recordings, a sharp, brief event.

Spindle. Muscle receptor organs (*which see*) of a spindle shape; generally used for any such organ in vertebrates. Also the configuration of slow, smooth brain waves characterized by a succession of gradually growing and gradually declining amplitude.

Spontaneous. Acting from internal tendency, self-acting. The essence is not independence from environmental influence, either constraints or sources of energy, but internally determined timing; within the permissive external conditions, the system has its own intrinsic mechanisms that set off the action.

Stalk. *See* Stem process.

Statocyst. A sense organ containing statoliths (*which see*) in one or more cavities; usually a receptor of gravity, acceleration, or vibration. The word is generally reserved for invertebrates but only because the corresponding structures in vertebrates are associated with the ear and are hence called otocysts (without implying a hearing or sound-receiving function).

Statolith. A heavy inclusion in a statocyst, not fixed firmly but free to fall at least enough to stimulate sensory structures; not distinguishable from otoliths except that the latter are part of that complex vertebrate structure called the ear.

Stem process. The part of a unipolar neuron leading from the cell body to the point where dendrites and axon can be recognized; the stalk.

Stimulus. In physiology, any agent, state, or change external to the excited cell that is capable of influencing the activity of a cell. *See* Adequate stimulus.

Stomatogastric nervous system. A term equivalent and alternative to stomodeal nervous system (*which see*); preferred in the literature of some groups perhaps because it has been thought that the innervation extends farther posteriorly in the gut.

Stomodeal nervous system. Pertaining to the ectodermal invagination at the anterior end of the alimentary canal. The stomodeal nervous system is that part of the peripheral nervous system of higher invertebrates that has its main connections to the central nervous system with the brain and anterior ventral ganglia and is distributed to the anterior part of the alimentary canal; the stomatogastric system. Sometimes, but inappropriately, called a sympathetic system.

Streptoneury. Equivalent to chiastoneury (*which see*).

Subesophageal ganglion. The ganglion or fused ganglia lying at the anterior end of the ventral nerve cord; it usually operates the mouth parts.

Summation. The addition of separate responses that are adjacent in time or space. The former is called temporal summation; the latter, spatial.

Supraesophageal ganglion. The main ganglion lying morphologically anterior and dorsal to the mouth. This is usually the ganglion that integrates excitation from head sense organs. It is nearly synonymous with cerebral ganglion or brain; only in a few animals is the cerebral ganglion not above the anterior end of the alimentary canal, and in a few animals the supraesophageal ganglion is not sufficiently developed to be called a brain.

Sympathetic nervous system. That division of the autonomic nervous system (*which see*) of vertebrates which has its outflow from the central nervous system in the thoracolumbar segments, most of its ganglia in a pair of chains near the spinal column (called paravertebral ganglia), and has among its actions acceleration of the heart, dilation of the pupils, constriction of visceral blood vessels, and decrease of activity in digestive and reproductive organs; its effector junctions usually adrenergic. Not appropriately used for invertebrates, although formerly used as a general term for visceral nerves.

Synapse. A functional connection between distinct neurons accomplished by contact or near contact of their membranes; by extension, the term is applied to neuromuscular junctions and to junctions between specialized nonnervous sense cells and afferent nerve endings. (*See* p. 29 for discussion of this definition.)

Synaptic vesicle. A membrane-bound bag, spherical or slightly flattened, seen in electron micrographs to be concentrated in number at sites near the cell membrane believed to be the presynaptic side of a synapse. (See p. 41.)

Syncytium. A multinucleate mass of protoplasm enclosed in one cell membrane, resulting either from fusion or from incomplete cell division.

Taxis. A form of behavior in which the animal first orients toward a fixed angle relative to a directional stimulus, usually 0° or 180°, and then moves in that direction more or less steadily.

Telodendria. The arborizing terminals of an axon.

Tetanus. In physiology, the state of muscles undergoing maintained contraction, commonly resulting from stimulation at a frequency high enough for individual twitches to summate to a smooth tension.

Threshold. Equivalent to limen or sill, a threshold exists if there is a clear and abrupt transition from one state to another. It is applied in neurophysiology to those states of excitation of excitable elements characterized by a discontinuous function of response on strength of stimulus. A threshold stimulus is often operationally defined as that intensity which produces a response on 50% of equivalent trials or some other arbitrary criterion to take account of variability.

Time constant (τ). A measure of the rate of decay of a passive potential or an equivalent process proceeding logarithmically; the time from the start of decay until the amplitude has declined to $1/e$ or about 37% of its initial height.

Tone. Often used in the Latin form *tonus* (from a Greek word meaning "stretch"). In physiology, a degree of tension, firmness, or maintained contraction in a muscle, and by extension, a degree of maintained nerve impulse traffic in the nervous system (as in central tone, vagal tone).

Tonic. Continuing for a relatively long time; a tonic receptor or fraction of a receptor discharge is one with no adaptation.

Torsion. In gastropods, a process in evolution and in ontogeny resulting in 180° anticlockwise rotation of the mantle complex, carrying anus, ctenidia, excretory openings, and mantle cavity from their primitive posterior position to a new position over the head and at the same time twisting the visceral loop of ganglia and connectives so that the (new) supraintestinal ganglion is on the left and its connective to the right pleural ganglion passes above the gut, the (new) subintestinal ganglion is on the right and its connective to the left pleural ganglion passes under the gut.

Tract. A bundle of nerve fibers between parts of the central nervous system; usually used for bundles all of whose fibers start in the same place or in serially homologous places and end in a common destination or series of homologous places.

Train. A series of nerve impulses of limited duration and therefore distinct from a run (*which see*), but lasting too long to be called a burst (*which see*).

Transducer. A device that changes energy from one form to another.

Transfer function. Sometimes used in a rigorous engineering sense for a mathematical function, sometimes more loosely for the input-output relation of a locus of integration (*which see*); the coupling function. It is determined in nervous systems both by the structural connectivity and by the properties of elements, such as noise, spontaneity, transmitter release, binding and decay, threshold, facilitation, accomodation, adaptation, and aftereffects.

Transmission. In neurophysiology, the passage of excitation or inhibition across synapses; distinguished from conduction and propagation, which occur within a neuron and do not involve junctions. A transmitter is an agent that mediates transmission.

Trigger. An arrangement of parts in a system, or an energy state in a system, such that proper application of an increment of energy (commonly but not necessarily small) will precipitate a change whose time course (commonly but not necessarily rapid) and magnitude are essentially independent of those of the energy increment. The trigger concept in biology and some examples are discussed in Bullock (1957).

Undershoot. A term applied when an internal microelectrode shows a positive or hyperpolarizing afterpotential following a spike.

Unipolar. In neurohistology, said of nerve cells having only one process; preferred to its synonym, monopolar.

Utilization time. Latency of response of an axon to a square-fronted electrical stimulus above threshold; measured at the site of stimulation and therefore without conduction time. This term is now little used because this latency is better understood and not always equivalent; it is partly determined by the time constant of the axonal membranes and others under the electrodes and partly by local response in some cases.

Volley. A simultaneous discharge of a single impulse in each of many nerve fibers in a bundle; also the impulses so discharged.

White matter. That part of the central nervous system of vertebrates consisting of axons and glia, without cell bodies, dendrites, or endings; historically, that part appearing white to the naked eye because of its greater concentration of myelin; contrasted with gray matter (*which see*).

Zygoneury. The condition of the nerve cords in some gastropods in which a connective (zygosis) between the supraintestinal ganglion and the pleural ganglion of the same (left) side, and between the subintestinal and the pleural ganglion of the same (right) side are both present. Believed to be a relatively derived or recent condition.

REFERENCES CITED

Included in this bibliography are the references cited in figure legends and in the text. Each item, therefore, is included for a special reason, and the list does not comprehend the subjects represented. Sources listed in Suggested Readings at the end of each chapter are not repeated here.

Ábrahám, A. 1940. Die Innervation des Darmkanals der Gastropoden. *Z. Zellforsch.*, A30:273–296.

Adey, W. R., and D. O. Walter. 1963. Application of phase detection and averaging techniques in computer analysis of EEG records in the cat. *Exp. Neurol.*, 7:186–209.

Adrian, E. D. 1928. *The Basis of Sensation: The Action of Sense Organs.* Christophers, London.

Adrian, E. D. 1932. *The Mechanism of Nervous Action.* Univ. Pennsylvania Press, Philadelphia.

Adrian, E. D., and Y. Zotterman. 1926a. The impulses produced by sensory nerve-endings. Part 2. The response of a single end-organ. *J. Physiol.*, 61:151–171.

Adrian, E. D., and Y. Zotterman. 1926b. The impulses produced by sensory nerve endings. Part 3. Impulses set up by touch and pressure. *J. Physiol.*, 61:465–483.

Akert, K., and R. B. Livingston. 1973. Morphological plasticity of the synapse. In L. Laitinen and K. E. Livingston, eds., *Surgical Approaches in Psychiatry.* Medical & Technical Publ. Co., St. Leonard's Gate, Lancaster, England.

Akert, K., K. Pfenninger, C. Sandri, and H. Moor. 1972. Freeze etching and cytochemistry of vesicles and membrane complexes in synapses of the central nervous system. In G. D. Pappas and D. P. Purpura, eds., *Structure and Function of Synapses.* Raven Press, New York.

Albuquerque, E. X., and R. J. McIsaac. 1970. Fast and slow mammalian muscles after denervation. *Exp. Neurol.*, 26:183–202.

Alexandrowicz, J. S. 1932. The innervation of the heart of the crustacea. I. Decapoda. *Quart. J. Micr. Sci.*, 75:181–249.

Alexandrowicz, J. S. 1951. Muscle receptor organs in the abdomen of *Homarus vulgaris* and *Palinurus vulgaris. Quart. J. Micr. Sci.*, 92:163–199.

Altman, P. L., and D. S. Dittmer. 1974. *Biology Data Book.* 1142–1149.

Anderson, P., and J. C. Eccles. 1965. Locating and identifying postsynaptic inhibitory synapses by the correlation of physiological and histological data. *Symp. Biol. Hung.*, 5:219–242.

Anderson, M. J., and M. W. Cohen. 1974. Fluorescent staining of acetylcholine receptors in vertebrate skeletal muscle. *J. Physiol.*, 237:385–400.

518

References Cited

Angevine, J. B., Jr. 1970. Critical cellular events in the shaping of neural centers. In F. O. Schmitt, ed., *The Neurosciences: Second Study Program*. Rockefeller Univ. Press, New York.

Angevine, J. B., Jr., and R. L. Sidman. 1961. Autoradiographic study of cell migration during histogenesis of cerebral cortex in the mouse. *Nature.* 192:766–768.

Attardi, D. G., and R. W. Sperry. 1963. Preferential selection of central pathways by regenerating optic fibers. *Exp. Neurol.*, 7:46–64.

Atwood, H. L. 1967. Crustacean neuromuscular mechanisms. *Amer. Zool.*, 7:527–551.

Atwood, H. L. 1973. An attempt to account for the diversity of crustacean muscles. *Amer. Zool.*, 13:357–378.

Balss, H. 1940–1957. Decapoda (Zehnfüsser). In *Bronn's Klassen, und Ordnungen des Tierreichs.* Akademische Verlagsgesellschaft Geest und Portig K.-G., Leipzig. 5:1:7:1–669, 1285–1770.

Bang, T. 1917. Zur Morphologie des Nervensystems von *Helix pomatia* L. *Zool. Anz.*, 48:281–292.

Barker, D. 1948. The innervation of the muscle-spindle. *Quart. J. Micr. Sci.*, 89:143–186.

Barr, M. L. 1972. *The Human Nervous System.* Harper & Row, New York.

Batham, E. J., and C. F. A. Pantin. 1950. Phases of activity in the sea anemone *Metridium senile* (L.), and their relation to external stimuli. *J. exp. Biol.*, 27:377–399.

Batham, E. J., C. F. A. Pantin, and E. A. Robson. 1960. The nerve-net of the sea anemone *Metridium senile*: The mesenteries and the column. *Quart. J. Micr. Sci.*, 101:487–510.

Bekesy, G. von. 1967. *Sensory Inhibition.* Princeton Univ. Press, Princeton, New Jersey.

Bennett, M. V. L., and H. Grundfest. 1961. The electrophysiology of electric organs of marine electric fishes. III. The electroplaques of the stargazer, *Astroscopus y-graecum. J. Gen. Physiol.* 44:819–843.

Bennett, M. V. L., Y. Nakajima, and G D. Pappas. 1967. Physiology and ultrastructure of electrotonic junctions. III. Giant electromotor neurons of *Malapterurus electricus. J. Neurophysiol.*, 30:209–235.

Berndt, W. 1903. Die Anatomie von *Cryptophialus striatus* Berndt (Cirripedia, Acrothoracica). *S.B. Ges. naturf. Fr., Berl.*, 1903:436–444.

Berry, M., A. W. Rogers, and J. T. Eayrs. 1964. Pattern of cell migration during cortical histogenesis. *Nature*, 203:591–593.

Bethe, A. 1897. Das Nervensystem von *Carcinus maenas.* Ein anatomisch-physiologischer Versuch. I. *Arch. mikr. Anat.*, 50:460–546, 589–639.

Blakemore, C., and G. E. Cooper. 1970. Development of the brain depends on the visual environment. *Nature*, 228:477–478.

Blinkov, S. M., and I. I. Glezer. 1968. *The Human Brain in Figures and Tables.* Basic Books, New York.

Bodian, D. 1942. Cytological aspects of synaptic function. *Physiol. Rev.*, 22:146–169.

Bodian, D. 1947. Nucleic acid in nerve cell regeneration. In *Symposium of the Society for Experimental Biology No. 1, Nucleic Acid.* Cambridge Univ. Press, London.

Bodian, D. 1952. Introductory survey of neurons. *Cold Spr. Harb. Symp. Quant. Biol.*, 17:1–13.

Bodian, D. 1967. Neurons, circuits, and neuroglia. In G. C. Quarton, T. Melnechuk, and F. O. Schmitt, eds., *The Neurosciences: A Study Program.* Rockefeller Univ. Press, New York.

Bodian, D. 1972. Neuron junctions: A revolutionary decade. *Anat. Rec.*, 174:73–82.

Boeckh, J., C. Sandri, and K. Akert. 1970. Sensorische Eingange und synaptische Verbindungen im Zentralnervensystem von Insekten. *Z. Zellforsch.*, 103:429–446.

Bogoraze, D., and P. Cazal. 1944. Recherches histologiques sur le système nerveux du poulpe: Les neurones, le tissu interstitiel et les éléments neurocrines. *Arch. Zool. exp. gén.*, 83:412–444.

Bouvier, E.-L. 1837. Système nerveux, morphologie générale et classification des gastéropodes prosobranches. *Ann. Sci. Nat. (Zool.)*, (7)3:1–510.

Boyd, I. A., and A. R. Martin. 1956. The endplate potential in mammalian muscle. *J. Physiol.*, 132:74–91.

Braitenberg, V. 1970. Ordnung und Orientierung der Elemente im Sehsystem der Fliege. *Kybernetik*, 7:235–242.

Brazier, M. A. B. 1960. *The Electrical Activity of the Nervous System* (2nd ed.). Macmillan, New York.

Brazier, M. A. B. 1961. *A History of the Electrical Activity of the Brain.* Pitman, London.

Brazier, M. A. B. 1968. *The Electrical Activity of the Nervous System* (3rd ed.). Pitman and Sons, Ltd., London.

Brodmann, K. 1909. *Vergleichende Lokalisationlehre der Grosshirnrinde in ihren Prinzipien dargestellt auf Grund des Zellenbaues.* J. A. Barth, Leipzig.

Buddenbrock, W. von. 1922. Mechanismus der phototropen Bewegungen. *Wiss. Meeresuntersuch. N. F. Abt. Helgoland*, 15:1–19.

Buddenbrock, W. von. 1956. *The Love of Animals.* Muller, London.

Buller, A. J., J. C. Eccles, and R. M. Eccles. 1960. Differentiation of fast and slow muscles in the cat hind limb. *J. Physiol.,* 150:399–416.

Bullock, T. H. 1943. Neuromuscular facilitation in scyphomedusae. *J. Cell. Comp. Physiol.,* 22:251–272.

Bullock, T. H. 1944. Oscillographic studies on the giant nerve fiber system in *Lumbricus. Biol. Bull.,* 87:159–160.

Bullock, T. H. 1957. *Physiological Triggers and Discontinuous Rate Processes.* Amer. Physiol. Soc., Washington, D. C.

Bullock, T. H. 1961a. The origins of patterned nervous discharge. *Behaviour,* 17:48–59.

Bullock, T. H. 1961b. On the anatomy of the giant neurons of the visceral ganglion of *Aplysia.* In E. Florey, ed., *Nervous Inhibition.* Pergamon Press, New York.

Bullock, T. H. 1968. Representation of information in neurons and sites for molecular participation. *Proc. Nat. Acad. Sci.,* 60:1058–1068.

Bullock, T. H. 1975. Are we learning what actually goes on when the brain recognizes and controls? In *New Directions in Comparative Physiology and Biochemistry.* Amer. Soc. Zool., *J. Exp. Zool.,* 194:13–33.

Bullock, T. H., and G. A. Horridge. 1965. *Structure and Function in the Nervous Systems of Invertebrates.* 2 vols. W. H. Freeman and Company, San Francisco.

Bullock, T. H., R. H. Hamstra, and H. Scheich. 1972. The jamming avoidance response of high frequency electric fish. *J. Comp. Physiol.,* 77:1–22.

Bumke, O., and O. Foerster. 1935–1937. *Handbuch der Neurologie.* 17 vols. Springer-Verlag, Berlin.

Bunge, R. P. 1968. Glial cells and the central myelin sheath. *Physiol. Rev.,* 48:197–251.

Bunge, R. P. 1970. Structure and function of neuroglia: Some recent observations. In F. O. Schmitt, ed., *The Neurosciences: Second Study Program.* Rockefeller Univ. Press, New York.

Bunge, M. B., R. P. Bunge, and H. Ris. 1961. Ultrastructural study of remyelination in an experimental lesion in adult cat spinal cord. *J. Cell. Biol.,* 10:67–94.

Burnstock, G. 1969. Evolution of the automatic innervation of visceral and cardiovascular systems in vertebrates. *Pharmac. Rev.,* 21(4):247–324.

Burnstock, G., and M. Costa. 1973. Inhibitory innervation of the gut. *Gastroenterology,* 64(1):141–144.

Burnstock, G., and T. Iwayama. 1971. Fine-structural identification of autonomic nerves and their relation to smooth muscle. *Prog. Brain Res.,* 34:389–404.

Cajal, S. R. 1894. *Die Retina der Wirbelthiere.* Bergmann. Wiesbaden.

Cajal, S. R. 1909–1911. *Histologie du système nerveux de l'homme et des vertébrés.* (French edition revised and updated by the author. Translated from Spanish by L. Azoulay. 2 vols. Maloine, Paris. Republished 1952 by Consejo Superior de Investigaciones Cientificas, Madrid.)

Cajal, S. R. 1917. Contribución al conocimiento de la retina y centros ópticos de los cefalópodos. *Trab. Lab. Invest. biol. Univ. Madr.,* 15:1–82.

Cajal, S. R. 1928. *Degeneration and Regeneration of the Nervous System.* Oxford Univ. Press.

Cajal, S. R. 1933. *Histology.* Translated by M. Fernán-Núñez. William Wood & Company, Baltimore.

Cajal, S. R. 1954. *Neuron Theory or Reticular Theory?* (Transl. by M. Ubeda Purkiss and C. A. Fox from the Spanish version, 1934.) Inst. Ramón y Cajal, C.S.I.S., Madrid.

Cajal, S. R. 1966. *Recollections of My Life.* (Republication of the 1937 translation by E. H. Craigie from the Spanish original, 1932.) Mass. Inst. Technol. Press, Cambridge.

Cajal, S. R., and D. Sánchez. 1915. Contribución al conocimiento de los centros nerviosos de los insectos. *Trab. Lab. Invest. Biol. Univ. Madr.,* 13:1–164.

Caldwell, P. C., and R. D. Keynes. 1957. The utilization of phosphate bond energy for sodium extrusion from giant axons. *J. Physiol.,* 137:12P.

Campos–Ortega, J. A., and N. J. Strausfeld. 1973. Synaptic connections of intrinsic cells and basket arborizations in the external plexiform layer of the fly's eye. *Brain Res.,* 59:119–136.

Chagas, C., and A. P. de Carvalho. 1961. *Bioelectrogenesis.* Elsevier, New York.

Clarke, E. S., and C. D. O'Malley. 1968. *The Human Brain and Spinal Cord: A historical study.* Univ. Calif. Press, Berkeley.

Cleland, B. G., and W. R. Levick. 1974. Properties of rarely encountered types of ganglion cells in the cat's retina and an overall classification. *J. Physiol.,* 240:457–492.

Cohen, M. J. 1970. A comparison of invertebrate and vertebrate central neurons. In F. O. Schmitt, ed., *The Neurosciences: Second Study Program.* Rockefeller Univ. Press, New York.

References Cited

Cole, K. S., and H. J. Curtis. 1938. Electrical impedance of the squid giant axon during activity. *J. Gen. Physiol.,* 22:649–670.

Colosanti, G. 1876. Anatomische und physiologische Untersuchungen über den Arm der Kephalopoden. *Arch. Anat. Physiol.,* 1876:480–500.

Conel, J. L. 1959. *The postnatal development of the human cerebral cortex.* Harvard Univ. Press, Cambridge.

Coombs, J. S., J. C. Eccles, and P. Fatt. 1955. The specific ionic conductances and the ionic movements across the motoneuronal membrane that produce the inhibitory post-synaptic potential. *J. Physiol.,* 155:326–373.

Couteaux, R. 1958. Morphological and cytochemical observations on the post-synaptic membrane at motor end-plates and ganglionic synapses. *Exp. Cell Res.* 5 (Suppl.):294–322.

Cowan, W. M., and T. P. S. Powell. 1963. Centrifugal fibres in the avian visual system. *Proc. Roy. Soc. B,* 158:232–252.

Curtis, B. A., S. Jacobson, and E. M. Marcus. 1972. *An Introduction to the Neurosciences.* Saunders, Philadelphia.

Dampier, W. C. 1948. *A History of Science* (4th ed.). Cambridge Univ. Press, London.

Davis, W. J. 1970. Motoneuron morphology and synaptic contacts: Determination by intracellular dye injection. *Science,* 168:1358–1360.

Debaisieux, P. 1944. Les yeux de crustacés; Structure développement, réactions à l'éclairement. *Cellule,* 50:9–122.

Dehorne, A. 1935. Sur le trophosponge des cellules nerveuses géantes de *Lanice conchylega* Pallas. *C. R. Soc. Biol., Paris,* 120:1188–1190.

Del Castillo, J., and B. Katz. 1954. Quantal components of the end-plate potential. *J. Physiol.,* 124:560–573.

Delage, Y. 1886. Études histologiques sur les planaires rhabdocoeles acoeles. *Arch. Zool. exp. gén.,* (2), 4:109–160.

De Lorenzo, A. J. 1960. The fine structure of synapses in the ciliary ganglion of the chick. *J. Biophys. Biochem. Cytol.,* 7:31–36.

Dennis, M. J., A. J. Harris, and S. W. Kuffler. 1971. Synaptic transmission and its duplication by focally applied acetylcholine in parasympathetic neurons in the heart of the frog. *Proc. Roy. Soc. B,* 177:509–539.

Diamond, J., E. G. Gray, and G. M. Yasargil. 1970. The function of the dendritic spine: An hypothesis. In P. Andersen and J. K. S. Jansen, eds., *Excitatory Synaptic Mechanisms.* Universitetsforlaget, Oslo.

Dijkgraaf, S. 1952. Bau und Funktionen der Seitenorgane und des Ohrlabyrinths bei Fischen. *Experientia,* 8:205–244.

Dijkgraaf, S. 1962. The functioning and significance of the lateral-line organs. *Biol. Rev.,* 38:51–105.

Dijkgraaf, S. 1967. Biological significance of the lateral-line organs. In P. Cahn, ed., *Lateral Line Detectors.* Indiana Univ. Press, Bloomington.

Dowling, J. E. 1970. Organization of vertebrate retinas. *Invest. Ophthalmol.,* 9:655–680.

Dreyer, Th. F. 1910. Über das Blutgefäss- und Nervensystem der *Aeolididae* und *Tritoniadae. Z. wiss. Zool.,* 96:373–418.

Droz, B. 1973. Renewal of synaptic proteins. *Brain Res.,* 62:383–394.

Droz, B., H. L., Koenig, and L. Di Giamberardino. 1973. Axonal migration of protein and glycoprotein to nerve endings. I. Radioautographic analysis of the renewal of protein in nerve endings of chicken ciliary ganglion after intracerebral injection of [^{3}H] lysine. *Brain Res.,* 60:93–127.

Dudel, J., and S. W. Kuffler. 1961. Presynaptic inhibition at the crayfish neuromuscular junction. *J. Physiol.,* 155:543–562.

Eakin, R. M. 1963. Lines of evolution of photoreceptors. In D. Mazia and A. Tyler, eds., *General Physiology of Cell Specialization.* McGraw-Hill, New York.

Eakin, R. M. 1965. Differentiation of rods and cones in total darkness. *J. Cell Biol.,* 25:162–165.

Eakin, R. M. 1968. Evolution of photoreceptors. In T. Dobzhansky, M. K. Hecht, and W. C. Steere, eds., *Evolutionary Biology,* vol. 2. Appleton-Century-Crofts, New York.

Easton, T. A. 1972. On the normal use of reflexes. *Amer. Sci.,* 60:591–599.

Ebbesson, S. O. E. 1970. On the organization of central visual pathways in vertebrates. *Brain Behav. Evol.,* 3:178–194.

Eccles, J. C. 1964. *The Physiology of Synapses.* Springer-Verlag, Berlin.

Eccles, J. C., M. Ito, and J. Szentágothai. 1967. *The Cerebellum as a Neuronal Machine.* Springer-Verlag, New York.

Eckert, R. 1963. Electrical interaction of paired ganglion cells in the leech. *J. Gen. Physiol.*, 46:575–587.

Economo, C. F. von. 1929. *The Cytoarchitectonics of the Human Cerebral Cortex*. Oxford Medical Publications, London.

Elazar, Z., and W. R. Adey. 1967. Spectral analysis of low frequency components in the electrical activity of the hippocampus during learning. *Electroenceph Clin. Neurophysiol.*, 23:225–240.

Engström, H., and J. Wersäll. 1958. The ultrastructural organization of the organ of Corti and of the vestibular sensory epithelia. *Exp. Cell Res. (Suppl.)*, 5:460–492.

Erickson, R. P. 1963. Sensory neural patterns and gustation. In Y. Zotterman, ed., *Olfaction and Taste*. Pergamon Press, London.

Erlanger, J., and H. S. Gasser. 1937. *Electrical Signs of Nervous Activity*. Univ. Pennsylvania Press, Philadelphia.

Eskin, A. 1971. Properties of the *Aplysia* visual system. In vitro entrainment of the circadian rhythm and certrifugal regulation of the eye. *Z. vergl. Physiol.*, 74:353–371.

Fatt, P., and B. Katz. 1951. An analysis of the end-plate potential recorded with an intra-cellular electrode. *J. Physiol.*, 115:320–370.

Fatt, P., and B. Katz. 1952. Spontaneous subthreshold activity at motor nerve endings. *J. Physiol.*, 117:109–128.

Fearing, F. 1970. *Reflex action: A study in the History of Physiological Pyschology*, Mass. Inst. Technol. Press, Cambridge.

Fields, H. L., and D. Kennedy. 1965. Functional role of muscle receptor organs in crayfish. *Nature*, 206:1232–1237.

Flock, Å. 1965. Transducing mechanisms in the lateral line canal organ receptors. *Cold Spr. Harb. Symp. Quant. Biol.*, 30:133–144.

Florey, E. 1957. Chemical transmission and adaptation. *J. Gen. Physiol.*, 40:533–545.

Fraenkel, G. 1931. Die Mechanik der Orientierung der Tiere im Raum. *Biol. Rev.*, 6:36–87.

Frazier, W. T., E. R. Kandel, I. Kupfermann, R. Waziri, and R. E. Coggeshall. 1967. Morphological and functional properties of identified neurons in the abdominal ganglion of *Aplysia californica*. *J. Neurophysiol.*, 30:1288–1351.

French, J. D., and H. W. Magoun. 1952. Effects of chronic lesions in central cephalic brain stem of monkeys. *Arch. Neurol. Psychia.* 68:591–604.

Frisch, K. von. 1948. Gelöste und ungelöste Rätsel der Bienensprache. *Naturwiss.*, 35:12–23, 38–43.

Fulton, J. F. 1949. *Physiology of the Nervous System* (3rd ed.). Oxford Univ. Press, New York.

Galambos, R. 1962. *Nerves and Muscles*. Doubleday, Garden City, New York.

Galambos, R., J. Schwartzkopff, and A. Rupert. 1959. Microelectrode study of superior olivary nuclei. *Am. J. Physiol.*, 197:527–536.

Gardner, E. 1968. *Fundamentals of Neurology* (5th ed.). Saunders, Philadelphia.

Gardner, E. J. 1965. *History of Biology* (2nd ed.). Burgess Minneapolis.

Gaze, R. M., M. Jacobson, and G. Székely. 1963. The retinotectal projection in *Xenopus* with compound eyes. *J. Physiol.*, 165:484–499.

Görner, P. 1961. Beitrag zum Bau und zur Arbeitsweise des Seitenorgans von *Xenopus laevis*. *Verh. dtsch. zool. Ges. Saarbrücken*, 1961:193–198.

Gray, E. G. 1960. The fine structure of the insect ear. *Phil. Trans. Roy. Soc. B*, 243:75–94.

Green, J. D., and A. A. Arduini. 1954. Hippocampal electrical activity in arousal. *J. Neurophysiol*, 17:533–557.

Greenfield, N. S., and R. S. Sternbach. 1972. *Handbook of Psychobiology*. Holt, Rinehart & Winston, New York.

Grillner, S. 1975. Locomotion in vertebrates: Central mechanisms and reflex interaction. *Physiol. Rev.*, 55:247–304.

Grundfest, H. 1957. Electrical inexcitability of synapses and some of its consequences in the central nervous system. *Physiol. Rev.*, 37:337–361.

Grüsser, O.-J., and U. Grüsser-Cornehls. 1972. Comparative physiology of movement-detecting neuronal systems in lower vertebrates. (*Anura* and *Urodela*.) *Bibl. ophthal.*, 82:260–273.

Gustafson, G. 1930. Anatomische Studien über die Polychäten-Famillien Amphinomidae und Euphrosynidae. *Zool. Bidr. Uppsala*, 12:305–471.

Hadenfeldt, D. 1929. Das Nervensystem von *Stylochoplana maculata* und *Notoplana atomata*. *Z. wiss. Zool.*, 133:586–638.

Ham, A. W. 1974. *Histology* (7th ed.). Lippincott, Philadelphia.

522

References Cited

Hamburger, V., and E. L. Keefe. 1944. The effects of peripheral factors on the proliferation and differentiation in the spinal cord of chick embryos. *J. Exp. Zool.*, 96:223–242.

Hamburger, V., M. Balaban, R. Oppenheim, and E. Wenger. 1965. Periodic motility of normal and spinal chick embryos between 8 and 17 days of incubation. *J. Exp. Zool.*, 159:1–14.

Hamlyn, L. H. 1962. The fine structure of the mossy fiber endings in the hippocampus of the rabbit. *J. Anat.* 96:112–120.

Hámori, J. 1973. Developmental morphology of dendritic postsynaptic specializations. In *Recent Developments of Neurobiology in Hungary.* Vol. 4. Akademiai Kiado, Budapest.

Hámori, J., and J. Szentágothai. 1965. The Purkinje cell baskets: Ultrastructure of an inhibitory synapse. *Acta biol. Acad. Sci. Hung.*, 15:465–479.

Hanström, B. 1924. Untersuchungen über das Gehirn, insbesondere die Sehganglien der Crustaceen. *Ark. Zool.*, 16(10):1–119.

Hanström, B. 1925. Über die sogenannten Intelligenzsphären des Molluskengehirns und die Innervation des Tentakels von *Helix. Acta Zool. Stockh.*, 6:183–215.

Hanström, B. 1926. Das Nervensystem und die Sinnesorgane von *Limulus polyphemus. Acta Univ. Lund., Avd.* 2, 22(5):1–79.

Hanström, B. 1928a. Die Beziehungen zwischen dem Gehirn der Polychäten und der der Arthropoden. *Z. Morph. Ökol. Tiere*, 11:152–160.

Hanström, B. 1928b. *Vergleichende Anatomie des Nervensystems der wirbellosen Tiere unter Berücksichtigung seiner Funktion.* Springer, Berlin.

Harlow, H. F., and C. N. Woolsey. 1958. *Biological and Biochemical Bases of Behavior.* Univ. Wisconsin Press, Madison.

Hartline, H. K. 1938a. The response of single optic nerve fibers of the vertebrate eye to illumination of the retina. *Amer. J. Physiol.*, 121:400–415.

Hartline, H. K. 1938b. The discharge of impulses in the optic nerve of *Pecten* in response to illumination of the eye. *J. Cell. Comp. Physiol.*, 11:465–478.

Hassenstein, B. 1951. Ommatidienraster und Afferente Bewegungsintegration. (Versuche an dem Rüsselkäfer *Chlorophanus viridis.*) *Z. vergl. Physiol.*, **33**:301–326.

Hassenstein, B., and W. Reichardt. 1959. Wie sehen Insekten Bewegungen? *Umschau*, 10:302–305.

Heiligenberg, W. 1973. Electrolocation of objects in the electric fish *Eigenmannia* (Rhamphichthyidae, Gymnotoidei), *J. Comp. Physiol.*, 87:137–164.

Hertwig, O., and R. Hertwig. 1878. Das Nervensystem und die Sinnesorgane der Medusen. Vogel, Leipzig.

Hess, E. H. 1956. Space perception in the chick. *Sci. Amer.*, 195:71–80.

Hess, W. N. 1925a. Nervous system of the earthworm, *Lumbricus terrestris* L. *J. Morph.*, 40:235–259.

Hess, W. N. 1925b. The nerve plexus of the earthworm, *Lumbricus terrestris. Anat. Rec.*, 31:335–336.

Heuser, J. E., and T. S. Reese. 1973. Evidence for recycling of synaptic vesicle membrane during transmitter release at the frog neuromuscular junction. *J. Cell Biol.*, 57:315–344.

Heuser, J. E., and T. S. Reese. 1974. Morphology of synaptic vesicle discharge and reformation at the frog neuromuscular junction. In M.V.L. Bennett, ed., *Synaptic Transmission and Neuronal Interaction.* Raven Press, New York.

Hibbard, E. 1965. Orientation and directed growth of Mauthner's cell axons from duplicated vestibular nerve roots. *Exp. Neurol.*, 13:289–301.

Hilton, W. A. 1934. Nervous system and sense organs. LI. Crustacea. Branchiopoda. *J. Ent. Zool.*, 26:35–45.

Hinde, R. A. 1954. Factors governing the changes in strength of a partially inborn response, as shown by the mobbing behavior of the chaffinch (*Fringilla coelebs*). I and II. *Proc. Roy. Soc. B*, 142:306,331, 331–358.

Hinde, R. A. 1958. The nest building behavior of domesticated canaries. *Proc. Zool. Soc. Lond.*, 131:1–48.

Hines, M. 1929. The brain of *Ornithorhynchus anatinus. Phil. Trans. Roy. Soc. B*, 217:155–288.

Hirano, A., and H. M. Dembitzer. 1967. A structural analysis of the myelin sheath in the central nervous system. *J. Cell Biol.*, 34:555–567.

Hodgkin, A. L. 1938. The subthreshold potentials in a crustacean nerve fiber. *Proc. Roy. Soc. B.*, 126:87–121.

Hodgkin, A. L. 1948. The local electric changes associated with the repetitive action in a non-medullated axon. *J. Physiol.*, 107:165–181.

Hodgkin, A. L. 1958. Ionic movements and electrical activity in giant nerve fibres. *Proc. Roy. Soc. B.* 148:1–37.

Hodgkin, A. L. 1964. *The Conduction of the Nervous Impulse.* Liverpool Univ. Press.

Hodgkin, A. L. and A. F. Huxley, 1952a. Currents carried by sodium and potassium ions through the membrane of the giant axon Loligo. *J. Physiol.*, 116:449–472.

Hodgkin, A. L. and A. F. Huxley, 1952b. A quantitative description of membrane current and its application to conduction and excitation in nerve. *J. Physiol.*, 117:500–544.

Hodgkin, A. L., and B. Katz. 1949. The effect of sodium ions on the electrical activity of the giant axon of the squid. *J. Physiol.* 108:37–77.

Holmes, W. 1942. The giant myelinated nerve fibres of the prawn. *Phil. Trans. Roy. Soc. B.*, 321:293–311. 311.

Holmgren, N. 1916. Zur vergleichenden Anatomie des Gehirns von Polychoren, Onychophoren, Xiphosuren, Arachniden, Crustaceen, Myriapoden und Insekten. *K. svenska Ventensk Akad. Handl.*, 56:1–303.

Holst, E. von. 1950a. Die Arbeitsweise des Statolithenapparates bei Fischen. *Z. vergl. Physiol.*, 32:60–120.

Holst, E. von. 1950b. Quantitative Messung von Stimmungen im Verhalten der Fische. *Symp. Soc. Exp. Biol.*, 4:143–172.

Holst, E. von, and U. von Saint Paul. 1960. Vom Wirkungsgefüge der Triebe. *Naturwiss.*, 47:409–422.

Hoog, E. G. Van't. 1918. Über Tiefenlokalisation in der Grosshirnrinde. *Psychiat. Neurol. Bladen*, 22:281–298.

Horridge, G. A. 1955. The nerves and muscles of medusae. II. *Geryonia proboscidalis* Escholtz. *J. Exp. Biol.*, 32:555–568.

Horridge, G. A. 1956. The nervous system of the ephyra larva of *Aurellia aurita*. *Quart. J. Micr. Sci.*, 97:59–74.

Horridge, G. A. 1966. Study of a system, as illustrated by the optokinetic response. *Symp. Soc. Exp. Biol.*, 20:179–198.

Horridge, G. A. 1968. *Interneurons*. W. H. Freeman and Company, San Francisco.

Hoyle, G. 1964. Exploration of neuronal mechanisms underlying behavior in insects. In R. F. Reiss, ed., *Neural Theory and Modeling*. Stanford Univ. Press.

Hsü, F. 1938. Étude cytologique et comparée sur les sensilla des insectes. *Cellule*, 47:7–60.

Hubel, D. H. 1963. The visual cortex of the brain. *Sci. Amer.*, 209:54–62.

Hubel, D. H., and T. N. Wiesel. 1962. Receptive fields, binocular interaction and functional architecture in the cat's visual cortex. *J. Physiol.*, 160:106–154.

Hubel, D. H., and T. N. Wiesel. 1963. Shape and arrangement of columns in the cat's striate cortex. *J. Physiol.*, 165:559–568.

Hubel, D. H., and T. N. Wiesel. 1972. Laminar and columnar distribution of geniculocortical fibers in the macaque monkey. *J. Comp. Neurol.*, 146:421–450.

Hubel, D. H., and T. N. Wiesel. 1974. Sequence, regularity, and geometry of orientation columns in the monkey striate cortex. *J. Comp. Neurol.*, 158:267–293.

Huber, F. 1960. Untersuchungen über die Funktion des Zentralnerven systems und insbesondere des Gehirnes bei der Fortbewegung und der Lauterzeugung der Grillen. *Z. vergl. Physiol.* 44:60–132.

Hughes, A. 1961. Cell degeneration in the larval ventral horn of *Xenopus laevis* (Daudin). *J. Embryol. Exp. Morph.*, 9:269–284.

Hunt, C. C., and S. W. Kuffler. 1951. Further study of efferent small-nerve fibres to mammalian muscle spindles. Multiple spindle innervation and activity during contraction. *J. Physiol.*, 113:283–315.

Huxley, A. F., and R. Stämpfli. 1949. Evidence for saltatory conduction in peripheral myelinated nerve fibres. *J. Physiol.*, 108:315–339.

Huxley, A. F., and R. Stämpfli. 1951. Effect of potassium and sodium on resting and action potentials of single myelinated nerve fibres. *J. Physiol.*, 112:496–508.

Jacobson, M. 1970. *Developmental Neurobiology*. Holt, Rinehart & Winston, New York.

Jänig, W. 1975. The autonomic nervous system. In R. F. Schmidt, ed., *Fundamentals of Neurophysiology*. Springer-Verlag, New York.

Jansen, J., and A. Brodal. 1958. Das Kleinhirn. *In* W. von Möllendorff and W. Bargmann (eds.), *Handbuch der mikroskopischen Anatomie des Menschen*. IV/8. Springer-Verlag, Berlin.

Jasper, H. H. 1941. Electroencephalography. In W. Penfield and T. C. Erickson, eds., *Epilepsy and Cerebral Localization*. Charles C Thomas, Springfield, Illinois.

Jennings, H. S. 1906. *Behavior of the Lower Organisms*. Columbia Univ. Press, New York.

Jerison, H. J. 1973. *Evolution of the Brain and Intelligence*. Academic Press, New York.

Johnson, G. E. 1924. Giant nerve fibers in crustaceans with special reference to *Cambarus* and *Palaemonetes*. *J. Comp. Neurol.*, 36:323–373.

Johnston, J. B. 1902. The brain of *Petromyzon*. *J. Comp. Neurol.*, 7:2–82.

Johnston, J. B. 1906. *Nervous System of Vertebrates*. Blakiston's Son & Co., Philadelphia.

Josephson, R. K. 1961a. The responses of a hydroid to weak water-borne disturbances. *J. Exp. Biol.*, 38:17–27.

524

References
Cited

Josephson, R. K. 1961b. Colonial responses of hydroid polyps. *J. Exp. Biol.*, 38:559–577.

Kandel, E. R. 1974. An invertebrate system for the cellular analysis of simple behaviors and their modifications. In F. O. Schmitt and F. G. Worden, eds., *The Neurosciences: Third Study Program.* Mass. Inst. Technol. Press, Cambridge.

Kappers, C. U. Ä. 1929. *The Evolution of the Nervous System in Invertebrates, Vertebrates and Man.* De Erven F. Bohn, Haarlem.

Kappers, C. U. Ä., G. C. Huber, and E. C. Crosby. 1936. *The Comparative Anatomy of the Nervous System of Vertebrates, Including Man.* Macmillan, New York. (1967 reprinting, published by Hafner Co., New York.)

Kater, S. B., and C. H. F. Rowell. 1973. Integration of sensory and centrally programmed components in generation of cyclical feeding activity of (*Helisoma trivolvis*). *J. Neurophysiol.*, 36:142–155.

Katsuki, Y. 1958. Electrical responses of auditory neurons in cat to sound stimulation. *J. Neurophysiol.* 21:569–588.

Katz, B. 1962. The transmission of impulses from nerve to muscle, and the subcellular unit of synaptic action. *Proc. Roy. Soc. B,* 155:455–477.

Katz, B., and R. Miledi. 1964. The measurement of synaptic delay, and the time course of acetylcholine release at the neuromuscular junction. *Proc. Roy. Soc. B,* 161:483–495.

Katz, B., and R. Miledi. 1967. A study of synaptic transmission in the absence of nerve impulses. *J. Physiol.,* 192:407–436.

Keynes, R. D. 1958. The nerve impulse and the squid. *Sci. Amer.,* 199:83–90.

Kirkpatrick, J. B., J. J. Bray, and S. M. Palmer. 1972. Visualization of axoplasmic flow in vitro by Nomarski microscopy. Comparison to rapid flow of radioactive proteins. *Brain Res.,* 43:1–10.

Kölliker, A. 1896. *Handbuch der Gewebelehre des Menschen.* Band 2, Engelmann, Leipzig.

Korn, H. 1958. Vergleichend-embroyologische Untersuchungen an *Harmothoë* Kinberg (Polychaeta, Annelida). *Z. wiss. Zool.,* 161:346–443.

Kruijt, J. P. 1964. Ontogeny of social behavior in Burmese red jungle fowl (*Gallus gallus spadiceus*) Bonnaterre. *Behaviour,* 12(suppl.):1–201.

Kuffler, S. W., and C. Eyzaguirre. 1955. Synaptic inhibition in an isolated nerve cell. *J. Gen. Physiol.,* 39:155–189.

Kupfermann, I., and E. R. Kandel. 1969. Neuronal controls of a behavioral response mediated by the abdominal ganglion of *Aplysia. Science,* 164:847–850.

Kusano, K. 1966. Electrical activity and structural correlates of giant nerve fibers in Kuruma shrimp (*Penaeus japonicus*). *J. Cell Physiol.,* 68:361–384.

La Croix, P. 1935. Recherches cytologiques sur les centres nerveux chez les invertébrés. I. *Helix pomatia. Cellule,* 44:5–42.

Lang, E., and H. L. Atwood. 1973. Crustacean neuromuscular mechanisms. *Amer. Zool.* 13:337–355.

Langedon, F. E. 1895. The sense organs of *Lumbricus agricola,* Hoffm. *J. Morph.,* 11:193–234.

Larsell, O. 1932. The cerebellum of reptiles: Chelonians and alligator. *J. Comp. Neurol.,* 56:299–345.

Laurent, P. 1957. L'Innervation auriculaire du coeur des téléostéens. *Arch. Anat. micr. Morph. exp.,* 46:503–520.

Lenhossék, M. von. 1895. *Der feinere Bau des Nervensystems im Lichte neuester Forschungen* (2nd ed.). Aufl. Fischer, Berlin.

Levi-Montalcini, R., and P. U. Angeletti. 1968. Biological aspects of the nerve growth factor. In G. E. W. Wolstenholme and M. O'Connor, eds., *Growth of the Nervous Systems.* Ciba Fdn., Little, Brown & Company, Boston.

Liley, A. W. 1956. The quantal components of the mammalian end-plate. *J. Physiol.,* 133:571–587.

Lindemann, W. 1955. Uber die Jugendentwicklung beim Luchs (*Lynx l. lynx* Kerr.) und bei der Wildkatze (*Felis s. silvestris.* Schreb.). *Behaviour,* 8:1–45.

Lissman, H. W. 1946. The neurological basis of the locomotory rhythm in the spinal dogfish (*Scyllium canicula, Acanthias vulgaris*). II. The effect of de-afferentation. *J. Exp. Biol.,* 23:162–176.

Livanow, N. A. 1904. Untersuchungen zur Morphologie der Hirudineen. II. Das Nervensystem des vorderen Körperendes und seine Metamerie. *Zool. Jahrb. (Anat.),* 20:153–226.

Lφmo, T., and Rosenthal, J. 1972. Control of ACh sensitivity by muscle activity in the rat. *J. Physiol.,* 221:493–513.

Lorente de Nó, R. 1949. General structural plan of cerebral cortex. In J. F. Fulton, ed., *Physiology of the Nervous System* (3rd ed.). Oxford Univ. Press, New York.

Lorenz, K. Z. 1954. *Man Meets Dog.* (Transl. from the German by M. K. Wilson.) Methuen, London.

Lucas, K. 1917. *The conduction of the nervous impulse.* Longmans, Green, London.

Macagno, E. R., V. Lopresti, and C. Levinthal. 1973. Structure and development of neuronal connections in isogenic organisms: Variations and similarities in the optic system of *Daphnia magna. Proc. Nat. Acad. Sci.,* 70:57–61.

McGaugh, J. L. 1971. *Psychobiology: Behavior from a Biological Perspective.* Academic Press, New York.

McHenry, L. C. 1969. *Garrison's History of Neurology.* Charles C Thomas, Springfield, Illinois.

MacLean, P. D. 1970. The triune brain, emotion, and scientific bias. In F. O. Schmitt, ed., *The Neurosciences: Second Study Program.* Rockefeller Univ. Press, New York.

MacLean, P. D., and J. M. R. Delgado. 1953. Electrical and chemical stimulation of frontotemporal portion of the limbic system in the waking animal. *Electroenceph. Clin. Neurophysiol.,* 5:91–100.

McMahan, U. J., and S. W. Kuffler. 1971. Visual identification of synaptic boutons on living ganglion cells and of varicosities in postganglionic axons in the heart of the frog. *Proc. Roy. Soc. B,* 177:485–508.

Mark, R. 1974. *Memory and Nerve Cell Connections.* Clarendon Press, Oxford.

Mast, S. O. 1911. *Light and the Behavior of Organisms.* Wiley, New York.

Matthews, B. H. C. 1931a. The response of a single end-organ. *J. Physiol.,* 71:64–110.

Matthews, B. H. C. 1931b. The response of a muscle spindle during active contraction of a muscle. *J. Physiol.,* 72:153–74.

Maturana, H. R., and S. Frenk. 1965. Synpatic connections of the centrifugal fibers in the pigeon retina. *Science,* 150:359–361.

Maturana, H. R., J. Y. Lettvin, W. S. McCulloch, and W. H. Pitts. 1960. Anatomy and physiology of vision in the frog (*Rana pipiens*). *J. Gen. Physiol.,* 43:129–175.

Maynard, D. M. 1955. Direct inhibition in the lobster cardiac ganglion. Ph.D. Dissertation, Univ. of California, Los Angeles.

Marton, P. A. 1972. How we control the contraction of our muscles. *Sci. Amer.,* 226:30–37.

Michael, C. R. 1969. Retinal processing of visual images. *Sci. Amer.,* 220:105–114.

Miledi, R. 1973. Transmitter release induced by injection of calcium ions into nerve terminals. *Proc. Roy. Soc. B,* 183:421–425.

Miledi, R., and C. R. Slater. 1970. On the degeneration of rat neuromuscular junctions after nerve action. *J. Physiol.,* 207:507–528.

Miller, N. E. 1956. Effects of drugs on motivation: the value of using a variety of measures. *Ann. N.Y. Acad. Sci.,* 65:318–333.

Milne, L. J., and M. J. Milne. 1959. Photosensitivity in invertebrates. In H. W. Magoun, ed., *Handbook of Physiology,* Sect. 1, Vol. I. Williams and Wilkins, Baltimore.

Miner, N. 1956. Integumental specification of sensory fibers in the development of cutaneous local sign. *J. Comp. Neurol.,* 105:161–170.

Mittelstaedt, H. 1950. Physiologie des Gleichgewichtssinnes bei fliegenden Libellen. *Z. vergl. Physiol.,* 32:422–463.

Mittelstaedt, H. 1961. Die Regelungstheorie als methodisches Werkzeug der Verhaltensanalyse. *Naturwiss.* 48:246–254.

Mittelstaedt, H. 1964a. Basic solutions to a problem of angular orientation. In R. F. Reiss, ed., *Neural Theory and Modeling.* Stanford Univ. Press. Stanford, California.

Mittelstaedt, H. 1964b. Basic control patterns of orientational homeostasis. *Symp. Soc. Exp. Biol.,* 18:365–385.

Mulloney, B. 1970. Structure of giant fibers of earthworms. *Science,* 168:994–996.

Murray, R. G. 1973. Ultrastructure of taste buds. In I. Friedmann, ed., *Ultrastructure of Sensory Organs.* North Holland Publishing Co., Amsterdam.

Naitoh, Y., and R. Eckert. 1969. Ionic mechanisms controlling behavioral responses of paramecium to mechanical stimulation. *Science.* 164:963–965.

Netter, F. H. 1962. *Nervous System.* CIBA Collection of Medical Illustrations. Summit, New Jersey.

Nicholls, J. G., and D. A. Baylor. 1968. Specific modalities and receptive fields of sensory neurons in CNS of the leech. *J. Neurophysiol.,* 31:740–756.

Nicol, J. A. C. 1955. Observations on luminescence in *Renilla* (Pennatulacea). *J. Exp. Biol.,* 32:299–320.

Nicolson, G. L. 1974. The interactions of lectins with animal cell surfaces. *Int. Rev. Cytol.,* 39:89–190.

Nieuwenhuys, R. 1969. The cerebellum. In C. A. Fox vol. 25, Elsevier, Amsterdam.

Noback, C. R. 1967. *The Human Nervous System.* McGraw-Hill, New York.

Nordenskiöld, E. 1935. *The History of Biology.* Tudor Publishing Company, New York.

526

References Cited

Ochs, S. 1972. Fast transport of materials in mammalian nerve fibers. *Science*, 176:252–260.

Ogawa, F. 1939. The nervous system of earthworm (*Pheretima communissima*) in different ages. *Sci. Rep. Tôhoku Univ.*, 13:395–488.

Orkand, R. 1971. Neuron-glia relations and control of electrolytes. In B. K. Slesjo and S. C. Sorensen, eds., *Ion Homeostasis of the Brain*. Munksgaard, Copenhagen.

Orkand, R., S. W. Kuffler and J. G. Nicholls. 1966. Effect of nerve impulses on the membrane potential of glial cells in the central nervous system of amphibia. *J. Neurophysiol.*, 29:788–806.

Ottoson, D., and G. M. Shepherd. 1971. Transducer properties and integrative mechanisms in the frog's muscle spindle. In W. R. Loewenstein, ed., *Handbook of Sensory Physiology* (Vol. 1). Springer-Verlag, New York.

Palay, S. L., and V. Chan-Palay. 1974. *Cerebellar Cortex: Cytology and Organization*. Springer-Verlag, New York.

Pantin, C. F. A. 1935. The nerve net of the Actinozoa. *J. Exp. Biol.*, 12:139–155.

Pantin, C. F. A., and M. Vianna Dias. 1952. Excitation phenomena in an actinian (*Bunodactis* sp?) from Guanabara Bay. *Ann. Acad. Bras. Sci.*, 24:335–349.

Patton, H. D. 1965. Reflex regulation of movement and posture. In T. C. Ruch, H. D. Patton, J. W. Woodbury, and A. L. Towe, eds., *Neurophysiology*. Saunders, Philadelphia.

Pearson, K. G., and J. F. Iles. 1970. Central programming and reflex control of walking in the cockroach. *J. Exp. Biol.*, 56:173–193.

Pelseneer, P. 1888. Sur la valeur morphologique des bras et la composition du système nerveux central des céphalopodes. *Arch. Biol.*, 8:723–756.

Pelseneer, P. 1892. Le système nerveux streptoneure des hétéropodes. *C. R. Acad. Sci.*, 111:775–777.

Perkel, D. H., and B. Mulloney. 1974. Motor pattern production in reciprocally inhibitory neurons exhibiting postinhibitory rebound. *Science*, 185:181–183.

Peters, A., S. L. Palay, and H. F. de Webster. 1970. *The Fine Structure of the Nervous System: The Cells and Their Processes*. Springer-Verlag, New York.

Picton, T. W., S. A. Hillyard, H. I. Krausz, and R. Galambos. 1973. Human auditory evoked potentials. 1. Evaluation of components. *Electroenceph. Clin. Neurophysiol*, 36:179–190.

Poggio, G. F., and V. B. Mountcastle. 1960. A study of the functional contributions of the lemniscal and spinothalamic systems to somatic sensibility. Central nervous mechanisms in pain. *Bull. Johns Hopkins Hosp.*, 106:266–316.

Poritsky, R. 1969. Two and three dimensional ultrastructure of boutons and glial cells on the motoneuronal surface in the cat spinal cord. *J. Comp. Neurol.*, 135:423–452.

Prosser, C. L. 1973. *Comparative Animal Physiology* (3rd ed.). Saunders, Philadelphia.

Ramón, P. 1891. El encéfalo de los reptiles. Zaragoza.

Ramon-Moliner, E. 1968. The morphology of dendrites. In G. H. Bourne, ed., *The Structure and Function of Nervous Tissue*. Vol. I, Structure I. Academic Press, New York.

Reid, J. V. O. 1969. The cardiac pacemaker: Effects of regularly spaced nervous input. *Amer. Heart J.*, 78:58–64.

Reisinger, E. 1926. Untersuchungen am Nervensystem der *Bothrioplana semperi* Braun. *Z. Morph. Ökol. Tiere*, 5:119–149.

Retzius, G. 1890. Zur Kenntnis des Nervensystems der Krustaceen. *Biol. Untersuch.*, N.F. 1:1–50.

Retzius, G. 1892. Das Nervensystem der Lumbriciden. *Biol. Untersuch.*, N.F. 3:1–16.

Retzius, G. 1895. Zur Kenntnis des Gehirnganglions und des sensiblen Nervensystems der Polychaeten. *Biol. Untersuch.*, N.F. 7:6–11.

Retzlaff, E., and J. Fontaine. 1960. Reciprocal inhibition as indicated by a differential staining reaction. *Science*. 131:104–105.

Ripley, S. H., and C. A. G. Wiersma. 1953. Spaced stimulation in the crayfish. *Physiol. Comp. Oecol.* 3:4, 1–17.

Roberts, A. 1971. The role of propagated skin impulses in the sensory system of young tadpoles. *Z. vergl. Physiol.*, 75:388–401.

Roberts, A., and C. A. Stirling. 1971. The properties and propagation of a cardiac-like impulse in the skin of young tadpoles. *Z. vergl. Physiol.*, 71:295–310.

Robertson, J. D. 1961. Ultrastructure of excitable membranes of the crayfish median-giant synapse. *Ann. N. Y. Acad. Sci.*, 94:339–389.

Robertson, J. D. 1962. The membrane of the living cell. *Sci. Amer.*, 206: 64–72.

Rodieck, R. W. 1973. *The Vertebrate Retina*. W. H. Freeman and Company, San Francisco.

Roeder, K. D. 1948. Organization of the ascending giant fiber system in the cockroach (*Periplaneta americana*) *J. Exp. Zool.*, 108:243–261.

Romanes, G. J. 1885. Jellyfish, starfish, and sea urchins, being a research on the primitive nervous systems. *Internat. Sci. Ser.* Appleton, New York.

Rose, J. E., and V. B. Mountcastle. 1959. Touch and kinesthesis. In H. W. Magoun, ed., *Handbook of Physiology*, Sect. 1., Vol. I. Williams and Wilkins, Baltimore.

Rose, J. E., R. Galambos, and J. R. Hughes. 1959. Microelectrode studies of the cochlear nuclei of the cat. *Bull. Johns Hopkins Hosp.*, 104:211–251.

Ross, L. S. 1922. Cytology of the large nerve cells of the crayfish (*Cambarus*). *J. Comp. Neurol.*, 34:37–71.

Ross, W. N., B. M. Salzberg, L. B. Cohen, and H. V. Davila. 1974. A large change in dye absorption during the action potential. *Biophys. J.*, 14:983–986.

Rovainen, C. M. 1967. Physiological and anatomical studies on large neurons of central nervous system of the sea lamprey (*Petromyzon marinus*). 1. Müler and Mauthner cells. *J. Neurophysiol.*, 30:1000–1023.

Rowland, V. 1967. Steady potential phenomena of cortex. In G. C. Quarton, T. Melnechuk, and F. O. Schmitt, eds., *The Neurosciences: A Study Program*. Rockefeller Univ. Press, New York.

Russell, G. V. 1961. Interrelationship within the limbic and centrencephalic systems. In D. E. Sheer, ed., *Electrical Stimulation of the Brain*. Univ. Texas Press, Austin.

Sánchez, D. 1912. El sistema nerviosa de los hirudineos. *Trab. Lab. Invest. biol. Univ. Madr.*, 10:1–143.

Sauer, F. C. 1935. Mitosis in the neural tube. *J. Comp. Neurol.*, 62:377–405.

Schaper, A. 1893. Zur feineren Anatomie des Kleinhirns der Teleostier. *Anat. Anz.*, 8:705–720.

Schaper, A. 1898. The finer structure of the selachian cerebellum (*Mustelus vulgaris*) as shown by chrome silver preparations. *J. Comp. Neurol.*, 8:1–20.

Scharrer, B. 1974. The spectrum of neuroendocrine communication. In K. Lederis and K. E. Cooper, eds., *Recent Studies of Hypothalamic Function*, Int. Symp. Calgary 1973. Karger, Basel. pp. 8–16.

Scheibel, M. E., and A. B. Scheibel. 1958. A symposium on dendrites. Formal discussion. *Electroenceph. Clin. Neurophysiol.* 10 Suppl.:43–50.

Scheibel, M. E., and A. B. Scheibel. 1963. Some structuro-functional correlates of development in young cats. *Electroenceph. Clin. Neurophysiol.* Suppl. 24:235–246.

Scheibel, M. E., and A. B. Scheibel. 1968a. The brain stem reticular core—An integrative matrix. In M. D. Mesarović, ed., *Systems Theory and Biology.* Springer-Verlag, New York.

Scheibel, M. E., and A. B. Scheibel. 1968b. On the nature of dendritic spines: Report of a workshop. *Communications in Behav. Biol.*, Part A, 1:231–265.

Scheibel, M. E., and A. B. Scheibel. 1970a. Elementary processes in selected thalamic and cortical subsystems—the structural substrates. In F. O. Schmitt, ed., *The Neurosciences: Second Study Program.* Rockefeller Univ. Press, New York.

Scheibel, M. E., and A. B. Scheibel. 1970b. Of pattern and place in dendrites. *Int. Rev. Neurobiol.*, 13:1–26.

Scheich, H., and T. H. Bullock. 1974. The detection of electric fields from electric organs. In A. Fessard, ed., *Handbook of Sensory Physiology*. Vol. III/3. Springer-Verlag, New York.

Schewiakoff, W. 1889. Beiträge zur Kenntnis des Acalephenauges. *Morph. Jahrb.*, 15:21–60.

Schmalz, E. 1914. Zur Morphologie des Nervensystems von *Helix pomatia*. *Z. wiss. Zool.*, 111:506–568.

Schneider, D. 1970. Olfactory receptors for the sexual attractant (bombykol) of the silk moth. In F. O Schmitt, ed., *The Neurosciences: Second Study Program.* Rockefeller Univ. Press, New York.

Schulman, J. 1969. Information transfer across an inhibitor to pacemaker synapse at the crayfish stretch receptor. Ph.D. dissertation, Univ. Calif., Los Angeles.

Scriban, I. A., and H. Autrum. 1932. Hirudinea. In *Kükenthal's Handb. Zool.*, 2(8):119–352.

Selverston, A. I., and B. Mulloney. 1972. Synaptic and structural analysis of a small neural system. In F. O. Schmitt and F. G. Worden, eds., *The Neuro-*

sciences: Third Study Program. Mass. Inst. Technol. Press, Cambridge.

Shantha, T. R., and G. H. Bourne. 1968. The perineural epithelium: A new concept. In G. H. Bourne, ed., *The Structure and Function of Nervous Tissue.* vol. I. Academic Press, New York.

Sherrington, C. S. 1906. *The Integrative Action of the Nervous System.* Yale Univ. Press, New Haven.

Shlaer, R. 1971. Shift in binocular disparity causes compensatory change in the cortical structure of kittens. *Science,* 173:638–641.

Simroth, H. 1896–1907. Gastropoda Prosobranchia. *Bronn's Klassen und Ordnungen des Tierreichs.* Akademische Verlags gessellschaft Geest und Portig K.-G., Leipzig. 3:2:1:1–1056.

Singer, C. J. 1959. *History of Biology* (3rd ed.). Abelard-Schuman, New York.

Singer, M. 1958. The regeneration of body parts. *Sci. Amer.,* 199:79–88.

Sirks, M. J., and C. Zirkle. 1964. *The Evolution of Biology.* The Ronald Press Company, New York.

Smith, J. E. 1957. The nervous anatomy of the body segments of nereid polychaetes. *Phil. Trans. Roy. Soc. B,* 240:135–196.

Speidel, C. C. 1933. Studies of living nerves. II. Activities of ameboid growth cones, sheath cells, and myelin segments, as revealed by prolonged observation of individual nerve fibers in frog tadpoles. *Amer. J. Anat.,* 52:1–79.

Spengel, J. W. 1881. Die Geruchsorgane und das Nervensystem der Mollusken. Ein Beitrag zur Erkenntnis der Einheit des Molluskentypus. *Z. wiss. Zool.,* 35:333–383.

Sperry, R. W. 1951. Mechanisms of neural maturation. In S. S. Stevens, ed., *Handbook of Experimental Psychology.* Wiley, New York.

Spoendlin, H. 1968. Ultrastructure and peripheral innervation pattern of the receptor in relation to the first coding of the acoustic message. In A.V.S. De Reuck and J. Knight, eds., *Hearing Mechanisms in Vertebrates.* Little, Brown & Company, Boston.

Steiger, U. 1967. Über den Feinbau des Neuropils im Corpus pedunculatum der Waldameise. *Z. Zellforsch.,* 81:511–536.

Stein, P. S. G. 1971. Intersegmental coordination of swimmeret motoneuron activity in crayfish. *J. Neurophysiol.* 34:310–318.

Steinbrecht, R. A. 1969. Receptor mechanisms. In C. Pfaffmann, ed., *Olfaction and Taste.* Rockefeller Univ. Press, New York.

Stotler, W. A. 1953. An experimental study of the cells and connections of the superior olivary complex of the cat. *J. Comp. Neurol.,* 98:401–432.

Strausfeld, N. J. 1970. Golgi studies on insects. Part II. The optic lobes of Diptera. *Phil. Trans. Roy. Soc. B.* 258:135–223.

Strausfeld, N. J. 1971. The organization of the insect visual system (light microscopy). I. Projections and arrangements of neurons in the lamina ganglionaris of Diptera. *Z. Zellforsch.,* 121:377–441.

Strausfeld, N. J. 1976. Atlas of an Insect Brain. Springer-Verlag, New York.

Strumwasser, F. 1973. Neural and humoral factors in the temporal organization of behavior. *The Physiologist,* 16:9–42.

Sutton, S., P. Tueting, J. Zubin, and E. R. John. 1967. Information delivery and the sensory evoked potential. *Science.* 155:1436–1439.

Swazey, J., and F. G. Worden. 1975. *Paths of Discovery in the Neurosciences.* Mass. Inst. Technol. Press, Cambridge.

Szabo, T. 1974. Anatomy of the specialized lateral line organs of electroreception. In A. Fessard, ed., *Handbook of Sensory Physiology.* Vol. III/3. Springer-Verlag, New York.

Székely, G., G. Setalo, and Gy. Lazar. 1973. Fine structure of the frog's optic tectum: Optic fibre termination layers. *J. Hirnforsch.,* 14:189–225.

Szentágothai, J. 1961. Anatomical aspects of inhibitory pathways and synapses. In E. Florey, ed., *Nervous Inhibition.* Pergamon Press, New York.

Szentágothai, J. 1963. Ujabb adatok a synapsisok funcionális anatómiájához. (New data on the functional anatomy of synapses). *Magy. Tud. Akad. Orv. Tud. Osztal. Közl.,* 6:217–227.

Szentágothai, J. 1970. Glomerular synapses, complex synaptic arrangements, and their operational significance. In F. O. Schmitt, ed., *The Neurosciences: Second Study Program.* Rockefeller Univ. Press, New York.

Takeuchi, A., and N. Takeuchi. 1960. On the permeability of end-plate membrane during the action of transmitter. *J. Physiol.,* 154:52–67.

Takeuchi, A., and N. Takeuchi. 1972. Actions of transmitter substances on the neuromuscular junctions of vertebrates and invertebrates. *Adv. Biophys.*, 3:45–95.

Tasaki, I. 1959. Conduction of the nerve muscle. In H. W. Magoun, ed., *Handbook of Physiology*, Sect. 1, Vol. I. Williams and Wilkins, Baltimore.

Taylor, A. C., and P. Weiss. 1965. Demonstration of axonal flow by the movement of tritium-labeled protein in mature optic nerve fibers. *Proc. Nat. Acad. Sci.*, 54:1521–1527.

Thesleff, S. 1960. Supersensitivity of skeletal muscle produced by botulinum toxin. *J. Physiol.*, 151:598–607.

Thore, S. 1942. Ein Beitrag zur Kenntnis der Sehzentren im Gehirn von *Sepia*. *Acta Univ. Lund*, Avd. 2, N.F. 38:1–18.

Thorpe, W. H. 1961. *Bird Song: The Biology of Vocal Communication and Expression in Birds*. Cambridge Univ. Press, Cambridge.

Thurm, U. 1969. General organization of sensory receptors. *Rendiconti della Scuola Internazionale di Fisica "E. Fermi"*, XLIII Corso: 44–68.

Tinbergen, N. 1951. *The Study of Instinct.* Clarendon Press, Oxford.

Tomita, T. 1965. Electrophysiological study of the mechanisms subserving color coding in the fish retina. *Cold Spr. Harb. Symp. Quant. Biol.*, 30:559–566.

Truex, R. C., and M. G. Carpenter. 1969. *Human Neuroanatomy* (6th ed.). Williams and Wilkins, Baltimore.

Tweedle, C. D., R. M. Pitman, and M. J. Cohen. 1973. Dendritic stability of insect central neurons subjected to axotomy and de-afferentation. *Brain Res.* 60:471–476.

Vayssière, A. 1879–1880. Recherches anatomiques sur les mollusques de la famille des Bullides. *Ann. Sci. Nat. (Zool.)*, ser. 6, 9:1–123.

Verplanck, W. S. 1957. A glossary of some terms used in the objective science of behavior. *Psychol. Rev.*, 64 (Suppl.):1–42.

Vogel, H. 1923. Über die Spaltsinnesorgane der Radnetzspinnen. *Jena. Z. Naturw.*, 59:171–208.

Walls, G. L. 1942. *The vertebrate eye and its adaptive radiation.* Cranbrook Institute of Science, Bloomfield Hills, Michigan.

Weed, L. H. 1923. The absorption of cerebrospinal fluid into the venous system. *Amer. J. Anat.* 31:190–221.

Weight, F. F., and J. Votava. 1970. Slow synaptic excitation in sympathetic ganglion cells: Evidence for synaptic inactivation of potassium conductance. *Science*, 170:755–758.

Weiss, P. 1950. Experimental analysis of co-ordination by the disarrangement of central-peripheral relations. *Symp. Soc. Exp. Biol.*, 4:92–111.

Weiss, P., and H. B. Hiscoe. 1948. Experiments on the mechanism of nerve growth. *J. Exp. Zool.*, 197:315–396.

Weller, W. L. 1972. Barrels in somatic sensory neocortex of the marsupial *Trichosurus vulpecula* (brush-tailed possum). *Brain Res.*, 43:11–24.

Wells, M. J. 1962. *Brain and Behavior in Cephalopods.* Stanford Univ. Press.

Whitsel, B. L., D. A. Dreyer, and J. R. Roppolo. 1971. Determinants of body representation in postcentral gyrus of macaques. *J. Neurophysiol.*, 34:1018–1034.

Whittaker, V. P., and E. G. Gray. 1962. The synapse: Biology and morphology. *Brit. Med. Bull.*, 18:223.

Wiersma, C. A. G. 1958. On the functional connections of single units in the central nervous system of the crayfish, *Procambolus clarkii*, Girard. *J. Comp. Neurol.*, 110:421–471.

Wiersma, C. A. G., and S. H. Ripley. 1952. Innervation patterns of crustacean limbs. *Physiol. Comp. Oecol.*, 2:391–405.

Wiesel, T. N., and D. H. Hubel. 1965. Comparison of the effects of unilateral and bilateral eye closure on cortical unit responses in kittens. *J. Neurophysiol.*, 28:1029–1040.

Willis, W. D., Jr., and R. G. Grossman. 1973. *Medical Neurobiology.* C. V. Mosby Company, St. Louis.

Wilson, D. M. 1968. The flight control system of the locust. *Sci. Amer.*, 218:83–90.

Woolsey, C. N. 1958. Organization of somatic sensory and motor areas of the cerebral cortex. In H. F. Harlow and C. N. Woolsey, eds., *Biological and Biochemical Bases of Behavior*. Univ. Wisconsin Press, Madison.

Woolsey, T. H., and H. Van der Loos. 1970. The structural organization of layer IV in the somatosensory region (SI of mouse cerebral cortex). *Brain Res.*, 17:205–242.

Wyman, R. J. 1966. Multistable firing patterns among several neurons. *J. Neurophysiol.* 29:807–833.

References Cited

Yoon, M. 1971. Reorganization of retinotectal projection following surgical operations on the optic tectum in goldfish. *Exp. Neurol.,* 33:395–411.

Young, J. Z. 1936a. The structure of nerve fibres in cephalopods and crustacea. *Proc. Roy. Soc. B.,* 121:319–337.

Young, J. Z. 1936b. The giant nerve fibre and epistellar body of cephalopods. *Quart. J. Micr. Sci.,* 78:367–386.

Young, J. Z. 1939. Fused neurons and synaptic contacts in the giant nerve fibres of cephalopods. *Phil. Trans. Roy. Soc. B.,* 229:465–503.

Young, J. Z. 1971. *The Anatomy of the Nervous System of Octopus vulgaris.* Clarendon Press, Oxford.

Zawarzin, A. 1924. Zur Morphologie der Nervenzentren. Das Bauchmark der Insekten. Ein Beitrag zur vergleichenden Histologie. (Histologische Studien über Insekten VI.) *Z. wiss. Zool.,* 122:323–424.

Zawarzin, A. 1925. Der Parallelismus der Strukturen als ein Grundprinzip der Morphologie. *Z. wiss. Zool.,* 124:118–212.

Zotterman, Y. 1939. Touch, pain and tickling: An electrophysiological investigation on cutaneous sensory nerves. *J. Physiol.,* 95:1–28.

Zucker, R. S. 1972. Crayfish escape behavior and central synapses. I. Neural circuit exciting lateral giant fiber. *J. Neurophysiol.,* 35:599–620.

INDEX

Page numbers in *italics* refer to figures and tables.